Gravity, Gauge Theories and Quantum Cosmology

Fundamental Theories of Physics

A New International Book Series on the Fundamental Theories of Physics: Their Clarification, Development and Application

Gravity, Gauge Theories and Quantum Cosmology

by

Jayant V. Narlikar

and

T. Padmanabhan

Tata Institute of Fundamental Research,
Bombay, India

D. Reidel Publishing Company

A MEMBER OF THE KLUWER ACADEMIC PUBLISHERS GROUP

Dordrecht / Boston / Lancaster / Tokyo

Library of Congress Cataloging-in-Publication Data

Narlikar, Jayant Vishnu, 1938–
Gravity, gauge theories, and quantum cosmology.

(Fundamental theories of physics)
Includes bibliographies and index.
1. Quantum gravity. 2. Cosmology. 3. Gauge fields (Physics) 4. Quantum field theory. 5. General relativity (Physics) I. Padmanabhan, T. (Thanu), 1957– II. Title. III. Series.
QC178.N26 1986 530.1 86–3282
ISBN 90–277–1948–9

Published by D. Reidel Publishing Company,
P.O. Box 17, 3300 AA Dordrecht, Holland

Sold and distributed in the U.S.A. and Canada
by Kluwer Academic Publishers,
101 Philip Drive, Assinippi Park, Norwell, MA 02061, U.S.A.

In all other countries, sold and distributed
by Kluwer Academic Publishers Group,
P.O. Box 322, 3300 AH Dordrecht, Holland

Printed in The Netherlands

Table of Contents

Preface

For several decades since its inception, Einstein's general theory of relativity stood somewhat aloof from the rest of physics. Paradoxically, the attributes which normally boost a physical theory – namely, its perfection as a theoretical framework and the extraordinary intellectual achievement underlying it – prevented the general theory from being assimilated in the mainstream of physics. It was as if theoreticians hesitated to tamper with something that is manifestly so beautiful.

Happily, two developments in the 1970s have narrowed the gap. In 1974 Stephen Hawking arrived at the remarkable result that black holes radiate after all. And in the second half of the decade, particle physicists discovered that the only scenario for applying their grand unified theories was offered by the very early phase in the history of the Big Bang universe. In both cases, it was necessary to discuss the ideas of quantum field theory in the background of curved spacetime that is basic to general relativity.

This is, however, only half the total story. If gravity is to be brought into the general fold of theoretical physics we have to know how to quantize it. To date this has proved a formidable task although most physicists would agree that, as in the case of grand unified theories, quantum gravity will have applications to cosmology, in the very early stages of the Big Bang universe. In fact, the present picture of the Big Bang universe necessarily forces us to think of quantum cosmology.

In this book we have highlighted these interdisciplinary problems, ending with a simplified version of quantum gravity that appears to provide useful inputs into the problems of the early universe. In particular, the difficulties associated with the classical models that have recently gained prominence under the heads of 'singularity', 'particle horizon' and 'flatness' are resolved in this picture of the quantum universe.

But this is a beginning, not the end of the quest for a general theory of quantum gravity. Indeed, none of the topics discussed in the various parts of this book is free from problems yet to be resolved: thus giving a lie to the belief shared in some circles that the end of physics is almost at hand.

Our discussion of classical general relativity and cosmology is self-contained and reasonably complete. As regards field theory, we have concentrated on those aspects which are of relevance to quantum field theory in curved spacetime and to quantum gravity. We have therefore assumed that the reader has elementary knowledge of quantum mechanics and field theory, for which excellent texts exist anyway. Illustrative solved examples (138 in number) sprinkled throughout the book are, however, expected to make the text easier to follow. We do admit to a lack of expertise and enthusiasm for the many formal approaches to quantum gravity. In a text on a frontier area, the authors, we believe, are entitled to their share of prejudice!

We have gained considerably, especially in Part IV, from discussions with colleagues too numerous to mention individually. We are grateful to our Institute for making its facilities available to get the manuscript of the book ready in time. In particular it is a pleasure to thank Mrs M. Benjamin for her cooperation in preparing the highly mathematical typescript with our numerous corrections, deletions and insertions.

Theoretical Astrophysics Group,
Tata Institute of Fundamental Research,
Homi Bhabha Road,
Bombay-400 005, India

Jayant V. Narlikar
T. Padmanabhan

September, 1984

Notation

We set out below the convention that will be followed in this book with regard to the presentation of the various physical and mathematical expressions. The reader is advised to refer to the list below as and when necessary.

(1) *Equations*. These are serially numbered in each chapter. However, while referring to an equation in another chapter, say Equation (11) in Chapter 4, we shall write it as Equation (4.11).

(2) *Physical constants*. The fundamental constants that are most frequently encountered in our discussions are, of course, the gravitational constant G, the speed of light c and the Planck constant/2π, $\hbar$. We shall often put one or more of these constants equal to unity when the relevant discussion does not explicitly require the knowledge of their magnitudes. While this procedure compactifies the various analytical expressions, it also has some drawbacks. In practical terms, the largeness or smallness of a given expression is not immediately apparent if some of these constants are taken to be unity. In conceptual terms limits such as

$$\lim_{\hbar \to 0} [\text{Quantum theory}] = \text{Classical theory}$$

$$\lim_{c \to \infty} [\text{Relativistic mechanics}] = \text{Newtonian mechanics}$$

$$\lim_{G \to 0} [\text{General relativity}] = \text{Special relativity}$$

cannot be taken if we set the relevant quantities equal to unity. In such circumstnaces we shall explicitly restore the constants.

(3) *Spacetime coordinates*. It is impossible to satisfy everybody in the matter of choice of coordinate labels. In general we shall use Latin superscripts to denote the four spacetime coordinates. Thus x^i denotes the quartet (x^0, x^1, x^2, x^3, of which *usually* x^0 is timelike and x^1, x^2, x^3 spacelike. In such circumstances the triplet x^1, x^2, x^3 will be abbreviated as x^μ. Thus, the Greek superscripts take values 1, 2, and 3.

If x^1, x^2, x^3 are Cartesian coordinates we may sometimes revert to the three-dimensional notation of writing the triplet (x^1, x^2, x^3) as **r** or **x**.

(4) *Spacetime metric*. Again, different conventions exist in literature. We shall adopt the rule that timelike intervals are real and spacelike ones, imaginary. This rule corresponds to the signature $(+, -, -, -)$ for the spacetime metric. We shall use $\eta_{ik} \equiv \text{diag}\,(1, -1, -1, -1)$ to denote the metric for Minkowski spacetime in rectangular coordinates.

(5) *Summation convention*. We shall use the standard convention that a quantity in which the same index appears twice – once as a superscript and once a subscript – will be automatically summed over all values of the index. Thus,

$$A^iB_i \equiv A^0B_0 + A^1B_1 + A^2B_2 + A^3B_3,$$

$$P^\mu_\mu = P^1_1 + P^2_2 + P^3_3.$$

Where this convention does not apply, the operating rule will be explicitly stated.

(6) *Differentiation*. Ordinary differentiation with respect to a coordinate x^i will either be denoted by the operator ∂_i or by the subscript i following a comma. Thus

$$\frac{\partial \varphi}{\partial x^i} \equiv \partial_i \varphi \equiv \varphi_{,i} \equiv \varphi_i.$$

The last identity implies further abbreviation of notation which will be used only for scalar quantities φ.

In Part II we shall encounter *covariant* derivatives which are denoted by subscripts following a semicolon. e.g. $A^i_{;k}$ to distinguish them from ordinary derivatives. On rare occasions we shall also use ∇_i to denote the covaraint derivative with respect to x^i. The wave operator $\eta^{ik}\,\partial_i\,\partial_k$ in Minkowski spacetime and its covariant generalization in Riemannian spacetime will be denoted by $\Box$.

(7) *Integration*. Ordinary integrals over all four spacetime coordinates x^i will be written as

$$I = \int \ldots \mathrm{d}^4x,$$

while those over the three spatial coordinates will be written as

$$I = \int \ldots \mathrm{d}^3\mathbf{r},$$

Path integrals, or functional integrals in general, will be written in the form

$$I[f(t)] = \int \ldots \mathcal{D} f(t).$$

(8) *Operators*. Quantum operators, or operators in general (where the operator nature *needs* to be emphasized), will be written in bold, as, for example $\mathbf{0}$, $\mathbf{T}_{ik}$, etc. We shall, on occasions, use Dirac's notation of bra, ket and *c*-numbers in writing quantum amplitudes, expectation values, etc.

(9) *Wave functions*. If ψ is a wave function, $\psi^\dagger$ will denote its Hermitian conjugate. For a 4-spinor ψ we shall adopt the commonly used notation $\bar{\psi} = \psi^\dagger\gamma_0$, where γ_i are the Dirac gamma matrices. ψ^* is the complex conjugate of ψ.

(10) *Delta functions*. We shall follow the usual convention of writing the Kronecker deltas as δ_i^k, δ_λ^μ, etc. As far as the continuum delta function is concerned, we shall use the convention of writing

$$\delta_3(\mathbf{r}) = \delta(x^1)\, \delta(x^2)\, \delta(x^3)$$

for Cartesian coordinates x^μ. If, however, we wish to emphasize the delta function as a two-point function we write it as $\delta_4(P_1, P_2)$. Here δ_4 is the four-dimensional delta function which vanishes unless $P_1 \equiv P_2$. In the neighbourhood of P_1 we can choose rectangular coordinates and define δ_4 in the usual manner.

For Heaviside function we use Θ. Thus

$$\Theta(x) = \begin{cases} 1 & \text{for } x \geqslant 0, \\ 0 & \text{for } x < 0. \end{cases}$$

(11) *Electromagnetism*. We use Heaviside's units in which the Coulomb field due to a charge at origin is $(Q/4\pi r^2)$. Our electromagnetic Lagrangian is

$$\mathrm{L} = -\tfrac{1}{4}\, F_{ik}F^{ik} - J_i A^i,$$

where J^i is the current 4-vector.

Chapter 1

Introduction

1.1. Historical Background

Looking back at the documented history of physics we find that there have been discontinuous changes of basic ideas on a few selected occasions. These changes can be traced to a handful of people, yet they were important enough to influence the direction of growth of the subject in a very major way, and for a long time to come. The proper description of such a change of course is by the word 'revolution'. Unlike political revolutions, which came with mixed benefits, the revolutions in physics have invariably led to a dramatically improved understanding of nature.

The first of these revolutions can be identified with Aristotle (384–321 BC). It is not known what, if any, coherent approach to a scientific study of nature existed prior to Aristotle. That one can formulate a set of rules of universal validity to describe the behaviour of the material content of the universe seems to have been appreciated by Aristotle. By hindsight we can now say that the rules postulated by Aristotle were wrong. That, however, is not the point. The universal applicability of physical laws which forms the basis of modern theoretical physics dates back to Aristotle.

The modern theoretical physicist looking for symmetry in nature and, more importantly, for instances of breakdown of that symmetry, may find some sympathy for Aristotle who believed that the paths of matter moving *naturally* would be straight lines or circles. These curves have one special symmetry: any portion of the curve can be taken out and placed anywhere else on it congruently. This symmetry is broken in the case of external (unnatural or man-made) forces, which lead to different paths. Aristotle called such motion *violent* motion as opposed to the natural motion in circles or straight lines (which are circles of zero curvature).

The geocentric theory of Hipparchus and Ptolemy was based on the Aristotelian concept of circular paths. Noticing that the planets *do not* move on circles round the Earth (as Aristotle would have suggested), the Greek astronomers constructed their elaborate epicyclic theory. In this theory a typical planet moves on a circle whose centre moves on another circle whose centre moves on another circle . . . until the system finally leads to the Earth as the centre of everything. These circles (called *epicycles*) were no different from the modern theoretician's attempts at parameter fitting in order to make the observations agree with his pet theory.

Again, wrong though the geocentric approach was, the appreciation that planets do not move arbitrarily but according to some pattern was the major contribution of Greek astronomy. In fact the heliocentric theory of Copernicus made considerable use of the epicyclic theory, although its major achievement was to correctly identify the Sun as the nucleus of the planetary system.

This brings us to the second revolution in physics, a revolution inspired by the Copernican theory but which can be properly attributed to Galileo and Newton. In his *Dialogues*, Galileo gave masterly arguments to demolish the foundations of the Aristotelian system. Galileo's arguments were not simply philosophical wranglings but consisted of thought experiments as well as experimental demonstrations. It was Galileo who correctly identified the role of the external force. Force is not proportional to velocity as Aristotle believed but is responsible for producing *changes* in velocity. This discovery of Galileo was later expressed in a precise mathematical form by Newton in his *Principia*.

By stating the law of gravitation, of course, Newton began the modern version of theoretical physics. Gravity is one of the basic interactions of nature and Newton deserves as much credit for recognizing the signature of gravity in a host of natural phenomena ranging from the (legendary) falling apple to the motions of planets and satellites, as he does for the actual mathematical statement of the law of gravitation. The successes of the Newtonian system include the correct prediction of the arrival of Halley's comet, the discovery of Neptune and Laplace's monumental work embodied in his five-volume *Mecanique-Celeste*.

If the Aristotelian system became established by religious fiat, the Newtonian system owed its popularity to such successful demonstrations. The next revolution in physics which challenged the Newtonian system was therefore even more profound in its implications for the subsequent growth of the subject. This revolution came at the turn of the century, on two fronts: quantum theory and the theory of relativity.

Quantum theory dealt a fatal blow to the Newtonian deterministic view of physics. Given the initial conditions and the equations of motion (of the relevant physical variables) the pre-quantum physicist confidently expected to be able to determine the future behaviour of a physical system. The work of Max Planck, Neils Bohr, Erwin Schrödinger, Max Born, Werner Heisenberg, and Paul Dirac showed that there are limitations both on what the physicist can measure and on what he can predict. Out of these considerations emerged the fundamental constant of nature $\hbar$, which characterizes the 'quantum of action'. No matter how precise his measuring technique, the physicist cannot simultaneously measure the position q and the momentum p of a particle with arbitrary accuracy. The uncertainties Δq and Δp in q and p are subject to the condition

$$\Delta q \, \Delta p \gtrsim \hbar . \tag{1}$$

The above relation is a typical example of quantum uncertainty in physical measurements. The smallness of $\hbar$ shows that macroscopic physics is hardly affected by quantum ideas. Indeed, if $\hbar$ were zero the classical deterministic physics *à-la*-Newton would apply rigorously.

The second front against the Newtonian system was opened by Albert Einstein in 1905. The special theory of relativity questioned the Newtonian postulates of absolute space and absolute time. By postulating a finite speed-limit on all physical interactions, Einstein was led to the notion of the observer-dependent space and time. Again, the limiting speed c, the speed of light, is so high that differences introduced into Newtonian kinematics and dynamics are hardly noticeable in the motion of macroscopic bodies. Like $\hbar$, c also came to be recognized as a fundamental constant of nature.

The same consideration applies to general relativity, which is Einstein's theory of gravity. The conditions under which the difference between the predictions of general relativity and Newtonian gravity becomes marked, hardly come about in the Solar System. Indeed, very sophisticated observations are needed to distinguish between the two theories. Small though these differences are for most practical purposes, their existence is sufficient to warrant change from old to new ideas.

Apart from the gravitational effects in the Solar System, general relativity made the bold attempt to bring 'cosmology', the study of the large-scale structure of the universe, within the purview of science. In this attempt it succeeded beyond expectation. In 1918 Einstein produced the first (static) model of the universe. Four years later A. Friedmann obtained expanding models from Einstein's field equations. The phenomenon of expansion was observationally confirmed in 1929 by the work of E. Hubble and M. Humason, and cosmology was launched as a respectable branch of science.

It is interesting to note the difference in the ways in which the new revolution came about via quantum theory and relativity. Quantum theory became necessary from the experimental considerations of spectroscopy, the photo-electric effect and the frequency-intensity distribution of the black-body radiation. The inadequacy of classical (Newton–Maxwell) electrodynamics was clearly brought out by these experiments even in the late nineteenth century. To explain these experiments quantum theory was invented. It had the appearance of a collection of *ad hoc* rules somewhat hastily brought together to explain the odd behaviour of nature. From this rather shabby appearance, the theory slowly evolved into a more aesthetically pleasing form. Even today it is not believed to be in a finished form and its foundations are subject to considerable discussion.

By contrast both the special and general theory were based on aesthetic aspects of symmetry and general principles. They were presented as finished products rather than as a set of rules put together pragmatically. As theories they are more beautiful than quantum theory; however, their contact with experiments and observations has not been as direct as in quantum theory. For special relativity we need particles moving very close to the speed of light, while for general relativity we need strong gravitational fields. The former conditions are achieved in high-energy particle accelerators while for the latter we have to resort to astrophysics and cosmology.

Although both the quantum theory and relativity represent revolutionary advances in physics, their coexistence has not been very smooth. Indeed it is not an exaggeration to say that the marriage of quantum theory and special relativity has not been a happy one and therefore doubts are being cast on the legitimacy of their offspring – the quantum field theory. Going one step further, we can say that the marriage between quantum theory and general relativity is yet to take place and so their offspring – quantum gravity – is certainly illegitimate.

The present book is devoted to the attempts at legitimizing this offspring. While we will review some of the many efforts in this direction, in the last analysis our emphasis will be on one particular approach to quantum gravity; partly because we have been concerned with its development and partly because it is simple to understand and execute in practice. Moreover, on the cosmological front the quantum ideas offer fresh insights into outstanding problems of classical gravity.

1.2. What This Book is About

In our efforts to make this book as self-contained as possible we have first discussed the relevant ideas that are believed to be useful towards the understanding of quantum gravity. Thus, Part I outlines the developments in quantum mechanics, quantum field theory, and gauge theories. Part II introduces the reader to classical gravity as interpreted by general relativity and its applications to astrophysics and cosmology. Part III is devoted to the problem of quantizing fields *other than gravity* in the presence of gravity. This description necessarily takes us to the current ideas of black-hole radiation and to the grand unified theories in the early universe. We come to quantum gravity *per se* in the final part, by which stage the reader will have become familiar with the complexity of the problem. At the outset it is fair to warn him that Part IV does not provide the master solution to all the difficulties. Rather it suggests one way along which some glimmer of understanding can be achieved. Let us briefly review here the problems and the tools that will be needed to deal with them.

What is the exact connection between classical and quantum physics? From our general remarks it is clear that the former is expected to follow from the latter in the limit $\hbar \to 0$. How does this limiting process operate in practice? Of all the different introductions to quantum mechanics, we find the path integral approach of Feynman intuitively easy to grasp. As we shall see in Chapter 2, the quantum mechanical propagator $K[f\,;\,i]$ giving the amplitude for a system to be in the final state f subject to its being earlier in the initial state i, gives in the limit $\hbar \to 0$ the result that the system would follow the classical behaviour specified by stationarizing the action S:

$$\delta S = 0. \tag{2}$$

While this method works very well in relating the classical behaviour of mechanical systems to their quantum behaviour, its intuitive advantage is not so apparent when describing the quantum behaviour of a 'field'. Unlike a mechanical system a field is characterized by a continuum of degrees of freedom. In Chapter 3 we develop methods that help us tackle problems which would arise in Chapter 4 when we undertake field quantization.

An unsolved problem to date in quantum theory of interacting fields is the computation of $K[f\,;\,i]$ in a finite and closed form. The perturbation theory invariably produces divergent integrals at each order of the expansion. In a limited class of field theories, pseudo-respectable techniques exist for extracting finite answers out of such infinities. Fortunately the field that is experimentally most accessible to us – viz. the electromagnetic field – belongs to this class. The measurements of Lamb shift, anomalous magnetic moment of the electron, etc., are instances where experiments have fully confirmed the finite answers extracted by these techniques.

By the 1970s it became clear that the property shared by fields to which these techniques of *renormalization* apply is that of gauge-invariance. Although the electromagnetic field is a 4-vector field, the vector A_i is not unique. The physical properties of the field are unchanged by the transformation

$$A_i \to A_i + \varphi_{,i}, \tag{3}$$

where φ is a scalar and $\varphi_{,i}$ is its gradient. Such a transformation is called *gauge transformation*. Chapter 5 discusses properties of gauge-invariant theories which also include the Salam–Weinberg theory of electro-weak interactions.

An important aspect of gauge-invariant field theories that finds use in cosmology is, of course, the phenomenon of spontaneous breakdown of symmetry. If we define vacuum as the state of lowest energy then it can happen that the state of lowest energy of a field system may change with temperature and so a state that represented vacuum at high temperature may not have that property at low temperature. The changeover may occur sharply at one critical temperature or, more interestingly, over a range of temperatures in analogy with the phase transition of a supercooled system.

As far as flat spacetime field theories are concerned, the phase transition manifests itself as a spontaneous breakdown of symmetry in the isospin space.

The new field configuration can acquire mass in such a transition whereas the original configuration had massless fields. To the uninitiated this may represent black magic or sleight of hand, but it is in fact the consequence of respectable mathematics, as we shall find in Chapter 5.

While these ideas have received backing in the recent experiments which have detected massive vector bosons that arise in the spontaneous symmetry breaking in the electro-weak formalism, their implications for cosmology are seen to be significant in the context of phase transitions in grand unified theories. To understand these it is necessary first to introduce the reader to classical general relativity and cosmology.

Accordingly, Part II begins with an elementary discussion in Chapter 6 of tensor calculus in Riemannian spacetimes and the arguments leading to general relativistic field equations. Chapter 7 discusses applications of Einstein's field equations to various situations in which massive distributions of matter generate curved spacetimes. The important difference between gravity and any other interaction studied by physicists is precisely demonstrated through such examples. In the latter case, the spacetime geometry is flat, regardless of the intensity of the field. In the former case the strength of gravity determines the geometry of spacetime. The dual role played by spacetime geometry – one of giving the background for describing physical reality and the other of giving the strength of gravity through its non-Euclidean nature – is mainly responsible for making the quantization of gravity so difficult.

While Chapter 7 describes phenomena ranging from the weak gravity cases of our Solar System to the strong gravity cases of black holes, Chapter 8 deals with cosmology. There is a basic difference between the two descriptions. In Chapter 7 we are concerned with the local effects of gravitating sources and these are characterized by geometries which are highly non-Euclidean near the sources but are asymptotically flat (i.e. of special relativity) away from them. In cosmology, on the other hand, we are interested in global (large-scale) effects of all pervading gravitating matter. Thus the models will exhibit global non-Euclidean character; but at the same time in the observer's local region they are presumed to be almost Euclidean. (An analogy is the arc of a circle which can be presumed to be 'straight' when its length is small compared to the radius of the circle.)

Reality, of course, combines both these properties but exact solutions of such situations are unfortunately not available. Being a nonlinear theory, relativity does not permit a simple combination of solutions of Chapter 7 with solutions of Chapter 8. The simplifying postulates of homogeneity and isotropy in cosmology therefore ignore local inhomogeneities like stars, galaxies, black holes, etc. We give in Chapter 8 a review of the present state of art in classical cosmology, ending with a series of problems that pose difficulties for the standard model, often glorified as the 'Big Bang model'. Because these problems do not seem to find solution at the classical level we have to invoke

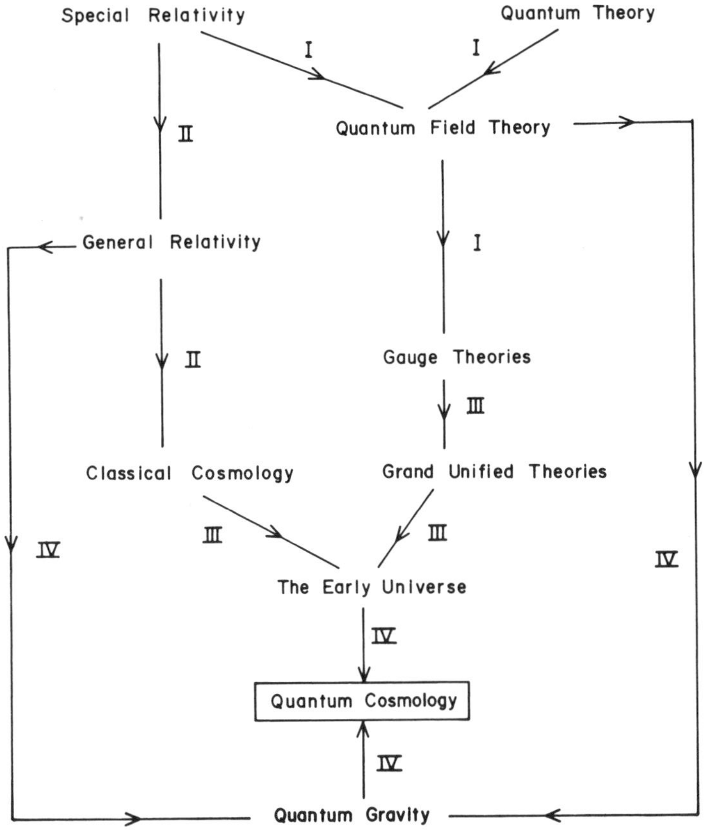

Fig. 1.1. Flow diagram showing the contents and interrelationship of the four parts of the book.

quantum cosmology.

Thus Parts I and II provide the necessary background for tackling problems of quantum gravity and quantum cosmology. We assume elementary knowledge of special relativity and quantum mechanics. While our discussion of most topics is sufficient for the reader to follow what comes next, the references given will help him to sharpen these tools of the trade further. In particular, a first course on quantum field theory will enable him to appreciate Part I better. Indeed, since quantum gravity is an interdisciplinary area between particle physics and general relativity it is very likely that a reader may be familiar with either Part I or Part II, but not with both.

The flow diagram shown in Figure 1.1 exhibits the developments leading to quantum cosmology. The roman figure besides each arrow indicates the part of this book in which the corresponding discussion is to be found. We have already described the developments that are to be discussed in Parts I and II. Here we briefly indicate how the discussions in Parts III and IV proceed.

Chapter 9 in Part III considers how quantum fields are described in curved spacetime. At first sight it might appear that with the application of the strong principle of equivalence (see Chapter 6) the techniques of flat space quantization can be trivially extended to the curved spacetime background. Not so! Although fields *per se* are local their interactions necessarily make us take note of the global properties of spacetime. Even the notion of vacuum becomes conceptually more difficult. And, to make matters worse, the notion of absolute energy presents new problems. It is not energy but differences of energy that matter in flat space quantization. However, in the curved spacetime the notion of energy turns out to be nontrivial: for, in general relativity the *total* energy density in a region of spacetime directly affects the geometry of that region. Thus 'infinities' in energy cannot be pushed under the rug any more.

With all these difficulties, only a modest success has been achieved in this area. Yet as the Hawking effect described in Chapter 9 indicates, the injection of quantum ideas seems to alter dramatically the classical gravitational picture of a black hole.

In Chapter 10 we return to the 'early universe' and see how the physical description of high-temperature matter in the grand unified theories alters the simple picture of the radiation universe. We discuss there the concept of the inflationary universe which is specifically related to the phase transition and spontaneous breakdown of symmetry in gauge theories. The idea certainly provides an important input to cosmology. It helps in solving some outstanding problems of the standard model but generates new problems of its own. Hence, searches for new cures of the problems of classical cosmology should certainly be kept on and it is here that quantum cosmology provides one likely avenue.

Thus in the final Part IV we come to quantum gravity and quantum cosmology. In Chapter 11 we outline some formal approaches to quantum gravity which are basically the logical extensions of the concepts and techniques of flat space field theory. It is not our purpose in this book to give the reader a comprehensive account of all such work on quantum gravity to date. Thus the ideas described in Chapter 11 merely indicate to the reader the flavour of the various important approaches. Our main emphasis, as mentioned earlier, is on our own approach which is described in Chapters 12 and 13.

It is a debatable issue whether one would gain more at this stage by adopting a highly formal stance or by being less ambitious and pragmatic. Had quantum theory waited on the sidelines until it acquired a formally fault-free and aesthetically beautiful formalism it would be waiting there even now. Likewise, in investigating such a complex phenomenon as quantum gravity we feel that tangible progress can be made by following the somewhat empirical and pragmatic approach of the early pioneers of quantum mechanics. In the last analysis, however, the final judgement rests with the reader.

Part I
Quantum Theory

Chapter 2

Path Integrals

2.1. Action in Classical Physics

The basic laws of classical physics are usually expressed as ordinary or partial differential equations. Although the form of the equations varies from law to law, there is a common characteristic shared by most of them: they are all derivable from a principle of stationary action. Let us begin by examining this principle in the context of the oldest established laws of physics: the laws of motion. We consider the simple example of a point particle moving in a one-dimensional space.

Let $q(t)$ denote the position of the particle at time t, m denote the mass of the particle, and let $V(q, t)$ be the potential in which the particle moves. The 'action' for the particle is given by the integral

$$S[q(t)] \equiv \int_{t_i}^{t_f} [\tfrac{1}{2}m\dot{q}^2 - V(q, t)]\, \mathrm{d}t; \quad \dot{q} = \frac{\mathrm{d}q}{\mathrm{d}t}\,. \tag{1}$$

We shall refer to the function $q(t)$ as the 'path' of the particle. Clearly, the integral in (1) can be evaluated for any path $q(t)$. Suppose that the particle was at q_i at time t_i and at q_f at time t_f. Consider the values of the action $S[q(t)]$ for all paths that satisfy the boundary conditions $q(t_i) = q_i$ and $q(t_f) = q_f$. The principle of stationary action can be stated as follows: the actual trajectory followed by the particle will be the one that makes the action S stationary for small variations of $q(t)$. In other words, if the trajectory followed by the particle is denoted by $q_c(t)$, then the value of S does not change for first-order variations in $q(t)$ as taken from $q_c(t)$. Writing

$$S = \int_{t_i}^{t_f} L(q, \dot{q}, t)\, \mathrm{d}t, \tag{2}$$

we get, for the variation in S,

$$\delta S = \int_{t_i}^{t_f} \left\{ \frac{\partial L}{\partial q}\, \delta q + \frac{\partial L}{\partial \dot{q}}\, \delta \dot{q} \right\} \mathrm{d}t$$

$$= \int_{t_i}^{t_f} \delta q \left\{ \frac{\partial L}{\partial q} - \frac{\mathrm{d}}{\mathrm{d}t}\left(\frac{\partial L}{\partial \dot{q}}\right)\right\} \mathrm{d}t + \frac{\partial L}{\partial \dot{q}} \, \delta q \, \Big|_{t_i}^{t_f} \quad . \tag{3}$$

In using the principle of stationary action we shall always consider variations of paths which vanish at the endpoints: $\delta q(t_i) = \delta q(t_f) = 0$. Then the second term in (3) vanishes. Therefore, $\delta S = 0$ implies

$$\frac{\mathrm{d}}{\mathrm{d}t}\left(\frac{\partial L}{\partial \dot{q}}\right) = \frac{\partial L}{\partial q} \; ; \tag{4}$$

that is, the trajectory q_c satisfies the equation

$$m \, \frac{\mathrm{d}^2 q_c}{\mathrm{d}t^2} = - \frac{\partial V}{\partial q_c} \tag{5}$$

which is the familiar Newton's law of motion. Among all solutions of (5), we have to choose the one that satisfies the boundary conditions $q(t_i) = q_i$ and $q(t_f) = q_f$.

This analysis shows that, mathematically, the principle of stationary action is completely equivalent to Newton's law of motion. All the same, physically, the two approaches treat the dynamics of the system on different footings.

The Newtonian law, crudely speaking, is a 'local' statement. Suppose that instead of specifying the positions at two different instants t_i and t_f, we had specified the position and velocity ($\dot{q}$) at $t = t_i$. Since (5) fixes $\ddot{q}$ at t_i, we can determine $\dot{q}$ at $(t_i + \delta t)$ and thus q at $t_i + \delta t$. This process can be iteratively repeated to construct the entire trajectory. The particle responds at each instant to the potential 'felt' by it at that instant. We do not require for our discussion any path $q(t)$ other than the physical trajectory $q_c(t)$.

Compared to this step-by-step analysis, the principle of stationary action works in a somewhat mysterious manner. The 'mechanism' by which the particle selects the physical trajectory of stationary action is not at all clear. As boundary conditions, we have specified the initial and final positions. The initial velocity is *not* given, so that the particle will not 'know' in which direction to start off and how fast to go. It is not clear how the particle can 'feel out' all trajectories and 'choose' the stationary one. It should be kept in mind that classical physics does not recognize any path other than the stationary path. Thus, out of a whole set of 'nonphysical' paths, introduced *a priori*, the classical principle of stationary action selects a unique physical trajectory through some mechanism which is not readily apparent.

Example 2.1. The action S can be evaluated for any path $q(t)$. Mathematically, one says that S is a functional of $q(t)$. Now consider the value of S for the classical trajectory $q_c(t)$ that connects (q_i, t_i) and (q_f, t_f). Let $S_c \equiv S[q_c]$. We can treat S_c as an ordinary function of (q_i, t_i) and (q_f, t_f):

$$S_c(q_f t_f; q_i t_i) = \int_{q_i, t_i}^{q_f, t_f} L(\dot{q}_c, q_c, t) \, \mathrm{d}t.$$

Suppose we vary q_f and t_f by infinitesimal amounts Δq and δt. We get

$$\delta S_c = L\,\delta t + \int_{q_i,t_i}^{q_f+\Delta q,t_f} \left(\frac{\partial L}{\partial \dot{q}}\,\delta\dot{q} + \frac{\partial L}{\partial q}\,\delta q \right) \mathrm{d}t$$

$$= L\,\delta t + \frac{\partial L}{\partial \dot{q}}\,\Delta q + \int_{q_i,t_i}^{q_f+\Delta q,t_f} \left\{ \frac{\partial L}{\partial q} - \frac{\mathrm{d}}{\mathrm{d}t}\left(\frac{\partial L}{\partial \dot{q}} \right) \right\}_{q=q_c} \mathrm{d}t.$$

The last term vanishes because of the equations of motion. A variation δt in t introduces a change $\dot{q}\ \delta t$ in q. Thus the total intrinsic variation in the q-coordinate is $\delta q = \Delta q + \dot{q}\ \delta t$. Writing δS_c in terms of δq, we get

$$\delta S_c = \frac{\partial L}{\partial \dot{q}}\,\delta q - \left(\dot{q}\,\frac{\partial L}{\partial \dot{q}} - L \right) \delta t.$$

We see that the momentum and energy of the particle can be expressed as the derivatives of S_c with respect to the end points:

$$p \equiv \frac{\partial L}{\partial \dot{q}} = \frac{\partial S_c}{\partial q_f}; \qquad H \equiv \dot{q}\,\frac{\partial L}{\partial \dot{q}} - L = -\frac{\partial S_c}{\partial t_f}.$$ □

Example 2.2. The action in (1) represents a particle which is nonrelativistic. For a particle which moves relativistically we have to modify this expression. Consider a free (relativistic) particle. The action for this particle can be taken to be

$$S = -mc \int \mathrm{d}s$$

where $\mathrm{d}s^2 = c^2\,\mathrm{d}t^2 - \mathrm{d}\mathbf{x}^2$, is the square of the spacetime interval between the events $(t, \mathbf{x})$ and $(t + \mathrm{d}t, \mathbf{x} + \mathrm{d}\mathbf{x})$. Writing $\mathrm{d}s = c\,\mathrm{d}t\,\sqrt{1 - (\mathbf{v}^2/c^2)}$, we get

$$S = -mc^2 \int \sqrt{1 - \frac{v^2}{c^2}}\,\mathrm{d}t \cong \int \left(-mc^2 + \frac{1}{2}mv^2 + O\left(\frac{v^4}{c^4} \right) \right) \mathrm{d}t$$

which has the correct nonrelativistic limit except for the $(-mc^2)$ term. This term contributes the 'rest mass energy' of the particle. Since,

$$\eta_{ik}\,\frac{dx^i}{\mathrm{d}s}\,\frac{\mathrm{d}x^k}{\mathrm{d}s} = \frac{\mathrm{d}s^2}{\mathrm{d}s^2} = 1,$$

we can also write S as

$$S = -mc \int \eta_{ik} \left(\frac{\mathrm{d}x^i}{\mathrm{d}s} \right) \left(\frac{\mathrm{d}x^k}{\mathrm{d}s} \right) \mathrm{d}s$$

$$= -\frac{mc}{2} \int \left[\eta_{ik} \left(\frac{\mathrm{d}x^i}{\mathrm{d}s} \right) \left(\frac{\mathrm{d}x^k}{\mathrm{d}s} \right) + 1 \right] \mathrm{d}s.$$

The first term in S is a four-dimensional generalization of $\frac{1}{2}m\mathbf{v}^2$ and the second term may be considered to be a constant potential energy. We shall have occasion to use this form later. □

2.2. Action in Quantum Physics

Quantum theory provides the 'mechanism' behind the principle of stationary action. What is more, the action S acquires a lot more significance in the quantum theory than it has in the classical theory.

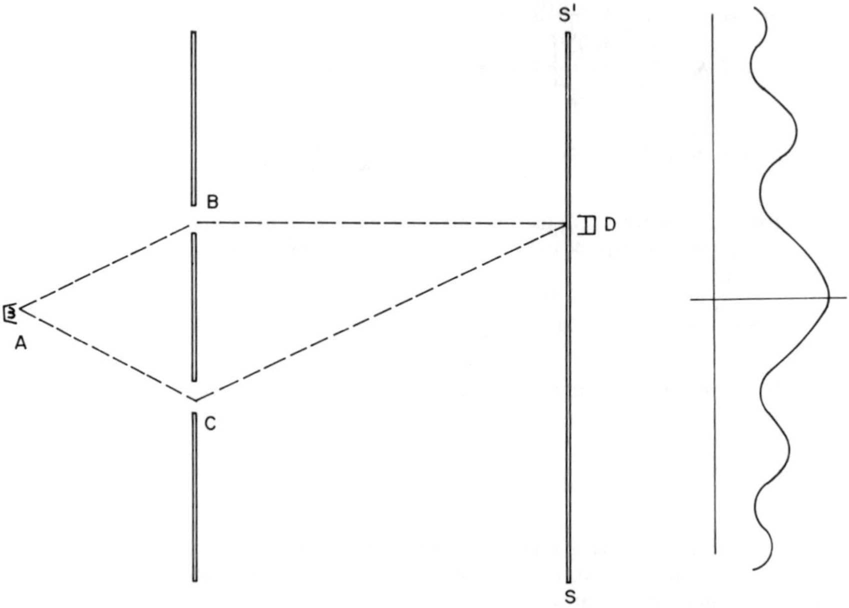

Fig. 2.1. The two slit interference pattern. The electrons emitted at A are detected by the detector D at various points in the screen SS'. The detected pattern (shown at far right) has maxima and minima indicating the interference of quantum amplitudes.

Let us consider again the example of the point particle moving in a potential $V(q, t)$. It is immediately obvious that the naïve iterative principle behind Newton's law (5) breaks down in quantum theory. This iteration depends on the simultaneous knowledge of position and velocity of the particle; such a knowledge is forbidden by the uncertainty principle in quantum theory. In fact the concept of a well-defined trajectory demands the same knowledge and hence cannot be used in quantum theory. We can still specify the end points q_i and q_f but cannot associate any specific physical trajectory connecting (q_i, t_i) and (q_f, t_f).

In other words, a quantum mechanical particle 'going' from (q_i, t_i) to (q_f, t_f) is actually free to choose all paths between (q_i, t_i) and (q_f, t_f). This idea is beautifully illustrated by the so-called 'two slit experiment' shown in Figure 2.1. The electrons emitted by the electron-gun at A are detected by the detector at D. Between A and D we have kept a screen with two slits B and C. The interference pattern observed by D cannot be explained by the known concepts of classical physics. This pattern clearly shows that every electron has a *probability amplitude* to go via slit B or via slit C and that these *amplitudes* add up to give the interference pattern.

Thus we are led to the following description: quantum mechanically, a particle may proceed from (q_i, t_i) to (q_f, t_f) by any path whatsoever connecting these points. There is a probability amplitude $A[q(t)]$ associated with each path $q(t)$. The total probability amplitude for the particle to 'propagate' from (q_i, t_i) to (q_f, t_f) is given by the sum over individual amplitudes $A[q(t)]$. Purely symbolically, we can write:

Probability amplitude for the particle to go from

$$(q_i, t_i) \text{ to } (q_f, t_f) \equiv K(q_f, t_f; q_i, t_i) = \sum_{\text{paths}} A[q(t)]. \tag{6}$$

Two points must be stressed about this equation. First, we have associated with each path $q(t)$ a complex number $A[q(t)]$ which is the 'probability amplitude' for the particular path. It is necessary that we work with probability amplitude (rather than probabilities) in order to produce the interference pattern of Figure 2.1. Second, note that all the paths that occur in the sum in (6) satisfy the boundary conditions $q(t_i) = q_i$ and $q(t_f) = q_f$. These are exactly the paths which were considered in the classical principle of stationary action. At this stage, we shall not bother to define the 'sum over paths' in (6) rigorously.

What can we say about the individual amplitudes $A[q(t)]$? The modulus $|A|^2$ denotes the *probability* for choosing the path $q(t)$; since quantum physics cannot grant special status to any particular path, this modulus must be a constant independent of $q(t)$. It follows that $A[q(t)]$ must have the form

$$A[q(t)] = N \exp i\, \theta[q(t)] \tag{7}$$

with N independent of $q(t)$. Therefore,

$$K(q_f, t_f; q_i, t_i) = N \sum_{\text{paths}} \exp i\, \theta[q(t)]. \tag{8}$$

Now the classical limit of quantum theory *does* single out one particular path $q_c(t)$ [for which $\delta S = 0$] from among all $q(t)$. In other words, all other terms that occur in the sum in (8) must cancel one another in the limit $\hbar \to 0$. This can be achieved most simply by taking $\theta = S/\hbar$, making

$$K(q_f, t_f; q_i\, t_i) = \sum_{\text{paths}} N \exp\left\{ i\, \frac{S[q(t)]}{\hbar} \right\}. \tag{9}$$

When $\hbar \to 0$, the phase $(S/\hbar)$ becomes very large and the amplitude oscillates rapidly even for small variations in the path, thus cancelling out the contributions from most of the paths. The only contributions that survive arise from the paths for which S does not vary (in first order) even when the paths are changed slightly. In other words, in the limit $\hbar \to 0$, only the path of stationary action contributes to K and we recover the classical theory.

One property of $K(q_f, t_f; q_i, t_i)$ can be obtained directly from the definition (9). Consider the class of all paths connecting (q_i, t_i) to (q_f, t_f). Let t be an instant in the interval (t_i, t_f). Suppose q is the value of $q(t)$ at that instant t for a particular path. Then the sum over all paths is equivalent to the following two summations: (i) sum over all paths from (q_i, t_i) to (q, t) and all paths from

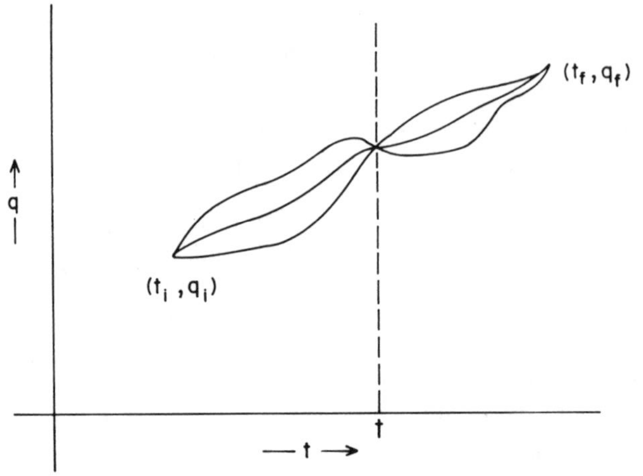

Fig. 2.2. Splitting up of all paths from (q_i, t_i) to (q_f, t_f). At the intermediate value of t $(t_i < t < t_f)$ one should integrate over all q.

(q, t) to (q_f, t_f); (ii) sum over all values of q(see Figure 2.2). We therefore get the integral relation

$$K(q_f, t_f; q_i, t_i) = \int_{-\infty}^{+\infty} K(q_f, t_f; q, t) K(q, t; q_i, t_i) \, \mathrm{d}q \tag{10}$$

which must be satisfied by the two-point function $K(q_f, t_f; q_i, t_i)$. This relation comes in handy in many cases.

It is usual to describe quantum mechanics of a point particle by the 'wave function' $\psi(q, t)$. The relation between ψ and K also follows immediately from the definition of K. Suppose the probability amplitude for the particle to be at q_i at time t_i is given by $\psi(q_i, t_i)$. Then, from the principle of addition of probability amplitudes, we get

$$\psi(q_f, t_f) = \int_{-\infty}^{+\infty} K(q_f, t_f; q_i, t_i) \psi(q_i, t_i) \, \mathrm{d}q_i. \tag{11}$$

In other words, given the initial configuration $\psi(q_i, t_i)$ the quantity K completely determines the future dynamics of the system. For this reason K is called the 'kernel' or 'propagator'.

In arriving at (9) we have used the terminology of a point particle. However, our discussion can be easily extended to any physical system described by an action S. The classical evolution of the system is described by the principle of stationary action $\delta S = 0$; the quantum mechanical evolution is described by the probability amplitude in Equation (9). Thus we have in (9) a general procedure for constructing the quantum theory of a dynamical system from its classical action, a procedure which tells us why the principle of stationary action 'works' in classical physics.

In order to proceed further we have to define precisely the notion of 'sum over paths' that appears in Equations (6), (8), and (9). We shall now turn our attention to this problem, which happens to be quite nontrivial.

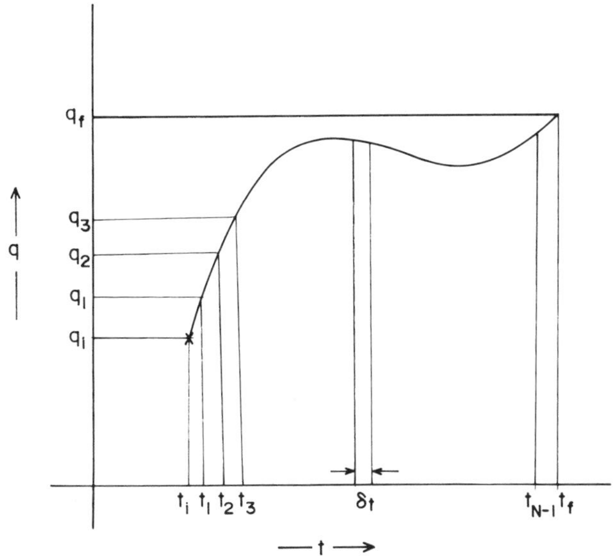

Fig. 2.3. Time slicing. A continuous path $q(t)$ can be approximated by a sequence of values $q_1, q_2, \ldots, q_{N-1}$, where $q_j = q(t_i + j\ \delta t)$. This approximation becomes exact $\delta t \to 0$.

2.3. The Path Integral

In order to give a meaning to the concept of sum over paths, we must first characterize and label the paths in some systematic way. Suppose we consider all paths with end points (q_i, t_i) and (q_f, t_f). Let us divide the time interval (t_i, t_f) into N equal parts of interval δt, such that

$$N\ \delta t = t_f - t_i.$$

A particular path $q(t)$ will take values $(q_1, q_2, \ldots, q_{N-1})$ at the intermediate instants $(t_1, t_2, \ldots, t_{N-1})$, where $t_A = t_i + A\delta t$. Any continuous path can be approximated arbitrarily well by the discrete set of points $(q_1, \ldots, q_{N-1})$, for sufficiently large N and small δt. This procedure is illustrated in Figure 2.3.

The kernel $K(q_f, t_f; q_i, t_i)$ can now be written, with the help of the transitivity relation (10), as

$$\begin{aligned} K(q_f, t_f; q_i, t_i) &= \int K(q_f, t_f; q_{N-1}, t_{N-1}) K(q_{N-1}, t_{N-1}; q_i, t_i)\, \mathrm{d}q_{N-1} \\ &= \iint K(q_f, t_f; q_{N-1}, t_{N-1}) K(q_{N-1}, t_{N-1}; q_{N-2}, t_{N-2}) \\ &\qquad K(q_{N-2}, t_{N-2}; q_i, t_i)\, \mathrm{d}q_{N-2}\, \mathrm{d}q_{N-1} \\ &= \iint \ldots \int K(f; N-1)\, K(N-1; N-2)] \ldots K(2; 1) \\ &\qquad K(1; i)\, \mathrm{d}q_{N-1}\, \mathrm{d}q_{N-2} \ldots \mathrm{d}q_1, \end{aligned} \tag{12}$$

where we have introduced a condensed notation

$$K(A;\, B) \equiv K(q_A,\, t_A;\, q_B,\, t_B). \tag{13}$$

Expression (12) becomes in the limit $N \to \infty$, $\delta t \to 0$ with $N\,\delta t$ constant:

$$K(f;i) = \lim_{\substack{N\to\infty \\ \delta t\to\infty}} \iint \ldots \int K(f;\, N\text{–}1) \ldots K(1,i)\, \mathrm{d}q_{N-1}\, \mathrm{d}q_{N-2} \ldots \mathrm{d}q_1 \tag{14}$$

All the kernels that appear on the right-hand side of (14) connect two events which are separated infinitesimally in time. Once this infinitesimal form of the kernel is known, (14) can be evaluated in a straightforward manner.

As a simple illustration, consider the case of a free particle with

$$S = \tfrac{1}{2}\int_{t_i}^{t_f} m\dot{q}^2\, \mathrm{d}t. \tag{15}$$

It can easily be shown that when t_A and t_B differ by an infinitesimal amount (with $t_A - t_B = \delta t$), $K(A;\, B)$ is given by,

$$K(q_A,\, t_A;\, q_B,\, t_B) = \left[\frac{m}{2\pi i\hbar\, \delta t}\right]^{1/2} \exp\left[\frac{im}{2\hbar\, \delta t}(q_A - q_B)^2\right]. \tag{16}$$

Example 2.3. When the two points $(q_A,\, t_A)$ and $(q_B,\, t_B)$ are separeted by an infinitesimal quantity, the only path that contributes in (9) is the straight line path. Then we have (denoting the normalization constant by $\bar{N}$)

$$K(q_A,\, t_A;\, q_B,\, t_B) = \bar{N} \exp\left\{\frac{im}{2\hbar}\int_{t_B}^{t_A} \frac{(q_A - q_B)^2}{(t_A - t_B)^2}\right\} \mathrm{d}t$$

$$= \bar{N} \exp\left\{\frac{im}{2\hbar}\frac{(q_A - q_B)^2}{t_A - t_B}\right\}.$$

In order to fix $\bar{N}$ consider the condition (with $t_A > t > t_B$)

$$K(q_A,\, t_A;\, q_B,\, t_B) = \int_{-\infty}^{+\infty} K(A;\, q,\, t)\, K(q,\, t;\, B)\, \mathrm{d}q,$$

that is,

$$\bar{N}(t_A,\, t_B) \exp\left\{\frac{im}{2\hbar}\frac{(q_A - q_B)^2}{(t_B - t_B)}\right\}$$

$$= \int_{-\infty}^{+\infty} \bar{N}(t_A,\, t)\bar{N}(t,\, t_B) \exp\frac{im}{2\hbar}\left[\frac{(q_A - q)^2}{t_A - t} + \frac{(q - q_B)^2}{t - t_B}\right] \mathrm{d}q$$

$$= \bar{N}(t_A,\, t)\bar{N}(t,\, t_B) \left\{\frac{2\pi i\hbar}{m}\frac{(t_A - t)(t - t_B)}{t_A - t_B}\right\}^{1/2} \exp\left\{\frac{im}{2\hbar}\frac{(q_A - q_B)^2}{(t_A - t_B)}\right\};$$

which implies the relation

$$\bar{N}(t_A,\, t_B) = \bar{N}(t_A,\, t)\bar{N}(t,\, t_B) \left(\frac{2\pi i\hbar}{m}\right)^{1/2} \frac{(t_A - t)^{1/2}(t - t_B)^{1/2}}{(t_A - t_B)^{1/2}}.$$

This relation can be satisfied by the choice

$$\bar{N}(x, y) = \left(\frac{m}{2\pi i\hbar} \right)^{1/2} \frac{1}{(x-y)^{1/2}},$$

thereby leading to Equation (16) of the text. □

This result allows us to perform the integrals in (12) in the following simple way. Let us begin with the q_1 integration which appears in the form

$$I_1 = \int_{-\infty}^{\infty} \left(\frac{2\pi i\hbar\, \delta t}{m} \right)^{-1/2} \left(\frac{2\pi i\hbar\, \delta t}{m} \right)^{-1/2} \exp \frac{im}{2\hbar\, \delta t} \left\{ (q_2 - q_1)^2 + (q_1 - q_i)^2 \right\} \mathrm{d}q_1. \tag{17}$$

Using the identity

$$\int_{-\infty}^{\infty} \exp[a(x - x_1)^2 + b(x - x_2)^2]\, \mathrm{d}x = \left(\frac{-\pi}{a+b} \right)^{1/2} \exp\left(\frac{ab}{a+b)} \right) (x_1 - x_2)^2 \tag{18}$$

we can write I_1 as

$$I_1 = \left(\frac{2\pi i\hbar\, \delta t}{m} \right)^{-2/2} \left(\frac{2i\acute{\pi}\,\hbar\delta t}{2m} \right)^{1/2} \exp \frac{im}{2\hbar(2\delta t)} (q_2 - q_i)^2$$

$$= \left(\frac{2\pi i\hbar\cdot 2\delta t}{m} \right)^{-1/2} \exp \frac{im}{2\hbar(2\delta t)} (q_2 - q_i)^2. \tag{19}$$

Notice that this expression is essentially the same as (16) with δt replaced by $(2\delta t)$. Doing the q_2 integration now, we get

$$I_2 \equiv \left(\frac{2\pi i\hbar\, \delta t}{m} \right)^{-1/2} \left(\frac{2\pi i\hbar 2\delta t}{m} \right)^{-1/2} \times \int_{-\infty}^{+\infty} \exp \frac{im}{2\hbar} \left\{ \frac{(q_2 - q_i)^2}{2\delta t} + \frac{(q_3 - q_2)^2}{\delta t} \right\} \mathrm{d}q_2 \tag{20}$$

$$= \left(\frac{2\pi i\hbar\cdot 3\delta t}{m} \right)^{-1/2} \exp \frac{im}{2\hbar} \frac{(q_3 - q_i)^2}{3\delta t}. \tag{21}$$

The pattern is now clear. After the $(N-1)$th integration over q_{N-I} in (14), we shall be left with

$$I_{N-I} \equiv \iint \ldots \int K(f; N-1) \ldots K(1; i)\, \mathrm{d}q_1\, \mathrm{d}q_2 \ldots \mathrm{d}q_{N-1}$$

$$= \left(\frac{2\pi\iota\hbar\cdot N\, \delta t}{m} \right)^{-1/2} \exp \frac{im}{2\hbar} \frac{(q_f - q_i)^2}{N\delta t}. \tag{22}$$

To obtain the kernel, we have to take the limit of (22) as $N \to \infty$, $\delta t \to 0$ with $N\, \delta t$ remaining constant at $(t_f - t_i)$. We thus arrive at the final result:

$$K[q_f, t_f; q_i, t_i) = \left[\frac{m}{2\pi i\hbar(t_f - t_i)}\right]^{1/2} \exp \frac{im}{2\hbar}\frac{(q_f - q_i)^2}{(t_f - t_i)}. \tag{23}$$

We have, therefore, succeeded in giving a meaning to the concept of 'sum over paths' at least for a free particle. The next logical step, obviously, would be to give a definition for a more general case. Unfortunately, it turns out to be an exceedingly difficult task if $S[q(t)]$ is an arbitrary functional. We shall not discuss this most general situation. However, when, $S[q(t)]$ has the form

$$S = \int [\tfrac{1}{2} m\dot{q}^2 - V(q, t)]\, \mathrm{d}t \tag{24}$$

then the following simple generalization of (14) exists for $K(f; i)$:

$$K(f; i) \equiv \lim_{\substack{N\to\infty \\ \delta t\to 0}} \iint \cdots \int \frac{1}{B} \exp \frac{i}{\hbar} \sum_{A=0}^{N-1} \left\{ \frac{m}{2}\frac{(q_{A+1} - q_A)^2}{\delta t} - \right.$$

$$\left. - V\left(\frac{q_{A+1} + q_A}{2}, t_A \right) \delta t \right\} \frac{\mathrm{d}q_{N-1}}{B} \cdot \frac{\mathrm{d}q_{N-2}}{B} \cdots \frac{\mathrm{d}q_1}{B}, \tag{25}$$

with

$$B = \left(\frac{2\pi i\hbar\, \delta t}{m} \right)^{1/2}. \tag{26}$$

For an arbitrary potential, it will not be possible to evaluate the integrals in (25) in a closed form. However, we shall show later that the expression in (25) does lead to the correct Schrödinger equation for a particle in a potential $V(q, t)$. This is the only rigorous justification that can be given to (25). We shall hereafter condense the notation in (25) and write

$$K(q_f, t_f; q_i, t_i) \equiv \int_{(q_i, t_i)}^{(q_f, t_f)} \exp \frac{i}{\hbar} S[q(t)]\, \mathcal{D}q(t). \tag{27}$$

The symbol $\mathcal{D}q(t)$ represents the series of integrations followed by the limit in (25). In analogy to ordinary integration, we shall call this expression, the 'path integral'.

Example 2.4. The path integral for the free particle presents an interesting analogy with the physics of one-dimensional Brownian motion. Consider a one-dimensional random walk in which all steps are of length Δx and take place at intervals of Δt. Let the probability for going to the right be p and the probability for going to the left be $q = 1-p$. Let $t = N\,\Delta t$ and $x = M\,\Delta x$. Let $A(M, N)$ be the probability for the particle to be at x at an instant t. If this was achieved after S_R steps to the right and S_L steps to the left, then we must have

$$S_R - S_L = M; \qquad S_R + S_L = N,$$

and from the binomial probability rule

$$A(M, N) = \binom{N}{S_R} p^{S_R} q^{S_L}, \quad \text{for } N - M = \text{even}$$
$$= 0, \qquad \text{for } N - M = \text{odd}.$$

Now, using Sterling's formula

$$n! \cong (2\pi n)^{1/2} \left(\frac{n}{e} \right)^n$$

we get, for large (M, N),

$$A(M, N) = \left\{ \frac{2}{\pi N} \right\}^{1/2} \exp \left\{ - \frac{(M - \bar{M})}{2N} \right\}^2$$

with $\bar{M} = N(p - q)$. Suppose that an ensemble of particles were performing the random walk. Then the density $\varrho(x, t)$ of particles per unit x interval, is just $A(M, N)/(2\Delta x)$. Therefore,

$$\varrho(x, t) = \left\{ \frac{\Delta t}{2\pi t(\Delta x)^2} \right\}^{1/2} \exp \left\{ - \frac{1}{2t} \left[\frac{\Delta t}{(\Delta x)^2} \right] (x - \bar{M}\ \Delta x) \right\}.$$

The continuum (diffusion) limit of Brownian motion is achieved by letting Δx, Δt, and $(p - q)$ go to zero while keeping

$$D \equiv \frac{(\Delta x)^2}{2(\Delta t)}, \qquad v \equiv \frac{\Delta x}{\Delta t}\ (p - q)$$

constant. In this limit

$$\varrho(x, t) = \left\{ \frac{1}{4\pi D t} \right\}^{1/2} \exp \left\{ - \frac{(x - vt)^2}{4Dt} \right\}.$$

In kinetic theory D is called the diffusion constant and v the drift velocity. If we replace t by $(-it)$ and identify $(\hbar/2m)$ with D, then the kernel in (23) becomes $\varrho(x, t)$ with zero drift velocity. Quantum mechanics with an imaginary time coordinate is thus quite similar to ordinary statistical mechanics. We shall say more about this analogy in our later chapters. □

2.4. The Quadratic Action

It is clear that the evaluation of K using (25) will be an extremely messy (if not impossible) job when $V \neq 0$. Fortunately, there exists a class of action functionals for which the path integral can be evaluated by an alternative method. We shall use this method repeatedly in various chapters of this book.

Consider the class of action functionals of the form

$$S = \int L\ \mathrm{d}t; \quad L = a(t)\dot{q}^2 + b(t)\dot{q} + c(t)q\dot{q} + d(t)q^2 + e(t)q + f(t). \qquad (28)$$

In other words, L contains terms up to second degree in $\dot{q}$ and q. We have to evaluate

$$K[f\,; i] = \int_{(q_i,\, t_i)}^{(q_f,\, t_f)} \exp \left\{ \frac{i}{\hbar} \int \mathrm{d}t\ L(\dot{q}, q, t) \right\} \mathcal{D}q(t). \qquad (29)$$

Let $q_c(t)$ denote the classical path that satisfies the principle of stationary action $(\delta S/\delta q) = 0$. Any arbitrary path $q(t)$ can be written as

$$q(t) = q_c(t) + \eta(t) \qquad (30)$$

with $\eta(t_i) = \eta(t_f) = 0$. From our basic definitions (25), (27) it follows that

$\mathcal{D}q = \mathcal{D}\eta$. Therefore,

$$K[f; i] = \int_{(0,\, t_i)}^{(0,\, t_f)} \left\{ \exp \frac{i}{\hbar} S[q_c + \eta] \right\} \mathcal{D}\eta(t). \tag{31}$$

We can expand S around the classical path $q_c(t)$ as

$$S = S_c(q_c) + S_1 + S_2, \tag{32}$$

where S_c depends only on q_c (i.e. terms independent of η), S_1 contains terms linear in η, and S_2 contains the part that is quadratic in η. But since q_c is the classical path that satisfies the $\delta S = 0$ condition, the linear term S_1 vanishes. Thus for (28) we are left with

$$S[q_c + \eta] = S_c(q_c) + \int_{t_i}^{t_f} [a(t)\dot{\eta}^2 + c(t)\eta\dot{\eta} + d(t)\eta^2]\, \mathrm{d}t, \tag{33}$$

so that

$$\begin{aligned} K(f; i) &= \int_{(0,\, t_i)}^{(0,\, t_f)} \exp \frac{i}{\hbar} \left\{ S_c(q_c) + \int_{(0,\, t_i)}^{t_f} \mathrm{d}t[a\dot{\eta}^2 + c\eta\dot{\eta} + d\eta^2] \right\} \mathcal{D}\eta\ (t) \\ &= \exp\left\{ \frac{i}{\hbar} S_c(q_c) \right\} \times \\ &\quad \times \int_{(0,\, t_i)}^{(0,\, t_f)} \exp\left[\frac{i}{\hbar} \int_{t_i}^{t_f} \mathrm{d}t \left\{ a\dot{\eta}^2 + c\eta\dot{\eta} + d\eta^2 \right\} \right] \mathcal{D}\eta(t). \end{aligned} \tag{34}$$

The crucial point to note about this expression is that all the (q_f, q_i) dependence is contained the $S_c(q_c)$ part. The remaining path integral is just a function of t_i and t_f. Calling this function $F(t_f, t_i)$ we write,

$$K(f; i) = F(t_f, t_i) \exp\left\{ \frac{i}{\hbar} S_c(q_f, t_f; q_i, t_i) \right\}. \tag{35}$$

In other words, *all the q-dependence of the kernel for a quadratic action can be determined completely by the value of the action for the classical trajectory.*

It is possible to express this function $F(t_f, t_i)$ as a determinant (usually called the Van Vleck determinant). But in the cases that we shall consider in this chapter, F can be determined by using (10). Techniques for evaluating this determinant will be discussed in later chapters as the need arises.

The free particle considered in Section 2.3 is, of course, a special case of the quadratic action. The classical solution for the free particle is

$$q_c(t) = q_i + \frac{(q_f - q_i)}{(t_f - t_i)} (t - t_i). \tag{36}$$

Substituting this into $S[q(t)]$ we get

$$S_c \equiv S[q_c] = \int_{t_i}^{t_f} \tfrac{1}{2} m \frac{(q_f - q_i)^2}{(t_f - t_i)^2} \mathrm{d}t = \tfrac{1}{2} m \frac{(q_f - q_i)^2}{(t_f - t_i)}\,. \tag{37}$$

We see that expression (35), along with (37), agrees with the form of the free particle kernel (23) derived via our detailed analysis.

Example 2.5. In a classical theory, equations of motion are more important than the action. The action in a classical theory is determined only up to an arbitrary function, in the following sense: two Lagrangians, L and

$$\bar{L} \equiv L + \frac{\mathrm{d}}{\mathrm{d}t} f(q, t),$$

will lead to the same equations of motion; but to different values of action, viz. S and

$$\bar{S} = S + \int \frac{\mathrm{d}f(q, t)}{\mathrm{d}t} \, \mathrm{d}t = S + f(q_f, t_f) - f(q_i, t_i).$$

It seems that the kernels for S and $\bar{S}$ will have entirely different dependences on (q_i, q_f) because

$$\bar{K}(q_f, t_f; q_i, t_i) = \mathrm{e}^{(i/\hbar)f(q_f, t_f)} K(q_f, t_f; q_i, t_i) \, \mathrm{e}^{-(i/\hbar)f(q_i, t_i)} .$$

However, this change can be completely neutralized by changing the wave functions $\psi(q, t)$ to

$$\bar{\psi}(q, t) = \mathrm{e}^{(i/\hbar)f(q, t)} \psi(q, t),$$

so that the evolution equation (11) retains the original form. To understand the physical meaning of such a phase change, notice that L and $\bar{L}$ have different canonical momenta, p and

$$\bar{p} = \frac{\partial \bar{L}}{\partial \dot{q}} = p + \frac{\partial f}{\partial q} .$$

In quantum theory the momentum operator is $(-ih \, \partial/\partial q)$ and we expect the above equation to become an operator equation

$$-i\hbar \frac{\partial \bar{\psi}}{\partial q} = -i\hbar \frac{\partial \psi}{\partial q} + \frac{\partial f}{\partial q} \psi.$$

The phase change accomplishes just this. Thus quantum physics does not change under the addition of a total time derivative to L. □

Example 2.6. There exist action functions which are not quadratic but can be transformed to the quadratic form by a change of variables. Consider for example the action

$$S = \int \dot{q}^2 \cdot f^2(q) \, \mathrm{d}t,$$

which is not quadratic. However, a simple transformation to the coordinate Q defined as,

$$Q = \int f(q) \, \mathrm{d}q$$

changes the action to

$$S = \int \dot{Q}^2 \, \mathrm{d}t,$$

which can be evaluated by our techniques. It is true that $\mathcal{D}q$ and $\mathcal{D}Q$ are related by a complicated factor; but $\mathcal{D}Q$ is the one that is defined via (25) (note that definition (25) is valid only for Lagrangians in the form (24)) and hence we can evaluate the path integral in the Q coordinate. (This choice is equivalent to a choice of factor ordering in the original action.) □

2.5. The Schrödinger Equation

In conventional approaches to quantum mechanics, the time evolution of the wave function $\psi(q, t)$ is determined by the equation

$$i\hbar \frac{\partial\psi}{\partial t} = -\frac{\hbar^2}{2m}\frac{\partial^2\psi}{\partial q^2} + V(q, t)\,\psi. \tag{38}$$

Given the initial wave function $\psi(q, t_i)$, this equation can be integrated forward in time, for all $t > t_i$. On the other hand, the path integral kernel does the same job through the equation

$$\psi(q, t) = \int_{-\infty}^{+\infty} K(q, t; q_i, t_i)\psi(q_i, t_i)\,\mathrm{d}q_i. \tag{39}$$

We have to make sure that the dynamics described by these two approaches is the same.

To do this, we proceed as follows: consider the value of ψ, determined via (39) at two infinitesimally separated instances t and $t + \varepsilon$. Using the fact that

$$K(q, t{+}\varepsilon; q_i, t) \cong \frac{1}{B}\exp\left[\frac{i}{\hbar}\varepsilon L\left(\frac{q + q_i}{2}, \frac{q - q_i}{\varepsilon}, t\right)\right], \tag{40}$$

(we have approximated $\dot{q}$ by $(q - q_i)/\varepsilon$ and q by $(q + q_i)/2$) we can write

$$\begin{aligned}\psi(q, t + \varepsilon) &= \int_{-\infty}^{\infty} \psi(q_i, t)\exp\left\{\frac{i}{\hbar}\varepsilon L\right\}\frac{\mathrm{d}q_i}{B}\\ &= \int_{-\infty}^{\infty} \psi(q_i, t)\exp\left\{\frac{i}{\hbar}\frac{m(q - q_i)^2}{\varepsilon}\right\}\times\\ &\quad\times\exp\left\{-\frac{i}{\hbar}\varepsilon V\left(\frac{q + q_i}{2}, t\right)\right\}\frac{\mathrm{d}q_i}{B}\end{aligned} \tag{41}$$

We write $q = q_i + \xi$ and expand both sides in a Taylor series in ξ retaining up to first order in ε:

$$\begin{aligned}\psi(q_i,t) + \varepsilon\frac{\partial\psi}{\partial t} &= \int_{-\infty}^{\infty}\exp\left\{\frac{im\xi^2}{2\hbar\varepsilon}\right\}\left(1 - \frac{i\varepsilon}{\hbar}V\right)\left(\psi + \xi\frac{\partial\psi}{\partial q_i} + \right.\\ &\quad\left. + \tfrac{1}{2}\xi^2\frac{\partial^2\psi}{\partial q_i^2}\right)\frac{\mathrm{d}\xi}{B} = \psi(q_i, t)\int_{-\infty}^{\infty}\exp\left\{\frac{im\xi^2}{2\hbar\varepsilon}\right\}\frac{\mathrm{d}\xi}{B} + \\ &\quad + \int_{-\infty}^{\infty}\xi\exp\left\{\frac{im\xi^2}{2\hbar\varepsilon}\right\}\frac{\partial\psi}{\partial q_i}\left(1 - \frac{i\varepsilon}{\hbar}V\right)\frac{\mathrm{d}\xi}{B} - \\ &\quad - \frac{i}{\hbar}\varepsilon\int_{-\infty}^{\infty}\exp\left\{\frac{im\xi^2}{2\hbar\varepsilon}\right\}\left[V\psi + \frac{V}{2}\xi^2\frac{\partial^2\psi}{\partial q_i^2}\right]\frac{\mathrm{d}\xi}{B} + \end{aligned}$$

$$+\int_{-\infty}^{\infty} \exp\left\{\frac{im\xi^2}{2\hbar\varepsilon}\right\} \tfrac{1}{2}\,\xi^2 \frac{\partial^2\psi}{\partial q_i^2}\cdot\frac{d\xi}{B} \quad . \tag{42}$$

Equating the zeroth-order term gives

$$\begin{aligned}\psi(q_i, t) &= \psi(q_i, t)\,\frac{1}{B}\int_{-\infty}^{+\infty} \exp\left\{\frac{im\xi^2}{2\hbar\varepsilon}\right\} d\xi \\ &= \frac{1}{B}\left(\frac{2\pi i\hbar\varepsilon}{m}\right)^{1/2} \psi(q_i, t).\end{aligned} \tag{43}$$

This equation is identically satisfied because of our choice of B. (In fact, it is this constraint which has forced us to use the particular form of B in (25)!)

The term involving $(\partial\psi/\partial q_i)$ vanishes identically because the integral vanishes. From an examination of the original expression, it is clear that first order of smallness in ε is equivalent to second order of smallness in ξ. Thus the term with $\varepsilon\xi^2$ is of one order higher than ε and can be dropped. In the remaining term, using

$$\int_{-\infty}^{\infty} \xi^2 \exp\left\{\frac{im\xi^2}{2\hbar\varepsilon}\right\}\frac{d\xi}{B} = \frac{i\hbar\varepsilon}{m} \tag{44}$$

we get the equality

$$\varepsilon\,\frac{\partial\psi}{\partial t} = -\,\frac{i\varepsilon}{\hbar}\,V\psi - \frac{\varepsilon\hbar}{2im}\,\frac{\partial^2\psi}{\partial q_i^2} \tag{45}$$

which is the same as

$$i\hbar\,\frac{\partial\psi}{\partial t} = -\,\frac{\hbar^2}{2m}\,\frac{\partial^2\psi}{\partial q^2} + V(q, t)\psi, \tag{46}$$

demonstrating the equivalence of the path integral and the Schrödinger equation approaches.

Though mathematically these two approaches are equivalent, from the physical point of view there is an interesting difference. The function $\psi(q, t)$ depends both on the dynamics and on the initial condition $\psi(q, t_i)$. In (46), there is no way of readily separating out these two aspects. There exist physical situations in which we would like to study the dynamics of the system without commiting ourselves to any particular initial conditions. The kernel is most suitable for such cases, since (39) clearly separates out the initial conditions from the dynamics. In short, kernel is independent of the initial conditions and represents the dynamics while the wave function $\psi(q, t)$ depends on both the dynamics and the initial conditions.

On the other hand, for many potentials V, the Schrödinger equation is easier to tackle than the corresponding path integral. Therefore, it would have been very convenient if we can find a way of disentangling the dynamics from the initial conditions in the Schrödinger equation itself. This is possible whenever the potential $V(q, t)$ is independent of time: $V(q, t) \equiv V(q)$. For such potentials, the Schrödinger equation becomes

$$\left[-\frac{\hbar^2}{2m} \frac{\mathrm{d}^2}{\mathrm{d}q^2} + V(q) \right] \varphi_E(q) = E\varphi_E(q), \tag{47}$$

where we have used the separation

$$\psi_E(q, t) = \mathrm{e}^{-iEt/\hbar}\, \varphi_E(q). \tag{48}$$

Equation (47) is the familiar eigenvalue equation for the Hamiltonian. From elementary quantum mechanics we know that φ_E forms a complete set of functions, when all eigenvalues of E are considered. We shall now show that the energy eigenstates $\{\varphi_E\}$ are related to the kernel in a simple manner.

Suppose the initial state of the system (at $t = 0$, say) was $\psi(q, 0)$. Since $\{\varphi_E\}$ forms a complete basis we can expand $\psi(q, 0)$ in terms of φ_E as

$$\psi(q, 0) = \sum_{\text{all } E} C_E \varphi_E(\mathrm{q}), \quad C_E = \text{const.} \tag{49}$$

From the orthonormality of $\{\varphi_E\}$ we know that

$$C_E = \int_{-\infty}^{\infty} \psi(q, 0)\, \varphi_E^*(q)\, \mathrm{d}q. \tag{50}$$

Now the energy eigenstates evolve in time with an $\exp(-iEt/\hbar)$ factor. Therefore, ψ at any other time, $\psi(q, t)$ is given by

$$\psi(q, t) = \sum_{\text{all } E} C_E \varphi_E(q) \exp\left(-\frac{i}{\hbar} Et \right). \tag{51}$$

Using (50) in (51) we get

$$\psi(q, t) = \sum_E \varphi_E(q) \exp\left(-\frac{i}{h} Et \right) \int_{-\infty}^{\infty} \psi(\bar{q}, 0) \varphi_E^*(\bar{q})\, \mathrm{d}\bar{q}$$

$$= \int_{-\infty}^{\infty} \psi(\bar{q}, 0) \sum_E \varphi_E(q)\, \varphi_E^*(\bar{q}) \exp\left(-\frac{i}{\hbar} Et \right) \mathrm{d}q. \tag{52}$$

Comparing with (39) we see that

$$K(q, t; \bar{q}, 0) = \sum_E \varphi_E(q)\, \varphi_E^*(\bar{q}) \exp\left(-\frac{iEt}{\hbar} \right). \tag{53}$$

In other words, for a potential which is independent of time, the energy eigenstates $\varphi_E(q)$ completely determine the kernel. In order to emphasize the time-independence of V we shall use the term 'stationary states' to describe $\psi_E(q, t)$ or $\varphi_E(q)$. Thus stationary states contain all the dynamical information about the system and are, of course, determined without any reference to initial conditions.

Our discussions have been modelled after the example of a point particle in one dimension. However, one can easily be convinced that the concepts introduced here have much wider domain of validity. For any classical system described by an action functional of the form (24) we can define path integral kernel, wave function, Schrödinger equation and – when, V is time independent – the stationary states. The quantity q may refer to any dynamical degree of freedom of the system.

2.6. The Spreading of Wave Packets

In this section we shall use the free particle kernel to study a purely quantum mechanical phenomenon which has come to be called the 'spreading of the wave packet'. Let us suppose that, at $t = 0$, the system variable q has the value a, say. Quantum mechanically, we expect the probability $|\psi(q, 0)|^2$ to be peaked about $q = a$. Ideally one would like to have

$$|\psi(q, 0)|^2 = (2\pi)^{-1}\, \delta(q-a). \tag{54}$$

However, such a completely certain description of q will make $\dot{q}$ completely undetermined at $t = 0$. In other words, the future evolution cannot be determined at all. To circumvent this difficulty, let us take the initial state to be such that $|\psi|^2$ is a Gaussian error function peaked at $q = a$. Let

$$\psi(q, 0) = \left[\frac{1}{2\pi\sigma^2(0)}\right]^{1/4} \exp\left\{-\frac{(q-a)^2}{4\sigma^2(0)}\right\} \quad (\text{at } t = 0), \tag{55}$$

denote such a wave function with dispersion $\sigma(0)$. In the limit of $\sigma(0) \to 0$, $|\psi(q, 0)|^2$ does go over to (54).

What is the wave function of the system at any other instant t? We know that

$$\psi(q, t) = \int_{-\infty}^{\infty} K(q, t; q_i, 0) \quad \psi(q_i, 0)\, dq_i. \tag{56}$$

Using the free particle kernel (23) and (55) we get

$$\psi(q, t) = \left(\frac{m}{2\pi i\hbar t}\right)^{1/2} \left[\frac{1}{2\pi\sigma^2(0)}\right]^{1/4} \times$$
$$\times \int_{-\infty}^{\infty} \exp\left\{\frac{im}{2\hbar}\frac{(q-q_i)^2}{t}\right\} \exp\left\{-\frac{(q_i-a)^2}{4\sigma^2(0)}\right\} dq_i. \tag{57}$$

The integral is again of the form in (18) and gives the result

$$\psi(q, t) = \left[\frac{m}{2\pi i\hbar t}\right]^{1/2} \left\{\frac{1}{2\pi\sigma^2(0)}\right\}^{1/4} \left\{\frac{4\pi\hbar\sigma^2(0)ti}{\hbar it + 2\sigma^2(0)m}\right\}^{1/2} \times$$
$$\times \exp\left\{-\frac{(q-a)^2}{4\sigma^2(0) + (2\hbar it/m)}\right\}. \tag{58}$$

Taking the modulus, we obtain

$$|\psi(q, t)|^2 = \left[\frac{1}{2\pi\sigma^2(t)}\right]^{1/2} \exp\left\{-\frac{(q-a)^2}{2\sigma^2(t)}\right\}, \tag{59}$$

where

$$\sigma^2(t) = \sigma^2(0) \left[1 + \frac{\hbar^2}{4m^2\sigma^4(0)} t^2\right]. \tag{60}$$

Thus at any time t, the probability is still peaked at the original (i.e. the $t = 0$) value $q = a$. However, the dispersion has increased from $\sigma^2(0)$ to $\sigma^2(t)$. So a Gaussian wave packet 'spreads' in time, even for a free particle.

The $\hbar$-dependent term in (60) has a simple interpretation based on the uncertainty principle. At $t = 0$, the dynamical variable q was known within an uncertainty of $\sigma(0)$. Therefore, the conjugate variable $\dot{q}$ is known with an uncertainty $\sim (\hbar/m\sigma(0))$. In a time interval t, the uncertainty in $\dot{q}$ manifests itself as an extra uncertainty in q of the order $(\hbar/m\sigma(0)t)$. Except for a factor of 2, this is the $\hbar$-dependent term that appears in (60).

The spreading of the wave packet, described by Equation (60), depends critically on the nature of the Lagrangian. In the free particle case, we see from (60) that $\sigma(t)$ is finite at all finite t and grows linearly with t for large t. As long as $\sigma(t)$ is finite, the mean value for q has a physical meaning; we may say that, in the above case, the mean value is a at all finite times.

It is possible to construct model Lagrangians in which the spread diverges at finite t. For example, consider the action

$$S = \tfrac{1}{2}\int mf(t)\dot{q}^2\, \mathrm{d}t. \tag{61}$$

By making the transformation to

$$T = \int_0^t \frac{\mathrm{d}t}{f(t)}, \tag{62}$$

this action S becomes

$$S = \tfrac{1}{2}\int m\left(\frac{\mathrm{d}q}{\mathrm{d}T}\right)^2 \mathrm{d}T, \tag{63}$$

which is just the action for a free particle. Repeating the analysis and transforming back to the t-coordinate, we find that

$$\sigma^2(t) = \sigma^2(0) \left[1 + \frac{\hbar^2}{4m^2\sigma^4(0)} \left\{\int_0^t \frac{\mathrm{d}\bar{t}}{f(\bar{t})}\right\}^2\right]. \tag{64}$$

It is easy to construct functions $f(t)$ such that $\int_0^t \mathrm{d}u/f(u)$ diverges at some finite $t = t_0$. In other words, $\sigma(t)$ will diverge at finite t_0. The mean value of the Gaussian then ceases to be of any physical interest as $t \to t_0$. Though such situations do not arise in ordinary quantum mechanics, they do occur in some quantum cosmological models that we shall consider in later chapters.

A much simpler example, in which the wave packet exhibits an interesting behaviour, is provided by a harmonic oscillator. The physics of the harmonic oscillator plays an important role throughout this book. We shall discuss various features of the harmonic oscillator in the next section.

Example 2.7. There exists an infinite class of functions which are peaked at a. What is special about the Gaussian? We can show that the Gaussian wave packet mimics the classical evolution very well in the limit of $\hbar \to 0$. Suppose that at $t = t_1$; the particle was described by

$$\psi_{q_1p_1}(q, t_1) = (2\pi\Delta^2)^{-1/4} \exp\left\{ -\frac{(q-q_1)^2}{4\Delta^2} + \frac{ip_1q}{\hbar} \right\}.$$

(Thus the expectation values of position and momentum are q_1 and p_1, respectively.) At a later time, let us assume that ψ had the same form with q_1 and p_1 replaced by q_2 and p_2. The transition amplitude between these two states is given by

$$A = \iint \psi^*_{q_2p_2}(x, t_2)\; K(x, t_2; y, t_1)\; \psi_{q_1p_1}(y, t_1)\, \mathrm{d}x\, \mathrm{d}y.$$

In the semiclassical limit, we consider only the classical path in the path integral and write

$$K(x, t_2; y, t_1) \cong N \exp\left\{ \frac{i}{\hbar} S_c(x, t_2; y, t_1) \right\}.$$

Then

$$A = (2\pi\Delta^2)^{-1/4} \iint N \exp\left[-\frac{(x-q_2)^2}{4\Delta^2} - \frac{ip_2x}{\hbar} + \right.$$
$$\left. + \frac{i}{\hbar} S_c(x, t_2; y, t_1) - \frac{(y-q_1)^2}{4\Delta^2} + \frac{ip_1y}{\hbar} \right] \mathrm{d}x\, \mathrm{d}y.$$

Now notice that in the semiclassical limit of $\hbar \to 0$, $\Delta \to 0$, most of the contribution arises from the regions where the argument of the exponential in the integrand is stationary with respect to x and y. That is, the contribution arises from points that satisfy the relations

$$q_2 = x, \qquad q_1 = y; \qquad -p_2 + \frac{\partial S_c}{\partial y} = 0; \qquad -p_1 + \frac{\partial S_c}{\partial x} = 0.$$

In other words (q_2, p_2) and (q_1, p_1) are related to each other by a canonical transformation generated by the action S_c, which is nothing but the action for the classical trajectory. □

2.7. The Harmonic Oscillator

The harmonic oscillator is described by the action of the form

$$S = \int \tfrac{1}{2} m(\dot{q}^2 - \omega^2 q^2)\, \mathrm{d}t, \tag{65}$$

where $m, \omega^2 > 0$. Since the action is quadratic the kernel can be written as

$$K(q_f, T; q_i, 0) = F(T) \exp\left\{ \frac{i}{\hbar} S_c[q_f, T; q_i, 0] \right\}, \tag{66}$$

where S_c is the classical value of the action. The classical equation of motion

$$\ddot{q} + \omega^2 q = 0 \tag{67}$$

has the solution (with $q_c(0) = q_i$ and $q_c(T) = q_f$),

$$q_c(t) = q_i \frac{\sin \omega(T-t)}{\sin \omega T} + q_f \frac{\sin \omega t}{\sin \omega T}, \tag{68}$$

Substituting (68) into (65) we get

$$S_c = \frac{m\omega}{2 \sin \omega T} \left\{ (q_i^2 + q_f^2) \cos \omega T - 2q_i q_f \right\}. \tag{69}$$

The quantity $F(T)$ can be determined by invoking (10). We get

$$F(T) = \left(\frac{m\omega}{2\pi i\hbar \sin \omega T} \right)^{1/2} \tag{70}$$

leading to the kernel

$$K(q_f, T, q_i, 0) = \left(\frac{m\omega}{2\pi i\hbar \sin \omega T} \right)^{1/2} \times$$
$$\times \exp\left[\frac{i}{\hbar} \frac{m\omega}{2 \sin \omega T} \left\{ (q_i^2 + q_f^2) \cos \omega T - 2q_i q_f \right\} \right]. \tag{71}$$

Example 2.8. Equation (10) implies that

$$K(q_f, T; q_i, 0) = \int_{-\infty}^{\infty} F(T-t)F(t) \times$$
$$\times \exp \frac{i}{\hbar} \left\{ S_c(q_f, T; q, t) + S_c(q, t; q_i, 0) \right\} dq.$$

Now

$$S_c(q_f, T; q, t) + S_c(q, t; q_i, 0) = \frac{m\omega}{2\hbar} \left\{ (q_f^2 + q^2) \cot \omega(T-t) - 2q_f q \operatorname{cosec} \omega (T-T) + \right.$$
$$\left. + (q^2 + q_i^2) \cot \omega t - 2qq_i \operatorname{cosec} \omega t \right\}$$
$$= \frac{m\omega}{2\hbar} \left\{ q_f^2 \cot \omega(T-t) + q_i^2 \cot \omega t \right\} +$$
$$+ \frac{m\omega}{2\hbar} q^2 \left\{ \cot \omega (T-t) + \cot \omega t \right\} -$$
$$- m\omega q \left\{ q_f \operatorname{cosec} \omega(T-t) + q_i \operatorname{cosec} \omega t \right\}$$
$$= A + Bq^2 + Cq \text{ (say).}$$

Therefore,

$$F(T) \exp \frac{i}{\hbar} S_c[q_f, T; q_i, 0]$$
$$= F(T-t)\, F(t) \exp \frac{i}{\hbar} A \left(\frac{2i\pi\hbar}{B} \right)^{1/2} \exp\left(-\frac{i}{\hbar} \frac{C^2}{4B} \right)$$

$$= F(T-t)\,F(t)\left\{\frac{2i\pi\hbar\sin\omega\,(T-t)\sin\omega t}{m\omega\sin\omega t}\right\}^{1/2}\quad \exp\frac{i}{\hbar}S_c(q_f,T;q_i,0),$$

where we have used the fact that

$$A-\frac{C^2}{4B}=S_c(q_f,T;q_i,0).$$

The equation for F,

$$F(T)=F(T-t)F(t)\quad\left(\frac{\sin\omega(T-t)\sin\omega t}{\sin\omega T}\cdot\frac{2i\pi\hbar}{m\omega}\right)^{1/2}$$

is solved by the choice

$$F(x)=\left(\frac{m\omega}{2\pi i\hbar\sin\omega x}\right)^{1/2}$$

thereby leading to (70). □

We shall now consider the various physical features of the system. Let us begin by looking at the evolution of a Gaussian wave packet, which at $t=0$ may be taken to be

$$\psi(q_i,0)=\left[\frac{1}{2\pi\sigma^2(0)}\right]^{1/4}\exp\left\{-\frac{(q_i-a)^2}{4\sigma^2(0)}\right\}. \tag{72}$$

We know that ψ at any later time is given by

$$\psi(q,T)=\int K(q,T;q_i,0)\psi(q_i,0)\,\mathrm{d}q_i. \tag{73}$$

Using (71) and (72),

$$\psi(q,T)=\int_{-\infty}^{\infty}F(T)\,[2\pi\sigma^2(0)]^{-1/4}\exp\left\{\frac{i}{\hbar}\,\frac{m\omega q^2}{2\tan\omega T}\right\}\times$$

$$\times\exp\left[\frac{i}{\hbar}\,\frac{m\omega}{2\sin\omega T}\,(q_i^2\cos\omega T-2qq_i)-\frac{(q_i-a)^2}{4\sigma^2(0)}\right]\mathrm{d}q_i$$

Performing the Gaussian integral and taking the modulus, we get

$$|\psi(q\;T)|^2=\left[\frac{1}{2\pi\sigma^2(T)}\right]^{1/2}\exp\left\{-\frac{(q-a\cos\omega T)^2}{2\sigma^2(T)}\right\}, \tag{74}$$

where

$$\sigma^2(T)=\sigma^2(0)\cos^2\omega T+\frac{1}{\sigma^2(0)}\left(\frac{\hbar}{2m\omega}\right)^2\sin^2\omega T. \tag{75}$$

The mean value of q oscillates in time as $a\cos\omega T$, which is the expected classical evolution for a harmonic oscillator. But notice that the spread σ^2 also oscillates in time between lower and upper bounds; this is in sharp contrast with (60) which represents a spread monotonically increasing with time.

From (75) we can also derive another important result. If we choose the spread of the initial state (carefully!) to be

$$\sigma^2(0) = \left(\frac{\hbar}{2m\omega}\right), \tag{76}$$

then we find that

$$\sigma^2(T) = \left(\frac{\hbar}{2m\omega}\right)[\sin^2 \omega T + \cos^2 \omega T] = \left(\frac{\hbar}{2m\omega}\right) = \sigma^2(0). \tag{77}$$

In other words, there exists a set of states for the harmonic oscillator characterized by only one parameter a, which do *not* spread in time. Using (76) in (74) we find

$$|\psi_a(q, T)|^2 = \left(\frac{m\omega}{\pi\hbar}\right)^{1/2} \exp\left\{-\frac{m\omega}{\hbar}(q - a\cos\omega T)^2\right\}. \tag{78}$$

For a given value of a the mean value of q oscillates in time as $a \cos \omega T$. Such states are called 'coherent states' and they describe very well a harmonic oscillator in a near-classical state.

Example 2.9. The quantum mechanics of the harmonic oscillator is usually discussed in terms of the creation and annihilation operators:

$$a^\dagger = \left(\frac{m\omega}{2\hbar}\right)^{1/2}\left(q - i\frac{p}{m\omega}\right); \qquad a = \left(\frac{m\omega}{2\hbar}\right)^{1/2}\left(q + i\frac{p}{m\omega}\right).$$

From the commutation rules $[q, p] \equiv qp - pq = i\hbar$, it follows that $[a, a^\dagger] = 1$. The ground state $|0\rangle$ is defined by

$$a\,|0\rangle = 0,$$

and the nth stationary state by

$$|n\rangle = \frac{(a^\dagger)^n}{\sqrt{n!}}\,|0\rangle.$$

The coherent states are still simpler to define if we use a and $a^\dagger$. A coherent state $|\alpha\rangle$ is just an eigenstate of a with the eigenvalue α. That is,

$$a\,|\alpha\rangle = \alpha\,|\alpha\rangle.$$

When we use the representation $q = q$, $p = -i\hbar(\partial/\partial q)$, this equation becomes (with $\psi_\alpha(q) = \langle q|\alpha\rangle$)

$$\sqrt{\frac{m\omega}{2\hbar}}\left(q + \frac{\hbar}{m\omega}\frac{\mathrm{d}}{\mathrm{d}q}\right)\psi_\alpha(q) = \alpha\psi_\alpha(q).$$

Integrating this equation and normalizing properly, we get

$$\psi_\alpha(q) = \left(\frac{m\omega}{\pi\hbar}\right)^{1/4}\exp\left\{-\frac{m\omega}{2\hbar}(q - \bar{\alpha})^2\right\}; \quad \bar{\alpha} = \sqrt{\frac{2\hbar}{m\omega}}\,\alpha.$$

An alternative definition of coherent state is

$$|\alpha\rangle = \exp(\alpha a^{\dagger} - \alpha^* a)|\,0\,\rangle.$$

Using $[a, a^{\dagger}] = 1$, it is easy to verify that this state is an eigenstate of a. □

The transition amplitude between two coherent states labelled by a and b proves to be of great use. This can again be evaluated by straightforward Gaussian integrations:

$$A(b, a) \equiv \int_{-\infty}^{\infty}\int_{-\infty}^{\infty} \psi_b^*(x, T)K(x, T; y, 0)\psi_a(y, 0)\,\mathrm{d}x\,\mathrm{d}y, \tag{79}$$

We get

$$A(b, a) = \exp\left[-\frac{i\omega T}{2} - \frac{m\omega}{4\hbar}(a^2 + b^2 - 2ab\,\mathrm{e}^{-i\omega T})\right]. \tag{80}$$

On the other hand since the harmonic oscillator potential is time-independent there must exist a complete set of stationary states $\{\varphi_n(q)\}$. Expanding the coherent states as

$$\psi_a(x) = \Sigma\, f_n(a)\varphi_n(x); \qquad \psi_b(x) = \Sigma\, f_n(b)\varphi_n(x) \tag{81}$$

we can also write (79) in the form

$$A(b, a) = \sum_n f_n^*(b)\, f_n(a) \exp\left(-\frac{iE_n\mathrm{T}}{\hbar}\right). \tag{82}$$

Expanding the exponential in (80) and comparing with (82) gives

$$E_n = \tfrac{1}{2}\hbar\omega(2n + 1) \tag{83}$$

and

$$f_n(a) = \left(\frac{m\omega}{2\hbar}\right)^{n/2} \frac{a^n}{\sqrt{n!}} \exp\left(-\frac{m\omega a^2}{4\hbar}\right). \tag{84}$$

Using the known form ψ_a and (84) in (81) we can immediately obtain the stationary states

$$\varphi_n(x) = (2^n n!)^{-1/2}\left(\frac{\omega}{\pi\hbar}\right)^{1/4} H_n\left(x\,\frac{m\omega}{\hbar}\right) \exp -\frac{m\omega x^2}{2\hbar}, \tag{85}$$

where $H_n(x)$ is defined through the expansion

$$\exp(-t^2 + 2ty) \equiv \sum_{n=0}^{\infty} H_n(y)\,\frac{t^n}{n!} \tag{86}$$

(the details of the algebra are given in Example 2.10). Equations (83) and (85) give the familiar energy eigenvalues and the eigenfunctions for the harmonic oscillator.

Example 2.10. Expanding (80) as a power series and comparing with (82) we have

$$\sum_n f_n^*(b)\, f_n(a) \exp - \frac{i}{\hbar} E_n T$$

$$= \exp - \frac{m\omega}{4\hbar}(a^2 + b^2) \times$$

$$\times \sum_{n=0}^{\infty} \left(\frac{m\omega}{2\hbar}\right)^n \frac{a^n b^n}{n!} \exp -i\omega T(n + \tfrac{1}{2})$$

$$= \sum_{n=0}^{\infty} \left(\frac{m\omega}{2\hbar}\right)^{n/2} \frac{b^n}{\sqrt{n!}} \exp - \frac{m\omega}{4\hbar} b^2 \times$$

$$\times \left(\frac{m\omega}{2\hbar}\right)^{n/2} \frac{a^n}{\sqrt{n!}} \exp - \frac{m\omega}{4\hbar} a^2 \times$$

$$\times \exp -i\omega T(n + \tfrac{1}{2})$$

Therefore,

$$E_n = \hbar\omega(n + \tfrac{1}{2}); \quad f_n(a) = \left(\frac{m\omega}{2\hbar}\right)^{n/2} \frac{a^n}{\sqrt{n!}} \exp - \frac{m\omega}{4h} a^2 \cdot$$

Equation (81) can now be written as

$$\psi_a(x) = \left(\frac{m\omega}{\pi\hbar}\right)^{1/4} \exp\left(- \frac{m\omega}{2\hbar}(x - a)^2 \right)$$

$$= \sum f_n \varphi_n(x) = \sum_{n=0}^{\infty} \left(\frac{m\omega}{2\hbar}\right)^{n/2} \exp\left(- \frac{\mu\omega}{4\hbar} a^2 \right) \frac{a^n}{\sqrt{n!}} \varphi_n(x).$$

Let

$$\varphi_n(x) \equiv (2^n n!)^{-1/2} \exp - \frac{m\omega x^2}{2\hbar} \cdot \left(\frac{m\omega}{\pi\hbar}\right)^{1/4} B_n(x),$$

where $B_n(x)$ are yet to be determined. Using this form we get

$$\exp - \frac{m\omega}{2\hbar}\left(\frac{a^2}{2} - 2ax\right) = \sum_{n=0}^{\infty} \left(\frac{m\omega}{2\hbar}\right)^{n/2} \frac{1}{2^{n/2} n!} B_n(x) a^n.$$

Putting

$$s = \frac{m\omega}{\hbar} \cdot \frac{a}{2} \quad \text{and} \quad \xi = x \cdot \frac{m\omega}{\hbar},$$

we have

$$\exp\left(- \frac{m\omega}{\hbar} \cdot \frac{a^2}{4} + \frac{m\omega}{\hbar} ax \right) = \exp(-s^2 + 2s\xi).$$

The Hermite polynomials are *defined* by the expansion

$$\exp\left(-s^2 + 2s\xi \right) = \sum_{n=0}^{\infty} \frac{H_n(\xi)}{n!} s^n.$$

Therefore,

$$\sum_{n=0}^{\infty} \frac{H_n(\xi)}{n!} s^n = \sum_{n=0}^{\infty} s^n \frac{1}{n!} B_n(x),$$

allowing the identification: $B_n = H_n$. Substituting back into $\varphi_n(x)$ leads to (85) of the text. □

A simple variant of the harmonic oscillator which will play an important role in our future discussions is described by the action

$$S = \int \{\tfrac{1}{2} m(\dot{q}^2 - \omega^2 q^2) + f(t)q(t)\}\ \mathrm{d}t \tag{87}$$

in which $f(t)$ is an external 'forcing' function. Since the action is still quadratic, the path integral kernel can be written in the form

$$K(q_f, t_f; q_i, t_i) = F(t_f, t_i) \exp \frac{i}{\hbar} S_c(q_f, t_f; q_i, t_i)\,. \tag{88}$$

The function $F(x)$ in (88) *must* be the same as the one in (66). To see this result, notice that F is determined by the path integral in (34) (see Section 2.4). This path integral does *not* depend on the term which is linear in $q(t)$ in the action, and depends only on the quadratic terms. Since the quadratic terms for (87) and (65) are the same, $F(t_f, t_i)$ is still given by (70).

To determine the classical action we have to solve the equation

$$\ddot{q} + \omega^2 q = m^{-1} f(t) \tag{89}$$

subject to the boundary conditions $q(t_i) = q_i$ and $q(t_f) = q_f$. We shall see in later chapters how Green's functions can be used to solve equations of this type. In one dimension, however, it is easy to verify directly that a particular solution to (89) is

$$q(t) = \int_0^t \frac{f(u)}{m\,\omega} \sin \omega(t-u)\ \mathrm{d}u. \tag{90}$$

We have

$$\begin{aligned} \frac{\mathrm{d}^2}{\mathrm{d}t^2} q &= \frac{\mathrm{d}}{\mathrm{d}t} \left\{ \int_0^t \frac{1}{m} \mathrm{f}(u) \cos \omega(t-u)\ \mathrm{d}u \right\} \\ &= m^{-1} f(t) - \int_0^t \frac{\omega}{m} f(u) \sin \omega(t-u)\ \mathrm{d}u \\ &= m^{-1} f(t) - \omega^2 q, \end{aligned} \tag{91}$$

which is the same as (89). A general solution to (89) is found by adding to (90) any solution of the homogeneous equation (i.e. Equation (89) with $f = 0$). Thus, we may take

$$q_c(t) = A \sin \omega t + B \cos \omega t + \int_0^t \frac{1}{m\omega} f(u) \sin \omega(t-u)\ \mathrm{d}u, \tag{92}$$

where A and B are determined by the conditions $q\ (t_i) = q_i$ and $q(t_f) = q_f$.

The classical action S_c can now be found by substituting (92) into (87). We get

$$S_c = S_c(f=0) + q_f\int_{t_i}^{t_f} f(u)\frac{\sin\omega(u-t_i)}{\sin\omega T}\,du + q_i\int_{t_i}^{t_f} f(u)\,\frac{\sin\omega(t_f-u)}{\sin\omega T}\,du$$
$$-\frac{1}{m\omega\sin\omega T}\int_{t_i}^{t_f} du_1\int_{t_i}^{u_1} f(u_1)f(u_2)\sin\omega(t_f-u_1)\sin\omega\,(u_2-t_i)\,du_2, \tag{93}$$

where $S_c(f=0)$ is the classical action for the harmonic oscillator (69). We shall have occasion to use this result later.

The external force $f(t)$ acts as a driving term in the classical picture. Quantum mechanically, $f(t)$ acts as a time-dependent perturbation, causing transitions between the harmonic oscillator energy levels. If the initial state at $t=0$ is taken to be the ground state of the harmonic oscillator,

$$\psi_0(q_i) = \left(\frac{m\omega}{\pi\hbar}\right)^{1/4}\exp\left(-\frac{m\omega q_i^2}{2\hbar}\right), \tag{94}$$

then we can ask for the probability that at $t=T$, the system is still in the ground state. If the probability is less than unity, it is clear that excitations to higher energy levels are taking place. The probability amplitude in question is given by

$$I = \int_{-\infty}^{+\infty}\int_{-\infty}^{+\infty} \psi_0^*(q_f,T)\,K(q_f,T;q_i,0)\,\psi_0(q_i,0)\,dq_i\,dq_f. \tag{95}$$

Performing the Gaussian integrations, we get (see Example 2.11)

$$I = \exp\left[-\frac{1}{2m\omega\hbar}\int_0^T f(t)\int_0^t f(s)\,e^{-i\omega(t-s)}\,ds\,dt\right] \tag{96}$$

which is clearly less than unity. The ground state of the harmonic oscillator plays the role of the vacuum in quantum field theory, and the quantity I represents the probability amplitude for the vacuum to remain a vacuum in the presence of external perturbations.

Example 2.11. The evaluation of (95) is straightforward but tedious. When written out in full (95) reads as

$$I = \left(\frac{m\omega}{\pi\hbar}\right)^{1/2}\left(\frac{m\omega}{2\pi i\hbar\sin\omega T}\right)^{1/2}\int_{-\infty}^{+\infty} dx\int_{-\infty}^{+\infty} dy\,\exp\frac{m\omega}{2\hbar}\times$$
$$\times\left\{C(x^2+y^2) + \frac{2i}{\sin\omega T}(By + Ax - xy) + D\right\},$$

where

$$C = i\,e^{i\omega T}\operatorname{cosec}\omega T,$$

$$A = \frac{1}{m\omega}\int_0^T f(t)\sin\omega t\,dt,$$

$$B = \frac{1}{m\omega} \int_0^T f(t) \sin \omega(T - t) \, \mathrm{d}t,$$

$$D = \frac{i}{m^2\omega^2} \int_0^T \int_0^t f(t)f(s) \frac{\sin \omega (s) \sin \omega (T - t)}{\sin \omega T} \, \mathrm{d}t \, \mathrm{d}s.$$

The integration over x, y can be performed by using

$$\int_{-\infty}^{+\infty} \mathrm{d}x \, \mathrm{e}^{ax^2 + bx} = \sqrt{\frac{\pi}{-a}} \exp\left(-\frac{b^2}{4a}\right)$$

This will give

$$I = \exp \frac{m\omega}{2\hbar} \left\{ D + \frac{1}{(\sin \omega T)} \frac{(A^2 + B^2)C \sin \omega T + 2\, iAB}{(1 + C^2 \sin \omega T)} \right\}$$

Substituting the expressions for A, B, C, D leads to expression (96) given in the text, provided $T \gg \omega^{-1}$.

The above result can be generalized to give the probability amplitude for the oscillator to be found at the nth energy level (at $t = T$) starting at the ground state at $t = 0$. The result in this case is

$$I_n = I_0(n!)^{-1/2}[i\tilde{f}(\omega)]^n,$$

where

$$\tilde{f}(\omega) = \frac{1}{m\sqrt{2\omega}} \int_0^T f(t) \, \mathrm{e}^{-i\omega t} \, \mathrm{d}t. \qquad \square$$

Notes and References

1. The best exposition of the subject of path integrals can be found in:

   Feynman, R. P., and Hibbs, A. R.: 1965, *Quantum Mechanics and Path Integrals*, McGraw-Hill, New York.

A discussion of topics which are somewhat more 'modern' can be found in:

   Schulman, L. S.: 1981, *Techniques and Applications of Path Integration*, Wiley, New York.

This book also gives an extensive list of references.

2. We assume that the reader is familiar with the concepts of mechanics at the level of:

   Landau, L. D., and Lifshitz, E. M.: 1973, *Mechanics*, Pergamon, London.

   and quantum mechanics at the level of:

   Schiff, L.: 1968, *Quantum Mechanics*, McGraw-Hill, New York.

3. A more mathematical discussion of functional integrals, measure, etc., can be found in:

   Albeverio, S. A., and Hough-Kohn R. J.: 1976, 'Mathematical theory of Feynman Path Integrals', *Lecture Notes in Math.* **523** (Springer-Verlag, Heidelberg).

4. More detailed discussion of the connection between probability theory and path integrals can be found in:

   Kac, M.: 1959, *Probability and Related Topics in the Physical Sciences*, Interscience, New York.

5. The evaluation of F in (34) is given in Schulman (1981) (see Note 1 above). Further details of the harmonic oscillator can also be found in the books cited in Note 1 above.

Chapter 3

En Route to Quantum Field Theory

3.1. The Field as a Dynamical System

The path integral approach described in the previous chapter is the simplest and most natural way of going over from classical mechanics to quantum mechanics. As discussed in Section 2.2, classical mechanics follows as the limit of quantum mechanics when $\hbar \to 0$. It is natural, therefore, to ask whether the same approach can give us the *quantum* theory of fields starting from what we know about the *classical* field theory, in a way that the latter follows from the former when $\hbar \to 0$.

The answer to the above question is "Yes, but . . .". The quantization of field brings in new issues both conceptual and computational, with the result that the transition from a dynamical system with finite degrees of freedom to a field system is nontrivial. In this chapter we will pave the way of a description of the quantum field theory by discussing some additional concepts. These concepts will be needed in the next chapter when we come to field quantization proper.

To begin with, we note that the typical field may be looked upon as a dynamical system with uncountably infinite degrees of freedom. Take, for example, a massive scalar field ϕ satisfying the wave equation

$$\Box\phi + m^2\phi = 0. \tag{1}$$

At first sight such a field looks quite different from, say, a harmonic oscillator whose displacement $q(t)$ satisfies the dynamic equation

$$\ddot{q} + \omega^2 q = 0. \tag{2}$$

However, let us look at (1) more closely before comparing with (2). Suppose we subject ϕ to Fourier analysis in the usual way:

$$\phi(t, \mathbf{r}) = \int q(\mathbf{k}, t) \exp(-i\mathbf{k} \cdot \mathbf{r}) \frac{d^3\mathbf{k}}{(2\pi)^3} . \tag{3}$$

Then substitution into (1) gives us the following condition on $q(\mathbf{k}, t)$:

$$\ddot{q}(\mathbf{k}, t) + (k^2 + m^2)\, q(\mathbf{k}, t) = 0, \quad k = |\mathbf{k}| . \tag{4}$$

Clearly, a similarity of (4) with (2) is now obvious and we can make it complete by identifying

$$\omega^2(\mathbf{k}) = k^2 + m^2. \tag{5}$$

What does all this mean? First, the integral (3) describes any ϕ as a linear combination of the exponentials $\exp(-i\mathbf{k} \cdot \mathbf{r})$ which are the base functions in Fourier analysis. Thus, in terms of such a basis all the dynamics of $\phi(t, \mathbf{r})$ is invested in the functions $q(\mathbf{k}, t)$. The q's are therefore the dynamical coordinates and being labelled by the continuous vector $\mathbf{k}$, their number is uncountably infinite.

Thus the path integral method, when applied to fields, must formulate all its techniques for a continuum of dynamical variables rather than for a finite discrete set of them. The complications caused by this circumstance will become apparent gradually as we approach our study of field theory.

At the same time we can take comfort in the fact that most physical fields satisfy straightforward wave equations like (1) and hence their dynamical behaviour is simulated to some extent by the simple harmonic oscillator which is a very well understood quantum mechanical system. The one important exception to this statement is, of course, the gravitational field whose quantization is considerably more difficult. We shall turn our attention to that problem in the last part of this book.

We begin our discussion by studying further properties of the harmonic oscillator.

3.2. Harmonic Oscillator in an External Potential

In Section 2.7 we discussed the harmonic oscillator with the help of path integrals. Starting with the Lagrangian

$$L(q, \dot{q}) = \tfrac{1}{2}(\dot{q}^2 - \omega^2 q^2) \tag{6}$$

for an oscillator of unit mass we arrived at the kernel

$$K(q_f, t_f; q_i, t_i) = \left(\frac{\omega}{2\pi i\hbar \sin\omega T} \right)^{1/2} \exp\left\{ \frac{i\omega}{2\hbar \sin \omega T} \times \right.$$
$$\left. \times \left[(q_f^2 + q_i^2) \cos \omega T - 2q_i q_f \right] \right\}, \tag{7}$$

where $T = t_f - t_i$.

However, such a 'free' harmonic oscillator has only limited applications, especially in field theory. In real life we wish to study the behaviour of interacting systems and hence it would be more relevant to know how a harmonic oscillator interacts with other disturbances. We would then be able to study the quantum theory of interacting fields.

Accordingly, let us add an external potential term to the Lagrangian of (6) and write

$$L = \tfrac{1}{2}(\dot{q}^2 - \omega^2 q^2) - \varepsilon V(q) = L_0 - \varepsilon V(q), \tag{8}$$

where $\varepsilon \ll 1$, to begin with. Such an assumption implies that the external disturbance is small.

Nevertheless, the modified path integral

$$K(q_f, t_f; q_i, t_i) = \int \exp\left(i \int_{t_i}^{t_f} \{\tfrac{1}{2}(\dot{q}^2 - \omega^2 q^2) - \varepsilon V(q)\}\, \mathrm{d}t \right) \mathcal{D}\, q(t) \tag{9}$$

cannot be evaluated in closed form for a general $V(q)$. [We shall henceforth set $\hbar = 1$ unless otherwise stated.]

Two approximate procedures are available to us. In the *perturbative* approximation we expand the exponential in (9) as a power series in $\varepsilon V(q)$ and compute K to any desired finite order in the power series for ε. In the *nonperturbative* method we try to evaluate the path integral by the saddle-point method or similar techniques. We shall discuss both approaches in this chapter.

Example 3.1. Some aspects of these path integral evaluations can be understood in comparison with ordinary integrals. Suppose we need to evaluate

$$I = \int_{q_1}^{q_2} \exp\{-q^2 + \varepsilon V(q)\}\, \mathrm{d}q.$$

For small ε, we can expand $\exp \varepsilon V(q)$ and write

$$I = \sum_{n=0}^{\infty} \frac{\varepsilon^n}{n!} \int_{q_1}^{q_2} [V(q)]^n \exp(-q^2)\, \mathrm{d}q$$

$$\cong \int_{q_1}^{q_2} \left\{ 1 + \varepsilon V(q) + \frac{\varepsilon^2}{2!} V^2(q) + \cdots \right\} \exp(-q^2)\, \mathrm{d}q.$$

For well-behaved $V(q)$ and sufficiently small ε the integral can be approximately by the first few terms of the above expansion.

Similar expansion exists for the path integral. Writing

$$\exp\left[-i\varepsilon \int_{t_1}^{t_2} V(q)\, \mathrm{d}t \right] = \sum_{n=0}^{\infty} \frac{(-i\varepsilon)^n}{n!} \left\{ \int_{t_1}^{t_2} V(q)\, \mathrm{d}t \right\}^n$$

the path integral becomes

$$K(q_2, t_2; q_1, t_1) = \sum_{n=0}^{\infty} \frac{(-i\varepsilon)^n}{n!} \int \left\{ \int_{t_1}^{t_2} V(q)\, \mathrm{d}t \right\}^n \exp i \int_{t_1}^{t_2} \{\tfrac{1}{2}(\dot{q}^2 - \omega^2 q^2\}\, \mathrm{d}t\, \mathcal{D}q$$

$$\cong \int \left\{ 1 - i\varepsilon \int_{t_1}^{t_2} V(q)\, \mathrm{d}t + \frac{(i\varepsilon)^2}{2!} \int_{t_1}^{t_2} \int_{t_1}^{t_2} V(q(t))\, V(q(t'))\, \mathrm{d}t\, \mathrm{d}t' \right\} \times$$

$$\times \exp\left(i \int L_0\, \mathrm{d}t \right) \mathcal{D}q + O(\varepsilon^3).$$

For small ε we can approximate the kernel by the first few terms. We shall see later how these terms can be evaluated in a systematic fashion. For example, using the fundamental definition of path integral as a product of ordinary integrals (see Equation (2.25)) we can write

$$\int (-i\varepsilon) \int_{t_1}^{t_2} V[q]\,\mathrm{d}t \exp\left(i \int_{t_1}^{t_2} L_0\,\mathrm{d}t\right) \mathcal{D}\, q(t)$$

$$= -i\varepsilon \int_{t_1}^{t_2} \int K^0(q_2, t_2; q, t)\, V[q]\, K^0(q, t; q_1, t_1)\,\mathrm{d}q\,\mathrm{d}t$$

where K^0 is the kernel for the harmonic oscillator, as given by (7).

Similar expression exist for higher order terms. This perturbation expansion is expressed in diagrammatic fashion in Figure 3.1. Each straight line from $(q_1, t_1$ to $q_2, t_2)$

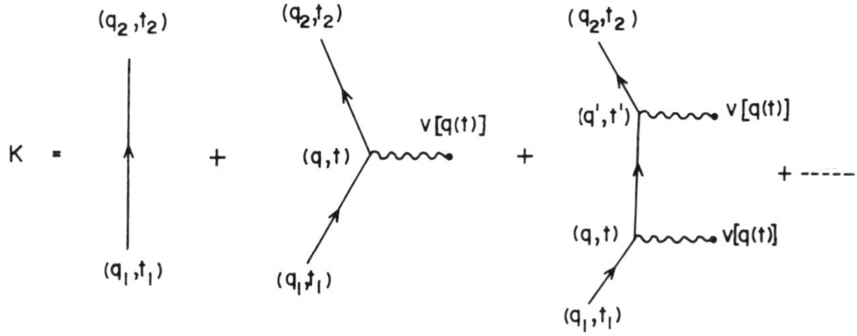

Fig. 3.1. Feynman diagrams for quantum mechanics. The successive terms in the perturbation expansion for the kernel can be represented by a series of diagrams with simple intuitive interpretation for each term.

corresponds to a term $K^0(q_2, t_2; q_1, t_1)$. Each wavy line represents the potential term $(-i\varepsilon V)$ and it is assumed that the coordinates at which the lines meet are integrated over the entire range. □

Let us first consider the perturbative approach. If we take the zeroth-order term in the ε power series we simply get the kernel K for the free harmonic oscillator given by (7). The higher order terms in the expansion can be used to answer two questions:

(i) How does the probability amplitude to go from (q_i, t_i) to (q_f, t_f) get altered by the presence of $\varepsilon V(q)$?

(ii) What is the probability that the potential causes a transition of the system from one stationary state to another?

In ordinary quantum mechanics we are often more interested in the second question than the first. For example, we are interested in the probability of an electron making transitions between two energy levels of a hydrogen atom when bombarded by radiation, rather than in knowing the likelihood of locating the electron in any particular place. The probability of transition is expressed as the square of modulus of a *transition amplitude* $\langle f|i\rangle$ between prescribed initial and final states given by wave functions ψ_i and ψ_f. Thus

$$\langle f|i\rangle = \iint \psi_f^*(q_f, t_f)\, K(q_f, t_f; q_i, t_i)\psi_i(q_i, t_i)\, \mathrm{d}q_i\, \mathrm{d}q_f. \tag{10}$$

We shall use the following compact notation in future to write the above transiton element as

$$\langle f|i\rangle \equiv \int_i^f \exp\left\{ i\int_{t_i}^{t_f} L\, \mathrm{d}t \right\} \bar{\mathcal{D}}\, q(t) \tag{11}$$

where the overbar on $\bar{\mathcal{D}}\, q$ implies that after summing over paths we have integrated over endpoints with the probability amplitudes ψ_f^* and ψ_i.

We shall next compute the transition amplitude for a 'ground state to ground state' transition. For reasons which become clear as we proceed, this amplitude is often called the *vacuum persistence amplitude*.

3.3. The Vacuum Persistence Amplitude

Applying (10) to harmonic oscillator we get

$$\begin{aligned}\langle 0, t_f \mid 0, t_i\rangle_V &= \int_i^f \exp\left\{ -i\varepsilon \int_{t_i}^{t_f} V(q)\, \mathrm{d}t \right\} \times \\ &\times \exp\left\{ \frac{i}{2}\int_{t_i}^{t_f} (\dot{q}^2 - \omega^2 q^2)\, \mathrm{d}t \right\} \bar{\mathcal{D}}\, q(t),\end{aligned} \tag{12}$$

for the probability amplitude that the harmonic oscillator will remain in ground state $|0\rangle$ at $t = t_f$ when it was known to be in the ground state $|0\rangle$ at $t = t_i$. As shown in Example 3.1, we can expand the first exponential in power series in ε. Thus, we have to evaluate terms like

$$I_n \sim \frac{(-i\varepsilon)^n}{n!}\int_i^f \left[\int_{t_i}^{t_f} V(q)\, \mathrm{d}t \right]^n \exp\left\{ \frac{i}{2}\int_{t_i}^{t_f} (\dot{q}^2 - \omega^2 q^2)\, \mathrm{d}t \right\} \bar{\mathcal{D}}\, q(t). \tag{13}$$

If we also assume that $V(q)$ can be expanded as a Taylor series in q, then a typical term in I_n can be

$$Q_{mn} \sim (\text{const}) \times \int_i^f \left[\int_{t_i}^{t_f} q^m\, \mathrm{d}t \right]^n \exp\left\{ \frac{i}{2}\int_{t_i}^{t_f} (\dot{q}^2 - \omega^2 q^2)\, \mathrm{d}t \right\} \bar{\mathcal{D}}\, q(t), \tag{14}$$

for integers $m = 0, 1, 2, \ldots$

We now show that the transition element (12) for an arbitrary potential can be obtained iteratively from the transition element

$$\langle 0, t_f \mid 0, t_i\rangle_J \equiv \int_i^f \exp\left[i\int_{t_i}^{t_f} \{ \tfrac{1}{2}(\dot{q}^2 - \omega^2 q^2) - J(t)\, q(t) \}\, \mathrm{d}t \right] \bar{\mathcal{D}}\, q(t). \tag{15}$$

The reader will recognize that the above expression is just the kernel for the forced harmonic oscillator discussed in Section 2.7, with $J(t)$ as the forcing term. This kernel can be evaluated exactly.

Example 3.2. Equations (14) and (15) can also be understood in the context of ordinary integrals introduced in Example 3.1. We saw that

$$I = \int_{q_1}^{q_2} \exp(-q^2 + \varepsilon V(q))\, dq = \sum_{n=0}^{\infty} \frac{\varepsilon^n}{n!} \int_{q_1}^{q_2} [V(q)]^n \, e^{-q^2} \, dq.$$

Now suppose $[V(q)]^n$ can be expanded in a Taylor series as

$$[V(q)]^n = \sum_{m=0}^{\infty} c_m^n q^m, \quad c_m^n = \frac{1}{m!}\left[\frac{\partial^m V^n}{\partial q^m} \right]_{q=0}.$$

Then

$$I = \sum_{n,\, m=0}^{\infty} \frac{c_m^n \varepsilon^n}{n!} \int_{q_1}^{q_2} q^m \, e^{-q^2} \, dq.$$

This expression can be cast into a much more useful form by using the concept of 'generating functions'. Suppose we want to evaluate a set of integrals of the form,

$$I_n = \int_{q_1}^{q_2} q^n F(q)\, dq, \quad q_1, q_2 \geq 0.$$

Consider the 'generating function' $G(\lambda)$ defined as

$$G(\lambda) \equiv \int_{q_1}^{q_2} F(q)\, e^{-q\lambda}\, dq.$$

Therefore,

$$\frac{d^n G}{d\lambda^n} = \int_{q_1}^{q_2} (-1)^n q^n F(q)\, e^{-q\lambda}\, dq = (-1)^n \int_{q_1}^{q_2} q^n F(q)\, e^{-q\lambda}\, dq.$$

Thus

$$I_n = (-1)^n \left(\frac{d^n G}{d\lambda^n} \right)\Bigg|_{\lambda=0}.$$

Using this result, we can write (with $G = \int_{q_1}^{q_2} \exp(-q^2 - q\lambda)\, dq$)

$$I = \sum_{n=0}^{\infty} \sum_{m=0}^{\infty} \frac{\varepsilon^n}{n!} c_m^n (-1)^m \frac{d^m}{d\lambda^m} G\Bigg|_{\lambda=0} = \sum_{n=0}^{\infty} \frac{\varepsilon^n}{n!} \left[\left\{ V\left[-\frac{d}{d\lambda} \right] \right\}^n G(\lambda) \right]_{\lambda=0}$$

$$= \left[\left\{ \exp \varepsilon V \left[-\frac{d}{d\lambda} \right] \right\} G(\lambda) \right]_{\lambda=0}.$$

Thus, our integral I can be evaluated by operating on the generating function $G(\lambda)$ by the operator $\exp \varepsilon V[-d/d\lambda]$. In the text, we generalize this concept to path integrals. □

To proceed further we introduce the concept of *functional derivative*. Suppose $F[x(t)]$ is a functional of $x(t)$ over the range $[t_i, t_f]$. Let the change in F for a small change $\eta(t)$ in $x(t)$ be expressed as

$$F[x(t) + \eta(t)] - F[x(t)] = \int_{t_i}^{t_f} Q(t)\, \eta(t)\, dt, \tag{16}$$

where terms of order η^2 and higher are neglected. Then we write the functional derivative of F as

$$\frac{\delta F}{\delta x(t)} = Q(t). \tag{17}$$

Example 3.3. An alternative definition for the functional derivative along the lines of the ordinary derivative is

$$\frac{\delta F(x(\bar{t}))}{\delta x(t)} \equiv \lim_{\varepsilon \to 0} \frac{F[x(\bar{t}) + \varepsilon\delta(\bar{t} - t)] - F(x(\bar{t}))}{\varepsilon}.$$

As an example consider the case, where

$$F(x(t)) = \exp \int J(t)x(t)\, \mathrm{d}t.$$

$$\begin{aligned} F(x(t) + \varepsilon\, \delta(t - t')) - F(x(t)) &= \exp \int J(t)[x(t) + \varepsilon\, \delta(t - t')]\, \mathrm{d}t - \exp \int Jx\, \mathrm{d}t \\ &= \left\{ \exp \int Jx\, \mathrm{d}t \right\} \cdot \exp[\varepsilon J(t')] - \exp \int Jx\, \mathrm{d}t \\ &\cong \left(\exp \int Jx\, \mathrm{d}t \right) \cdot [1 + \varepsilon J(t')] - \exp \int Jx\, \mathrm{d}t + O(\varepsilon^2) \\ &= \varepsilon\, J(t') \exp \int Jx\, \mathrm{d}t + O(\varepsilon^2) \end{aligned}$$

Therefore,

$$\lim_{\varepsilon \to 0} \frac{F(x + \varepsilon\delta(t - t')) - F(x)}{\varepsilon} = J(t') \exp \int J(t)x(t)\, \mathrm{d}t = \frac{\delta F(x(t'))}{\delta x(t)}.$$

The same definition can be used to proved the 'chain rule' for functional differentiation

$$\frac{\delta F}{\delta x(t)} = \int \frac{\delta F}{\delta f[x(t')]} \cdot \frac{\delta f(x(t'))}{\delta x(t)}\, \mathrm{d}t'.$$

This rule is very useful in evaluating more complicated functional derivatives. □

In our case if $F[q(t)]$ is given by (15), that is,

$$\langle 0, t_f \mid 0, t_i \rangle_J \equiv F[J(t)] = \int_i^f \exp\left[i \int_{t_i}^{t_f} \{ L_0 - J(t)q(t) \}\, \mathrm{d}t \right] \bar{\mathcal{D}}\, q(t), \tag{18}$$

then for t_1, t_2 in the range $[t_i, t_f]$ we can write

$$\frac{\delta F}{\delta J(\mathrm{t}_2)} = \int_i^f -iq(t_2) \exp\left[i \int_{t_i}^{t_f} \{ L_0 - J(t)q(t) \}\, \mathrm{d}t \right] \bar{\mathcal{D}}\, q(t) \tag{19}$$

and

$$\frac{\delta^2 F}{\delta J(t_1)\, \delta J(t_2)} = -\int_i^f q(t_1)q(t_2) \exp\left[i \int_{t_i}^{t_f} \{ L_0 - J(t)q(t) \}\, \mathrm{d}t \right] \bar{\mathcal{D}}\, q(t). \tag{20}$$

Apart from the minus sign, the right-hand side is just the transition element of the product $q(t_1)q(t_2)$, provided $t_1 > t_2$. If $t_2 > t_1$ we can express the same as the transition element of $q(t_2)q(t_1)$. In other words, if we define a time-ordering operator T by

$$T[q(t_1)q(t_2)] = \begin{cases} q(t_1)q(t_2) & \text{for } t_1 > t_2 \\ q(t_2)q(t_1) & \text{for } t_2 > t_1, \end{cases}$$

then (20) gives

$$\langle 0, t_f|\ T[q(t_1)\ q(t_2)]\ |0, t_i\ \rangle_J = (i)^2\ \frac{\delta^2 F}{\delta J(t_1)\ \delta J(t_2)}$$

$$= (i)^2\ \frac{\delta^2}{\delta J(t_1)\ \delta J(t_2)}\ \langle 0, t_f\ |\ 0, t_i\rangle_J. \qquad (21)$$

It is evident that we can generalize this result to the transition element of the time-ordered product of any number of q's by introducing $i\delta/\delta J$ operators as many times on the right-hand side. Thus, if $f[q(t)]$ is any functional of q we get

$$\langle 0, t_f|\ Tf[q(t)]\ |0, t_i\ \rangle_J = f[\ i\ \frac{\delta}{\delta J(t)}\ ]\ \langle 0, t_f\ |\ 0, t_i\rangle_J, \qquad (22)$$

provided a power series expansion of f in terms of q is available.

We now return to (12) and note that we have there an expression which can formally be expressed in the above notation. Thus we have,

$$\langle 0, t_f\ |\ 0, t_i\rangle_V = \int_i^f\ \exp\left\{\int_{t_i}^{t_f}\ -i\varepsilon V(q)\ \mathrm{d}t + i\int_{t_i}^{t_f}\ L_0\ \mathrm{d}t\right\}\ \mathcal{D}q(t)$$

$$= \lim_{J\to 0}\int_i^f\ \exp\left\{\int_{t_i}^{t_f}\ -i\varepsilon V(q)\ \mathrm{d}t\right\}\ \times$$

$$\times\ \exp\left\{\ i\int_{t_i}^{t_f}\ (L_0 - Jq)\ \mathrm{d}t\right\}\ \mathcal{D}q(t)$$

$$= \lim_{J\to 0}\ \exp\left\{-i\varepsilon\int_{t_i}^{t_f}\ V\left(i\,\frac{\delta}{\delta J(t)}\right)\mathrm{d}t\right\}\ \langle 0, t_f\ |\ 0, t_i\rangle_J, \qquad (23)$$

by using (22). The meaning of the functional differentiation operator is given by the power series expansion:

$$\exp\left\{-i\varepsilon\int_{t_i}^{t_f}\ V\left(i\,\frac{\delta}{\delta J}\right)\mathrm{d}t\right\} = \sum_{n=0}^{\infty}\ \frac{(-i\varepsilon)^n}{n!}\ \left[\int_{t_i}^{t_f}\ V\left(i\,\frac{\delta}{\delta J}\right)\mathrm{d}t\right]^n. \qquad (24)$$

It is convenient to introduce the quantities $Z_0(J)$ and $Z_v(J)$ to express these results:

$$Z_0(J) = \langle 0, t_f\ |\ 0, t_i\rangle_J, \qquad (25)$$

$$Z_V(J) = \exp\left\{-i\varepsilon\int_{t_i}^{t_f}\ V\left(i\,\frac{\delta}{\delta J}\right)\mathrm{d}t\right\}\ Z_0(J). \qquad (26)$$

$Z_0(J)$ is the vacuum to vacuum transition amplitude for the forced harmonic oscillator while $Z_V(J)$ is the same amplitude for the forced harmonic oscillator *in the* presence of the external potential εV. By using arguments similar to those given above, we get the result analogous to (21), viz.:

$$\langle 0, t_f \mid T\left[q(t_1)q(t_2)\right] \mid 0, t_i\rangle_V = (i)^2 \frac{\delta^2}{\delta J(t_1)\,\delta J(t_2)} Z_V(J). \tag{27}$$

When discussing the quantum theory of free and interacting fields we shall have recourse to formulae similar to these.

3.4. Euclidean Time

The above formal results suggest that $Z_0(J)$, the vacuum persistence amplitude for the forced harmonic oscillator, plays a key role in perturbation techniques. We now consider methods of actually evaluating this quantity. Some of the methods, of course, have wider applications (in field theory) which go beyond the computation of $Z_0(J)$.

(i) First we note that from the results of Section 2.7 the kernel K_J of the forced harmonic oscillator can be computed in a straightforward manner. If $\phi_0(q, t)$ is the ground state of the system, straightforward computation gives us

$$\begin{aligned} Z_0(J) &\equiv \int\!\!\int \phi_0^*(q_f, t_f)\, K_J(q_f, t_f; q_i, t_i)\, \phi_0(q_i, t_i)\, \mathrm{d}q_f\, \mathrm{d}q_i \\ &= \exp\left\{-\frac{1}{2\omega}\int_{t_i}^{t_f} \mathrm{d}t \int_{t_i}^{t} J(t)J(s)\, \mathrm{e}^{-i\omega(t-s)} \mathrm{d}s\right\}. \end{aligned} \tag{28}$$

(This calculation was carried out in Example 2.11.)

(ii) We now compute $Z_0(J)$ by another method which is shorter and free from the ambiguities which are present in (i) in the limiting cases of $t_i \to -\infty$, $t_f \to \infty$ when the exponentials $\mathrm{e}^{\pm i\omega t}$ oscillate. In practice we are interested in precisely such situations: we know the initial state of the system long before the perturbing force acted on it and we wish to know its final state long after the force ceased to act on it.

The 'trick' involved here is first to make time an imaginary quantity by writing

$$t = -i\tau, \qquad \tau = it \tag{29}$$

and then to compute the relevant path integrals (see Figure 3.2). For the forced harmonic oscillator the action changes from

$$S = \int_{t_i}^{t_f} \{\tfrac{1}{2}(\dot{q}^2 - \omega^2 q^2) - J(t)q(t)\}\, \mathrm{d}t \tag{30}$$

to the following when the transformation (29) is effected:

$$S^{\mathrm{E}} = i\int_{\tau_i}^{\tau_f} \{\tfrac{1}{2}(\dot{q}^2 + \omega^2 q^2) + J(\tau)q(\tau)\}\, \mathrm{d}\tau, \tag{31}$$

where the dot on q denotes $\mathrm{d}/\mathrm{d}\tau$ and the superscript 'E' on S indicates that we have made time a'Euclidean' coordinate on par with spatial distance.

To evaluate the path integral

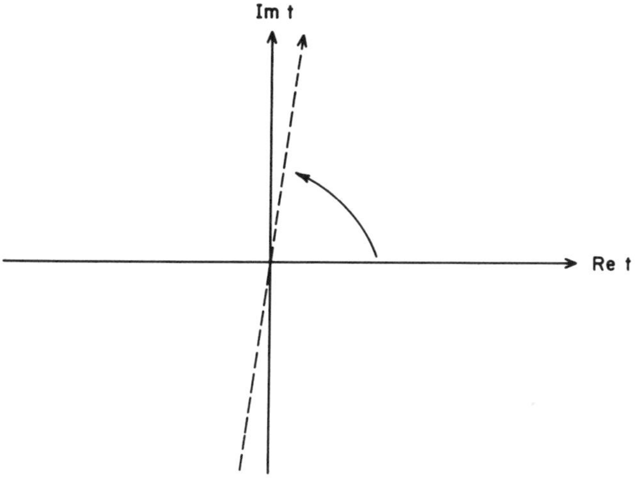

Fig. 3.2. The 'Euclidean rotation' from real to complex values of t.

$$K^{\mathrm{E}}[q_f, \tau_f; q_i, \tau_i] = \int \exp\left\{ -\int_{\tau_i}^{\tau_f} \left[\tfrac{1}{2}(\dot{q}^2 + \omega^2 q^2) + Jq \right] \mathrm{d}\tau \right\} \mathcal{D}\, q(\tau) \tag{32}$$

in the Euclidean space we shall discuss the case where $\tau_i \to -\infty$, $\tau_f \to \infty$ and $q \to 0$ at these times. Write

$$q(\tau) = q_c(\tau) + \eta(\tau), \tag{33}$$

where $q_c(\tau)$ is the 'classical' path satisfying the equation

$$\ddot{q}_c - \omega^2 q_c = J(\tau). \tag{34}$$

The boundary conditions on $q_c(\tau)$ being the same as on $q(\tau)$ at $\tau = \pm\infty$, $\eta(\tau)$ also vanishes at these times. Following arguments similar to those in Section 2.4 we arrrive at this limiting form of K^{E}:

$$I(J) \equiv K^{\mathrm{E}}[0, \infty; 0, -\infty] = N \exp\{ -S^{\mathrm{E}}[q_c(\tau)]\}, \tag{35}$$

where N is a normalizing constant independent of J.

To evaluate $S^{\mathrm{E}}[q_c(\tau)]$ we need to solve (34) with the appropriate asymptotic boundary conditions. For this purpose we use the Green's function $D^{\mathrm{E}}(\tau)$ which satisfies the equation

$$\left[\frac{\mathrm{d}^2}{\mathrm{d}\tau^2} - \omega^2 \right] D^{\mathrm{E}}(\tau) = \delta(\tau), \tag{36}$$

where $D^{\mathrm{E}}(\tau) \to 0$ as $|\tau| \to \infty$. Example 3.4 illustrates how this is done.

Example 3.4. To solve Equation (36), introduce the Fourier transform of $D^{\mathrm{E}}(\tau)$ as

$$D^{\mathrm{E}}(\nu) \equiv \int_{-\infty}^{+\infty} D^{\mathrm{E}}(\tau)\, \mathrm{e}^{-i\nu\tau}\, \mathrm{d}\tau.$$

Substituting into Equation (36) we get,

$$-(\nu^2+\omega^2)D^{\mathrm{E}}(\nu)=1.$$

Therefore,

$$D^{\mathrm{E}}(\tau)=\int_{-\infty}^{+\infty}D^{\mathrm{E}}(\nu)\,\mathrm{e}^{i\nu\tau}\,\frac{\mathrm{d}\nu}{2\pi}=-\int_{-\infty}^{+\infty}\frac{\mathrm{d}\nu}{2\pi}\,\frac{\mathrm{e}^{i\nu\tau}}{(\nu^2+\omega^2)}.$$

This integral can be evaluated by the method of contours. The poles are at $\nu=\pm i\omega$. For $\tau>0$ we can close to contour in the upper half plane, and for $\tau<0$ the contour can be closed in the lower half. Thus we get,

$$D^{\mathrm{E}}(\tau)=-\theta(\tau)\left\{2\pi i\cdot\frac{1}{2\pi}\cdot\frac{1}{2i\omega}\cdot\mathrm{e}^{-\omega\tau}\right\}+\left\{2\pi i\cdot\frac{-1}{2\pi}\cdot\frac{1}{2i\omega}\cdot\mathrm{e}^{+\omega\tau}\right\}\theta(-\tau)$$

$$=-\frac{1}{2\omega}\exp(-\omega|\tau|).$$

□

Thus the answer may be written simply as

$$D(\tau)=-\frac{\mathrm{e}^{-\omega|\tau|}}{2\omega} \tag{37}$$

and

$$q_c(\tau)=\int_{-\infty}^{\infty}D^{\mathrm{E}}(\tau-\tau')\,J(\tau')\,\mathrm{d}\tau'. \tag{38}$$

With the help of (37) and (38) we can evaluate $S^{\mathrm{E}}[q_c(\tau)]$ in a straightforward manner. The final answer is

$$I(J)=N\exp\left\{-\frac{1}{2\omega}\int_{-\infty}^{\infty}\int_{-\infty}^{\infty}J(\tau)D^{\mathrm{E}}(\tau-\sigma)J(\sigma)\,\mathrm{d}\tau\,\mathrm{d}\sigma\right\}. \tag{39}$$

This integral gives the expression for $Z_0(J)$ on analytic continuation on to the real time axis. The constant N is not yet determined, but in most cases we need ratios $I(J)/I(0)$ and the constant N drops out.

$Z_0(J)$ is the amplitude that the ground state remains a ground state even in the presence of the forcing function $J(t)$. The reader may wonder why in the Euclidean-time method we did not have to use the information about $\phi_0(q,t)$, the ground state wave functions, in deriving $Z_0(J)$. Instead, we have used the 'classical ground state' conditions $q_\cdot=0$ at $\tau=\pm\infty$. This feature, that classical solutions in Euclidean time contain information about quantum solutions in real time, will also be found elsewhere.

We also notice that, in Example 3.4, the Green's function $D(\tau)$ was evaluated fairly simply by using the boundary conditions $q\to 0$ at $|\tau|\to\infty$. Had we tackled a similar Green's function evaluation in real time we would be forced to evaluate the Fourier transform

$$D(t)=-\int_{-\infty}^{\infty}\frac{\mathrm{e}^{-i\alpha t}}{\alpha^2-\omega^2}\frac{\mathrm{d}\alpha}{2\pi}\ . \tag{40}$$

This integral evidently diverges at $\alpha = \pm\omega$, and to evaluate it we have to go to the complex $-\alpha$ plane and treat $\alpha = \pm\omega$ as poles which are to be avoided in suitably chosen contours. Example 3.5 illustrates how this is done and how different Green's functions are thereby obtained.

Example 3.5. In order to give meaning to (40) we must specify a contour in the complex α plane which avoids the poles in the real axis occurring at $\alpha = \pm\omega$. Essentially there are three different ways of avoiding the poles which are shown in Figure 3.3. They lead to three different definitions of Green's functions.

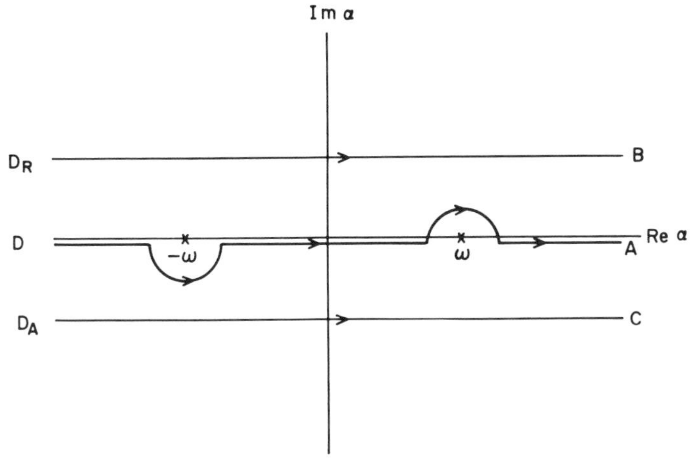

Fig. 3.3 The contours for A, B, C. Note that the contours in (3.40) are closed in the upper half plane for $t < 0$ and in the lower half plane for $t > 0$.

(i) *Contour A*. Here the contour is completed in the upper half plane for $t > 0$ and in the lower half plane for $t < 0$. In either case one of the poles contributes and we get (on the lines of Example 3.4)

$$D(t) = \frac{1}{2i\omega}\left[\theta(t)\,\mathrm{e}^{-i\omega t} + \theta(-t)\,\mathrm{e}^{+i\omega t}\right].$$

On analytic continuation this becomes the $D^{\mathrm{E}}(\tau)$ discussed previously.

(ii) *Contour B*. In this case $D(t) = 0$ for $t < 0$ and is given by

$$D(t) = \frac{1}{2\omega}\sin\omega t \quad \text{for} \quad t > 0.$$

Since such a Green's function takes support only from the values of $J(t)$ at earlier times, it is called the 'retarded' Green's function.

(iii) *Contour C*. Here $D(t) = 0$ for $t > 0$ and,

$$D(t) = -\frac{1}{2\omega}\sin\omega t \quad \text{for} \quad t < 0.$$

In analogy with the previous case, this is called the 'advanced' Green's functions.

The choice of contours can be incorporated into the definition of the integrals by a simple trick.

Consider the definition

$$D(t) = \lim_{\varepsilon\to 0} \int_{-\infty}^{+\infty} \frac{e^{-i\alpha t}}{\alpha^2 - \omega^2 + i\varepsilon} \frac{d\alpha}{2\pi} .$$

There is no problem in evaluating this integral because the $i\varepsilon$ in the denominator has shifted the poles off the real axis. This definition is therefore completely equivalent to the definition based on contour A.

Similarly, one can provide the following '($i\varepsilon$) prescriptions' for the retarded and advanced Green's functions.

$$D_R(t) = \int_{-\infty}^{+\infty} \frac{e^{-i\alpha t}}{\alpha^2 - \omega^2 + i\varepsilon\alpha} \cdot \frac{d\alpha}{2\pi}$$

$$D_A(t) = \int_{-\infty}^{+\infty} \frac{e^{-i\alpha t}}{\alpha^2 - \omega^2 - i\varepsilon\alpha} \cdot \frac{d\alpha}{2\pi} .$$

Following normal practice we have not bothered to express the limit $\varepsilon \to 0$ explicitly. It is always assumed that, in these expressions, the limit $\varepsilon \to 0+$ must be taken at the end of the calculations, keeping ε positive. □

3.5. The Double-Hump Potential

Nonperturbative methods are needed when, for various reasons, the perturbation approach fails. This can happen, for example, when the perturbation expansion is known not to converge even for very small ε. We shall encounter such situations in quantum field theories in Chapter 4. Another situation, which we consider now, arises when even for small ε the nature of the problem is completely changed. An example of this type is provided by the addition of a potential $-\lambda q^4/4$ to the potential of the free harmonic oscillator.

The potential is now

$$V = \tfrac{1}{2}\omega^2 q^2 - \tfrac{1}{4}\lambda q^4. \tag{41}$$

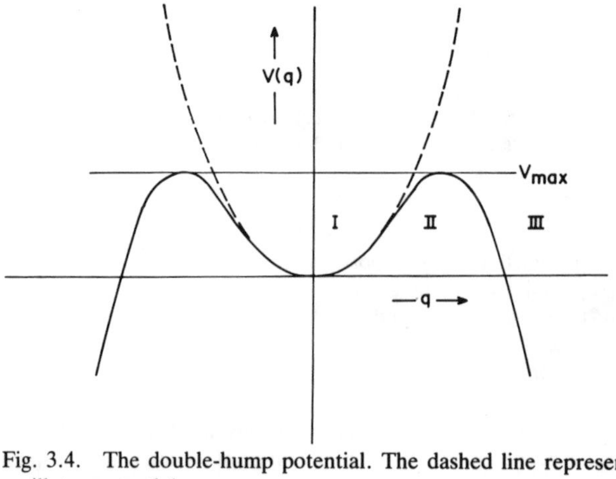

Fig. 3.4. The double-hump potential. The dashed line represents the $\lambda = 0$ harmonic oscillator potential.

As illustrated in Figure 3.4, the solid line curve corresponding to (41) differs fundamentally from the 'free' case $\lambda = 0$ shown by the dashed line. No matter how small λ is, the potential has a finite maximum on both sides of the minimum at $q = 0$, $V = 0$. There is thus a finite tunnelling probability for a particle to go from the 'bound' region I out to the 'unbound' region III of Figure 3.4. In other words, the low-energy bound states ($E < V_{max}$) which exist are unstable. Clearly, this kind of result would not be apparent from any perturbation expansion for small λ, which treats the dashed curve as the zeroth-order potential. For large enough q the discrepancy between the dotted and the solid curves is too large to sustain the reliability of the perturbation approach.

Example 3.6. The double-hump potential becomes infinitely negative for large $|q|$ and is thus not bounded from below. There is a simple variant of this potential obtained by reversing the signof $V(q)$:

$$\begin{aligned} V\,(\text{double well}) &\equiv -V\,(\text{double hump}) \\ &= -\tfrac{1}{2}\omega^2 q^2 + \tfrac{1}{4}\lambda q^4. \end{aligned}$$

Notice that ω can no longer be considered to be a frequency. This potential has the shape shown in Figure 3.5(a).

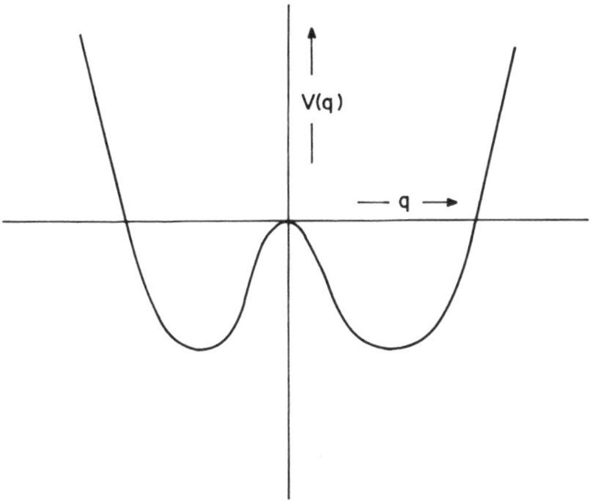

Fig. 3.5(a). The double-well potential in real (physical) time coordinate.

Obviously any perturbation about $q = 0$ is unstable, even classically. At first sight, one may think that well-defined perturbation theory can exist about the classical minimum, say $q = +\sqrt{\omega/\lambda}$. However, quantum mechanics allows for tunneling between classical minima, at $\pm\sqrt{\omega/\lambda}$. This tunnelling, as we shall see in the next section, introduces nonperturbative features into the problem. □

Example 3.7. Some interesting aspects of this potential can again be explored by considering ordinary integrals rather than path integrals. Let us look at the following integral:

$$I \equiv \int_0^\infty \exp i(q^2 - \lambda q^4)\, dq.$$

Consider the evaluation of I by an expansion in λ followed by a term by term integration. This will give

$$I = \int_0^\infty e^{iq^2} \sum_{n=0}^\infty \frac{(-i\lambda)^n q^{4n}}{n!}\, dq = \sum_{n=0}^\infty \frac{(-i\lambda)^n}{n!} \int_0^\infty q^{4n} e^{iq^2}\, dq.$$

Using the fact that

$$\int_0^\infty e^{-\alpha x^2} x^{2n-1}\, dx = \frac{\Gamma(n)}{2\alpha^n},$$

we can *formally* write this as,

$$I = \tfrac{1}{2}\sqrt{i} \sum_{n=0}^\infty \frac{(i\lambda)^n}{n!} \Gamma(2n + \tfrac{1}{2}).$$

The expression has only a formal significance because the series is divergent. This is a reflection of the fact that our original integral is ill defined for large $|q|$.

Suppose we do the expansion not about $q = 0$ but around the minimum. Substituting $q = x/\sqrt{\lambda}$, the integral becomes

$$I = \frac{1}{\sqrt{\lambda}} \int_0^\infty \exp \frac{i}{\lambda} (x^2 - x^4)\, dx.$$

The critical points are at $x = 0$, and $x = (\sqrt{2})^{-1}$. Expansion about $x = 0$ will lead to the previous result. But we can argue that as $\lambda \to 0$, $\lambda^{-1} \to \infty$ and hence the integral can be approximated by its saddle-point value. This will give, on expanding in $u \equiv x - 1/\sqrt{2}$,

$$I^1 = e^{i/4\lambda} \frac{1}{\sqrt{\lambda}} \int_{-1/\sqrt{2}}^\infty e^{-2iu^2/\lambda} \sum_{n=0}^\infty \left(-\frac{i}{\lambda} \right)^n \frac{[u^3(2\sqrt{2} + u)]^n}{n!}\, du.$$

The leading contributions from both the saddle points $(0, 2^{-1/2})$ added together give

$$I \simeq \tfrac{1}{2} \sqrt{i\pi} \left(1 + \sqrt{2}\, i \exp \frac{i}{4\lambda} \right).$$

Notice that the second term is nonanalytic in λ. Thus no perturbation in λ will reproduce this term. In other words, expansion about the minimum value (saddle-point method) reveals more structure than a simple-minded perturbation in λ. In the text we shall extend this method to path integrals. □

How do we deal with such situations? One way is to consider that the path integral amplitude is made up of two components, one coming entirely from the classical path with $\delta S = 0$ and the other from nonclassical paths which give the 'quantum corrections'. Although simple in principle, this method does encounter complications when followed through in detail. The following discussion is intended only to capture the essential features of this approach. We shall use the method of Euclidean time in evaluating the path integrals.

Accordingly, use the new time $\tau = it$ and take (without loss of generality) the time interval to be $[-\tau_0/2, \tau_0/2]$. The classical equation of motion given by $\delta S = 0$ is

$$\frac{\mathrm{d}^2 q_c}{\mathrm{d}\tau^2} - V'(q_c) = 0. \tag{42}$$

As in (33) we expand the typical path near the classical path by writing $q = q_c + \eta$. Ignore terms of order η^3 or higher in the path integral. The transition amplitude becomes

$$\left\langle q_f, \frac{\tau_0}{2} \middle| q_i, -\frac{\tau_0}{2} \right\rangle = \exp\left\{ - S[q_c] \right\}$$
$$\times \int \exp\left[-\tfrac{1}{2} \int_{-\tau_{0/2}}^{\tau_{0/0}} \left\{ \dot{\eta}^{\,2} + V''(q_c)\,\eta^2 \right\} \mathrm{d}\tau \right] \mathcal{D}\,\eta(t). \tag{43}$$

For the harmonic oscillator alone V'' is constant and the coefficient of η^2 is independent of time. This is not so in the general case, or for $\lambda \neq 0$ in (41). Nevertheless, we can write the functional integral in (43) as a determinant which is often useful. We proceed as follows.

Integrate by parts to get

$$\tfrac{1}{2} \int_{-\tau_{0/2}}^{\tau_{0/2}} [\,\dot{\eta}^2 + V''\,\eta^2\,]\,\mathrm{d}\tau = \tfrac{1}{2} \int_{-\tau_{0/2}}^{\tau_{0/2}} \eta\,[\,-\ddot{\eta} + V''\eta\,]\,\mathrm{d}\tau \tag{44}$$

since η vanishes at the endpoints. Let the eigenvalues of the differential operator in the square brackets of the above integrand be $\{\lambda_n\}$, and the eigenfunctions $f_n(\tau)$, i.e.

$$\left[-\frac{\mathrm{d}^2}{\mathrm{d}\tau^2} + V''(q_c) \right] f_n(\tau) = \lambda_n f_n(\tau). \tag{45}$$

Since the eigenfunctions form a complete orthonormal set,

$$\int_{-\tau_{0/2}}^{\tau_{0/2}} f_n(\tau)\, f_m(\tau)\,\mathrm{d}\tau = \delta_{nm}; \qquad f_n\left(\pm \frac{\tau_0}{2} \right) = 0 \tag{46}$$

and any $\eta(\tau)$ can be expressed as an expansion

$$\eta(\tau) = \sum_{n=0}^{\infty} c_n f_n(\tau). \tag{47}$$

The integration over all paths $\eta(\tau)$ is equivalent to the ordinary integrations over all c_n. Thus,

$$\mathcal{D}\,\eta \propto \prod_{n=0}^{\infty} \mathrm{d}c_n. \tag{48}$$

Further, for η given by (47), the use of (46) gives

$$\tfrac{1}{2} \int_{-\tau_{0/2}}^{\tau_{0/2}} \eta\,[-\ddot{\eta} + V''\eta\,]\,\mathrm{d}\tau = \tfrac{1}{2} \int_{-\tau_{0/2}}^{\tau_{0/2}} \left\{ \sum_{n=0}^{\infty} c_n f_n(\tau) \sum_{m=0}^{\infty} \lambda_m c_m f_m(\tau) \right\} \mathrm{d}\tau$$
$$= \tfrac{1}{2} \sum_{n=0}^{\infty} \lambda_n c_n^2. \tag{49}$$

Thus

$$\int \exp\left[-\tfrac{1}{2}\int_{-\tau_0/2}^{\tau_0/2} (\dot{\eta}^2 + V''\eta^2)\,\mathrm{d}\tau\right] \mathcal{D}\eta = \prod_{n=0}^{\infty} \int_{-\infty}^{\infty} \exp\left(-\tfrac{1}{2}\lambda_n c_n^2\right) \mathrm{d}c_n$$
$$= \prod_{n=0}^{\infty} \sqrt{2\pi/\lambda_n}\ . \tag{50}$$

Putting all these results together we get

$$\left\langle q_f, \frac{\tau_0}{2} \,\middle|\, q_i, -\frac{\tau_0}{2} \right\rangle = N \exp\left\{-S\,[q_c(\tau)]\right\} \prod_n \lambda_n^{-1/2}$$
$$= N \exp\left\{-S\,[q_c(\tau)]\right\} \left[\det\left(-\partial_\tau^2 + V''\right)\right]^{-1/2} \tag{51}$$

Here we have absorbed the $\sqrt{2\pi}$ factors in (50) into the constant N. The determinant appearing in the final answer is thus the determinant of the operator which, in diagonalized form, is simply the diagonal matrix with $\{\lambda_n\}$ in the diagonal. The dimensionality of the matrix is, of course, infinite.

Thus the quantum corrections appear as the reciprocal of the square root of the above determinant. The result obtained above is quite general since the explicit form of $V(q)$ does not appear in the argument, except for the approximation of ignoring terms higher than η^2 in the path integral.

3.6. The Instanton Solutions

Let us apply the above method to the double-well potential. By adding a suitable constant and inverting the sign we shall write the potential (41) in the form

$$V(q) = \frac{\lambda}{4}(q^2 - a^2)^2. \tag{52}$$

Figure 3.5(a) shows this potential as a 'double-well' potential with the deepest points at $V = 0$, $q = \pm a$ and a hump in the middle at $V = \lambda a^4/4$, $q = 0$. In Euclidean time $\tau = it$, the classical equation of motion becomes

$$-\frac{\mathrm{d}^2 q}{\mathrm{d}\tau^2} + \frac{\partial V}{\partial q} = -\ddot{q} + \lambda q(q^2 - a^2) = 0. \tag{53}$$

In terms of τ-time the potential changes sign and becomes, as shown in Figure 3.5(b), a double-hump potential.

Consider the following interesting situation. We are interested in the transition amplitude to go from $q = -a$ to $q = +a$. Classically, in real time this is prevented by a potential barrier but quantum mechanically this tunnelling is possible. However, in the τ-time as can be seen from Figure 3.5(b),

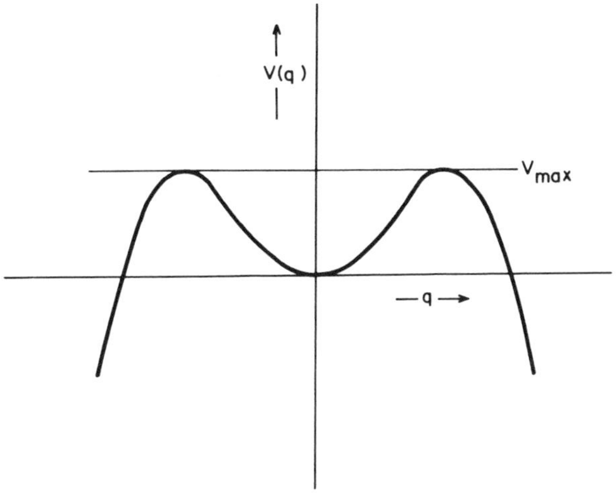

Fig. 3.5(b). The double-well potential in imaginary time. Note that it becomes a double-hump potential.

the particle can 'roll' along the trough (which was the hump in Figure 3.5(a)) from $-a$ to $+a$. *Thus classical motion in τ-time is able to tell us about quantum tunnelling in real time.* Such classical solutions are called 'instantons', for reasons which will soon become clear.

The simplest instanton solution describes a motion which begins at $q = -a$ at $\tau = -\infty$ and ends at $q = +a$ at $\tau = +\infty$. This solution is given by

$$q_c(\tau) = a \tanh\left(a \frac{\lambda}{2} \tau\right). \tag{54}$$

The classical action for this solution is

$$S\,[q_c(\tau)] = \frac{2\sqrt{2}}{3}\,(a^3\sqrt{\lambda}\,). \tag{55}$$

Neglecting the determinant in (51), the transition amplitude can be written as

$$\langle a, \infty \mid -a, -\infty \rangle \approx N \exp\left(-\tfrac{2}{3}\,\sqrt{2\lambda}\,a^3\right). \tag{56}$$

This amplitude could, of course, have been evaluated as the tunnelling amplitude in real time by using the WKB approximation method. Written in terms of ω and λ the amplitude becomes

$$\langle a, \infty \mid -a, -\infty \rangle \approx N \exp\left(-\frac{2\sqrt{2\omega^3}}{3\lambda}\right), \tag{57}$$

which shows that the transition amplitude is a nonanalytic function of λ near $\lambda = 0$ and could not have been obtained by the perturbation expansion near $\lambda = 0$ (cf. Example 3.7).

In general, instanton solutions are bounded and compact, as is the solution

(54). Since the tunnelling through a potential is instantaneous in real time, the solution has become known as 'instanton'.

What is the role of the quantum fluctuations represented by the determinant in (51)? At first sight it appears that the determinant vanishes, thus leading to a divergent result for (51). This conclusion arises because the differential equation (53) is invariant under time translation so that the eigenvalue equation (45) must also be satisfied for $\lambda_n = 0$. (If λ_n were not equal to zero for some n, then time translation invariant solution could not be accommodated. Indeed, it can be verified that

$$f_0(\tau) = S[q_c]^{-1/2} \frac{dq_c}{d\tau} \tag{58}$$

satisfies the eigenvalue equation with $\lambda_0 = 0$.) Example 3.8 below shows how this difficulty can be avoided.

Example 3.8. Consider the following function $q(b, \tau)$ defined by

$$q(b, \tau) \equiv q_c(\tau - b) = a \tanh a \frac{\lambda}{2} (\tau - b).$$

Because our potential is invariant under translations in time, $q(b, \tau)$ is also a solution to classical equation of motion and contributes $\exp(-S_c)$ to the path integral! In other words, the total contribution from *all* such classical solutions will be

$$\langle 0, +\infty, | 0, -\infty \rangle = \int_{-\infty}^{\infty} \exp(-S_c)\, db.$$

This expression is divergent because S_c is independent of b. Alternatively, because of time translation invariance, there exists a zero eigenvalue among the set $\{\lambda_n\}$. In (50) integration over c_i will give a divergent result if $\lambda_i = 0$.

In many physical situations, however, this divergence does not matter. For example, consider the symmetric and antisymmetric vacua defined as $|a\rangle + |-a\rangle$ and $|a\rangle - |-a\rangle$. (Here $|-a\rangle$ denote the states with particle localized around $\pm a$). We are interested in the energy difference ΔE between these states. First, from the *Euclidean* version of the real time formula

$$K(q_2, t_2; q_1, t_1) = \sum_n \phi_n^*(q_2)\phi_n(q_1)\, e^{-iE_n(t_2 - t_1)},$$

we get, for large T,

$$K(q_2, T; q_1, 0) = \phi_0^*(q_2)\phi_0(q_1) \exp(-E_0 T).$$

Thus,

$$\begin{aligned} -E_0 &= -(\text{Ground state energy}) \\ &= \lim_{T\to\infty} \frac{1}{T} \ln K(q_2, T; q_1, 0). \end{aligned}$$

So, in our case, if $E^{\pm}$ denote the energies of states $|a\rangle \pm |-a\rangle$,

$$E^+ - E^- \cong \lim_{T\to\infty} \left\{ \frac{2}{T} \frac{K(+a, T; -a, 0)}{K(+a, T; +a, 0)} \right\}.$$

Here we are calculating K for finite but arbitrarily large T. The integral over b above gives, in the interval $(0, T)$,

$$K(a, T; -a, 0) \cong T \exp(-S_c) \times (\text{const}).$$

Therefore, for large T,

$$K(-a, T; -a, 0) \propto T \exp(-S_c) \cdot K(+a, T; +a, 0)$$

giving

$$\Delta E \propto \lim_{T\to\infty} \frac{1}{T} \left\{ T \exp(-S_c) \right\} = \exp(-S_c).$$

Notice that the divergence of $K(T\ldots)$ as $T\to\infty$ is eliminated because of the existence of $(1/T)$ in E_0. In many situations of interest similar cancellations occur and the physical quantities are well defined even though kernel K itself may not be. □

3.7. The Concept of an Effective Action

The techniques described in the previous section can be used to develop a concept of effective action. Qualitatively, the effective action incorporates the effect of quantum fluctuations around the classical trajectory. Consider a theory in which the classical action for the path $q(t)$ is $S[q(t)]$. The *complete* quantum theory of the system is contained in the path integral

$$K = \int \exp iS[q(t)]\mathcal{D}\, q(t). \tag{59}$$

However, let us assume that the system is more or less classical and that the quantum fluctuations are small. Then the paths $q(t)$ in the path integral (59) will not fluctuate drastically from the classical path $q_c(t)$. We shall assume that we are only interested in those paths $q(t)$ which have the same average behaviour as $q_c(t)$. In other words, if $\eta(t) \equiv q(t) - q_c(t)$ denotes the quantum fluctuation from the classical trajectory, then we expect the time average of η to vanish (thus, on average, the fluctuations are evenly distributed about the classical path):

$$\langle\eta\rangle \equiv \int_0^T \frac{dt}{T}\eta(t) = 0 = \int_0^T [q(t) - q_c(t)]\, \frac{dt}{T}\,. \tag{60}$$

In considering the semiclassical limit of (59) we shall sum over only those paths $q(t)$ that satisfy (60). Thus, we define a new kernel, $\bar{K}(q_c)$ by the path integral:

$$\begin{aligned}
\bar{K}[q_c(t)] &\equiv \int \delta(\langle\eta\rangle) \exp iS[q]\, \mathcal{D}\, q(t) \\
&= \int \delta\left[\int_0^T \frac{dt}{T}\,(q - q_c)\right] \exp iS[q]\, \mathcal{D}\, q(t) \\
&= \int\int_{-\infty}^{+\infty} \frac{d\lambda}{2\pi} \exp i\left\{S[q] + \lambda\int_0^T \frac{dt}{T}\,(q - q_c)\right\} \mathcal{D}q(t) \\
&= \int_{-\infty}^{+\infty} \frac{d\lambda}{2\pi} \int \exp i\int_0^T dt\left\{L[q] + \frac{\lambda}{T}\,(q - q_c)\right\} \mathcal{D}q(t).
\end{aligned} \tag{61}$$

In arriving at this expression we have used the well-known representation of Dirac's delta function:

$$\delta(x) = \int_{-\infty}^{+\infty} \exp(i\lambda x) \frac{d\lambda}{2\pi}.$$

The effective action is defined to be that action which will lead to $\bar{K}[q_c]$ for the path $q_c(t)$. In other words,

$$\bar{K}[q_c] \equiv \exp iS_{\text{eff}}[q_c, T] \equiv \exp i\int_0^T L_{\text{eff}}(q_c, T)\, dt. \tag{62}$$

Or

$$S_{\text{eff}}[q_c, T] = -i \ln\int_{-\infty}^{\infty} \frac{d\lambda}{2\pi} \int \exp i\int_0^T dt\left\{L(q) + \frac{\lambda}{T}(q - q_c)\right\} \mathcal{D}\, q(t). \tag{63}$$

In practice, we shall avoid the 'edge effects' by considering $T \to \infty$. As an example, consider the case:

$$L = \tfrac{1}{2}m(\dot{q}^2 - \omega^2 q^2) - V(q).$$

Since the condition on $\langle\eta\rangle$ is meaningful only when quantum fluctuations are small, we shall expand the exponent up to quadratic terms in η for evaluating the path integral. Thus

$$\begin{aligned}
&\int \mathcal{D}q \exp i\int_0^T dt\left[L + \frac{\lambda}{T}\eta\right] \\
&= \int \exp i\int_0^T dt\left[L(q_c) + \tfrac{1}{2}\frac{\delta^2 L}{\delta q_c^2}\eta^2 + \frac{\lambda}{T}\eta\right]\mathcal{D}\eta \\
&= \int \exp i\int_0^T dt\left[\tfrac{1}{2}m(\dot{\eta}^2 - \omega^2\eta^2) - \tfrac{1}{2}\left(\frac{\partial^2 V}{\partial q^2}\right)_{q_c}\eta^2 + \right. \\
&\qquad \left. + \frac{\lambda}{T}\eta\right] \exp iS[q_c]\,\mathcal{D}\eta \\
&= \exp iS[q_c]\int \exp i\int_0^T dt\left\{\tfrac{1}{2}m(\dot{\eta}^2 - \nu^2\eta^2) + \frac{\lambda}{T}\eta\right\}\mathcal{D}\eta,
\end{aligned} \tag{64}$$

where we have put

$$\nu^2 \equiv \omega^2 + \frac{1}{m}\left(\frac{\partial^2 V}{\partial q^2}\right)_{q_c}. \tag{65}$$

The functional integral in (64) has to be evaluated with $\eta = 0$ at the boundaries. It is easier to work with the variable

$$y \equiv \left(\eta - \frac{\lambda}{mT\nu^2}\right)$$

which satisfies the boundary conditions $y = -\lambda/mTv^2$ at $t = 0, T$. In terms of y we have,

$$\bar{K} = \int_{-\infty}^{+\infty} \frac{d\lambda}{2\pi} \exp iS_c[q_c] \exp \frac{i\lambda^2}{2mTv^2} \times$$
$$\times \int \exp i \int_0^T dt \tfrac{1}{2} m [\dot{y}^2 - v^2 y^2] \mathcal{D}y, \tag{66}$$

The path integral over y is the same as that for a harmonic oscillator. Using the proper boundary conditions on y we get

$$\bar{K} = \int_{-\infty}^{+\infty} \frac{d\lambda}{2\pi} \exp iS_c[q_c] \times$$
$$\times \exp\left\{ \frac{i\lambda^2}{2mTv^2}\left[1 - \frac{2}{vT} \tan \frac{vT}{2}\right]\right\}\left(\frac{mv}{2\pi i \sin vT}\right)^{1/2}. \tag{67}$$

The integral over λ is Gaussian and can be done trivially, giving:

$$\begin{aligned}\bar{K} &= \left(\frac{mv}{2\pi i \sin vT}\right)^{1/2} \left(\frac{2\pi m i T v^2}{1 - (2/vT)\tan(vT/2)}\right)^{1/2} \\ &= \exp\left\{\frac{1}{2} \ln \frac{m^2 v^3 T}{(\sin vT)(1 - (2/vT)\tan(vT/2))}\right\} \\ &= \exp i \int_0^T \frac{1}{2i} \ln\left\{\frac{m^2 v^3 T}{\sin vT(1 - (2/vT)\tan(vT/2))}\right\}\frac{dt}{T}.\end{aligned} \tag{68}$$

Using our definition of effective action (62), we see that

$$S_{\text{eff}}(q_c, T) \equiv \int_0^T L_{\text{eff}}(q_c, T)\, dt$$

with

$$L_{\text{eff}}(q_c, T) = L(q_c) + \frac{1}{2iT} \ln\left\{\frac{m^2 v^3 T}{\sin vT(1 - (2/vT)\tan(vT/2))}\right\}. \tag{69}$$

In order to avoid the effects of end conditions, we consider the limit of $T \to \infty$. If we take this limit naïvely, we face ambiguities in expressions like

$$A \equiv \lim_{T\to\infty} \sin vT.$$

These ambiguities arise because integrals like

$$I = \int_{-\infty}^{+\infty} e^{iv^2\eta^2}\, d\eta$$

do not exist in the strict sense. This problem can be tackled by assuming that the frequency v has a negative imaginary part [that is, $v = \text{Lim}\ \varepsilon \to \infty\ (v - i\varepsilon)$]. With this assumption, we take

$$\lim_{T\to\infty} \sin \nu T \equiv \lim_{T\to\infty} \cos \nu T \equiv \lim_{T\to\infty} \frac{1}{2i} \exp(i\nu T).$$

Using this in (69) we get our effective Lagrangian,

$$\begin{aligned} L_{\text{eff}}(q_c) &= \lim_{T\to\infty} L_{\text{eff}}(q_c, T) \\ &= L(q_c) - \tfrac{1}{2}\,\nu = L(q_c) - \tfrac{1}{2}\left(\omega^2 + \frac{1}{m}\left(\frac{\partial^2 V}{\partial q^2}\right)_{q_c}\right)^{1/2}. \end{aligned} \tag{70}$$

Thus the effect of quantum fluctuations is to replace the classical potential $V(q)$ by the 'effective potential'

$$\begin{aligned} V(q)_{\text{eff}} &= V(q)_c + \tfrac{1}{2}\left(\omega^2 + \frac{1}{m}\left(\frac{\mathrm{d}^2 V_c}{\mathrm{d}q^2}\right)\right)^{1/2} \\ &V(q)_c + \tfrac{1}{2}\,\hbar\nu(q) + O(\hbar^2). \end{aligned} \tag{71}$$

In the above equation we have reintroduced $\hbar$. The second term has the form of a zero-point energy. (In fact, if $V_c = 0$, then the second term will be just a constant addition $\frac{1}{2}\hbar\omega$). This expression is valid to first order in $\hbar$ and can be used to study quantum corrections around the classical trajectory. In the next chapter we shall discuss the concept of effective action for a field theoretic system which will prove to be very useful.

Example 3.9. We have developed the concept of effective action in a somewhat intuitive way in the text. The following is a more formal definition of effective action (which is useful in the field theoretical contexts).

For a given action functional $S(q(t))$, define, $Z[J(t)]$ and $W[J(t)]$ by

$$\begin{aligned} Z[J(t)] &\equiv \exp\left\{+ iW(J(t))\right\} \\ &\equiv \int \exp\left\{+ iS[q(t)] + i\int J(t)q(t)\,\mathrm{d}t\right\} \mathcal{D}q(t). \end{aligned}$$

We define a 'classical trajectory' by

$$q_c(t) \equiv \frac{\delta W[J]}{\delta J(t)}.$$

Assuming this equation is invertible we can express J in terms of q_c and thus, W in terms of q_c. The effective action is defined to be the Legendre transform

$$S_{\text{eff}}(q_c) = W(q_c) - J(q_c)q_c.$$

This definition will give the same answer as the method given in the text. However, formal manipulations are easier with the above definition than with the one developed in the text. □

The effective action is also useful when we have two interacting systems characterized by dynamical variables $q(t)$ and $Q(t)$, respectively. If we are interested only in one (q, say) we can study the joint path integral in which all paths of the other (Q) are integrated out. The resulting expression can be related to the effective action for q, along the lines of (61).

3.8. Quantum Mechanics at Finite Temperature

Let us consider the quantum mechanical description of a system which is interacting with an external heat bath maintained at temperature T. Suppose that the system could be in any one of a whole range of stationary states with wave functions $\{\psi_n(q)\}$ and energy levels $\{E_n\}$. Then the probability that the system is found in the nth state is proportional to the Boltzmann factor exp $(-E_n/kT)$, where k is the Boltzmann constant. The probability that the system is found in a range $(q, q+\mathrm{d}q)$ of the dynamical variable of q is $|\psi_n(q)|^2\,\mathrm{d}q$, if it is in the nth state. Combining the two probabilities we get the probability of finding the system of the range $(q, q+\mathrm{d}q)$ to be $P(q, T)\,\mathrm{d}q$, where

$$P(q, T) = \left(\sum_n \exp\left(-\frac{E_n}{kT}\right) |\psi_n(q)|^2\right) Z^{-1}, \tag{72}$$

and

$$Z = \sum_m \exp\left(-\frac{E_m}{kT}\right).$$

We shall now try to relate $P(q, T)$ to the quantum mechanical kernel at zero temperature.

We know that the kernel can be expressed in the form

$$K(q_f, t_f; q_i, t_i) = \sum_n e^{-iE_n(t_f - t_i)}\, \psi_n(q_f)\, \psi_n^*(q_i) \tag{73}$$

Setting $q_f = q_i = \bar{q}$ gives

$$K(\bar{q}, t_f; \bar{q}, t_i) = \sum_n e^{-iE_n(t_f - t_i)}\, |\psi_n(\bar{q})|^2 \tag{74}$$

A comparison of (72) with (74) gives

$$P(\bar{q}, T) = K(\bar{q}, t_i - i\beta; \bar{q}, t_i)\,.\,Z^{-1}, \tag{75}$$

where

$$\beta = \frac{1}{kT}. \tag{76}$$

The value of Z is determined by noticing that the $\psi_n(q)$ are normalized to unity. We have

$$\int_{-\infty}^{+\infty} K(q, t_i - i\beta; q, t_i)\,\mathrm{d}q = \sum_n \exp\left(-\frac{E_n}{kT}\right)\int_{-\infty}^{+\infty} |\psi_n(q)|^2\,\mathrm{d}q$$

$$= \sum_n \exp\left(-\frac{E_n}{kT}\right) = Z. \tag{77}$$

Thus P is obtained from K by analytic continuation into imaginary time axis. This result suggests that we can do finite temperature statistical mechanics through path integrals in Euclidean time. Thus we may write

$$P(\bar{q}, T) = \int \exp\left\{-i \int_{t_i}^{t_i - i\beta} L(q, \dot{q})\, \mathrm{d}t\right\} \mathcal{D}\, q(t)$$

$$= \int \exp\left\{-\int_0^{\beta} L\left(q, i\frac{\mathrm{d}q}{\mathrm{d}\tau}\right)\mathrm{d}\tau\right\} \mathcal{D}\, q(\tau), \tag{78}$$

where we have put $t = t_i - i\tau$. There is one restriction on the paths being summed over. They start and end with the same q value:

$$q(t_i) = q(t_i - i\beta) = \bar{q}.$$

These are periodic boundary conditions.

Example 3.10. Consider a harmonic oscillator in a heat bath at temperature T with $\beta = 1/kT$. Using the known form of $K(q_f, t_f; q_i, t_i)$ we immediately get

$$P(q, T) = \left(\frac{m\omega}{2\pi\hbar \sinh \beta h\omega}\right)^{1/2} \exp\left\{-\frac{m\omega q^2}{\hbar} \tanh \frac{\hbar\omega}{2kT}\right\} \cdot Z^{-1},$$

where we have reintroduced $\hbar$ and m denotes the mass of the particle. Using (77), we find Z to be

$$Z = \int_{-\infty}^{+\infty} \left(\frac{m\omega}{2\pi\hbar \sinh \beta\hbar\omega}\right)^{1/2} \exp\left\{-\frac{m\omega}{\hbar} \tanh \frac{\beta\hbar\omega}{2}\right\} q^2\, \mathrm{d}q$$

$$= \left(\frac{m\omega}{2\pi\hbar \sinh \beta\hbar\omega}\right)^{1/2} \left(\frac{\pi\hbar}{m\omega \tanh \beta\hbar\omega/2}\right)^{1/2} = \tfrac{1}{2}\cosh\left(\frac{\beta\hbar\omega}{2}\right).$$

Thus,

$$P(q, T) = \left(\frac{m\omega}{\pi\hbar} \tanh \frac{\beta\hbar\omega}{2}\right)^{1/2} \exp\left(-\frac{m\omega q^2}{\hbar} \tanh \frac{\hbar\omega}{2kT}\right)$$

This expression has simple interpretations when $\hbar\omega \ll kT$ or when $\hbar\omega \gg kT$. In the high-temperature limit $kT \gg \hbar\omega$.

$$P(q, T) \approx \left(\frac{m\omega^2}{2\pi kT}\right)^{1/2} \exp\left(-\frac{1}{2}\frac{m\omega^2 q^2}{kT}\right)$$

which represents the classical Boltzmann distribution. (Note that P is independent of $\hbar$). On the other hand, as $T \to 0$, P becomes

$$P \simeq \left(\frac{m\omega}{\pi\hbar}\right)^{1/2} \exp\left(-\frac{m\omega q^2}{\hbar}\right),$$

which is the ground state probability distribution for the harmonic oscillator. These are the results we would expect on intuitive grounds. □

We shall discuss in the next chapter an extension of these ideas to finite temperature field theory, and its applications.

This brings to an end our discussion of an assorted collection of techniques which will come in useful in the following chapter where we tackle the problems of field quantization. As mentioned at the beginning of this chapter, the field has an infinite number of degrees of freedom. This fact brings in additional difficulties of its own besides those we encountered here. Nevertheless, the similarity between an individual degree of freedom of a field and the harmonic oscillator helps in visualizing many properties of the quantized field.

Notes and References

1. For the use of path integral in perturbation theory see:

    Feynman, R. P. and Hibbs, A. R.: 1965, *Quantum Mechanics and Path Integrals*, McGraw-Hill, New York, chs 6 and 7.

2. Functional differentiation is discussed in most of the field theory texts. A discussion in the context of point quantum mechanics is given in ch. 7 of Feynman and Hibbs (above).

3. Physics in the Euclidean time is discussed, for example, in the article:

    Abers, E. S., and Lee, B. W.: 1973, 'Gauge Theories', *Phys. Repts* **9C** 67.

4. A more detailed discussion of various aspects of instanton physics (including double-well and double-hump potentials) can be found in the excellent article:

    Coleman, S.: 1979 'The Uses of Instantons', in *The Whys of Subnuclear Physics, Erice* 1977 (ed. A. Zichichi), Plenum, New York.

    Some (definitely *non*-representative) original references relevant to the instanton physics, double-well potential, etc., are:

    Langer, J. S.: 1967, *Ann. Phys.* **41**, 108.
    Collins, J. C., and Soper, D. E.: 1978, *Ann. Phys.* **112**, 209.
    Jaffe, A. M.: 1965, *Comm. Math. Phys.* **1**, 127.
    Bender, C. M., and Wu, T. T.: 1973, *Phys. Rev.* **D7**, 1620.

5. Our discussion of effective action parallels that which appears in pp. 280–285 of Feynman and Hibbs (1965) – see Note 1 above. See also:

    Coleman, S.: 1975, 'Secret Symmetry' in *The Laws of Hadronic Matter* (ed. A. Zichichi), Academic Press, New York).

6. Quantum statistical mechanics using path integrals is discussed in ch. 10 of Feynman and Hibbs (1965) – see Note 1 above.

Chapter 4

Quantum Field Theory

4.1. Classical Field Theory (General)

We mentioned in the previous chapter that a typical field can be looked upon as a dynamical system with infinite degrees of freedom. In this section we briefly review some dynamical properties of fields at the classical level to develop concepts that are analogous to those of Lagrangian mechanics.

To see the analogy the reader should keep in mind the following picture. In Lagrangian mechanics the N independent dynamical coordinates $q_1(t)$, $q_2(t), \ldots, q_N(t)$ specify a system with N degrees of freedom. Sometimes such coordinates cannot be identified and we have to use $N+m$ coordinates ($m > 0$) which are constrained by m relations. We shall ignore this additional complication for this discussion (which could be modified to take care of it). In the case of a field we may consider $q(\mathbf{r}, t)$ as a collection of dynamical variables where the space coordinate $\mathbf{r}$ is the label for the variable at point $\mathbf{r}$. We therefore have a continuum set of such coordinates.

For the mechanical system the Lagrangian L is a function of q_n, $\dot{q}_n$, and t. The Lagrange equations

$$\frac{\mathrm{d}}{\mathrm{d}t}\left(\frac{\partial L}{\partial \dot{q}_n}\right) - \frac{\partial L}{\partial q_n} = 0, \quad n = 1, \ldots, N \tag{1}$$

follow from a variation of the action

$$S = \int_{t_i}^{t_f} L \,\mathrm{d}t \tag{2}$$

defined over a specified interval (t_i, t_f). It is also possible to re-express (1) as Hamilton's equations:

$$-\frac{\partial H}{\partial q_n} = \dot{p}_n, \qquad \frac{\partial H}{\partial p_n} = \dot{q}_n, \tag{3}$$

where the Hamiltonian is defined by

$$H = \sum_{n=1}^{N} p_n \dot{q}_n - L \tag{4}$$

and is expressed as a function of $\{q_n\}$ and the conjugate momenta $\{p_n\}$ defined by

$$p_n = \frac{\partial L}{\partial \dot{q}_n} . \tag{5}$$

These concepts can be translated into field theory language in the following way. Since we are now dealing with a continuum rather than with a discrete set now, we define a *Lagrangian density* $\mathscr{L}$ which is a functional of $q(\mathbf{r}, t)$, $\dot{q}(\mathbf{r}, t)$, and t. We write

$$L = \int \mathscr{L}\, \mathrm{d}^3\mathbf{r} \tag{6}$$

as the Lagrangian. The action functional then becomes

$$S = \int L\, \mathrm{d}t = \int_{\mathcal{V}} \mathscr{L}\, \mathrm{d}^4x, \tag{7}$$

where the integration is over spacetime coordinates. The integration over $\mathrm{d}^3\mathbf{r}$ is the logical extension of the definition of the mechanical Lagrangian as a sum over the discrete set $1, \ldots, N$. $\mathcal{V}$ is the four-dimensional subspace of spacetime under consideration.

In addition, if we demand that our theory is Lorentz invariant in order to be consistent with the special relativistic description of nature, we need $\mathscr{L}$ to be a scalar density, and S given by (7) to be a scalar. This is ensured by requiring $q(\mathbf{r}, t)$ to be expressed in a spinorial or tensorial form. Likewise, Lorentz invariance also requires that not only the time derivatives be present in $\mathscr{L}$ but also the space derivatives. Thus,

$$\mathscr{L} = \mathscr{L}[q, q_{,i}]. \tag{8}$$

In addition, $\mathscr{L}$ may also be an explicit function of spacetime coordinates.

Lagrange equations obtained by $\delta S = 0$ are then easily seen (*vide* Example 4.1) to be

$$\frac{\partial}{\partial x^i}\left[\frac{\delta \mathscr{L}}{\delta q_{,i}}\right] - \frac{\delta \mathscr{L}}{\delta q} = 0. \tag{9}$$

Here the derivatives δ are functional derivatives as defined in Section 3.3. (Because of the repeated index i the summation convention operates.) In field theory jargon equations like (9) are called *field equations*. Example 4.1 shows how the field equation for a massive scalar field follows by the above prescription.

Example 4.1. Consider the variation of the action S under a variation, $\delta q(x^i)$ of the dynamical variables $q(x^i)$. We have

$$\delta S = \int_{\mathcal{V}} \left\{ \frac{\partial \mathscr{L}}{\partial q}\, \delta q + \frac{\partial \mathscr{L}}{\partial (q_{,i})}\, \delta(q_{,i}) \right\} \mathrm{d}^4x,$$

$$= \int_{\mathcal{V}} \left\{ \frac{\partial \mathcal{L}}{\partial q} \delta q + \frac{\partial \mathcal{L}}{\partial (q_{,i})} \frac{\partial}{\partial x^i} (\delta q) \right\} d^4x, \text{ (since } \delta(q_{,i}) = \partial_i(\delta q))$$

$$= \int_{\mathcal{V}} \left\{ \frac{\partial \mathcal{L}}{\partial q} \delta q + \frac{\partial}{\partial x^i} \left[\frac{\partial \mathcal{L}}{\partial (q_{,i})} \delta q \right] - \frac{\partial}{\partial x^i} \left(\frac{\partial \mathcal{L}}{\partial (q_{,\mathrm{i}}} \right) \delta q \right\} d^4x,$$

$$= \int_{\mathcal{V}} \left[\left\{ \frac{\partial \mathcal{L}}{\partial q} - \frac{\partial}{\partial x^i} \left(\frac{\partial \mathcal{L}}{\partial (q_{,i})} \right) \right\} \delta q + \frac{\partial}{\partial x^i} \left\{ \frac{\partial \mathcal{L}}{\partial (q_{,i})} \delta \mathrm{q} \right\} \right] d^4x.$$

Using Gauss's theorem, the 4 divergence part is converted to an integral on the surface $\partial \mathcal{V}$ bounding the 4-volume $\mathcal{V}$. As usual we shall assume the δq to vanish on this surface; so if n_i is the unit normal to $\partial \mathcal{V}$.

$$\int_{\mathcal{V}} \frac{\partial}{\partial x^i} \left\{ \frac{\partial L}{\partial (q_{,i})} \delta q \right\} d^4x = \int_{\partial \mathcal{V}} n_i \left\{ \frac{\partial L}{\partial (q_{,i})} \delta q \right\} d^3x = 0.$$

Thus, $\delta S = 0$ implies

$$\frac{\partial L}{\partial q} = \frac{\partial}{\partial x^i} \left(\frac{\partial L}{\partial (q_{,i})} \right).$$

As an example, consider

$$L(q, q_{,i}) = \tfrac{1}{2}(q_{,i}q^{,i} - m^2q^2)$$

$$\frac{\partial L}{\partial q} = -\, m^2 q; \qquad \frac{\partial L}{\partial q_{,i}} = q_{,i}.$$

Therefore,

$$\frac{\partial}{\partial x^i} (q_{,i}) = -\, m^2 q; \text{ i.e. } (\Box + m^2)\, q(x^i) = 0,$$

which is the wave equation for the scalar field. □

The analogues of (4) and (5) are also obtained in a straightforward manner. Corresponding to (5), we write the field momentum as the functional derivative

$$p^i = \frac{\delta \mathcal{L}}{\delta q_{,i}}. \tag{10}$$

Note that, because of our insistance on treating space and time on equal footing (as dictated by Lorentz invariance), we get a 4-vector for the momentum rather than a single quantity like p_n. Likewise, instead of the Hamiltonian H we get an energy–momentum tensor which is defined by

$$T^i{}_k = p^i q_{,k} - \delta^i_k \mathcal{L}. \tag{11}$$

4.1.1. *Conservation Laws*

According to Noether's theorem, a symmetry in the Lagrangian can lead to conservation laws for the dynamical system. If time does not occur explicitly in (4), i.e. if the Lagrangian is symmetric with respect to time translation, then H is a constant of motion. Likewise, if $\mathscr{L}$ is invariant under translation of coordinates $x'^i \rightarrow x^i + \varepsilon^i$, then we can show that

$$T^i_{k,i} \equiv 0 \tag{12}$$

The proof is briefly as follows.

Consider the change $\delta\mathscr{L}$ produced in $\mathscr{L}$ by the above displacement. We can write

$$\delta\mathscr{L} = \frac{\partial\mathscr{L}}{\partial x^i}\, \varepsilon^i. \tag{13}$$

However, $\mathscr{L}$ cannot have explicit dependence of x^i because of its translation invariance. Hence, the above $\delta\mathscr{L}$ is obtained through the changes in q and $q_{,i}$. Thus

$$\begin{aligned}\delta\mathscr{L} &= \frac{\delta\mathscr{L}}{\delta q}\,\delta q + \frac{\delta\mathscr{L}}{\delta q_{,i}}\,\delta q_{,i} \\ &= \frac{\delta\mathscr{L}}{\delta q}\, q_{,k}\, \varepsilon^k + \frac{\delta\mathscr{L}}{\delta q_{,i}}\, q_{,ik}\, \varepsilon^k.\end{aligned}$$

But from field equations (9) we know that

$$\frac{\delta\mathscr{L}}{\delta q} = \frac{\partial}{\partial x^i}\left(\frac{\delta\mathscr{L}}{\delta q_{,i}}\right).$$

Also, $q_{,ik} = q_{,ki}$. Hence, we can write

$$\delta\mathscr{L} = \frac{\partial}{\partial x^i}\left[\frac{\delta\mathscr{L}}{\delta q_{,i}}\, q_{,k}\right]\varepsilon^k = \frac{\partial}{\partial x^i}\,(p^i q_{,k})\,\varepsilon^k. \tag{14}$$

Equating (13) and (14) and arguing that ε^i are arbitrary, we get the required result.

Notice that (12) gives an energy tensor

$$T_{ik} = \eta_{i\ell} T^\ell_k = p_i q_{,k} - \delta_{ik}\,\mathscr{L} \tag{15}$$

which is *not* manifestly symmetric with respect to (i, k). In those cases where it is not symmetric it can be made so by adding to it an appropriate tensor of zero divergence. Later, in Chapter 6, we shall show how a manifestly symmetric T_{ik} follows from considerations of general spacetime transformations.

If Σ is a spacelike surface extending in all spacelike directions of infinity and n_i is its unit 4-normal then the total 4-momentum of the field is defined by

$$P_k = \int_\Sigma T^i_k n_i\, \mathrm{d}\Sigma. \tag{16}$$

If we foliate the space–time with such Σ surfaces then the conservation law

(12) leads to P_k = const from one Σ surface to another. This explains why (12) is called the law of conservation of 4-momentum.

If $n_i = (1, 0, 0, 0)$ we get

$$P_k = \int T_k^0 \, d\Sigma = \int T_k^0 \, d^3\mathbf{r}.$$

In particular,

$$P_0 = \int_\Sigma (p^0\dot{q} - \mathscr{L}) \, d\Sigma = \int_V (p_0\dot{q} - \mathscr{L}) \, d^3\mathbf{r}$$

$$= \int \mathscr{H} \, d^3\mathbf{r}, \tag{17}$$

where

$$\mathscr{H} = p_0\dot{q} - \mathscr{L} \equiv \frac{\partial \mathscr{L}}{\partial \dot{q}} \dot{q} - \mathscr{L} \tag{18}$$

is the *Hamiltonian density* for the field. Like its counterpart H in (4), $\mathscr{H}$ plays an important role in field quantization.

Example 4.2. The conservation of $T^i{}_k$ is only a special case of a general result. Suppose that an action is invariant under a set of transformations

$$x^i \to x'^i \equiv x^i + \delta x^i; \quad \psi_A(x) \to \psi'_A(x) \equiv \psi_A(x) + \bar{\delta}\psi_A(x).$$

Here $\psi_A(x)$, $A = 1, 2 \ldots$ denote different spinor/tensor components of field. Also note that since x^i has changed to x'^i, the change $\bar{\delta}\psi_A(x)$ actually denotes the change in the functional form of $\psi_A(x)$. This is to be contrasted with the local variation

$$\delta\psi_A \equiv \psi'_A(x') - \psi_A(x)$$
$$= \bar{\delta}\psi_A + \partial_i\psi_A(x)\delta x^i,$$

which measures the change in the field $\psi_A(x)$ at the same physical event x^i.

Since the action is invariant under the above changes, we can write

$$0 = \delta S = \int_{\mathcal{V}} \mathscr{L}(\psi + \bar{\delta}\psi, \partial_i\psi + \bar{\delta}\partial_i\psi) \, d^4x' - \int_{\mathcal{V}} \mathscr{L}(\psi, \partial_i\psi) \, d^4x$$
$$= \int_{\mathcal{V}} \left(\frac{\partial \mathscr{L}}{\partial \psi_A} \bar{\delta}\psi_A + \frac{\partial \mathscr{L}}{\partial(\partial_i\psi_A)} \partial_i(\bar{\delta}\psi_A) + \mathscr{L} \frac{\partial \delta x^i}{\partial x^i} \right) d^4x$$
$$= \int_{\mathcal{V}} \left\{ \frac{\partial \mathscr{L}}{\partial \psi_A} - \partial_i \left(\frac{\partial \mathscr{L}}{\partial(\partial_i\psi_A)} \right) \right\} \bar{\delta}\psi_A \, d^4x + \int_{\mathcal{V}} \partial_i \left\{ \frac{\partial \mathscr{L}}{\partial(\partial_i\psi_A)} \bar{\delta}\psi_A + \mathscr{L}\delta x^i \right\} d^4x,$$

The first integral vanishes when equations of motion ae used. In the second integral we introduce local variation $\delta\psi_A$ to arrive at the conservation law

$$0 = \int \partial_i J^i \, d^4x = \int_{t=t_1} J^0 \, d^3\mathbf{r} - \int_{t=t_2} J^0 \, d^3\mathbf{r},$$

where

$$J^i = \frac{\partial \mathscr{L}}{\partial(\partial_i\psi_A)} \delta\psi_A - \left(\frac{\partial \mathscr{L}}{\partial(\partial_i\psi_A)} \partial_k\psi_A - \delta^i_k \mathscr{L} \right) \delta x^k.$$

Let us apply this result to homogeneous Lorentz transformations,

$$\delta x^i = \varepsilon^{ik} x_k; \qquad \varepsilon^{ik} = -\varepsilon^{ki}$$

$$\delta \psi_A = \tfrac{1}{2}\, I^{AB}_{ik}\, \varepsilon^{ik} \psi_B,$$

where I^{AB}_{ik} depends on the Lorentz transformation property of the field (scalar, vector, spinor, etc.). Using our general result, we get ($\pi^i_A \equiv \partial L/(\partial(\partial_i \psi_A))$)

$$J^i = \pi^i_A (\tfrac{1}{2} I^{AB}_{mn}\, \varepsilon^{mn} \psi_B) - (\pi^i_A \partial_m \psi_A - \delta^i_m \mathscr{L})\, \varepsilon^{mn} x_n$$

$$= \tfrac{1}{2} \varepsilon^{mn} \left\{ \pi^i_A I^{AB}_{mn}\, \psi_B + \pi^i_A (x_m \partial_n - x_n \partial_m) \psi_A + (\delta^i_m x_n - \delta^i_n x_m)\, \mathrm{L} \right\}$$

$$\equiv \tfrac{1}{2} \varepsilon^{mn}\, M^i_{mn}$$

The quantity M^i_{mn} satisfies the equations

$$\partial_i M^i_{mn} = 0; \quad M_{mn} = -M_{mn}$$

and represents the *angular momentum density* of the field. The total conserved angular momentum at any time is given by

$$L_{mn} \equiv \int M^0_{mn}\, d^3\mathbf{r}.$$

For purely spatial components, $L_{\mu\nu}$ (μ, ν = 1, 2, 3), we have, for example,

$$L_{12} = \int \left\{ \pi_A I^{AB}_{12}\, \psi_B + \pi_A (x_1 \partial_2 - x_2 \partial_1)\, \psi_A \right\} \mathrm{d}^3\mathbf{r}.$$

The first term represents the intrinsic spin angular momentum and the second term gives the orbital angular momentum. □

4.2. Classical Field Theory (Specific Fields)

We next consider briefly three specific fields since they will feature frequently in our later discussions. These are (i) the massive scalar field, (ii) the electromagnetic field, and (iii) the spin half Dirac field. We are excluding from our discussion the gravitational field which forms the main part of this book and receives our attention in Parts II, III, and IV.

4.2.1. *The Scalar Field*

The scalar field will be denoted by a function $\varphi(t, \mathbf{r})$ of spacetime coordinates. At each spacetime point, φ is invariant under Lorentz transformation (which is why we use the adjective 'scalar'). A 'free' scalar field of mass m is specified by the action

$$S = \int (\tfrac{1}{2}\, \varphi_i \varphi^i - m^2 \varphi^2)\, \mathrm{d}^4 x, \tag{19}$$

and its field equation (9) takes the form

$$(\Box + m^2)\, \varphi = 0. \tag{20}$$

In the previous chapter we briefly discussed this field and showed in Equations (3.3) and (3.4) that its Fourier representation is equivalent to a continuum distribution of harmonic oscillators. Now let us study the interaction of φ with an external potential in form

$$V = \varphi(t, r)\, J(t, \mathbf{r}). \tag{21}$$

Writing

$$\mathcal{L} = \tfrac{1}{2}\varphi_i\varphi^i - \tfrac{1}{2}m^2\varphi^2 + \varphi J \tag{22}$$

we find that (20) is modified to

$$(\Box + m^2)\varphi = J. \tag{23}$$

To solve this equation we use a scalar Green's function $G(x, x')$ which is defined over two spacetime points x and x', and satisfies the wave equation

$$(\Box_x + m^2)\, G(x, x') = \delta_4(x, x'); \tag{24}$$

where $\delta_4(x, x')$ is the four-dimensional delta function that vanishes unless x and x' coincide. The solution of (23) is then

$$\varphi(x) = \varphi_0(x) + \int G(x, x')\, J(x')\, \mathrm{d}^4x, \tag{25}$$

where φ_0 is a solution of (20). Since

$$\delta_4(x, x') = \int \frac{\mathrm{e}^{-ip\cdot(x-x')}}{(2\pi)^4}\, \mathrm{d}^4p \tag{26}$$

is the Fourier representation of the delta function, we get the following representation for $G(x, x')$:

$$G(x, x') = \int \frac{\mathrm{e}^{-ip\cdot(x-x')}}{(-p^2 + m^2)} \frac{\mathrm{d}^4p}{(2\pi)^4}\ . \tag{27}$$

This integral is not well defined because of the poles in the path of integration. We had encountered this problem in Equation (3.40) in Section 3.4. We have to evaluate G by choosing a suitable contour in the complex p plane. The answer is not unique and Figure 4.1 illustrates the various combinations possible. Table 4.1 denotes the type of solution that emerges from the choice of each contour.

TABLE 4.1
Contours and Green's functions

Contour	Green's function	Comment
C_S	$G_S(x, x')$	The principal part of the integral in (27); symmetric in x, x'
C_R	$G_R(x, x')$	Retarded, i.e. vanishes for $x^0 < x'^0$
C_A	$G_A(x, x')$	Advanced, i.e. vanishes for $x^0 > x'^0$
C_F	$G_\mathrm{F}(x, x')$	Propagates positive frequencies forward and negative frequencies backward in time
C_D	$G_\mathrm{D}(x, x')$	Propagates positive frequencies backward and negative frequencies forward in time.

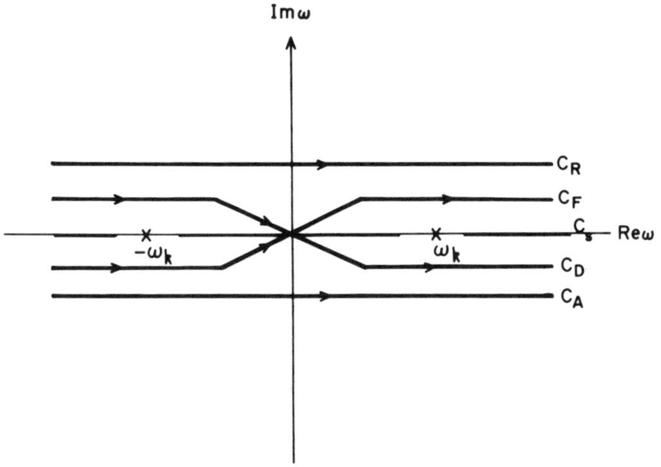

Fig. 4.1. The contours for various Green's functions given in Table 4.1. The retarded (advanced) Green's function propagates both positive and negative frequency components forward (backward) in time. The Feynman Green's function propagates positive (negative) frequency components forward (backward) in time.

The last two Green's functions, G_F and G_D, in the table are also called the 'Feynman propagator' and the 'Dyson propagator', respectively. We evaluate G_F in Example 4.3. below.

Example 4.3. Consider the evaluation of $G_F(x)$ using the contour C_F. It is clear that integrating along C_F is equivalent to integrating along the real axis after replacing m^2 by $m^2 - i\varepsilon$. Thus, we have to calculate

$$G_F(x) \equiv - \int \frac{e^{-ik \cdot x}}{k^2 - m^2 + i\varepsilon} \frac{d^4k}{(2\pi)^4} .$$

Since Gaussian integrals are easier to evaluate than the one above, write:

$$\frac{1}{k^2 - m^2 + i\varepsilon} = i^{-1} \int_0^\infty \exp i\lambda\,(k^2 - m^2 + i\varepsilon)\, d\lambda$$

$$G_F(x) = i \int_0^\infty d\lambda \int \exp(i\lambda k^2 - ik \cdot x) \exp i\lambda(-m^2 + i\varepsilon) \frac{d^4k}{(2\pi)^4}$$

$$= \frac{i}{16\pi^4} \int_0^\infty \left\{ \exp i\lambda(-m^2 + i\varepsilon) \left(\frac{\pi}{-i\lambda}\right)^{1/2} \exp -\frac{(-t^2)}{4i\lambda} \times \right.$$

$$\left. \times \left(\frac{\pi}{i\lambda}\right)^{3/2} \exp -\frac{(-|\mathbf{r}|^2)}{(-4i\lambda)} \right\} d\lambda$$

$$= - \int_0^\infty \left(\frac{\pi}{i\lambda}\right)^2 \exp i\lambda(-m^2 + i\varepsilon) \exp\left(-\frac{ix^2}{4\lambda}\right) \cdot \frac{1}{16\pi^4} d\lambda$$

$$= \frac{1}{16\pi^2} \int_0^\infty \frac{1}{\lambda^2} \exp -i\left[\lambda(m^2 - i\varepsilon) + \frac{x^2}{4\lambda}\right] d\lambda.$$

When $m = 0$, this expression simplifies considerably. We get

$$G_F(x)\Big|_{m=0} = \frac{1}{4\pi^2 i}\,\frac{1}{x^2 - i\varepsilon}\,.$$

For nonzero m, the integral can be expressed in terms of modified Bessel functions. Writing $M^2 \equiv m^2 - i\varepsilon$, $\lambda = (s/2M)\,\alpha$ where, $s^2 = x^i x_i$, we get

$$G_F(x) = \frac{1}{16\pi^2}\cdot\left(\frac{2M}{s}\right)\int_0^\infty \frac{1}{\alpha^2}\exp -\frac{iMs}{2}\left(\alpha + \frac{1}{\alpha}\right)\,d\alpha$$

$$= \frac{1}{16\pi^2}\left(\frac{2M}{s}\right)\int_0^\infty \exp -\frac{iMs}{2}\left(\eta + \frac{1}{\eta}\right)\,d\eta = \frac{1}{4\pi^2}\left(\frac{M}{s}\right)\,K_1(iMs).$$

In arriving at the last expression we have used the following integral representation for the modified Bessel function,

$$K_\nu(x) = \int_0^\infty \tfrac{1}{2}\lambda^{\nu-1}\exp\left\{-\tfrac{1}{2}x(\lambda + \lambda^{-1})\right\}\,d\lambda.$$

This formula is valid for $\mathrm{Re}(\nu) > -\frac{1}{2}$. Oscillating exponents are taken care of by the $i\varepsilon$ in the definition of M^2. The function, $K_\nu(x)$ has the asymptotic form,

$$K_\nu(x) \cong (\mathrm{const})x^{-1/2}\,e^{-x}, \quad x \gg 1.$$

Note that for spacelike separations $t = 0$, $x^i = (0, R)$, $s = iR$, with $mR \gg 1$, the Green's function behaves as

$$G_F(t = 0, R) \cong (\mathrm{const})R^{-3/2}\exp(-mR);\quad mR \gg 1.$$ □

These Green's functions are not all independent. For example,

$$G_S = \tfrac{1}{2}\left\{G_R + G_A\right\}. \tag{28}$$

4.2.2. *The Electromagnetic Field*

This is a zero mass vector field which may be characterized by the 4-vector potential A_i. The field itself is a second-rank antisymmetric tensor defined by

$$F_{mn} = A_{n,m} - A_{m,n}. \tag{29}$$

The Lagrangian density for the free field is given by

$$\mathcal{L} = -\tfrac{1}{4}F_{mn}F^{mn}, \tag{30}$$

and for a field coupled to electric charges and currents

$$\mathcal{L} = -\tfrac{1}{4}F_{mn}F^{mn} - A_i J^i, \tag{31}$$

where J^i is the 4-vector for current density.

The coupled case gives the field equations

$$F^{ik}{}_{,k} = J^i. \tag{32}$$

Besides, the definition (29) implies the identity

$$F_{ik,l} + F_{kl,i} + F_{li,k} \equiv 0. \tag{33}$$

The Maxwell equations are contained in (32) and (33) and can be reproduced by identifying A_0 with the scalar potential φ and A_μ with the 3-vector potential $\mathbf{A}$. In the usual notation

$$\mathbf{E} \equiv [F_{01}, F_{02}, F_{03}] = -\nabla\varphi - \frac{\partial \mathbf{A}}{\partial t},$$

$$\mathbf{H} = [F_{23}, F_{31}, F_{12}] = \nabla \times \mathbf{A}, \tag{34}$$

$$J^i = [\varrho, \mathbf{j}].$$

Equation (32) is expressed as a wave equation

$$\Box A_i - (A^k{}_{,k})_{,i} = J_i. \tag{35}$$

This equation can, however, be simplified by a 'gauge transformation'. The gauge transformation is defined by

$$A^*_i = A_i + \chi_{,i} \tag{36}$$

where χ is an arbitrary scalar function. It is easy to see that under (36) the field tensor remains unchanged.

$$F^*_{mn} = F_{mn}. \tag{37}$$

Since all physical effects of the electromagnetic nature are connected with the field tensor F_{ik} rather than with the 4-potential A_i, we say that the theory is invariant under gauge transformations. Such theories are called 'gauge theories' and we shall discuss such theories in the following chapter. For the present we only note that the electromagnetic theory is a gauge theory.

The immediate advantage of this property is that we can always choose χ to make

$$A^{*k}{}_{,k} = A^i{}_{,i} + \Box\chi = 0. \tag{38}$$

Hence the wave equation for A^*_k becomes simply

$$\Box A^*_k = J_k. \tag{39}$$

Henceforth we shall drop the asterisk and assume that (38) and (39) hold for A_k. We have thus 'gauged away' the quantity $A^i{}_{,i}$. This particular gauge is called the *Lorentz gauge*.

The Green's functions for (39) are obtainable from the Green's functions for the scalar field by setting $m = 0$ and multiplying by η_{ik} to include the vector effect. Thus a particular integral of (39) is

$$A_k(x) = \int G(x, x')\eta_{ik}J^i \, d^4x'. \tag{40}$$

In some cases it is preferable to have the Coulomb gauge in which it is possible to set $A_0 = 0$ and reduce (38) to $\nabla \cdot \mathbf{A} = 0$.

Example 4.4. It is possible to add to the electromagnetic Lagrangian terms that fix the choice of gauge. For example, consider

$$\mathcal{L}_\lambda \equiv -\tfrac{1}{4} F_{ik}F^{ik} + \frac{\lambda}{2}\,(\partial_i A^i)^2, \quad \lambda \neq 0.$$

The field equations read,

$$\Box A^i + (\lambda - 1)\,\partial^i(\partial_k A^k) = J^i.$$

Taking the divergence and assuming that $\partial_i A^i$ vanishes at infinity leads to the conclusion

$$\partial_i A^i \equiv 0,$$

which is the Lorentz gauge. □

4.2.3. *The Dirac Field*

A spin-$\frac{1}{2}$ particle of mass m is described by a 4-spinor wave function ψ whose equation can be derived from the Lagrangian density

$$\mathcal{L} = i\,\bar{\psi}\gamma^i\psi_{,i} - m\bar{\psi}\,\psi, \tag{41}$$

where

$$\bar{\psi} = \psi^\dagger\gamma^0. \tag{42}$$

Here $\psi^\dagger$ is the Hermitian conjugate of a 4×1 matrix ψ so that $\bar{\psi}$ is a 1×4 matrix. The 4×4 matrices γ^i satisfy the relation

$$\gamma^i\gamma^k + \gamma^k\gamma^i = 2\eta^{ik}; \tag{43}$$

and it is usual to chose the following as their representation:

$$\gamma^0 = \begin{bmatrix} I & 0 \\ 0 & -I \end{bmatrix}, \qquad \gamma^\mu = \begin{bmatrix} 0 & \sigma^\mu \\ \sigma^\mu & 0 \end{bmatrix} \tag{44}$$

where I is the 2×2 unit matrix and $\sigma^\mu$are the 2×2 Pauli spin matrices:

$$\sigma^1 = \begin{bmatrix} 0 & 1 \\ 1 & 0 \end{bmatrix}, \qquad \sigma^2 = \begin{bmatrix} 0 & i \\ -i & 0 \end{bmatrix}, \qquad \sigma^3 = \begin{bmatrix} 1 & 0 \\ 0 & -1 \end{bmatrix}. \tag{45}$$

The Feynman convention of writing

$$\not{A} \equiv \gamma^i A_i \tag{46}$$

will be followed here. Thus

$$\not{A}\,\not{A} = A^i A_i. \tag{47}$$

We shall also write

$$\not{\nabla} \equiv \gamma^i \frac{\partial}{\partial x^i}\,. \tag{48}$$

Thus the equation satisfied by ψ as given by the Lagrangian density (41) is

$$(i\nabla\!\!\!/ - m)\psi = 0. \tag{49}$$

(Note that in the variational problem $\delta S = 0$, $\bar{\psi}$ and ψ are treated as independent.)

Equation (49) is called the Dirac equation. It can be written in the form

$$i \frac{\partial \psi}{\partial t} = -i\gamma^0\gamma^\mu \frac{\partial \psi}{\partial x^\mu} + m\psi,$$

that is,

$$E\psi = \gamma^0\boldsymbol{\gamma} \cdot \mathbf{p}\psi + m\psi; \tag{50}$$

where $\mathbf{p}$ is the 3-momentum operator. Dirac obtained it first in this form in his attempt to describe a relativistic free particle of mass m. Squaring the operator relation

$$E = \gamma^0\boldsymbol{\gamma} \cdot \mathbf{p} + m \tag{51}$$

and using relations (43) we get the free particle relation

$$E^2 = p^2 + m^2.$$

The Green's function for (49) satisfies the equation

$$(i\nabla\!\!\!/ - m)_x S(x, x') = \delta_4(x, x'). \tag{52}$$

To obtain its form it is convenient to relate it to the Green's function $G(x, x')$ for the massive scalar field. Indeed, it is easy to verify that the substitution

$$S(x, x') = -(i\nabla\!\!\!/ + m)_x G(x, x')$$

converts (52) to (24). Like the various forms of G we therefore have various forms of S.

For a charged particle interacting with an electromagnetic field, the form of $\mathscr{L}$ is changed to

$$\mathscr{L} = i\,\bar{\psi}\gamma^i\psi_{,i} - m\,\bar{\psi}\psi - e\bar{\psi}\gamma^i A_i\psi, \tag{53}$$

where e is the particle charge. The Dirac equation then becomes

$$(i\nabla\!\!\!/ - eA\!\!\!/ - m)\psi = 0. \tag{54}$$

The reader may wonder whether (54) is invariant under the gauge transformation (36). In fact it is, if we make the change

$$\psi \to \psi \exp(ie\chi). \tag{55}$$

Since ψ has the interpretation of wave function, no change of physics results from (55) wherein ψ changes only its phase.

4.3. Quantization of the Scalar Field

Having introduced the classical version of field theory we now take on the more difficult problem of formulating the quantum version of field theory. Although one argument for quantizing fields could be: 'any classical dynamical system has a quantum version and hence we should study the quantum theory of fields', somewhat stronger reasons can also be given.

Take, for example, a massive relativistic particle. If it is of spin zero we can describe it by the Klein–Gordon equation

$$\Box\varphi + m^2\varphi = 0. \tag{56}$$

We have seen that the propagator for such a particle is described by one of the Green's functions of Table 4.1. If we assume that the wave function φ is made of positive energy states, then we need the G_F propagator. This is given by (Example 4.3)

$$G_F(x, x') = \frac{1}{4\pi^2}\left(\frac{m}{s}\right) K_1(ims). \tag{57}$$

The argument of the functions above is s, where

$$s^2 = \eta_{ik}(x^i - x'^i)(x^k - x'^k). \tag{58}$$

For timelike displacements s^2 is positive; for spacelike displacements it is negative. In the latter case, for $m|s| \gg 1$, we have

$$G_F \cong (\text{const})\,(m|s|)^{-3/2}\exp(-m|s|) \tag{59}$$

In other words, even for spacelike separations the propagation amplitude is nonzero. This result means that the particle cannot be localized when it is free. We are thus led to formulate a 'many-particle' picture.

The reason for this behaviour lies in the relativistic equation for energy,

$$E_p = \sqrt{\mathbf{p}^2 + m^2}.$$

From general principles of quantum mechanics, a single particle wave function will be a superposition,

$$\psi(\mathbf{r}, t) = \sum_{\mathbf{p}} c(\mathbf{p})\, e^{i(\mathbf{p}\cdot\mathbf{r} - E_p t)}.$$

At a fixed $\mathbf{r}$, ψ may be treated as a function of time alone. Taking $E_p > 0$, $\psi(t)$ is analytic in the lower half of the complex plane. Clearly $\psi(t)$ cannot vanish outside the light cone without vanishing identically. Thus, one must either (a) abandon single-particle description or (b) allow negative values of E_p.

When Dirac formulated his theory of relativistic spin-$\frac{1}{2}$ massive particle he was forced to postulate a vacuum filled with negative energy states. Thus an attempt at a consistent single-particle picture led him to a many-particle formalism.

In a properly *quantized* field theory a formalism for dealing with many-particle states does follow naturally. We illustrate this result by developing a quantized theory of the scalar field.

Suppose we want to know the probability amplitude that the field φ given by Equations (19) and (20) has a value $\varphi_f(\mathbf{r})$ at time t_f if it had value $\varphi_i(\mathbf{r})$ at the initial time t_i. Following almost mechanically the analogy of free particle nonrelativistic quantum mechanics of Chapter 2, we write the answer as

$$K\,[\varphi_f(\mathbf{r}),\, t_f;\, \varphi_i(\mathbf{r}),\, t_i] = \int \exp\,\{iS[\varphi]\}\,\mathcal{D}\varphi. \tag{60}$$

In the above path integral the sum is over all 'paths' from φ_i to φ_f, each path a function $\varphi(\mathbf{r}, t)$ which satisfies the above endpoint conditions. Since $S[\varphi]$ is quadratic in φ and $\varphi_{,i}$, we can use the reasoning of Section 2.4 to write the answer as

$$K = f(t_f,\, t_i)\, \exp\,\{iS[\varphi_c]\}. \tag{61}$$

φ_c is the solution of the classical equation (20) with the prescribed endpoint conditions and f is a function of t_f, t_i only.

Assuming that the path integral (60) can be so defined rigorously we have formally solved the problem of quantizing the scalar field. The propagator K can in principle describe the evolution of a quantum state in the form of a wave functional $\Psi_i[\varphi_i(\mathbf{r})]$:

$$\Psi_f\,[\varphi_f(\mathbf{r})] = \int K\,[\varphi_f(\mathbf{r}),\, t_f;\, \varphi_i(\mathbf{r}),\, t_i)]\;\Psi_i\,[\varphi_i(\mathbf{r})]\mathcal{D}\,\varphi_i(\mathbf{r}). \tag{62}$$

This description, however, has only formal value since it does not provide any physical insight into many-particle systems nor does it prove useful in the study of interacting fields. For this purpose we need another description which naturally leads us to the many-particle concept.

We recall the formulae (3.3), (3.4), (3.5) of the previous chapter wherein we established a representation of φ in terms of a continuum set of harmonic oscillators. Let us carry on from there and write down the transition amplitude for the **k**th harmonic oscillator to go from $q(\mathbf{k},\, t_i) = a(\mathbf{k})$ to $q(\mathbf{k},\, t_f) = b(\mathbf{k})$. From the formulae derived in Chapter 2, this quantity is

$$K\,[b(\mathbf{k}),\, t_f;\, a(\mathbf{k}),\, t_i] = N(\mathbf{k})\,\exp Q\,\{a(\mathbf{k}),\, b(\mathbf{k}),\, \omega(\mathbf{k}),\, T\}\,, \tag{63}$$

where $T = t_f - t_i$ and

$$Q\left\{\, a(\mathbf{k}), b(\mathbf{k}), \omega(\mathbf{k}), T \,\right\} = \frac{i\omega(\mathbf{k})}{2\sin\,(T\omega(\mathbf{k}))}\;\times$$
$$\times\left\{\, (|b(\mathbf{k})|^2 + |a(\mathbf{k})|^2)\cos(T\omega(\mathbf{k})) - 2a(\mathbf{k})\, b(\mathbf{k}) \,\right\}. \tag{64}$$

Since the oscillators are all independent we may express the transition amplitude for the entire field system in the form

$$K(\varphi_f, t_f; \varphi_i, t_i) = N \exp\left[\int Q\{a(\mathbf{k}), b(\mathbf{k}), \omega(\mathbf{k}), T\} \frac{d^3\mathbf{k}}{(2\pi)^3}\right]. \tag{65}$$

Here the initial and final states of the field are given by

$$\varphi_i(\mathbf{r}) = \int a(\mathbf{k}) \exp(-i\mathbf{k} \cdot \mathbf{r}) \frac{d^3\mathbf{k}}{(2\pi)^3}, \tag{66}$$

$$\varphi_f(\mathbf{r}) = \int b(\mathbf{k}) \exp(-\mathbf{k} \cdot \mathbf{r}) \frac{d^3\mathbf{k}}{(2\pi)^3}. \tag{67}$$

N in (65) is a normalizing constant.

Again, as yet we have made no connection with the many-particle representation that we were after. However, we have made some progress in recognizing that the transitions of a free field can be related to transitions of individual harmonic oscillators. Let us now consider the typical oscillator in one of its stationary states. From Chapter 2 we know that the stationary states of the oscillator form a complete orthonormal basis and that the nth state in the series (where n is an integer) corresponds to the energy level $E_n = (2n + 1)\omega/2$. We use that result now and suppose that $\psi[n(\mathbf{k}), \omega(\mathbf{k}), \mathbf{r}, t]$ denotes this state for the **k**th harmonic oscillator. Then the wave functional for the initial state of the field is given by

$$\Psi[\{n(\mathbf{k})\}, \mathbf{r}, t_i] = \prod_k \psi[n(\mathbf{k}), \omega(\mathbf{k}), \mathbf{r}, t_i], \tag{68}$$

These wave functionals $\{\Psi\}$ form a complete orthonormal basis for describing any arbitrary state of the field φ. Note that each member of this set is labelled by a continuum set of integers $\{n(\mathbf{k})\}$. The energy of the field in the state (68) is given by

$$\begin{aligned}\varepsilon[\{n(\mathbf{k})\}] &= \sum_{\mathbf{k}} \omega(\mathbf{k}) \{n(\mathbf{k}) + \tfrac{1}{2}\} \\ &= \int \sqrt{m^2 + |\mathbf{k}|^2} \{n(\mathbf{k}) + \tfrac{1}{2}\} \frac{d^3\mathbf{k}}{(2\pi)^3},\end{aligned} \tag{69}$$

This representation is called 'Fock representation', and the Ψ basis the 'Fock basis'. It appears from (69) that the field made of $\{n(\mathbf{k})\}$ particles of momentum **k** and energy $\sqrt{m^2 + |\mathbf{k}|^2}$, together with a ground state energy

$$\varepsilon_0 = \sum_{\mathbf{k}} \tfrac{1}{2} \omega(\mathbf{k}). \tag{70}$$

This latter energy corresponds to the state of zero particles, i.e. the vacuum state.

We are now confronted with the mathematical difficulty that we have so far ignored, namely the problem of uncountably infinite degrees of freedom of φ. When summed over all **k**, the expression (70) for ε_0 diverges. Why should the energy of a system in its lowest level state – associated with no 'particles' in any state – diverge? Although recognized from early days as a difficulty in field theory, no satisfactory solution has yet been found. The usual argument to get round this difficulty is that the absolute value of energy does not 'count', it is only the difference in the energy levels of two states that has physical significance. This difference

$$\varepsilon\,[\{n(\mathbf{k})\}] - \varepsilon_0 = \sum_{\mathbf{k}} n(\mathbf{k})\omega \tag{71}$$

is finite and can be interpreted (as discussed above) as the energy of $\{n(\mathbf{k})\}$ particles.

Example 4.5. It should have been clear from the above discussion that the relationship between 'fields' and 'particles' is rather subtle. In particular, $\varphi(t, \mathbf{r})$ does not represent any simple-minded 'wave function' for a particle. Consider, for example, the vacuum state without any particles. When $n_{\mathbf{k}} = 0$, the wave functional for the state is given by

$$\Psi_{\text{vac}}\,[\{q_{\mathbf{k}}\}] = \prod_{\mathbf{k}} \left(\frac{\omega_{\mathbf{k}}}{\pi}\right)^{1/4} \exp(-\tfrac{1}{2}\,\omega_{\mathbf{k}}\,|q_{\mathbf{k}}|^2)\ .$$

$$= N \exp\left\{-\tfrac{1}{2} \int \frac{d^3\mathbf{k}}{(2\pi)^3}\,\omega_{\mathbf{k}}\,|q_{\mathbf{k}}|^2\right\}$$

Note that (at $t = 0$, say),

$$\varphi(\mathbf{r}) = \int \frac{d^3\mathbf{k}}{(2\pi)^3}\, q_{\mathbf{k}}\, e^{i\mathbf{k}\cdot\mathbf{r}}$$

so that

$$\nabla\varphi(\mathbf{r}) = \int \frac{d^3\mathbf{k}}{(2\pi)^3}\, i\mathbf{k} q_k\, e^{i\mathbf{k}\cdot\mathbf{r}}.$$

giving

$$i\mathbf{k} q_{\mathbf{k}} = \int d^3\mathbf{x}\, \nabla\varphi\, e^{-i\mathbf{k}\cdot\mathbf{r}}.$$

Now, for a massless field, $\omega_{\mathbf{k}} = |\mathbf{k}|$ and

$$\int \frac{d^3\mathbf{k}}{(2\pi)^3}\omega_{\mathbf{k}}\,|q_{\mathbf{k}}|^2 = \int \frac{d^3\mathbf{k}}{(2\pi)^3} \frac{|\mathbf{k}|^2 |q_{\mathbf{k}}|^2}{|\mathbf{k}|}$$

$$= -\int \frac{d^3\mathbf{k}}{(2\pi)^3} \int d^3\mathbf{x} \int d^3\mathbf{y}\, \frac{1}{|\mathbf{k}|}\, \nabla_x\varphi \cdot \nabla_y\varphi\, e^{-i\mathbf{k}(\mathbf{x}-\mathbf{y})}$$

$$= \frac{1}{2\pi^2} \int d^3\mathbf{x} \int d^3\mathbf{y} \left\{\frac{\nabla_x\varphi \cdot \nabla_y\varphi}{|\mathbf{x}-\mathbf{y}|^2}\right\}$$

Therefore, the probability amplitude to observe a field configuration $\varphi(\mathbf{r})$ in the vacuum state is

$$\mathcal{P}\,[\varphi(\mathbf{r})] = |\,\Psi_{\text{vac}}\,[\varphi(\mathbf{r})]\,|^2$$

$$= N \exp\left\{-\frac{1}{2\pi^2} \iint d^3\mathbf{x}\, d^3\mathbf{y}\, \frac{\nabla_x\varphi \cdot \nabla_y\varphi}{|\mathbf{x}-\mathbf{y}|^2}\right\}$$

This expression is indicative of the fact that nontrivial field configurations can exist even in the absence of 'particles'. □

However, this argument is not the correct way out, if we are to take Einstein's general relativity seriously. As we shall see in Chapter 5–7, the absolute value of energy does have physical significance in that it generates spacetime curvature. Infinite energy densities, therefore, cause severe problems in relativity.

4.4. Canonical Quantization

For the present we shall ignore general relativity and continue to work in flat spacetime. Although in Section 4.3 we derived the essential features of field quantization through the path integral method, we now present another approach which uses operator formulation and arrives at the above results more directly. Again we illustrate the approach first through the example of the scalar field.

We start with the classical notions developed in Section 4.1 and write the Lagrangian density as

$$\mathcal{L} = \tfrac{1}{2}[\dot{\varphi}^2 - (\nabla\varphi)^2 - m^2\varphi^2]\ ; \tag{72}$$

and define the timelike component of momentum by

$$\pi(t, \mathbf{r}) = \frac{\partial \mathcal{L}}{\partial \dot{\varphi}} = \dot{\varphi}(t, \mathbf{r}). \tag{73}$$

The Hamiltonian density (18) is

$$\mathcal{H} = \tfrac{1}{2}[\pi^2 + (\nabla\varphi)^2 + m^2\varphi^2]\ . \tag{74}$$

For quantization we 'convert' φ and π into operators satisfying the commutation relations

$$[\ \boldsymbol{\varphi}(t, \mathbf{r}), \boldsymbol{\varphi}(t, \mathbf{r}')\ ] = [\ \boldsymbol{\pi}(t, \mathbf{r}), \boldsymbol{\pi}(t, \mathbf{r}')\ ] = 0,$$

$$[\ \boldsymbol{\varphi}(t, \mathbf{r}), \boldsymbol{\pi}(t, \mathbf{r}')\ ] = i\ \delta_3(\mathbf{r} - \mathbf{r}'). \tag{75}$$

The space coordinates $\mathbf{r}$, $\mathbf{r}'$, etc., here serve as labels for the independent field components. The field components themselves are operators in the Hilbert space of quantum states. The time development of these operators is governed by the Hamiltonian

$$\mathbf{H} = \int \mathcal{H}\, \mathrm{d}^3\mathbf{r}. \tag{76}$$

From (74) – (76) we get the Heisenberg equations of motion:

$$i\dot{\varphi} = [\mathbf{H}, \pi], \qquad i\pi = -[\mathbf{H}, \varphi] \tag{77}$$

These equations are consistent with (and are in fact equivalent to) the operator field equation

$$(\Box + m^2)\varphi = \mathbf{0}. \tag{78}$$

We next look for a basis of states of φ in terms of which $\mathbf{H}$ can be diagonalized. As in previous work we start with Fourier states and write

$$\varphi(t, \mathbf{r}) = \int \frac{1}{[2\omega(\mathbf{k})]^{1/2}} \times$$

$$\times \{\mathbf{a}(\mathbf{k})\, \mathrm{e}^{[i\mathbf{k}\cdot\mathbf{r} - i\omega(\mathbf{k})t]} + \mathbf{a}^{\dagger}(\mathbf{k})\, \mathrm{e}^{[-i\mathbf{k}\cdot\mathbf{r} + i\omega(\mathbf{k})t]}\} \frac{\mathrm{d}^3\mathbf{k}}{(2\pi)^3}. \tag{79}$$

Here $\mathbf{a}$ and $\mathbf{a}^{\dagger}$ are Hermitian conjugate operators, and

$$\omega(\mathbf{k}) = \sqrt{m^2 + |\mathbf{k}|^2}\,. \tag{80}$$

The commutation rules (75) imply that

$$[\mathbf{a}(\mathbf{k}), \mathbf{a}(\mathbf{k}')] = [\mathbf{a}^{\dagger}(\mathbf{k}), \mathbf{a}^{\dagger}(\mathbf{k}')] = 0,$$

$$[\mathbf{a}(\mathbf{k}), \mathbf{a}^{\dagger}(\mathbf{k}')] = \delta_3(\mathbf{k} - \mathbf{k}') \tag{81}$$

Straightforward substitution of (79) into (76) gives

$$\mathbf{H} = \tfrac{1}{2} \int [\mathbf{a}(\mathbf{k})\, \mathbf{a}^{\dagger}(\mathbf{k}) + \mathbf{a}^{\dagger}(\mathbf{k})\, \mathbf{a}(\mathbf{k})]\, \omega(\mathbf{k}) \frac{\mathrm{d}^3\mathbf{k}}{(2\pi)^3}\,. \tag{82}$$

A comparison with the case of the simple harmonic oscillator (see Example 2.9) tells us that the operator $\mathbf{a}^{\dagger}(\mathbf{k})$ converts the nth eigenstate into the $(n + 1)$th one while $\mathbf{a}(\mathbf{k})$ does the reverse. Following the clue of the preceding section we define the 'vacuum' state as the lowest energy state of all harmonic oscillators $\mathbf{k}$. Denote it by the ket vector $|0\rangle$ and write

$$\mathbf{a}(\mathbf{k})\, |0\rangle = 0 \quad \text{for all } \mathbf{k}, \tag{83}$$

to indicate this property. A state with $n(\mathbf{k})$ particles can be generated from $|0\rangle$ by the repeated applications of $\mathbf{a}^{\dagger}(\mathbf{k})$.

$$|n(\mathbf{k})\rangle = \frac{1}{\sqrt{n(k)!}} [\mathbf{a}^{\dagger}(\mathbf{k})]^{n(\mathbf{k})}\, |0\rangle\,. \tag{84}$$

For this property $\mathbf{a}^{\dagger}$ is called the *creation operator* and likewise $\mathbf{a}$ is called the *annihilation operator*.

The base states of the field are composed of products of the above states for all **k**. We next show that in terms of these the operator **H** is indeed diagonalized. To this end we compute the matrix elements of **H**:

$$\mathbf{H}[n'(\mathbf{k}), n(\ell)] = \langle n'(\mathbf{k})|\ \mathbf{H}\ |n(\ell)\rangle \tag{85}$$

From (81), (83), and (84) it is easy to verify that

$$\mathbf{a}^\dagger(\mathbf{k}')\,|\mathbf{n}(\mathbf{k})\rangle = (2\pi)^3\delta_3(\mathbf{k}) - \mathbf{k}')\ [n(\mathbf{k}) + 1]^{1/2}|n(\mathbf{k}) + 1\rangle, \tag{86}$$

$$\mathbf{a}(\mathbf{k}')|\mathbf{n}(\mathbf{k})\rangle = (2\pi)^3\delta_3(\mathbf{k} - \mathbf{k}')\ [n(\mathbf{k})]^{1/2}|n(\mathbf{k}) - 1\rangle\ . \tag{87}$$

Use these relations and their Hermitian conjugates in (82) to get

$$\begin{aligned} H[n'(\mathbf{k}), n(\ell)] &= \tfrac{1}{2}\int [\langle n'(\mathbf{k})|\ \mathbf{a}(\mathbf{k}')\mathbf{a}^\dagger(\mathbf{k}')\ |n(\ell)\rangle\ + \\ &\quad + \langle n'(\mathbf{k})|\ \mathbf{a}^\dagger(\mathbf{k}')a(\mathbf{k}')\ |n(\ell)\rangle]\omega(\mathbf{k}')\ \frac{\mathrm{d}^3\mathbf{k}'}{(2\pi)^3} \\ &= \tfrac{1}{2}\int [\delta(\ell - \mathbf{k}')\ \delta(\mathbf{k} - \mathbf{k}')\ \{n'(\mathbf{k}) + 1\}^{1/2} \times \\ &\quad \times \{n(\ell) + 1\}^{1/2} \times \langle n'(\mathbf{k}) + 1|\ n(\ell) + 1\ \rangle + \\ &\quad + \delta\ (\ell - \mathbf{k}')\ \delta(\mathbf{k} - \mathbf{k}')\ \{n'(\mathbf{k})n(\ell)\}^{1/2} \times \\ &\quad \times \langle n'(\mathbf{k}) - 1\ |\ n(\ell) - 1\rangle]\ (2\pi)^3 \cdot \omega(\mathbf{k}')\ \mathrm{d}^3\mathbf{k} \\ &= (2\pi)^3\ \delta(\ell - \mathbf{k}) \cdot [n(\mathbf{k}) + \tfrac{1}{2}]\ \omega(\mathbf{k})\delta_{nn'}\ . \end{aligned} \tag{88}$$

In other words, the matrix element vanishes unless the sandwiching states are the same. Thus **H** is diagonalized. Its trace is given by the sum over the diagonal elements

$$\begin{aligned} \mathrm{Tr}\ \mathbf{H} &= \int \mathbf{H}(\mathbf{k}, \mathbf{k})\ \frac{\mathrm{d}^3\mathbf{k}}{(2\pi)^3} \\ &= \int \{n(\mathbf{k}) + \tfrac{1}{2}\}\ \omega(\mathbf{k})\ \frac{\mathrm{d}^3\mathbf{k}}{(2\pi)^3}\ . \end{aligned} \tag{89}$$

This is precisely the same expression as (69), and denotes the energy of the field. As discussed earlier, this is infinite.

To extract a finite answer from this formally infinite expression, we introduce the concept of normal ordering. An operator **T** after normal ordering is written as :**T**: and contains all creation operators ($\mathbf{a}^\dagger$) to the left of all annihilation operators. Thus from (83) it follows that

$$\langle 0\ |\ :\mathbf{T}:\ |0\rangle = 0. \tag{90}$$

We also have,

$$:\mathbf{H}: = \int \mathbf{a}^{\dagger}(\mathbf{k})\, \mathbf{a}^{\dagger}(\mathbf{k})\, \omega(\mathbf{k})\, \frac{d^3\mathbf{k}}{(2\pi)^3}\,. \tag{91}$$

and it is easy to verify that

$$\mathrm{Tr}\ :\mathbf{H}: = \int n(k)\, \omega(\mathbf{k})\, \frac{d^3\mathbf{k}}{(2\pi)^3} \tag{92}$$

does not have the infinite vacuum contribution. This device of normal ordering has generally proved useful in regularizing infinite expressions appearing in quantum field theory.

Example 4.6. While discussing harmonic oscillator in Chapter 2 we showed that there exist 'coherent states' which are the closest approximation in quantum mechanics to the classical behaviour. Such coherent states can also be constructed for our scalar field. Define

$$|f(\mathbf{k})\rangle' = \exp \int \frac{d^3\mathbf{k}}{(2\pi)^3}\, f(\mathbf{k})\, \mathbf{a}^{+}(\mathbf{k})\, |0\rangle.$$

Such states are the eigenstates of the 'positive frequency' part of the field, i.e.

$$\varphi^{+}(x)\, |\, f(\mathbf{k})\rangle' \propto \int \frac{d^3\mathbf{k}}{(2\pi)^3}\, f(\mathbf{k})\, e^{-i(\omega_k t\, -\, \mathbf{k}\, \cdot\, \mathbf{r})}\, |\, f(\mathbf{k})\rangle',$$

where

$$\varphi^{+}(x) \equiv \int \frac{d^3\mathbf{k}}{(2\pi)^3}\, \mathbf{a}(\mathbf{k})\, e^{-i(\omega_k t\, -\, \mathbf{k}\, \cdot\, \mathbf{r})}.$$

Using the relation, $e^A e^B = e^{A + B + [A, B]/2}$ (valid whenever, $[A, B]$ commutes with A and B), we can see that

$$\langle f_1 \,|\, f_2 \rangle' = \exp \int \frac{d^3\mathbf{k}}{(2\pi)^3}\, f_1^{*}(\mathbf{k}) f_2(\mathbf{k}).$$

Therefore, correctly normalized coherent states are

$$|f(\mathbf{k})\rangle = \exp\left[-\tfrac{1}{2} \int |f(\mathbf{k})|^2\, \frac{d^3\mathbf{k}}{(2\pi)^3} \right] |f(\mathbf{k})\rangle'.$$

A classical field configuration can be specified by giving Fourier coefficients of $\varphi(\mathbf{x})$ [at $t = 0$, say]. The quantum state closest to such a classical description is the coherent state $|f(\mathbf{k})\rangle$, where $f(\mathbf{k})$ is the Fourier transform of $\varphi(\mathbf{x})$. □

4.4.1. *The Electromagnetic Field*

We discussed the quantization of the scalar field in some detail partly because it is the simplest example of field quantization and partly because we shall have future occasions in this book to refer to these ideas. Since we are not concerned here with details of quantum field theory in general (for which excellent texts exist; see references cited at the end of this chapter) we shall only highlight a few important ideas in the remainder of this chapter. We begin with the electromagnetic field.

First we recall from Section 4.2 the fact that the classical electromagnetic theory is gauge invariant. Thus, while the potential vector A_i has apparently four degrees of freedom, in reality the field F_{ik} has only two degrees of freedom. This 'discrepancy' shows up in many ways. For example, for $\mathscr{L}$ given by (30) the conjugate momenta

$$\pi^i = \frac{\partial \mathscr{L}}{\partial \dot{A}_i} \tag{93}$$

are such that $\pi^0 \equiv 0$. How can we impose commutation relations in this case?

To get out of this difficulty we have to pay the price of manifest coordinate and gauge invariance and restrict quantization to the specific choice of gauge

$$\nabla \cdot \mathbf{A} = 0, \qquad \varphi = 0. \tag{94}$$

Then the Lagrangian density becomes

$$\mathscr{L} = \tfrac{1}{2}[\dot{A}^2 - (\nabla \times \mathbf{A})^2]\,. \tag{95}$$

As in the case of the scalar field we expand $\mathbf{A}$ in a Fourier integral:

$$\mathbf{A}(t, \mathbf{r}) = \int \frac{1}{\sqrt{2\omega(\mathbf{k})}} \{\mathbf{a}(\mathbf{k})\, \mathrm{e}^{i[\mathbf{k}\cdot\mathbf{r} - \omega(\mathbf{k})t]} + \mathbf{a}^\dagger(\mathbf{k})\, \mathrm{e}^{-i[\mathbf{k}\cdot\mathbf{r} - \omega(\mathbf{k})t]}\} \frac{\mathrm{d}^3\mathbf{k}}{(2\pi)^3}\,. \tag{96}$$

The gauge condition (94), however, implies that $\mathbf{k} \cdot \mathbf{a}(\mathbf{k}) = 0$. Hence we express $\mathbf{a}$ in terms of two unit vectors $\mathbf{e}_\lambda(\mathbf{k})$, $\lambda = 1, 2$ perpendicular to $\mathbf{k}$:

$$\mathbf{a}(\mathbf{k}) = \sum_{\lambda=1,2} \mathbf{e}_\lambda(\mathbf{k}) a_\lambda(\mathbf{k}). \tag{97}$$

The commutation relations are

$$[\mathbf{A}(t, \mathbf{r}), \mathbf{A}(t, \mathbf{r}')] = 0, \qquad [\pi(t, \mathbf{r}), \pi(t, \mathbf{r}')] = 0,$$

$$[A_\mu(t, \mathbf{r}), \pi_\nu(t, \mathbf{r}')] = i\eta_{\mu\nu}\, \delta_3(\mathbf{r}, \mathbf{r}'). \tag{98}$$

Dropping the boldface notation for operators these become

$$[a_\lambda(\mathbf{k}), a_{\lambda'}(\mathbf{k}')] = [a_\lambda^\dagger(\mathbf{k}), a_{\lambda'}^\dagger(\mathbf{k}')] = 0, \tag{99}$$

$$[a_\lambda(\mathbf{k}), a_{\lambda'}^\dagger(\mathbf{k}')] = \delta_{\lambda\lambda'}\delta_3(\mathbf{k} - \mathbf{k}').$$

The two components a_λ of course denote the two polarization states of the plane electromagnetic wave.

The identification of the field as a continuum set of independent harmonic oscillators carries over from the scalar field case in an analogous manner. The Hamiltonian operator is given by

$$H = \tfrac{1}{2}[\boldsymbol{\pi}^2 + (\nabla \times \mathbf{A})^2] \,. \tag{100}$$

In terms of the operators $a_\lambda(\mathbf{k})$ and $a_\lambda^\dagger(\mathbf{k})$ we get

$$H = \int \frac{d^3\mathbf{k}}{(2\pi)^3} \left\{ \sum_{\lambda=1,2} (a_\lambda^\dagger(\mathbf{k})a_\lambda(\mathbf{k}) + \tfrac{1}{2})\omega(\mathbf{k}) \right\}. \tag{101}$$

The same zero-point energy problem exists and it is 'resolved' by introducing normal ordering.

Example 4.7. The following is one of the most striking demonstrations of the nontrivial nature of the vacuum: two parallel, uncharged plane conductors kept in the vacuum at a given separation will attract each other. We shall give below a brief derivation of this effect, known commonly as the *Casimir effect.*

Consider two large square conductors of size L kept perpendicular to the z axis at a separation a (with $L \gg a$). Consider the electromagnetic field modes labelled by $\mathbf{k} = (k_x, k_y, k_z)$. Since the field must vanish on the surfaces of the conductor, we get

$$k_z = \frac{n\pi}{a}\,, \quad n = 1, 2, \ldots$$

These modes have two degrees of freedom. When $k_z = 0$ the modes have only one degree of freedom. Therefore, the 'vacuum energy' *in the presence of the conductors* is

$$E \equiv \sum_{\mathbf{k},\lambda} \tfrac{1}{2}\omega_{\mathbf{k}} = \tfrac{1}{2}\int \frac{L^2}{(2\pi)^2}\left[(k_x^2 + k_y^2)^{1/2} + 2\sum_{n=1}^{\infty}\left(k_x^2 + k_y^2 + \frac{n^2\pi^2}{a^2}\right)^{1/2}\right] dk_x\, dk_y.$$

This expression, of course, is divergent. However, we must subtract from it the 'vacuum energy' in the absence of the plates, which is

$$E_0 \equiv \tfrac{1}{2}\int \frac{L^2\, dk_x\, dk_y}{(2\pi)^2}\int_{-\infty}^{+\infty} \frac{a}{\pi}\sqrt{k_x^2 + k_y^2 + k_z^2}\, dk_z$$

$$= \tfrac{1}{2}\int \frac{L^2\, dk_x\, dk_y}{(2\pi)^2}\int_0^{\infty} 2\sqrt{k_x^2 + k_y^2 + \frac{n^2\pi^2}{a^2}}\, dn\,.$$

The energy difference per unit area of the surface is

$$\Delta E \equiv \frac{E - E_0}{L^2} = \frac{1}{2\pi}\int_0^{\infty} k\left\{\frac{k}{2} + \sum_{n=1}^{\infty}\sqrt{k^2 + \frac{n^2\pi^2}{a^2}} - \int_0^{\infty} dn\sqrt{k^2 + \frac{n^2\pi^2}{a^2}}\right\} dk\,.$$

In order to make precise the subtraction of one infinity from the other, introduce a cut-off function $f(k)$ with the property, $f(k) = 0$ for $k > k_m$ and $f(k) \approx 1$ for $k < k_m$. Then, taking $x = a^2k^2/\pi$ we get

$$\Delta E = \frac{\pi^2}{4a^3}\int_0^{\infty}\left[\frac{\sqrt{x}}{2} f\left(\frac{\pi}{a}\sqrt{x}\right) + \sum_1^{\infty}\sqrt{x + n^2}\, f\left(\frac{\pi}{a}\sqrt{x + n^2}\right) - \right.$$

$$\left. - \int_0^{\infty} dn\sqrt{x + n^2}\, f\left(\frac{\pi}{a}\sqrt{x + n^2}\right)\right] dx = \frac{\pi}{4a^3}\left[\tfrac{1}{2}I(0) + I(1) + \cdots - \int_0^{\infty} dn\, I(n)\right]$$

with

$$I(n) \equiv \int_0^\infty \sqrt{x + n^2}\, f\left(\frac{\pi}{a}\sqrt{x + n^2}\right) dx.$$

Using the Euler–MacLaurin formula

$$\tfrac{1}{2}I(0) + I(1) + \cdots - \int_0^\infty dn\, I(n) = -\tfrac{1}{2}B_2 I'(0) - \frac{1}{4!}\, I'''(0) + \cdots,$$

where the B_n are defined through the expansion

$$y(e^y - 1)^{-1} = \sum_{n=0}^{\infty} B_n \frac{y^n}{n!},$$

we get (assuming $f(0) = 1$ and that all derivatives of f vanishes at the origin)

$$\Delta E = \frac{\pi^2}{a^3} \frac{B_4}{4!} = -\frac{\pi^2}{720} \frac{\hbar c}{a^3}.$$

In the last step we have used the fact that $B_4 = -\frac{1}{30}$ and reintroduced $\hbar$ and c. This energy density leads to a tiny, but detectable, force of attraction between the plates:

$$F = -\frac{\pi^2}{240} \frac{\hbar c}{a^4} \cong -\frac{0.01}{a^4_{(\mu m)}} \text{ dyne cm}^{-2}.$$ □

4.4.2. *The Dirac Field*

The Dirac field presents problems of a new kind when attempts are made to quantize it along the above lines. For $\mathscr{L}$ given by (41) we have

$$\pi \equiv \frac{\partial \mathscr{L}}{\partial \dot{\psi}} = i\bar{\psi}\gamma^0, \qquad \text{i.e.} \quad \bar{\psi} = -i\pi\gamma^0. \tag{102}$$

Therefore,

$$H = \pi\gamma^0\gamma^\mu\psi_{,\mu} - i\pi\gamma^0\psi. \tag{103}$$

Writing the Fourier integral for ψ as before (here $a(k)$ and $a^\dagger(k)$ are spinor operators)

$$\psi = \int \frac{d^3\mathbf{k}}{(2\pi)^3} \frac{1}{\sqrt{2\omega_k}} [a(k)\, e^{i(\mathbf{k}\cdot\mathbf{r} - \omega_k t)} + a^\dagger(k)\, e^{-i(\mathbf{k}\cdot\mathbf{r} - \omega_k t)}] \tag{104}$$

we can set up commutation relations for a, $a^\dagger$ based on the canonical commutation relations:

$$[\psi_\alpha(t, \mathbf{x}), \pi_\beta(t, \mathbf{y})] = i\delta_{\alpha\beta}\, \delta(\mathbf{x} - \mathbf{y})$$

$$[\psi_\alpha(t, \mathbf{x}), \psi_\beta(t, \mathbf{y})] = [\pi_\alpha(t, \mathbf{x}), \pi_\beta(t, \mathbf{y})] = 0. \tag{105}$$

However, we then find that the total energy of the system as computed from (103) is not positive definite. To make it so, we have to change the quantization rules from (105) to those involving the *anticommutation* relations

$$\{\psi_\alpha(t, \mathbf{x}), \pi_\beta(t, \mathbf{y})\} = i\delta_{\alpha\beta}\,\delta(\mathbf{x} - \mathbf{y})$$

$$\{\psi_\alpha(t, \mathbf{x}), \psi_\beta(t, \mathbf{y})\} = \{\pi_\alpha(t, \mathbf{x}), \pi_\beta(t, \mathbf{y})\} = 0, \tag{106a}$$

which lead to (suppressing the spinor indices)

$$\{a_k, a_p^\dagger\} = \delta_{kp};\ \{a_k, a_p\} = \{a_k^\dagger, a_p^\dagger\} = 0. \tag{106b}$$

These relations imply that the number operator $N = a^\dagger a$ can have only two eigenvalues, 0 and 1. For,

$$N^2 = a^\dagger a a^\dagger a = a^\dagger(1 - a^\dagger a)a = a^\dagger a = N. \tag{107}$$

So that

$$(N^2 - N)|n\rangle = (n^2 - n)|n\rangle = 0; \quad n = 0, 1.$$

This mathematical requirement is the realization of Pauli's exclusion principle that there can be at most one particle (of Dirac type) in a given quantum state. The energy operator H then has positive definite eigenvalues which can be made finite by normal ordering.

We now tackle more problems of the quantum theory of interacting fields which is where real physics begins!

Example 4.8. The switch over from commutation to anticommutation rules makes $\psi(t, \mathbf{r})$ a rather peculiar operator (e.g. as an operator, $\psi^2(t, \mathbf{r}) \equiv 0$). This requires a special formalism when integrations over $\psi(t, \mathbf{r})$ are performed, say, in a path integral.

Such a formalism involves introducing anticommuting c-numbers. Let us call them θ_i; $i = 1, 2, \ldots, N$. This set $\{\theta_i\}$ satisfies the conditions

$$\{\theta_i, \theta_i\} = 0.$$

The 'derivative' is defined through the rule

$$\left\{\frac{\partial}{\partial\theta_i}, \theta_j\right\} = \delta_{ij}; \qquad \left\{\frac{\partial}{\partial\theta_i}, \frac{\partial}{\partial\theta_j}\right\} = 0.$$

Any function $f(\theta_i)$ will become a polynomial of finite degree because of anticommutation rules. For example, any function of a single variable θ, $f(\theta)$ becomes

$$f(\theta) = a + b\theta,$$

where $b^2 = 0$ and a is an ordinary number. The integrations over θ's are defined through the relations,

$$\int d\theta_i = 0 \qquad \int d\theta_i\,\theta_i = 1 \quad \text{(no summation on } i\text{)}$$

when multiple integrations are required one proceeds in a step by step fashion: for example,

$$\iint d\theta_1\, d\theta_2\, \theta_1\theta_2 = -\iint d\theta_1(d\theta_2\, \theta_2)\theta_1 = -1.$$

In evaluating path integrals over fermionic variables one often requires evaluation of the following expression:

$$I_N(M) \equiv \int \ldots \int \mathrm{d}\theta_1 \ldots \mathrm{d}\theta_N \exp(-\theta^T M\theta),$$

where M is an $N \times N$ matrix. When $N = 2$, for example,

$$I_2(M) = \iint \mathrm{d}\theta_1\, \mathrm{d}\theta_2 [1 - 2M_{12}\, \theta_1\theta_2]$$
$$= 2(\det M)^{1/2}.$$

This result generalizes to any even N in a pairwise manner. We get

$$I_N(M) = 2^{N/2}(\det M)^{1/2}.$$

(In contrast, for bosonic variables, the expression goes as $\sim(\det M)^{-1/2}$.) The above discussion can easily be extended to cover the case of complex M. Treating θ and θ^* as independent variables we get

$$I_N(M) = \iint \mathrm{d}\theta\, \mathrm{d}\theta^* \exp(-\theta^* M\theta) = 2^N(\det M).$$ □

4.5. Scalar Field with Quartic Self-interaction

As we shall see later (in the context of gauge theories and the grand unification programme), an important theory of interacting scalar fields is the so-called $\lambda\varphi^4$ theory described by the Lagrangian density

$$\mathscr{L} = \tfrac{1}{2}[\varphi_i\varphi^i - m^2\varphi^2 - \tfrac{1}{2}\lambda\varphi^4]\,. \tag{108}$$

We may look upon this theory as that of a massive field interacting with itself in a nonlinear fashion. This theory forms the analogue of the quartically coupled harmonic oscillator discussed in Chapter 3 (*vide* Equation (3.41)). We shall begin by studying this interaction by the use of perturbation theory. To this end the machinery developed in Chapter 3 will be useful.

Accordingly, we first consider the vacuum to vacuum transition amplitude for the free field case ($\lambda = 0$) in the presence of an external source $J(x)$. Using the notation of Section 3.3 we write

$$\langle 0, t_f | 0, t_i\rangle_J$$
$$\equiv \int_i^f \exp\left[i \int_{t_i}^{t_f} \{\tfrac{1}{2}(\varphi_k\varphi^k - m^2\varphi^2) + J\varphi\}\,\mathrm{d}^4x\right] \overline{\mathscr{D}}\varphi(x). \tag{109}$$

This is the field analogue of Equation (3.15) with $\overline{\mathscr{D}}$ defined in the same way. We shall call this expression $Z_0(J)$.

Writing $V(\varphi) = \frac{1}{2}\lambda\varphi^4$, we use the analogue of relation (3.26) to write

$$Z_V(J) = \exp\left\{-i\int \mathrm{d}^4x\, V\left[\frac{\delta}{i\cdot\delta J(x)}\right]\right\} Z_0(J). \tag{110}$$

By expanding the exponential in power series we get the required perturbation series. But first we evaluate $Z_0(J)$.

4.5.1. *The Green's Function*

We shall use the technique of Euclidean time described in Chapter 3. Write the Euclidean path integral

$$Z_0(J) = \int \exp\{-S[\varphi]\}\, \mathcal{D}\varphi \equiv \exp\{-W_0(J)\}\,, \tag{111}$$

where

$$S[\varphi] = \tfrac{1}{2}(\varphi, A\varphi) - (J, \varphi), \tag{112}$$

$$A \equiv -\Box_E + m^2, \qquad (\psi, \chi) = \int \psi\chi\, d^4x. \tag{113}$$

Since the integrand is the exponential of a quadratic, we can use the technique of Section 3.4 to write

$$Z_0(J) = (\det A)^{-1/2} \exp\left(\frac{A^{-1}J, J}{2}\right), \tag{114}$$

where A^{-1} is easily seen to be the Green's function of the operator A.

The inverse of a differential operator is usually defined in the Fourier space. In our case,

$$\begin{aligned}(-\Box_E + m^2)^{-1} f(x) &= (-\Box_E + m^2)^{-1} \int e^{-ikx} \tilde{f}(k)\, \frac{d^4k}{(2\pi)^4} \\ &\equiv \int \frac{e^{-ik\cdot x}}{k_E^2 + m^2}\, \tilde{f}(k)\, \frac{d^4k}{(2\pi)^4} \\ &\equiv \iint \frac{e^{-ik\cdot x}}{k_E^2 + m^2}\, \frac{d^4k}{(2\pi)^4}\, f(y)\, e^{iky}\, dy \\ &\equiv i \int \Delta_F(x - y) f(y)\, d^4y\end{aligned}$$

where

$$\Delta_F(x) \equiv -i \int \frac{e^{-ik\cdot x}}{k_E^2 + m^2} \cdot \frac{d^4k}{(2\pi)^4}\,.$$

Also note that

$$A_x \Delta_F(x, x') \equiv (-\Box_x + m^2)\Delta_F(x, x') = i\delta_4(x, x'). \tag{115}$$

The reader may recognize Δ as the Euclidean version of G defined by Equation (24). The factor $-i$ has come on the right-hand side because of Euclideanization of time from t to τ. Like the many types of G in Table 4.1 we have many types of Δ. We shall shortly see that analytic continuation from Euclidean extension uniquely specifies the choice of G, in real time as G_F.

Thus (114) becomes

$$Z_0(J) = (\det A)^{-1/2} \exp\left[\tfrac{1}{2} i \int J(x)\, \mathrm{d}^4x \int \Delta_F(x, x')\, J(x')\, \mathrm{d}^4x'\right]$$

$$= (\det A)^{-1/2} \exp\left[\frac{i}{2} \iint \Delta_F(x, x')\, J(x)\, J(x')\, \mathrm{d}^4x\, \mathrm{d}^4x'\right]. \qquad (116)$$

So far we have not said anything about the factor $(\det A)^{-1/2}$ in (116). We choose it from the criterion that the amplitude for vacuum to remain should be unity in the absence of any external disturbance. Thus $Z_0(J = 0) = 1$. From (116) we therefore omit the factor $(\det A)^{-1/2}$ and write

$$Z_0(J) = \exp\left[\frac{i}{2} \iint \Delta_F(x, x') J(x) J(x')\, \mathrm{d}^4x\, \mathrm{d}^4x'\right]. \qquad (117)$$

Now we analytically continue (117) back to real time and write

$$Z_0(J) = \exp\left[+\frac{i}{2} \iint G_F(x, x') J(x) J(x')\, \mathrm{d}^4x\, \mathrm{d}^4x'\right]. \qquad (118)$$

Notice that in the process of analytic continuation we have made the decision to use the Feynman type Green's function G_F. This was the Green's function used by Feynman and it arises naturally in field theory from the requirement that positive energy states are propagated forward in time while the negative energy states are propagated backwards in time (and are interpreted as antiparticles)

Example 4.9. The Green's function G_F plays a central role in the theory of quantized scalar field. We shall now show that this Green's function can be obtained from a straightforward path integral approach. Consider a relativistic particle of mass m described by a classical action (see Example 2.2)

$$\mathcal{A} = -\frac{m}{2}\int_{\lambda_1}^{\lambda_2} \left(\frac{\mathrm{d}x_i\, \mathrm{d}x^i}{\mathrm{d}\lambda\, \mathrm{d}\lambda} + 1\right) \mathrm{d}\lambda.$$

The probability amplitude for the particle to go from x_2^i to x_2^i is given by the sum over amplitudes,

$$K(x_2, x_1) \equiv \int_0^\infty \mathrm{d}\lambda \int \exp i\mathcal{A}[x_2, \lambda; x_1, 0]\, \mathcal{D}x(\lambda).$$

(Note that we also have to integrate over all proper times λ). Performing the quadratic path integral,

$$K(x_2, x_1) = \int_0^\infty \left(\frac{m}{2\pi i\lambda}\right)^2 \exp -\frac{im}{2}\left\{\frac{(x_2 - x_1)^2}{\lambda} + \lambda\right\} \mathrm{d}\lambda$$

$$= \left(\frac{m}{2\pi i}\right)^2 \frac{1}{s}\int_0^\infty \alpha^{-2} \exp\left\{-\frac{ims}{2}\left(\alpha + \frac{1}{\alpha}\right)\right\} \mathrm{d}\alpha = -2mG(x_2, x_1),$$

where $s^2 = (x_2 - x_1)^2$. Notice that, except for the multiplicative constant, K is same as our Green's function G. This leads to the following physical interpretation for $G(x_2, x_1)$:

One may assume that a relativistic particle goes from x_1 to x_2 along all possible worldlines $x^i(\lambda)$ taking all possible proper time values λ. Each path is given an amplitude $\exp(i\mathcal{A})$. The crucial difference between this propagation and that of a nonrelativistic particle is that the paths $x^i(\lambda)$ may go backward in the coordinate time t. In other words, one can have propagation from $x_1^i = (t_1, \mathbf{x}_1)$ to $x_2^i = (t_2, \mathbf{x}_2)$ with $t_2 < t_1$. In general, one must include in the sum over paths in K paths like the one shown in Figure 4.2. For coordinate times $t_A < t < t_B$, notice that the 'particle' is at three different points in space. We are once given led to the conclusion that the single-particle description is not viable in relativistic quantum theory. □

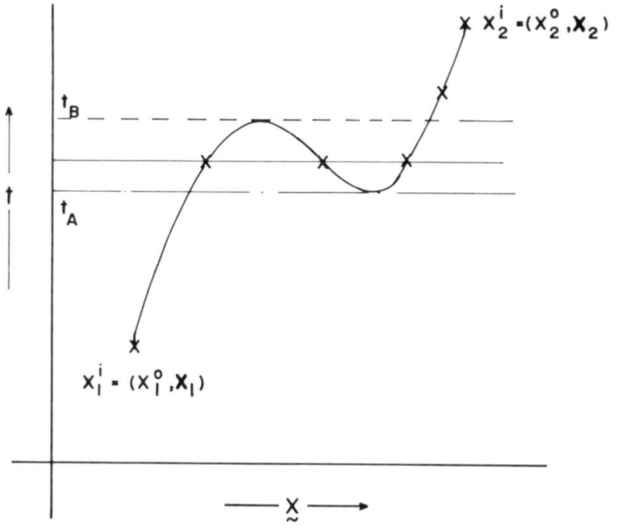

Fig. 4.2. One of the paths that contribute in the path integral for $K(x_2, x_1)$. Though the path corresponds to going forward in the proper time λ, it actually goes backward in time as far as $t = (x^0)$ is concerned. For $t_A < t < t_B$, the 'particle' following this trajectory will be at three different points in space.

Recalling relation (3.21) from Chapter 3 we write

$$-iG_F(x, x') = \frac{1}{i^2}\left[\frac{\delta^2 Z_0(J)}{\delta J(x)\,\delta J(x')}\right]_{J=0} = \langle 0|\, T\,\{\varphi(x)\varphi(x')\}\,|0\rangle\,. \quad (119)$$

4.5.2. *The Perturbation Expansion*

Next we go to (110) and rewrite it as

$$Z_V(J) \equiv \exp -iW_V(J) = \exp\left\{-i\int \frac{\lambda}{4}\left[\frac{\delta}{i\delta J(x)}\right]^4 \mathrm{d}^4x\right\} Z_0[\mathrm{J}]. \quad (120)$$

In real spacetime (111) is written as

$$Z_0(J) = \exp\{-iW_0(J)\}\ , \quad (121)$$

while we similarly define $W_V(J)$ by

$$Z_V(J) = \exp\left\{-iW_V(J)\right\}. \tag{122}$$

Substitute these expressions into (120) and take logarithms, remembering that we are dealing with operators in the exponential. After some algebra we get

$$W_V(J) = W_0(J) + i \ln (1 + \Delta), \tag{123}$$

where

$$\Delta(J) = \exp\{iW_0(J)\}\left[\exp \frac{-i\lambda}{4}\int\left(i\frac{\delta}{\delta J}\right)^4 \mathrm{d}^4x - 1\right]\times$$
$$\times \exp\left\{-iW_0(J)\right\}. \tag{124}$$

The reason for writing this relation in the above form is to argue that for small λ, the second term in (123) is small and hence Δ is small. This conclusion enables us to make the perturbation expansions

$$W_V(J) = W_0(J) + i\Delta(J) - \frac{i[\Delta(J)]^2}{2} + \cdots \tag{125}$$

and

$$\Delta(J) = \lambda\Delta_1(J) + \lambda^2\Delta_2(J) + \cdots \tag{126}$$

where Δ_1, Δ_2, etc., are known terms.

We may now define the Green's function $G_F^V(x, x')$ for the scalar field interacting via the $\lambda\varphi^4$ potential by

$$G_F^V(x, x') = i\left[\frac{\delta^2 Z_v(J)}{\delta J(x\ \delta J)x')}\right]_{J=0}, \tag{127}$$

and make the above substitutions. To order λ^2 the perturbation expansion becomes

$$G_F^V(x, x') = G_F(x, x') - \frac{\lambda}{2}\int G_F(x, x_1)G_F(x_1, x_1)G_F(x_1, x')\,\mathrm{d}^4x_1 +$$
$$+\frac{\lambda^2}{6}\iint G_F(x, x_1)G_F^3(x_1, x_2)G_F(x_2, x')\,\mathrm{d}^4x_1\,\mathrm{d}^4x_2 +$$
$$+\frac{\lambda^2}{4}\iint G_F(x, x_1)G_F(x_1, x_2)G_F(x_2, x_2)G_F(x_2, x') + \mathrm{d}^4x_1\,\mathrm{d}^4x_2.$$
$$+\frac{\lambda^2}{4}\iint G_F(x, x_1)G_F(x_1, x_1)G_F(x_1, x_2)G_F(x_2, x_2)$$
$$\times G_F(x_2, x')\,\mathrm{d}^4x_1\,\mathrm{d}^4x_2. \tag{128}$$

The presence of the $G_F(x, x)$ –type terms in the integrals is sufficient to show that we are dealing with a perturbation expansion in which individual terms diverge badly. In other words, we appear to have been carried away by formal aspects of the theory without having any indication that the formalism so far developed has any practical value.

The divergence of individual terms in the perturbation expansion was the most serious difficulty in the early years of field theory. This difficulty is probably another signal of the fact that our formalism for relativistic quantum theory is certainly incomplete, if not incorrect. No general solution to this difficulty is known. In other words, given a Lagrangian containing arbitrary interaction terms (even very realistic interactions like gravitational interactions), nobody knows how to produce a quantum theory which is perturbatively meaningful.

A much more limited goal, however, has been achieved. There exists a particular class of interactions for which these infinities can be absorbed into the definition of the parameters that appear in the original Lagrangian. The present theory happens to belong to this particular class, normally called 'renormalizable' theories, another important example being quantum electrodynamics.

A detailed discussion of 'renormalization' is beyond the scope of our book. (Interested readers may consult the literature cited at the end of this chapter.) Some basic aspects of the renormalization of the $\lambda\varphi^4$ theory are presented in Appendix A, for the sake of completeness.

4.6. Nonperturbative Methods

We next consider the application of nonperturbative methods developed in the previous chapter to the study of interacting quantum fields. For illustrative purposes we will use the $\lambda\varphi^4$ theory given by (108).

4.6.1 *The Effective Potential*

By a straightforward extension of the ideas developed in Section 3.6 we write (in Euclidean time)

$$\exp\{-W(J)\} \equiv \int \exp\left\{-\int (\mathscr{L}[\varphi] + J\varphi)\, \mathrm{d}^4x\right\} \mathscr{D}\varphi, \tag{129}$$

$$\varphi_c(x) \equiv \frac{\delta W}{\delta J}, \tag{130}$$

$$S_{\text{eff}}[\varphi_c] \equiv W(J) - \int J\varphi_c\, \mathrm{d}^4x. \tag{131}$$

It is assumed in (130) that J can be expressed in terms of φ_c so that writing (131) in terms of φ_c is possible. Similarly, by an analysis similar to the one leading to (3.64) we arrive at the result

$$S_{\text{eff}}[\varphi_c] = S[\varphi_c] + \tfrac{1}{2}\ln\det\{-(\Box_{\mathrm{E}} + V''(\varphi_c))\}. \tag{132}$$

Here $\Box_E$ denotes the Euclidean wave operator and

$$V(\varphi) = \tfrac{1}{2}m^2\varphi^2 + \tfrac{1}{4}\lambda\varphi^4. \tag{133}$$

To evaluate the determinant of

$$A = -\Box_E + V''(\varphi_c) \tag{134}$$

we use a technique called the zeta-function regularization.

Suppose an operator A has a spectrum $\{\lambda_n\}$ of eigenvalues. We define a 'zeta function' corresponding to that operator A to be

$$\zeta_A(s) = \sum_n \frac{1}{\lambda_n^s},$$

where the sum is over all eigenvalues $\{\lambda_n\}$ defined by the equation $Af_n(x) = \lambda_n f_n(x)$. From the definition of ζ_A we get

$$-\frac{d\zeta_A(s)}{ds} = \sum_n \frac{1}{\lambda_n^s} \ln \lambda_n$$

so that

$$-\left.\frac{d\zeta_A(s)}{ds}\right|_{s=0} = \sum_n \ln \lambda_n = \ln \det A.$$

Thus,

$$\det A = \exp\left\{-\left.\frac{d\zeta_A}{ds}\right|_{s=0}\right\}. \tag{135}$$

Now define the 'heat kernel K' for this operator as

$$K(x_2, x_1, \tau) = \sum_n e^{-\lambda_n\tau} f_n(x_2) f_n^*(x_1).$$

Then,

$$\int_0^\infty \tau^{s-1}\, d\tau \int_{-\infty}^{+\infty} K(x, x, \tau)\, dx$$

$$= \sum_n \int_0^\infty \tau^{s-1} e^{-\lambda_n\tau}\, d\tau \int_{-\infty}^{+\infty} f_n(x) f_n^*(x)\, dx$$

$$= \sum_n \frac{1}{\lambda_n^s} \Gamma(s) = \Gamma(s)\zeta_A(s).$$

Thus,

$$\zeta_A(s) = \Gamma(s)^{-1} \int_0^\infty \tau^{s-1}\, d\tau \int_0^{+\infty} K(x, x, \tau)\, dx. \tag{136}$$

The function K satisfies the 'conduction' equation

$$\frac{\partial K}{\partial \tau} = -\sum_n \lambda_n \quad e^{-\lambda_n \tau} f_n(x_2) f_n^*(x_1)$$
$$= -\sum_n e^{-\lambda_n \tau} A f_n(x_2) f_n^*(x_1)$$
$$= -AK; \tag{137}$$

and the boundary condition,

$$\lim_{\tau \to 0} K(x_2, x_1, \tau) = \delta(x_2 - x_1). \tag{138}$$

Thus, if we can solve Equation (137) for an operator K (satisfying the boundary condition (138)), then we can evaluate det A using (136) and (135).

Corresponding to Equation (137) we set up the conduction equation

$$\frac{\partial K}{\partial \tau} = (\Box_E - m^2 - 3\lambda \varphi_c^2)K.$$

As a first approximation we take $\varphi_c = \text{const}$. Then this equation has the integral

$$K(x, x', \tau) = \frac{\mu^4}{16\pi^2 \tau^2} \exp \frac{\mu^2 (x - x^{2\prime})^2}{4\tau} \exp\left\{ -(m^2 + 3\lambda \varphi_c^2) \frac{\tau}{\mu^2} \right\}, \tag{139}$$

where μ is a constant with dimensions of mass.

Following (136) we define

$$\zeta_A(s) = \frac{1}{\Gamma(s)} \int_0^\infty \tau^{s-1}\, d\tau \quad K(x, x, \tau)\, d^4x$$
$$= \frac{1}{s(s-1)} \int \frac{\mu^4}{16\pi^2} \left[\frac{m^2 + 3\lambda \varphi_c^2}{\mu^2} \right]^{2-s} d^4x.$$

From (135) we have

$$\det A = \exp\left\{ -\left. \frac{d\zeta_A}{ds} \right|_{s=0} \right\}.$$

Using these definitions and (132) we get

$$S_{\text{eff}}(\varphi_c) = S(\varphi_c) + \int \frac{1}{64\pi^2} (m^2 + 2\lambda \varphi_c^2)^2 \left(-\frac{3}{2} + \ln \frac{m^2 + 3\lambda \varphi^2}{\mu^2} \right) d^4x.$$

Thus the effective potential becomes[1]

$$V_{\text{eff}}(\varphi_c) = \tfrac{1}{2} m^2 \varphi_c^2 + \frac{\lambda}{4} \varphi_c^4 + \frac{\hbar}{64\pi^2} (m^2 + 3\lambda \varphi_c^2)^2 \times$$
$$\times \left(-\frac{3}{2} + \ln \frac{m^2 + 3\lambda \varphi_c^2}{\mu^2} \right). \tag{140}$$

[1] Notice that the $\left(-\frac{3}{2}\right)$ in the last term can be absorbed into μ.

For the sake of explaining the extent of quantum corrections we have restored $\hbar$ in (140). The corrections are therefore accurate to order $\hbar$. Why has the constant μ appeared in the final answer?

To answer this question consider the simpler case $m = 0$. Then we have

$$V_{\text{eff}}(\varphi_c) = \frac{\lambda}{4}\varphi_c^4 + \frac{36\hbar\lambda^2}{256\pi^2}\varphi_c^4\left(-\frac{3}{2} + \ln\frac{3\lambda\varphi_c^2}{\mu^2}\right) + O(\hbar^2). \tag{141}$$

Now the constant m^2 could be defined at the classical level by

$$m^2 = V''(\varphi = 0).$$

We see that to order $\hbar$, m^2 continues to be zero even after quantum corrections. However, the same is not true of λ. Classically, λ may be defined by

$$\lambda = \frac{1}{6}\frac{d^4V}{d\varphi^4}\bigg|_{\varphi=\text{const}} . \tag{142}$$

Suppose we choose the constant in (142) to be $\varphi = M$. Then the requirement

$$\lambda = \frac{1}{6}\frac{d^4V_{\text{eff}}}{d\varphi^4}\bigg|_{\varphi=M} \Rightarrow \frac{\lambda M^2}{2\mu^2} = -\frac{8}{3} . \tag{143}$$

Thus, in terms of M and λ the effective potential may be written as

$$V_{\text{eff}}(\varphi) = \tfrac{1}{4}\lambda\varphi^4 + \frac{36\lambda^2\varphi^4}{256\pi^2}\left[\ln\frac{\varphi^2}{M^2} - \frac{25}{6}\right] . \tag{144}$$

If, instead of the arbitrary M, we had chosen M', say, then the form of V_{eff} can be retained to be the same to order λ^2 if we transform λ to

$$\lambda' = \lambda + \frac{9\lambda^2}{16\pi^2}\ln\frac{M'}{M} . \tag{145}$$

The moral of all this is that, to order $\hbar$, the effective potential has the invariant form $V_{\text{eff}}(\varphi, \lambda, M)$ given by (144) providing M and λ transform as (145). The physics remain unchanged as long as we keep M nonzero.

Example 4.10. The above discussion emphasizes the need for operational definition of the coupling constants, masses, etc., that appear in the Lagrangian. We have seen that the theory does not possess a unique coupling constant λ but is actually described by a function $\lambda(\mu)$. This function specifies the value of the coupling λ to be used at the scale μ, which enters into the effective potential. It must be noted that any physical quantity A in the theory should be independent of the scale μ in the following sense:

$$A(\lambda(\mu), \mu) = A(\lambda(\mu'), \mu').$$

It turns out that the above requirement is quite nontrivial. The so called 'renormalization group approach' to field theory depends crucially on the above fact. □

4.6.2. *Effective Potential and Counter Terms*

We have seen that quantum corrections to the classical potential introduce an arbitrary scale into the theory. Because of the importance of this concept, we shall examine it from a different angle.

The scale μ appeared in Equation (139) in our attempt to define the determinant of an operator. Suppose we had chosen an alternative definition, viz.

$$\det A = \exp[\operatorname{Tr} \ln A]. \tag{146}$$

Then we would have arrived at the effective potential (using (132) and defining the logarithm in the Fourier space),

$$V(\varphi) = V_0(\varphi) + \tfrac{1}{2} \int \ln(k_E^2 + V_0''(\varphi)) \frac{d^4 k_E}{(2\pi)^4} . \tag{147}$$

This integral is divergent. In order to extract a meaningful answer, let us introduce a cutoff in the k integration at $k_E^2 = V^2$, (say). Then,

$$\int \ln(k_E^2 + V'') \frac{d^4 k_E}{(2\pi)^4} = \pi^2 \left[\Lambda^4 \left(\ln \Lambda - \frac{1}{4} \right) + \Lambda^2 V_0'' + \right.$$

$$\left. + \tfrac{1}{2} (V_0'')^2 \left(\ln \frac{V_0''}{\Lambda^2} - \frac{1}{2} \right) \right] + O\left(\frac{1}{\Lambda^2} \right) . \tag{148}$$

Dropping constant terms, the effective potential can be written as

$$V(\varphi) = V_0(\varphi) + V_1(\varphi),$$

$$V_1(\varphi) = \frac{\Lambda^2}{32\pi^2} V_0''(\varphi) + \frac{[V_0''(\varphi)]^2}{64\pi^2} \left[- \frac{1}{2} + \ln \frac{V_0''(\varphi)}{\Lambda^2} \right] . \tag{149}$$

At this stage, let us introduce a finite, but arbitrary, scale μ by the relation

$$\ln \frac{V_0''}{\Lambda^2} = \ln \frac{V_0''}{\mu^2} + \ln \left(\frac{\mu^2}{\Lambda^2} \right) , \tag{150}$$

so that

$$V_1(\varphi) = \frac{(V_0'')^2}{64\pi^2} \left(\ln \frac{V_0''}{\mu^2} - \frac{3}{2} \right) + \frac{1}{32\pi^2} \left[\Lambda^2 V_0'' - \tfrac{1}{2} ((V_0'')^2 \ln \frac{\Lambda^2}{\mu^2} \right] . \tag{151}$$

When the limit of $\Lambda \to \infty$ is taken, the first term remains finite while the second term diverges. However, for certain class of potentials $V_0(\varphi)$, this divergence can be completely absorbed in the constants that appear in $V_0(\varphi)$. For example, if $V_0(\varphi)$ has the form

$$V_0(\varphi) = a\varphi^2 + b\varphi^4, \tag{152}$$

then,

$$\Lambda^2 V_0'' - \left(\frac{1}{2}\ln\frac{\Lambda^2}{\mu^2}\right)(V_0'')^2 = \Lambda^2(2a + 12b\varphi^2) - \left(\frac{1}{2}\ln\frac{\Lambda^2}{\mu^2}\right)(2a + 12b\varphi^2)^2$$
$$= a'\varphi^2 + b'\varphi^4 + c; \qquad (153)$$

so that (up to a constant)

$$V_{\text{eff}}(\varphi) = (a + a')\varphi^2 + (b + b')\varphi^4 + \text{finite terms in } V_1(\varphi). \qquad (154)$$

In other words, a redefinition of the constants in the original Lagrangian takes care of the divergences!

We shall illustrate this procedure in the case of a potential given by

$$V_0(\varphi) = \frac{\lambda_0}{4}\varphi^4 - \frac{m_0^2}{2}\varphi^2. \qquad (155)$$

This potential differs from that in (108) as regards the sign of the φ^2 term, and will play a crucial role in our discussion of gauge fields in the next chapter. Since we are planning to cancel the divergent parts by redefinition of λ_0 and m_0^2, it is useful to write, purely arbitrarily,

$$\lambda_0 = \lambda + \delta\lambda; \qquad m_0^2 = m^2 + \delta m^2. \qquad (156)$$

We hope to choose $\delta\lambda$ and δm^2 in such a way as to cancel the divergent parts of the effective potential. The quantities λ and m have to be identified with the physical coupling constant and mass.

The divergent part of $V_1(\varphi)$ has the form,

$$V_1^{\text{divergent}} = \frac{1}{32\pi^2}\left\{\Lambda^2(3\lambda\varphi^2 - m^2) - \tfrac{1}{2}\left(\ln\frac{\Lambda^2}{\mu^2}\right)(3\lambda\varphi^2 - m^2)^2\right\}$$
$$= \frac{1}{32\pi^2}\left\{\left(3\lambda\Lambda^2 + 3\lambda m^2\ln\frac{\Lambda^2}{\mu^2}\right)\varphi^2 - \left(\frac{9}{2}\lambda^2\ln\frac{\Lambda^2}{\mu^2}\right)\varphi^2\right\} +$$
$$+ (\text{const}). \qquad (157)$$

Dropping constants independent of φ, the effective potential becomes

$$V_{\text{eff}}(\varphi) = \left(\frac{\lambda}{4}\varphi^4 - \frac{m^2}{2}\varphi^2\right) + \frac{1}{64\pi^2}(3\lambda\varphi^2 - m^2)^2\left\{\ln\frac{(3\lambda\varphi^2 - m^2)}{\mu^2} - \frac{1}{2}\right\} +$$
$$+ \varphi^2\left\{\frac{3\lambda}{32\pi^2}\left(\mathrm{V}^2 + 3m^2\ln\frac{\Lambda^2}{\mu^2}\right) - \frac{\delta m^2}{2}\right\} +$$
$$+ \varphi^4\left\{\frac{9}{64\pi^2}\lambda^2\ln\frac{\Lambda^2}{\mu^2} + \frac{\delta\lambda}{4}\right\}. \qquad (158)$$

We shall choose our counter terms δm^2, $\delta\lambda$ to be

$$\delta\lambda = -\frac{9}{16\pi^2}\lambda^2\ln\frac{\Lambda^2}{\mu^2}; \qquad \delta m^2 = \frac{3\lambda}{16\pi^2}\left(\Lambda^2 + 3m^2\ln\frac{\Lambda^2}{\mu^2}\right), \qquad (159)$$

thereby making $V_{\text{eff}}(\varphi)$ completely finite:

$$V_{\text{eff}}(\varphi) = \left(\frac{\lambda}{4}\varphi^4 - \frac{m^2}{2}\varphi^2 \right) + \frac{1}{64\pi^2}(3\lambda\varphi^2 - m^2)^2 \times$$
$$\times \left\{ \ln \frac{(3\lambda\varphi^2 - m^2)}{\mu^2} - \frac{1}{2} \right\}. \tag{160}$$

Notice that this expression is the same as (140) except for the change of sign of m^2.[2] The origin of μ in (160), is clearly from the need to separate out divergent parts of $V_{\text{eff}}(\varphi)$. However, as discussed before, μ is arbitrary; a change of μ to μ' can always be accommodated by a corresponding change of λ to λ', as in Equation (145).

If $V_0(\varphi)$ is a polynomial of degree n, V_0'' will have a degree $(n - 2)$. The divergent part of V_{eff} which involves (V_0''), will have a degree of $(2n - 4)$. Since $V_0(\varphi)$ is of degree n, we can at best introduce n counterterms into V_0. Thus our scheme for removing the divergences will work only if

$$2n - 4 \leq n; \quad \text{that is, } n \leq 4.$$

This explains why $\lambda\varphi^4$ coupling enjoys a special status of being the 'maximal' interaction compatible with renormalizability.

4.7. Quantum Theory in External Fields

In the later chapters of the book we shall consider the quantization of fields in a given background gravitational field. We shall then see that the externally specified classical field produces particle–antiparticle pairs. In this section we shall consider some simple models that illustrate this concept.

Consider, for example, the scalar field interacting with an external source $J(x)$. The system is described by the Lagrangian

$$\mathscr{L} = \tfrac{1}{2}(\varphi^i\varphi_i - m^2\varphi^2) + J(x)\varphi(x). \tag{161}$$

In Section 4.5 we worked out the 'vacuum persistence amplitude' in the presence of the source J to be (see Equation (118)),

$$\langle 0|0\rangle_J = Z_0(J) = \exp\left[\frac{i}{2}\iint G_F(x, x')J(x)J(x')\,\mathrm{d}^4x\,\mathrm{d}^4x'\right]$$
$$= \exp\left[-\frac{i}{2}\int \frac{\tilde{J}(p)\tilde{J}(-p)}{p^2 - m^2 + i\varepsilon}\frac{\mathrm{d}^4p}{(2\pi)^4}\right]$$
$$= \exp\left[-\frac{i}{2}\int \frac{|\tilde{J}(p)|^2}{p^2 - m^2 + i\varepsilon}\frac{\mathrm{d}^4p}{(2\pi)^4}\right], \tag{162}$$

[2] Note that the $(-\frac{1}{2})$ inside the curly brackets, in the last term of (160), can be absorbed into the definition of μ.

where we have used the fact that, for real $J(x)$,

$$\tilde{J}(p) = \int J(x)\, e^{ipx}\, d^4x = \tilde{J}^*(-p). \tag{163}$$

The *probability* for vacuum to remain a vacuum is given by

$$|\langle 0|0\rangle_J|^2 = \exp\left\{ \text{Im} \int \frac{|\tilde{J}(p)|^2}{p^2 - m^2 + i\varepsilon} \frac{d^4p}{(2\pi)^4} \right\}. \tag{164}$$

We now use the relation (where $\mathcal{P}$ denotes the principal value)

$$\frac{1}{x + i\varepsilon} = \mathcal{P}\left(\frac{1}{x}\right) - i\pi\, \delta(x)$$

to obtain

$$\begin{aligned} \text{Im} \int \frac{|\tilde{J}(p)|^2}{p^2 - m^2 + i\varepsilon} \frac{d^4p}{(2\pi)^4} &= -\pi \int |\tilde{J}(p)|^2\, \delta(p^2 - m^2) \frac{d^4p}{(2\pi)^4} \\ &= -\pi \int \frac{1}{2\omega_{\mathbf{p}}} \cdot \frac{1}{2\pi} \cdot 2|\tilde{J}(\mathbf{p}, \omega_{\mathbf{p}})|^2 \frac{d^3\mathbf{p}}{(2\pi)^3}. \end{aligned} \tag{165}$$

Thus we get,

$$|\langle 0|0\rangle_J|^2 = \exp\left\{ -\int \frac{1}{2\omega_{\mathbf{p}}} |\tilde{J}(\mathbf{p}, \omega_{\mathbf{p}})|^2 \frac{d^3\mathbf{p}}{(2\pi)^3} \right\}, \tag{166}$$

where $\omega_{\mathbf{p}}^2 = \mathbf{p}^2 + m^2$. This expression shows that in the presence of the source, the probability for the vacuum to remain a vacuum is less than 1. Suppose the probability for creating a pair of particles in the momentum range $(\mathbf{p}, \mathbf{p} + d\mathbf{p})$ is $\mathcal{A}(\mathbf{p})\, d^3\mathbf{p}$. Then, assuming that the creation events at various values of $\mathbf{p}$ are independent, the probability that *no* pairs were created at all is given by

$$\text{Probability for no creation} = \exp(-\int \mathcal{A}(\mathbf{p})\, d^3\mathbf{p}). \tag{167}$$

However, this is precisely the vacuum persistence probability $|\langle 0|0\rangle|^2$. Therefore, we get the expression for the probability for creation of a particle with momentum $\mathbf{p}$ to be,

$$\mathcal{A}(\mathbf{p}) = \frac{1}{2\omega_{\mathbf{p}}} |\tilde{J}(\mathbf{p}), \omega_{\mathbf{p}})|^2. \tag{168}$$

One may also interpret $\mathcal{A}(\mathbf{p})$ as the average number of particle pairs created at the mode $\mathbf{p}$. Since the creation of particles is a statistically random event, the probability for creating $n_{\mathbf{k}}$ particles with momentum $\mathbf{k}$ is given by the Poisson distribution,

$$P(n_{\mathbf{k}}) = \frac{\mathcal{A}_{\mathbf{k}}^{n_{\mathbf{k}}}}{n_{\mathbf{k}!}} \exp(-\mathcal{A}_{\mathbf{k}}). \tag{169}$$

Example 4.11. Because of the simple nature of the interaction, the above results can also be obtained by straightforward Fourier decomposition. Note that the action corresponding to (161) can be written, in Fourier space, as,

$$S = \int \mathscr{L}\, \mathrm{d}t\, \mathrm{d}^3\mathbf{r} = \int \frac{\mathrm{d}^3\mathbf{k}}{(2\pi)^3} \int \mathrm{d}t \left\{ \tfrac{1}{2}|\dot{q}_{\mathbf{k}}^2| - \tfrac{1}{2}\omega_{\mathbf{k}}^2 |q_{\mathbf{k}}|^2 + J_{\mathbf{k}}^* q_{\mathbf{k}} \right\}.$$

This is the action corresponding to a set of independent harmonic oscillators with forcing functions $J_{\mathbf{k}}^*(t)$. The probability amplitude for the **k**th oscillator to have made a transition from $|0\rangle$ at $t = -\infty$ to $|n_{\mathbf{k}}>$ at $t = +\infty$ is given (Chapter 2, Example 2.11) by

$$\mathcal{P}_{\mathbf{K}}(n_{\mathbf{k}}) = \frac{\mathcal{A}_{\mathbf{k}}(0)}{\sqrt{n_k!}} (i\beta_{\mathbf{k}})^{n_{\mathbf{k}}}$$

with,

$$\beta_{\mathbf{k}} = \frac{1}{\sqrt{2\omega_{\mathbf{k}}}} \int_{-\infty}^{+\infty} \mathrm{d}t\, J_{\mathbf{k}}(t)\, \mathrm{e}^{i\omega_k t} \equiv \frac{1}{\sqrt{2\omega_{\mathbf{k}}}} \bar{J}_{\mathbf{k}}(\omega_{\mathbf{k}}).$$

So we have

$$P_{\mathbf{k}}(n_{\mathbf{k}}) = |\mathcal{P}|^2 = \frac{|\mathcal{A}_{\mathbf{k}}(0)|^2}{n_{\mathbf{k}}!} \left\{ \frac{1}{2\omega_{\mathbf{k}}} |J_{\mathbf{k}}(\omega_{\mathbf{k}})|^2 \right\}^{n_{\mathbf{k}}}.$$

This is exactly the same as (169) because normalization of $P_{\mathbf{k}}$ guarantees $\Sigma(n_{\mathbf{k}})\ P_{\mathbf{k}}(n_{\mathbf{k}}) = 1$ so that $|A_{\mathbf{k}}(0)|^2$ is equal to $\exp(-1/2\omega_{\mathbf{k}})|J|^2)$. □

Next, let us consider a more complicated case of two interacting fields, one of which is treated classically and the other quantum mechanically. A charged spin-$\frac{1}{2}$ particle coupled to electromagnetic field is described by the action (see (53))

$$S = \int \mathrm{d}^4x\ \bar{\psi}(i\gamma^k\partial_k - qA\!\!\!/ - m)\psi - \tfrac{1}{4}\int \mathrm{d}^4x\ F_{ik}F^{ik}. \tag{170}$$

The full quantum theory of this system involves the functional integral,

$$Z \equiv \iiint \exp\Big[iS(\psi, \bar{\psi}, A_i)\Big]\ \mathcal{D}\psi\ \mathcal{D}\bar{\psi}\ \mathcal{D}A_i. \tag{171}$$

However, suppose that we are prepared to treat the electromagnetic field semiclassically. Then we would like to evaluate the effective action for the electromagnetic field, taking into account the quantum fluctuations of $(\psi, \bar{\psi})$. As usual, we define the effective action through the equation,

$$\exp\left\{iS_{\mathrm{eff}}(A_i)\right\} = \frac{\iint \exp[iS(\psi, \bar{\psi}, A_i)]\ \mathcal{D}\psi\ \mathcal{D}\bar{\psi}}{\iint \exp[iS(\psi, \bar{\psi}, 0)]\ \mathcal{D}\psi\ \mathcal{D}\bar{\psi}}, \tag{172}$$

where we have carefully normalized the expression to get $S_{\mathrm{eff}} = 0$ for $A_i = 0$.

In this case $(\exp iS_{\mathrm{eff}})$ can be given an alternative intepretation. Suppose

$A_i(x)$ is specified externally. Then the probability amplitude for the vacuum to remain a vacuum without the creation of electron–positron pairs is given by the expression in the right-hand side of (172). Therefore,

$$\langle 0|0\rangle_{A^i} = \exp[iS_{\text{eff}}(A_i)] \ . \tag{173}$$

Clearly, an imaginary part in the effective action will signal pair creation. Suprisingly, such an imaginary part can exist even for a time-independent, constant, electric field. Thus a constant electric field is capable of creating electron–positron pairs from the vacuum.

Using the result of Example 4.8, we can write

$$\exp(iS_{\text{eff}}) = \frac{\det[i\gamma^k\partial_k - qA - m]}{\det[i\gamma^k\partial_k - m]}\,, \tag{174}$$

so that

$$iS_{\text{eff}} = \text{Tr}\ln\left\{(\not{P} - q\not{A} - m + i\varepsilon)\,\frac{1}{\not{P} - m + i\varepsilon}\right\}\ . \tag{175}$$

Here we have written $\not{P}$ for $i\gamma^k\partial_k$ and have introduced the $i\varepsilon$ prescription. Using $\text{Tr}(AB) = \text{Tr}(BA)$ we can write (175) as

$$iS_{\text{eff}} = \tfrac{1}{2}\text{Tr}\ln\left\{[(\not{P} - q\not{A})^2 - m^2 + i\varepsilon]\,\frac{1}{P^2 - m^2 + i\varepsilon}\right\}\ . \tag{176}$$

This expression can be evaluated by a series of tricks described in Example 4.12 below. For the case of a pure electric field E the final answer is

$$S_{\text{eff}} = \int \mathscr{L}_{\text{eff}}\,\mathrm{d}^4x \tag{177}$$

with

$$\mathscr{L}_{\text{eff}} = \frac{1}{8\pi^2}\int_0^\infty \mathrm{e}^{-i(m^2 - i\varepsilon)s}\left[qE\coth(qEs) - \frac{1}{s}\right]\frac{\mathrm{d}s}{s^2}\ . \tag{178}$$

From equation (173), we find that the probability for the creation of electron–positron pairs per unit space–time volume is given by

$$W \equiv -\frac{1}{4\pi^2}\int_0^\infty \left[qE\coth(qEs) - \frac{1}{s}\right]\cdot \text{Re}(i\,\mathrm{e}^{-is(m^2 - i\varepsilon)})\,\frac{\mathrm{d}s}{s^2}\ . \tag{179}$$

The integral can be performed by taking a contour encircling negative imaginary axis. We thus get

$$W = \frac{q^2E^2}{\pi^2}\sum_{n=1}^{\infty}\frac{1}{n^2}\exp\left(-\frac{n\pi m^2}{|qE|}\right). \tag{180}$$

Note that the expression is a nonanalytic function of the coupling constant q. No amount of perturbation expansion in q would have led to the above result.

The leading term in W (i.e., the term with $n = 1$),

$$W \propto \exp\left(- \frac{\pi m^2}{|qE|}\right), \tag{181}$$

can be interpreted in terms of a Euclidean path integral for a charged particle moving in a constant electric field. The Minkowski action for a charged, relativistic particle in an electric field E,

$$S = - \int \left[m\sqrt{1 - \dot{x}^2} - qEx \right] \mathrm{d}t \tag{182}$$

can be analytically continued to the Euclidean domain,

$$S_{\mathrm{E}} = \int \left[m(1 + \dot{x}^2)^{1/2} - qEx \right] \mathrm{d}\tau. \tag{183}$$

The trajectory of a charged particle, in the Euclidean coordinates is a circle,

$$(x - x_0)^2 + (\tau - \tau_0)^2 = \ell^2; \ell = \frac{m}{|qE|} . \tag{184}$$

Substituting (184) into (183) we get

$$S_{\mathrm{E}} = \frac{\pi m^2}{|qE|} . \tag{185}$$

Therefore, the leading exponential in W is given by

$$W \propto \exp(-S_{\mathrm{E}}). \tag{186}$$

The nonperturbative nature of the result is reproduced in the Euclidean path integral expression.

Example 4.12. Below we indicate the derivation of Equation (178). A more detailed treatment can be found in the references cited at the end of the chapter.

Introduce a representation in which the operators X_i and P_k satisfy the relations

$$X_i |x\rangle = x_i |x\rangle, \qquad [X_i, P_j] = i\delta_{ij}.$$

Then, with the help of the identity

$$\ln\left(\frac{a}{b}\right) = \int_0^\infty \frac{1}{s} \left(\mathrm{e}^{is(b+i\varepsilon)} - \mathrm{e}^{is(a+i\varepsilon)}\right) \mathrm{d}s,$$

Equation (176) becomes the relation

$$iS_{\text{eff}} = \tfrac{1}{2} \int \mathrm{d}^4x \,\mathrm{tr} \int_0^\infty \frac{1}{s}\, \mathrm{e}^{-is(m^2 - i\varepsilon)} \left[\langle x| \,\mathrm{e}^{is(\not{P} - q\not{A})^2} |x\rangle - \langle x| \,\mathrm{e}^{isP^2} |x\rangle \right] \mathrm{d}s.$$

Here 'tr' refers to the trace over γ-matrices. Noting that

$$\langle x| \exp is(\not{P} - q\not{A})^2 |x\rangle = \left\langle x \right| \exp is\left[(P - qA)^2 + \frac{q}{2}\,\sigma_{ik}F^{ik}\right] \left| x \right\rangle,$$

where $\sigma_{ik} = \frac{1}{2} i[\gamma_i, \gamma_k]$, and that, for a pure electric field,

$$\mathrm{tr} \exp(isq\sigma_{ik}F^{ik}) = 4\cosh(qsE)$$

we can write

$$\text{tr}\,\langle x|\,\exp is[(P-qA)^2+\tfrac{1}{2}q\sigma_{ik}F^{ik}]\,|x\rangle$$
$$=\frac{2qE}{(4\pi^2)is}\cosh(qEs)\int_{-\infty}^{\infty}\langle\alpha|\,\exp i(P_0^2-q^2E^2x_0^2)\,|\alpha\rangle\,d\alpha.$$

The remaining integral is just the trace of the kernel for a harmonic oscillator with imaginary frequency, and is given by

$$\int_{-\infty}^{+\infty}d\alpha\,\langle\alpha|\,\exp i(P_0^2-q^2E^2x_0^2\,|\alpha\rangle=\frac{1}{2\sinh(qEs)}\,.$$

Putting these together, we get the result (178) of the text. □

4.8. Field Theory at Finite Temperature

In the last chapter we discussed a quantum mechanical system at finite temperature. The generalization to infinite degrees of freedom is straightforward in principle. Let us consider a field theoretic system with energy eigenstates $|E_\alpha\rangle$. We introduce finite temperature ($\beta^{-1}=kT$) by postulating that in thermodynamic equilibrium a state $|E_\alpha\rangle$ has a weightage of $\exp(-\beta E_\alpha)$. Thus the thermodynamic average of an operator is defined to be

$$\bar{A}\equiv\frac{\sum_\alpha\exp(-\beta E_\alpha)\langle\alpha\,|A|\,\alpha\rangle}{\sum_\alpha\exp(-\beta E_\alpha)}\,. \tag{187}$$

Such averages are usually written in terms of a density matrix $\varrho\equiv\exp(-\beta H)$, as

$$\bar{A}=\frac{\text{Tr}(\varrho A)}{\text{Tr}\,\varrho}\,. \tag{188}$$

There is a simple recipe for calculating the finite temperature averages from the zero temperature values (this is an extension of the method used in Section 3.8). Consider, for example, the finite temperature Green's function,

$$\overline{T(\psi(x)\psi(y))}=\frac{\text{Tr}[\varrho T(\varphi(x)\varphi(y))]}{\text{Tr}\,\varrho}\equiv G_\beta(x,y). \tag{189}$$

We have, in Euclidean time, the following relation:

$$\begin{aligned}G_\beta(\mathbf{x},x^0=0;y)&=(\text{Tr}\,\text{e}^{-\beta H})^{-1}\text{Tr}[\text{e}^{-\beta H}\,\varphi(y)\varphi(0,\mathbf{x})]\\&=(\text{Tr}\,\text{e}^{-\beta H})^{-1}\,\text{Tr}[\text{e}^{-\beta H}\,\text{e}^{\beta H}\,\varphi(0,\mathbf{x})\,\text{e}^{-\beta H}\,\varphi(y)]\\&=G_\beta(\mathbf{x},x^0=\beta;y).\end{aligned} \tag{190}$$

In arriving at (190) we have used the relations

$$\mathrm{Tr}(AB) = \mathrm{Tr}(BA),$$

and

$$\mathrm{e}^{iHt}\,\varphi(0, \mathbf{x})\,\mathrm{e}^{-iHt}\,\varphi = \varphi(t, \mathbf{x}).$$

In other words, the finite temperature Green's function is obtained from the periodic identification of the standard Euclidean Green's function with a period β in the Euclidean time coordinate. In momentum space, this result can be expressed as follows:

$$\overline{\hat{T}(\varphi(x)\varphi(y))} = \frac{1}{\beta}\sum_{n=-\infty}^{n=+\infty} \exp -2\pi i\frac{n}{\beta}(x^0 - y^0)\int \mathrm{e}^{i\mathbf{k}\cdot(\mathbf{x}-\mathbf{y})} \times$$

$$\times \frac{i}{k^2 - m^2}\frac{\mathrm{d}^2\mathbf{k}}{(2\pi)^3}$$

$$= \frac{1}{\beta}\sum_{k^0 = 2\pi i n/\beta} \mathrm{e}^{-i\mathbf{k}\cdot(\mathbf{x}-\mathbf{y})}\frac{i}{k^2 - m^2}\frac{\mathrm{d}^3\mathbf{k}}{(2\pi)^3}\,. \tag{191}$$

Thus the Fourier space integrations are replaced by summations in k^0:

$$\int (\;)\,\frac{\mathrm{d}^4k}{(2\pi)^4} \to \frac{1}{\beta}\sum_{k^0 = 2\pi i n/\beta}\int (\;)\,\frac{\mathrm{d}^3\mathbf{k}}{(2\pi)^3}\,. \tag{192}$$

One of the important uses of this rule is in the evaluation of finite temperature effective potentials. Consider, for example, the $\lambda\varphi^4$ theory. We saw in Section 4.6 that, at zero temperature, the effective potential involves calculating (see (147))

$$V_0(\varphi) \approx -\tfrac{1}{2}i\int \ln(k^2 - m^2 - \tfrac{1}{2}\lambda\bar{\varphi}^2)\,\mathrm{d}^4k. \tag{193}$$

The finite temperature result is obtained by the replacement in (192), viz.:

$$V_\beta(\varphi) = \frac{1}{2\beta}\sum_n\int \ln\left(-\frac{4\pi^2n^2}{\beta^2} - E^2\right)\frac{\mathrm{d}^3\mathbf{k}}{(2\pi)^2}\,, \tag{194}$$

where

$$E^2 = \mathbf{k}^2 + m^2 + \tfrac{1}{2}\lambda\bar{\varphi}^2. \tag{195}$$

We know that the original integral over d^4k is divergent and has to be regularized. The summation over n is also divergent. We shall extract the temperature dependent parts of V_β by the following trick: let us define $\upsilon(E)$ by

$$\upsilon(E) = \sum_n \ln\left(\frac{4\pi^2n^2}{\beta^2} + E^2\right) \tag{196}$$

$$\frac{\partial\upsilon}{\partial E} = \sum_n \frac{2E}{(4\pi^2n^2/\beta^2 + E^2)} = 2\beta\left(\frac{1}{2} + \frac{1}{\mathrm{e}^{\beta E} - 1}\right) \tag{197}$$

Integrating again,

$$\mathcal{V}(E) = 2\beta\left[\frac{E}{2} + \frac{1}{\beta}\ \ln(1 - e^{-\beta E})\right] + (\text{terms independent of } E). \quad (198)$$

Substituting back into $V_\beta(\varphi)$ we see that the thermal contribution is given by

$$V_\beta(\varphi)_{\text{thermal}} = \frac{1}{2\pi^2\beta^4}\int_0^\infty x^2 \ln(1 - \exp\{-(x^2 + \beta^2M^2)^{1/2}\})\,\mathrm{d}x \quad (199)$$

with $M^2 = m^2 + \frac{1}{2}\lambda\varphi^2$. At any finite temperature we should add the thermal contribution to the zero temperature effective potential. We shall have occasion to use this result in the future.

Notes and References

1. We assume that the reader has a brief acquaintance with relativistic wave equations at the level discussed in any quantum mechanics text. There exist many excellent texts in field theory which cover all the topics of this chapter in more detailed form. For example, the reader may consult:

   Itzykson, C., and Zuber, J. B.: 1980, *Quantum Field Theory*, McGraw-Hill, New York.
   Ramond, P.: 1981, *Field Theory – A Modern Primer*, Benjamin.

2. Detailed exposition of the quantization of Dirac and electromagnetic fields can be found in Itzykson and Zuber (1980), pp. 127–151 (see Note 1 above).
3. More on functional integrals over fermionic variables can be found in Itzykson and Zuber (1980), p. 439 and Ramond (1981), p. 214 (see Note 1 above) as well as in

   Nash, C.: 1978, *Relativistic Quantum Fields*, Academic Press, New York, p. 19.

4. A thorough discussion of the quantization of a self interacting scalar field (Feynman diagrams, renormalization, etc.) can also be found in Nash (1978), chs 1 and 2 (see Note 3 above).
5. The method of effective potential is discussed in detail in ch. 10 of

   Huang, K.: 1982, *Quarks, Leptons, and Gauge Fields*, World Scientific.

6. The pair creation in a constant electric field was discussed in:

   Schwinger, J.: 1954, *Phys. Rev.* **93**, 615; **94**, 1362.

   See also (4-3-3) of Itzykson and Zuber (1980) (see Note 1 above).

7. Quantum field theory at finite temperature is developed in detail in:

   Dolan, L., and Jackiw, R.: 1974, *Phys. Rev.* **D9**, 3320. Kirzhnits, D. A., and Linde, A. D. 1976, *Rep. Prog. Phys.* **20**, 195.

Chapter 5

Gauge Fields

5.1. Gauge Invariance – Electromagnetism

We saw in the last chapter that if we extend the quantization principles from a finite to an infinite number of degrees of freedom we encounter a new difficulty, viz. that of divergences in the perturbation expansion. We have two ways of dealing with this difficulty: (i) analyse the theory nonperturbatively or (ii) confine ourselves to only those theories which are perturbatively renormalizable.

At present, we do not possess sufficiently powerful methods of applied mathematics to tackle realistic field theories nonperturbatively. Thus, a considerable amount of emphasis has been placed in recent years on theories which are perturbatively renormalizable. A class of theories, loosely called 'gauge field theories', satisfy this requirement. The formalism of gauge fields has been used successfully in recent years to unify electromagnetic and weak interactions, and is invoked to unify strong interactions as well. Moreover, it seems possible that these models may have some relevance to the physics of the very early universe.

With the above motivation in mind, we shall introduce the reader to some basic concepts of gauge field theory in this chapter. Our discussion will be of an introductory nature and the interested reader is referred to books cited at the end of this chapter for more details.

The simplest kind of 'gauge field' is the Maxwellian electromagnetic field. With an eye on future generalizations, we shall introduce the familiar electromagnetic field via an unfamiliar route.

Let us assume that we were given a Dirac Lagrangian,

$$\mathscr{L} = i\bar{\psi}\gamma^i\partial_i\psi - m\bar{\psi}\psi \tag{1}$$

describing, say, electrons of mass m. We notice that this Lagrangian is invariant under the transformation

$$\psi \to \psi' \equiv \mathrm{e}^{i\alpha}\,\psi; \qquad \bar{\psi} \to \bar{\psi}' = \mathrm{e}^{i\alpha}\,\bar{\psi}, \tag{2}$$

where α is a constant. This symmetry – invariance under transformation (2) – leads to a conserved 'current' proportional to

$$J^i = \bar{\psi}\gamma^i\psi. \tag{3}$$

One may now ask the question: What happens to $\mathscr{L}$ under (2) when we allow α to be a function of x^i, i.e. $\alpha = \alpha(x)$? Clearly,

$$\begin{aligned}\mathscr{L}' &= i\,e^{-i\alpha}\,\bar{\psi}\gamma^i\,[\,e^{i\alpha}\,\partial_i\psi + i\,e^{i\alpha}\,\psi\partial_i\alpha\,] - m\bar{\psi}\psi \\ &= \mathscr{L} - J^i\,\partial_i\alpha.\end{aligned} \tag{4}$$

Thus $\mathscr{L}$ is not invariant under (2) if $\alpha = \alpha(x)$, but picks up a term $(-J^i\,\partial_i\alpha)$.

We can, of course, leave matters at that. However, it is usual for theories possessing a larger class of symmetries to have a richer dynamical structure. Motivated by this feeling we may ask: What is the minimum modification required in $\mathscr{L}$ so that our theory will be invariant under the transformations

$$\psi \to \psi' = [\exp i\alpha(x)]\,\psi; \qquad \bar{\psi}' = \exp[-i\alpha(x)]\,\bar{\psi} \quad ? \tag{5}$$

Such a modification is easily derived. Notice that J^i in (3) is invariant under (5), i.e.

$$J'^i \equiv \bar{\psi}'\gamma^i\,\psi' = J^i. \tag{6}$$

So if we take our modified Lagrangian to be

$$\mathscr{L}_1 \equiv \mathscr{L} + qA^iJ_i = \mathscr{L} + qA^i(x)J_i(x); \quad q = \text{constant} \tag{7}$$

and assume that, under the transformations (5), the quantities A^i transform as

$$A^i \to A^{i\prime} = A^i + \frac{1}{q}\,\partial^i\alpha, \tag{8}$$

then, using (4), (6), and (8), we get the desired result:

$$\begin{aligned}\mathscr{L}_1' &= \mathscr{L}' + qA'^iJ_i' = \mathscr{L} - J^i\,\partial_i\alpha + q\left[A^i + \frac{\partial^i\alpha}{q}\right]J_i \\ &= \mathscr{L} + qA^iJ_i = \mathscr{L}_1.\end{aligned} \tag{9}$$

In other words, the new Lagrangian $\mathscr{L}_1$ is invariant under the transformations (5). To achieve this property we had to introduce a new set of variables $A^i(x)$, which must transform as in (8). It is also clear that, since $\mathscr{L}_1$ is a scalar and J^i is a 4-vector, $A^i(x)$ must be a vector field. For historical reasons transformations like those in (2), (5), and (8) are called *gauge transformations*. Transformations like (2), in which the parameter α is a constant, are called *global* gauge transformations while (5) and (8) are called *local* gauge transformations.

To understand the physical nature of $A^i(x)$, notice that

$$\begin{aligned}\mathscr{L}_1 &= i\bar{\psi}\gamma^i\,\partial_i\psi - m\bar{\psi}\psi + q\bar{\psi}\gamma^iA_i\psi \\ &= i\bar{\psi}\,[\gamma^i(\partial_i - iqA_i)]\,\psi - m\bar{\psi}\psi.\end{aligned} \tag{10}$$

From the elementary theory of Dirac's equation we see that $A^i(x)$ is just the vector potential for the electromagnetic field and that q is the charge of the particle.

Though $\mathscr{L}_1$ is invariant under gauge transformation, it cannot be the complete Lagrangian for the system, because $\mathscr{L}_1$ does not contain a 'kinetic energy' term for $A^i(x)$. (As it stands, the variation of A^i in $\mathscr{L}_1$ will lead to the absurd result of $J_i = 0$.) Thus, we must add to $\mathscr{L}_1$ a term that is quadratic in the derivatives of A_i. Also, we must ensure that this term is invariant under gauge transformations. Obviously, the combination

$$F_{ik} \equiv \partial_i A_k - \partial_k A_i \tag{11}$$

is gauge invariant. Thus the additional term may be taken to be proportional to $F_{ik}F^{ik}$. The proportionally constant depends on the choice of units and we take it to be $(-\frac{1}{4})$.[1] Thus, we arrive at the complete gauge invariant Lagrangian:

$$L_{\text{ED}} = i\bar{\psi}\gamma^i(\partial_i - iqA_i)\psi - m\bar{\psi}\psi - \tfrac{1}{4}F_{ik}F^{ik}. \tag{12}$$

Let us summarize what we have done. We started with a Lagrangian which was invariant under a certain global transformation. We found that such a Lagrangian can be modified to be invariant under local transformations by introducing some extra degrees of freedom $A^i(x)$. Lastly, we completed the dynamics by adding the kinetic energy term for $A^i(x)$. It is usual to call the field $A^i(x)$ – introduced to maintain local gauge invariance – as a 'gauge field'.

Notice that the gauge field couples to the spinor field through the 'generalized derivative'

$$D_i \equiv \partial_i - iqA_i. \tag{13}$$

This prescription can be used to couple the gauge field to other fields like, say, a scalar field. We shall encounter such examples in the coming sections.

Thus, an electromagnetic field can be 'discovered' by starting with global symmetry and extending it to local symmetry. It is also known that electromagnetic interaction is a 'good', renormalizable interaction. This fact generates the hope that by starting with more general global symmetries and constructing their local versions we can describe the other interactions of nature. Fortunately, such a naïve hope of the physicist is realized to a great extent. We shall now consider these generalizations.

5.2. Gauge Invariance – Generalized

The global gauge transformation in the previous section was described by a single parameter α which was later taken to be a function of x^i. (Such a

[1] The sign of $F_{ik}F^{ik}$ must be negative because $(\partial_t \mathbf{A})^2$ should appear with a positive sign in the action.

transformation forms a representation for the one-parameter abelian group called U(1). For a simple review of group theory see Appendix B.) It is possible to consider transformations described by more than one parameter. As an example, consider a theory describing a doublet of complex scalar fields denoted by a column vector

$$\varphi \equiv \begin{pmatrix} \varphi_u \\ \varphi_d \end{pmatrix} . \tag{14}$$

We take the Lagrangian for this field to be

$$\mathscr{L} = \partial_i \varphi^\dagger \partial^i \varphi - \tfrac{1}{2}\mu^2 \varphi^\dagger \varphi - \tfrac{1}{2}\lambda(\varphi^\dagger\varphi)^2. \tag{15}$$

We may consider φ to be a vector in an abstract (internal symmetry) space with $\varphi^\dagger\varphi$ forming the analogue of 'length'. This generalization suggests that 'rotations' in this space may leave the Lagrangian invariant. This is indeed true. The Lagrangian $\mathscr{L}$ in (15) is invariant under the transformations

$$\varphi \to \varphi' = \exp\left(-i\frac{\tau^A}{2}\alpha_A\right)\varphi \equiv \varphi_\alpha. \tag{16}$$

Here, φ is treated as a 2×1 matrix. The quantities α_A, with $A = 1, 2, 3$, represent three parameters that characterize this transformation ('rotation'), and τ^A are three 2×2 matrices that satisfy the commutation relation

$$\left[\frac{\tau^A}{2}, \frac{\tau^B}{2}\right] = i\varepsilon^{ABC}\left(\frac{\tau_C}{2}\right), \tag{17}$$

where ε^{ABC} is the completely antisymmetric symbol.

Here, and in what follows, we shall denote the components in the internal symmetry space by capital letters A, B, etc. It is assume that repeated capital index is summed over. We do not distinguish between upper and lower indices (i.e. between τ^A and τ_A or α^A and α_A) and will use them interchangeably. The summation convention operates irrespective of the position of the index (in other words, we take $\tau_A = \delta_{AB}\tau^B$, etc.). Notice that (16) represents a matrix equation with the understanding that, for a matrix M,

$$\exp M = \sum_{n=0}^{\infty} \frac{(M)^n}{n!} . \tag{18}$$

This transformation forms a representation for the so-called SU(2) group. In the simplest sense, unitary 2×2 matrices of unit determinant under the operation of multiplication form a representation of this group.

Notice that the commutation rule in (17) is exactly the same as that of angular momentum generators in quantum mechanics. This is a physical motivation for considering the transformations in (16) as 'rotations'.

These transformations differ from the transformations discussed in Section 5.1 in one crucial respect, viz. two successive transformations do not commute. Suppose, for simplicity, that α^A and β^A were two infinitesimal parameters. Consider

$$\begin{aligned}\varphi_{(\alpha\beta)} &\equiv \exp\left(-\frac{i}{2}\tau_A\alpha^A\right)\exp\left(-\frac{i}{2}\tau_B\beta^A\right)\varphi\\ &\cong \left(1-\frac{i}{2}\tau_A\alpha^A\right)\left(1-\frac{i}{2}\tau_A\beta^A\right)\varphi\\ &\cong \left(1-\frac{i}{2}\tau_A\alpha^A-\frac{i}{2}\tau_A\beta^A-\tfrac{1}{4}(\tau_A\alpha^A)(\tau_B\beta^B)\right)\varphi.\end{aligned} \tag{19}$$

Thus

$$\begin{aligned}\varphi_{(\alpha\beta)}-\varphi_{(\beta\alpha)} &= -\tfrac{1}{4}\left[\tau_A\alpha^A\tau_B\beta^B-\tau_B\beta^B\tau_A\alpha^A\right]\varphi\\ &= -\tfrac{1}{4}\,\alpha^A\beta^B\left[\tau_B,\tau_B\right]\varphi = -\,\alpha^A\beta^B i\varepsilon_{ABC}\frac{\tau^C}{2}\,\varphi\end{aligned} \tag{20}$$

is nonzero. Groups for which (20) leads to a nonzero result are called *nonabelian groups*. Thus SU(2) is nonabelian. The gauge theories based on nonabelian groups are richer and more complicated, than the *Abelian* gauge theory of electromagnetism.

Let us now enlarge the symmetry by making the parameters α^A functions of x^i, thereby arriving at the local version of (16):

$$\varphi \rightarrow \varphi' = \exp\left[-\frac{i}{2}\tau_A\alpha^A(x)\right]\varphi. \tag{21}$$

The Lagrangian $\mathscr{L}$ in (15) is not invariant under (21). However, it can be made invariant by introducing *three* gauge fields $A_i^N (N = 1, 2, 3)$ into the Lagrangian. We first replace the ordinary derivative ∂_i in (15) by the generalized derivatives (where g is a constant somewhat analogous to q in (7) and (8)):

$$D_i \equiv \partial_i + i\,\frac{g}{2}\,\tau_N A_i^N(x). \tag{22}$$

Next, we have to choose the transformation law for A_i^N under (21) suitably so as to leave (15) invariant. This transformation law turns out to be

$$A_i' = UA_iU^{-1} - \frac{i}{g}\,U\,\partial_i U^{-1}, \tag{23}$$

where A_i and U are the matrices,

$$A_i \equiv \tfrac{1}{2}\tau_A A_i^A; \qquad U = \exp\left[-i\frac{\tau_A}{2}\,\alpha^A\right] \equiv \exp(-i\alpha). \tag{24}$$

Example 5.1. Let us verify that (23) is the 'correct' transformation law. We have

$$\begin{aligned} D_i'\varphi &= (\partial_i + igA_i')\,(U\varphi) \\ &= \{\partial_i + ig[\,U A_i U^{-1} - \frac{i}{g}\; U\,\partial_i U^{-1}]\,\}\,(U\varphi) \\ &= (\partial_i U)\,\varphi + U\partial_i\varphi + igUA_i\varphi + U(\partial_i U^{-1})\,U\varphi \\ &= U(\partial_i + igA_i)\,\varphi + [\,e^{-i\alpha}(-i\partial_i\alpha) + e^{-i\alpha}(+i\partial_i\alpha)\; e^{+i\alpha}\; e^{-i\alpha}]\,\varphi \\ &= U(\partial_i + igA_i)\,\varphi = U(D_i\varphi). \end{aligned}$$

Thus $(D_i\varphi)$ transforms covariantly under the local gauge transformations. □

Thus, enlarging the global symmetry to a local version leads to the Lagrangian

$$\mathscr{L}_1 = (D^i\varphi)^\dagger(D_i\varphi) - \tfrac{1}{2}\mu^2(\varphi^\dagger\varphi) - \frac{\lambda}{4}\,(\varphi^\dagger\varphi)^2. \tag{25}$$

Though $\mathscr{L}_1$ is gauge invariant, it suffers from the same defect as $\mathscr{L}_1$ in (10), viz. lack of a kinetic energy term for the gauge fields A_i^N. To remedy this situation, we need to construct a gauge invariant object which is quadratic in the derivatives of A_i^N. The simplest choice is given by

$$\mathscr{L}_A \equiv -\tfrac{1}{4}\; F_{ik}^N F_N^{ik}\;; \tag{26}$$

with

$$F_{ik}^N = \partial_i A_k^N - \partial_k A_i^N - g\varepsilon^{NPQ}A_{iP}A_{kQ}.$$

Or, in our matrix notation, with $F_{ik} \equiv \tfrac{1}{2}\;\tau_A F_{ik}^A$,

$$F_{ik} = \partial_i A_k - \partial_k A_i + ig[A_i, A_k]. \tag{27}$$

The complete gauge invariant Lagrangian is given by

$$\mathscr{L}_G = (D_i\varphi)^\dagger(D^i\varphi) - \tfrac{1}{2}\mu^2(\varphi^\dagger\varphi) - \frac{\lambda}{2}\,(\varphi^\dagger\varphi) - \tfrac{1}{4}\;F_{ik}^N F_N^{ik}. \tag{28}$$

Quite clearly, we have ended up with a rather complicated theory. The Lagrangian in (28) describes a world with (i) a pair of complex scalar fields (φ_u, φ_d) and (ii) a triplet of gauge vector fields (A_i^1, A_i^2, A_i^3). The scalar fields (φ_u, φ_d) have the same mass μ and interact through the self-coupling $\lambda(\varphi^\dagger\varphi)^2$. The vector fields A_i^N $(N = 1, 2, 3)$ are massless and interact among themselves through the last term in (27). This nonlinear interaction is the main point of departure from the electromagnetic case.

Example 5.2. Let us verify that F_{ik} in (27) does transform in a gauge invariant manner. Consider the quantity

$$\varepsilon_{ik}' \;\equiv\; \partial_i A_k' + igA_i'A_k' = \partial_i\;[\;UA_k U^{-1} - \frac{i}{g}\;U\,\partial_k U^{-1}\,] +$$

$$+ ig\,[\,UA_iU^{-1} - \frac{i}{g}\,U\,\partial_i U^{-1}\,]\,[\,UA_kU^{-1} - \frac{i}{g}\,U\,\partial_k U^{-1}\,] +$$

$$= (\partial_i U)A_k U^{-1} + U\,\partial_i A_k U^{-1} + UA_k\,\partial_i U^{-1} - \frac{i}{g}\,\partial_i U\,\partial_k U^{-1} - \frac{i}{g}\,U\partial_i\,\partial_k U^{-1}\,] +$$

$$+ ig\,[\,UA_iA_kU^{-1} - \frac{i}{g}\,UA_i\,\partial A_k U^{-1} - \frac{i}{g}\,U(\partial_i U^{-1})\,UA_kU^{-1} -$$

$$- \frac{1}{g^2}\,U(\partial_i U^{-1})U(\partial_k U^{-1})\,]$$

$$= U[\partial_i A_k + igA_iA_k]\,U^{-1} + [(\partial_i U)A_kU^{-1} + UA_k\,\partial_i U^{-1} + UA_i\,\partial_k U^{-1} +$$

$$+ U(\partial_i U^{-1})\,UA_kU^{-1}] +$$

$$+ [-\frac{i}{g}\,\partial_i U\,\partial_k U^{-1} - \frac{i}{g}\,U\,\partial_i\,\partial_k U^{-1} - \frac{i}{g}\,U(\partial_i U^{-1})U(\partial_k U^{-1})]\,.$$

Clearly, when we take the antisymmetric combination $(\varepsilon'_{ik} - \varepsilon'_{ki})$, only the first term on the right-hand side survives (remember that $U = \exp(-i\alpha)$). Thus we get

$$\varepsilon'_{ik} - \varepsilon'_{ki} = \partial_i A'_k - \partial_k A'_i\,ig\,[A'_iA'_k - A'_kA'_i] = F'_{ik}$$

$$= U[\,\partial_i A_k - \partial_k A_i + ig[A_i, A_k]\,]U^{-1}$$

$$= UF_{ik}U^{-1}.$$

This proves that F_{ik} transforms covariantly under gauge transformation. We have shown in Appendix B that the matrix generators can be chosen (in the adjoint representation) to satisfy the condition

$$\mathrm{Tr}(T_AT_B) = \tfrac{1}{2}\delta_{AB}.$$

Therefore, the Lagrangian can be written as

$$\mathcal{L}_A = -\tfrac{1}{4}\,F^A_{ik}\,F^{ik}_A = -\tfrac{1}{2}\,\mathrm{Tr}(F_{ik}F^{ik})$$

so that

$$\mathcal{L}'_A = -\tfrac{1}{2}\,\mathrm{Tr}(UF_{ik}U^{-1}UF^{ik}U^{-1}) = -\tfrac{1}{2}\,\mathrm{Tr}(F_{ik}F^{ik}) = \mathcal{L}_A.$$ □

5.3. General Formalism for Gauge Theory

In the previous sections we have considered gauge transformations from the physical point of view. The theory can be developed in a more compact form by using the concepts of Lie group theory. We describe here a gauge theory based on an arbitrary Lie group G, from an intuitive point of view. A formal development of the necessary definitions, results, etc., is given in Appendix B.

In this section we shall be concerned with groups that are 'transformation groups'. These are groups whose elements generate a well-defined transformation in a linear vector space. We shall write an arbitrary element of the group in the form

$$U = \exp(-i\lambda^A T_A).$$

Here λ^A are a set of N real numbers ($A = 1, 2, \ldots, N$) that characterize the element of the group. The operators T_A – called the generators of the transformation – produce the infinitesimal changes. We have parametrized our group so that $\lambda^A = 0$ corresponds to the identity element of the group. For small λ^A,

$$U \cong 1 - i\lambda^A T_A.$$

A central quantity of interest is the commutator $[T_A, T_B]$. Clearly, if two infinitesimal transformations $U(\lambda)$ and $U(\eta)$ are applied in the order $U(\lambda)U(\eta)$ or as $U(\eta)U(\lambda)$, it is the commutator $[T_A, T_B]$ that determines the net effect. In general, we have the relation

$$[T_A, T_B] = iC^M{}_{AB}\, T_M, \tag{29}$$

where the quantities $C^M{}_{AB}$ – called structure constants – determine all the local properties of the group. In any group, the structure constants satisfy the relations (see Appendix B)

$$C^A{}_{BM} = -C^A{}_{MB}\,,$$

$$C^N{}_{AB}C^D{}_{NM} + C^N{}_{BM}C^D{}_{NA} + C^N{}_{MA}C^D{}_{NB} = 0. \tag{30}$$

The theory of classification of Lie groups essentially boils down to classifying all possible $C^A{}_{BM}$ which satisfy the above conditions. When all C's vanish the group becomes Abelian. It should be noted that we can choose some other linearly independent set of T_A to describe the same group. This will change the numerical values of the structure constants.

Suppose there exists a choice of T_A in which $C^M{}_{AB}$ has the following form. The index A is divided into two sets such that, when A and B belong to two different sets, $C^M{}_{AB} = 0$. (In other words, all members of a subset of T_A's commute with all T_A's of the complementary subset.) If such a division is possible, the group is said to be 'factored'. A group which cannot be so factored is called 'simple'. A direct product of nonabelian groups is called 'semisimple'.

A set of n matrices (of any $p \times p$ dimension) which obey the same commutation rules as T_A is said to provide a matrix representation for the group. Clearly, if a matrix T_A satisfies the relation so does $T_A^\dagger$. Therefore, we can take the matrices in the representation to be Hermitian. The group elements then becomes unitary matrices.

There exists one particular representation which is completely decided by the structure constants themselves. This is called the adjoint representation and is given by

$$(T_A)_{BM} = -iC^M{}_{AB}.$$

In other words, the BMth element of the matrix that corresponds to the generator T_A is given by $-iC^M{}_{AB}$. The proof of this assertion can be found in Appendix B.

In order to construct a gauge theory based on a particular group, we must specify the representation under which the physical fields transform. In general, different physical fields transform under different dimensional representations. Let us denote by $\psi(x)$ an arbitrary, multicomponent field [$\psi = \{\psi_A\}$ $A = 1, \ldots, p$, say]. Under a general local transformation

$$\psi(x) \to U(x)\psi(x) \equiv \exp(-i\lambda^A(x)T_A)\psi(x).$$

Here (T_A) (and hence U) must be a $p \times p$ matrix and ψ transforms under the p-dimensional representation. The infinitesimal transformation is

$$\delta\psi = (-i\lambda^A\, T_A)\psi \equiv (-i\lambda(x))\psi(x).$$

In the last equation we have simplified the notation by assuming that $\lambda(x)$ denotes the $p \times p$ matrix $(\lambda^A\, T_A)$. It is easy to see that ordinary derivatives do not transform covariantly:

$$\delta(\partial_k\psi) = -i\lambda(x)\,\partial_k\psi - i(\partial_k\lambda(x))\psi(x).$$

We, therefore, define the gauge covariant derivative as

$$D^k\psi \equiv [\partial^k + igA^k(x)]\psi. \tag{31}$$

Here again $A^k(x)$ is the matrix

$$A^k(x) \equiv A^{kM}(x)T_M.$$

Thus we need N gauge fields ($M = 1, \ldots$ N-dimension of the group). It is easy to work out the transformation law for $A^k(x)$:

$$\delta A^k(x) = \frac{1}{g}\,\partial^k\lambda - i[\lambda(x, A^k(x)]. \tag{32}$$

In the component notation this is equivalent to

$$\delta A^k_A = \frac{1}{g}\,\partial^k\,\lambda_A(x) + C_{ABM}\lambda^B A^{kM}(x).$$

The gauge covariant field tensor can be similarly defined as

$$F^{ik} = \partial^i A^k - \partial^k A^i + ig[A^i, A^k],$$

that is,

$$F^{ik}_A = \partial^i A^k_A - \partial^k A^i_A - gC_{ABM}A^{Bi}A^{kM}. \tag{33}$$

It should be clear to the reader that we have merely generalized the discussion of the last section to an arbitrary group with N generators. (In particular, the proofs given in Examples 5.1 and 5.2 are also valid for the present case.) In Section 5.2 we dealt with the SU(2) group for which $C_{ABM} = \varepsilon_{ABM}$.

We wish to conclude this section with a mention of 'global properties of the group'. In quantum mechanics, it is well known that the spin matrices σ_A and the rotation matrices J_A satisfy the same communtation rules:

$$[J_A, J_B] = i\varepsilon_{ABC}J_C; \qquad [\sigma_A, \sigma_B] = i\varepsilon_{ABC}\sigma_C.$$

In the group theoretic language, J_A are the generators for the rotation group O(3) and σ_A are the generators for SU(2). Thus we see that SU(2) and O(3) have the same structure constants and hence the same local properties.

However, the global properties of these two groups are very different. For example, spin can have integral or half integral values while angular momentum can have only integral values.

In general, the local properties are decided by the generators of the group – or by what is called the 'Lie algebra'. There can be more than one Lie group corresponding to a given Lie algebra. In obtaining the group element U from the Lie algebra element T_A, we write

$$U(\lambda) = \exp(-i\lambda^A T_A).$$

The global structure of the group depends on the range of allowed $\{\lambda^A\}$. Different ranges will lead to different groups.

The space of possible values of λ^A is called the group manifold. A group is said to be simply connected if every closed path in the group manifold can be continuously deformed to a point. Otherwise, it is multiply connected.

Any rotation can be specified by a vector pointing in the direction of the axis of rotation with a magnitude equal to the angle of rotation. Thus the group manifold of O(3) is the region enclosed by a sphere of radius π, with diametrically opposite points on the surface identified. (This is because rotations of $\pm\pi$ about the same axis are physically the same.) Closed loops drawn within the sphere can be deformed to a point, but the diameters of the sphere are also 'closed curves' in the group manifold because of the identification of opposite points. These cannot be shrunk to a point. Thus O(3) is multiply connected.

On the other hand, a general element of SU(2) can be expressed as

$$U = \exp i\theta\hat{\mathbf{n}}\cdot\hat{\sigma}/2 = b_0 + i\mathbf{b}\cdot\sigma$$

$$b_0 = \cos\frac{\theta}{2}; \qquad \mathbf{b} = \mathbf{n}\sin\frac{\theta}{2}.$$

where σ are the Pauli matrices, i.e. $\sigma = \{\sigma_A\}$. Thus the group element is specified by four numbers $(b_0, \mathbf{b})$ with the condition

$$\sum_0^3 b_i^2 = 1.$$

Therefore, the group manifold is the surface of unit sphere in Euclidean 4-space which is simply connected. This example illustrates the difference between the global properties of two groups which have the same local structure.

5.4. Spontaneous Symmetry Breaking

In quantum electrodynamics, the gauge field $A_i(x)$ mediates interactions between electrically charged particles. Pursuing this simple analogy we may expect a quantum gauge field A_i^N to mediate strong and weak interactions. It turns out that the gauge field theories are good renormalizable theories, just like quantum electrodynamics. However, there is one crucial difference between the electromagnetic interaction and (say) the weak interaction. While the former has a long range and thus can adequately be described by a massless gauge field, the latter cannot. In other words, if weak interactions are mediated by a vector particle, then that particle must be massive. Thus we seem to have reached a dead end; renormalizability requires gauge invariance of the Lagrangian, thereby excluding the appearance of the mass term. However, the short-range nature of the interaction requires the mediating field to be massive. Thus, we have to find some means of giving mass to the gauge field without explicitly introducing it in the Lagrangian. Let us now see how this can be done.

In classical physics, the symmetries of physical observables are directly determined by the symmetries of the Lagrangian (or the Hamiltonian) because the Lagrangian itself is made out of physical parameters. In quantum theory the Lagrangian is constructed from the basic operators of the theory. These operators have definite expectation values, in any particular state chosen for the system. This leads to the following interesting possibility: the Lagrangian may respect a particular symmetry while the state vector of the system may not. Observed quantities, such as expectation values and transition elements, will then no longer exhibit the basic symmetry of the Lagrangian. Such a process in which the choice of the ground state breaks the symmetry of the Hamiltonian is often called 'spontaneous symmetry breaking' (SSB in brief).

Condensed matter physics is full of examples of such a phenomenon. Consider a 'Heisenberg ferromagnet' described by an interaction Hamiltonian of the form $J \Sigma \mathbf{S}_i \cdot \mathbf{S}_j$, where $\mathbf{S}_i$, $\mathbf{S}_j$ are the spins at the (i, j)th sites. Being a vector dot product this Hamiltonian is invariant under arbitrary rotations. However, the ground state of the ferromagnet (below the transition temperature) has all the spins aligned in one particular direction. This ground state clearly violates the rotational invariance of the Hamiltonian.

The simplest field theory exhibiting this phenomenon is that of a complex scalar field with the Lagrangian:

$$L = \partial_i\varphi^\dagger \partial^i\varphi - V(\varphi^\dagger\varphi), \tag{34}$$

where

$$V(\varphi^\dagger\varphi) = \tfrac{1}{2}\lambda^2 \, |\varphi|^4 - \tfrac{1}{2}\mu^2 \, |\varphi|^2 \tag{35}$$

If μ is imaginary, then the $-\frac{1}{2}\mu^4 \, |\varphi|^2$ term in $V(\varphi)$ can be interpreted as a mass term for the scalar field [2]. [This is the $\lambda\varphi^4$ theory discussed at length in the previous chapter.] However, the theory is renormalizable and well defined even when μ is real. In this case, of course, we cannot treat μ as the mass of the field quanta. We should interpret the Lagrangian as describing a massless complex scalar field with a peculiar self-interaction described by $V(\varphi)$. We also notice that the Lagrangian is invariant under the global transformation

$$\varphi \to \varphi' = e^{i\alpha} \, \varphi \quad (\text{constant } \alpha).$$

What is the ground state of such a system? We expect the physical ground state to minimize the energy of the system. The Hamiltonian for the above Lagrangian is

$$H = |\dot{\varphi}|^2 + |\nabla\varphi|^2 + V(\varphi).$$

Any constant value of φ minimizes the first two terms of the Hamiltonian. By choosing this constant to be the minimum of $V(\varphi)$ we can make the third term also minimum. For our potential, the Hamiltonian is thus minimized when

$$|\varphi|^2 = \frac{\mu^2}{2\lambda^2} \equiv |\varphi_0|^2.$$

If we separate φ into real and imaginary parts by writing

$$\varphi = \frac{1}{\sqrt{2}} \, (\varphi_1 + i\varphi_2), \tag{36}$$

then this conditions means that

$$\varphi_1^2 + \varphi_2^2 = \frac{\mu^2}{\lambda^2} \, ,$$

So far we have been using classical terminology. In the quantum version of the theory, we would expect the ground state (loosely called 'vacuum' and denoted by $|0\rangle$) to have a nonzero expectation value for the φ operator. In other words, we assume that

$$\langle 0| \, \varphi \, |0\rangle = \frac{1}{\sqrt{2}} \, \frac{|\mu|}{|\lambda|} \, e^{i\theta} \, .$$

Here θ is an arbitrary phase angle which is not determined by the minimum

[2] Although V is function of $\varphi^+\varphi$ we shall economize by writing it as $V(\varphi)$.

condition. With this choice, the Hamiltonian will have minimum expectation value in the ground state – as it should.

This choice for the ground state immediately leads to two important conclusions. First, our ground state does not respect the symmetry present in the Lagrangian. Clearly, ($\varphi' \equiv e^{i\alpha}\varphi$)

$$\langle 0|\,\varphi'\,|0\rangle = e^{i\alpha}\langle 0|\,\varphi\,|0\rangle = \frac{1}{\sqrt{2}}\,\frac{|\mu|}{|\lambda|}\quad e^{i(\theta+\alpha)} \neq \langle 0|\,\varphi\,|0\rangle. \tag{37}$$

Second, the field φ cannot describe the physical particles in the conventinal sense. To see this, assume that φ represents physical particles. Then, the state $\varphi\,|0\rangle$ will be proportional to a (physical) one-particle state. But a physical one-particle state must be orthogonal to the physical vacuum requiring $\langle 0|\,\varphi\,|0\rangle$ to be zero, which is contrary to our assumption. Actually, the physical excitations are to be found around the 'vacuum condensate'. In order to analyse the physical content of the theory we shall therefore proceed as follows:

From (35) we see that the minimum value of V only determines the modulus $|\,\varphi\,|^2$ and not the phase angle θ. This is a consequence of the phase invariance of the Lagrangian. Using the freedom, let us choose $\theta = 0$. In other words, take

$$\langle 0|\,\varphi_1\,|0\rangle = \frac{|\mu|}{|\lambda|}\,; \qquad \langle 0|\,\varphi_2\,|0\rangle = 0. \tag{38}$$

Now introduce two new field variables χ_1 and χ_2 by the relations

$$\varphi_1 = \frac{|\mu|}{|\lambda|} + \chi_1, \qquad \varphi_2 = \chi_2. \tag{39}$$

Clearly χ_1 and χ_2 have zero expectation values in the ground state and thus $\chi_1\,|0\rangle$ and $\chi_2\,|0\rangle$ can represent physical one-particle states. Substituting (39) into the original Lagrangian, we express our system in terms of the physical fields χ_1 and χ_2, as,

$$\begin{aligned} L = {} & \tfrac{1}{2}(\partial_i\chi_1\partial^i\chi_1 - \mu^2\chi_1^2) + \tfrac{1}{2}\,\partial_i\chi_2\,\partial^i\chi_2 - \\ & -\tfrac{1}{2}\,|\mu|\,\lambda\chi_1(\chi_1^2 + \chi_2^2) - \frac{1}{8}\lambda^2(\chi_1^2 + \chi_2^2)^2 + \frac{\mu^4}{8\lambda^2}\,. \end{aligned} \tag{40}$$

This Lagrangian describes two scalar fields χ_1, χ_2, one with mass μ(notice that it appears with the correct sign) and the other massless. The fields interact via a complicated coupling (see Figure 5.1).

Let us pause for a moment and take stock of the situation. We began with a Lagrangian for a complex scalar field which has two degrees of freedom. This Lagrangian contained a potential which is minimum for nonzero value of the field. This minimum corresponds to the vacuum state of the quantum theory.

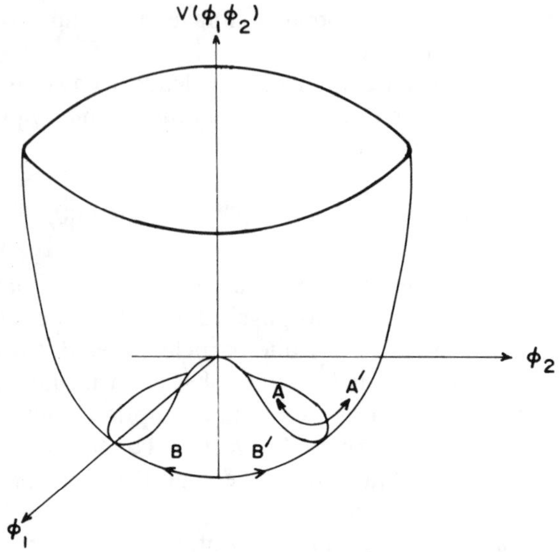

Fig. 5.1. The potential $V(\varphi)$ of (5.35) plotted against φ_1, φ_2 of Equation (5.36). The oscillatory degree of freedom along AA' appears as a massive field, while the 'rolling' degree of freedom along BB' (in the $\varphi_1 - \varphi_2$ plane) appears as a massless field.

In order to describe the physical interactions, we must know the nature of small excitations about the minimum. Instead of separating $\varphi(x)$ into real and imaginary parts, suppose that we had written

$$\varphi(x) = \alpha(x) \exp i\beta(x). \tag{41}$$

Clearly, the minimum condition does not say anything about $\beta(x)$. In other words, it is possible to have excitations around the 'valley' of the potential $V(\varphi)$ without changing the energy. These excitations correspond to the massless χ_2 field in (40). It is also possible to have 'radial' oscillations around the minimum. The energy required for this oscillation is proportional to the second derivative of the potential and appears as the massive χ_1 mode. In short, the nonzero expectation value for φ has led to the appearance of one massless and one massive degree of freedom in the theory.

Example 5.3. The existence of massless excitation in this example is just a particular case of a general result: whenever the ground state breaks a symmetry of the Hamiltonian, there will exist massless excitations in the theory. The general result can be proved in the following way:

Let J^i be the conserved current corresponding to the symmetry in question ($\partial_i J^i = 0$). The charge operator $\int J^0(\mathbf{x}, t)\, d^3\mathbf{x} = Q$ acts as the generator of the transformation. If φ is not invariant under this symmetry we shall have $\varphi(y) = [Q, \varphi'(y)]$ for some $\varphi'(y)$. When $\langle|\, \varphi(y)\, |0\rangle$ is not zero, we get

$$0 \neq \langle 0|\, \varphi(y)\, |0\rangle = \langle 0|\, [Q, \varphi'(y)]\, |0\rangle$$

$$= \int \langle 0|\ [J^0(x^0, \mathbf{x}), \varphi'(y^0, \mathbf{y})]|0\rangle_{x^0=y^0}\ d^3\mathbf{x}$$

$$= \int d^3\mathbf{x} \sum_n \left\{ \langle 0|\ J^0(x)\ |n\rangle \langle n|\ \varphi'(y)\ |0\rangle - \langle 0|\ \varphi'(y)\ |n\rangle \langle n|\ J^0(x)\ |0\rangle \right\}_{x^0=y^0}$$

$$= \sum_n \delta^3(\mathbf{P}_n) \Big[\langle 0|\ J^0(0)\ |n\rangle \langle n|\ \varphi'(y)\ |0\rangle\ e^{ip_n^0 x^0}$$

$$- \langle 0|\ \varphi'(y)\ |n\rangle \langle n|\ \varphi'(y)\ |0\rangle\ e^{-ip_n^0 x^0} \Big] .$$

In the last step we have used

$$J^0(x) = e^{-ip\cdot x}\ J^0(0)\ e^{ipx}$$

and integrated over $d^3\mathbf{x}$. Since Q is a conserved quantity we know that the right-hand side should actually be independent of x^0. At the same time, the right-hand side cannot vanish because the left-hand side does not. However, the right-hand side contains such terms as $\exp(ip_n^0 x^0)$ which do depend on x^0. Therefore, we must have states $|n\rangle$ with $P_n^0 \to 0$ as $\mathbf{P}_n \to 0$ (notice that $\delta(\mathbf{P}_n)$ restricts contributions to $\mathbf{P}_n = 0$ set). This is the only way of producing a nonzero value on the right-hand side which is independent of x^0. The states $|n\rangle$ with $(P^0)^2 \to 0$ as $|\mathbf{P}|^2 \to 0$ are precisely the massless states. (Note that $m^2 = (P^0)^2 - |\mathbf{P}|^2$.) □

5.5. SSB with an Abelian Gauge Field

The Lagrangian in (40) possesses the invariance only under the global phase transformation. We have seen in the previous section that we can enlarge the symmetry into a local invariance by introducing a gauge field. This will lead us to the Lagrangian

$$\mathcal{L} = [(\partial_i + iqA_i)\varphi^\dagger]\ \ [(\partial^i - iqA^i)\varphi] - V(\varphi) - \tfrac{1}{4} F_{ik}F^{ik}. \tag{42}$$

This Lagrangian now has four degrees of freedom, the two degrees of freedom of φ and the two degrees of freedom of the massless gauge field. What are the physical states now?

The ground state again corresponds to the choice

$$\langle 0|\ \varphi\ |0\rangle = \frac{1}{\sqrt{2}} \frac{|\mu|}{|\lambda|} \quad \text{and} \quad \langle 0|\ A_\mu\ |0\rangle = 0. \tag{43}$$

Once again we see that φ cannot be the physical field. In order to obtain the physical field, we could use the same decomposition as before. However, it is simpler to proceed as follows. Let us write

$$\varphi(x) = \frac{1}{\sqrt{2}} \left(\frac{|\mu|}{|\lambda|} + \alpha(x) \right) \exp \frac{i|\lambda|}{|\mu|}\ \beta(x). \tag{44}$$

Here $\alpha(x)$ represents the 'radial' mode and $\beta(x)$ represents the phase freedom. We have arranged things so that $\alpha(x)$ and $\beta(x)$ represent physical excitations, i.e.

$$\langle 0|\ \alpha\ |0\rangle = 0;\ \langle 0|\ \beta\ |0\rangle = 0;\ \langle 0|\ \varphi\ |0\rangle = \frac{1}{\sqrt{2}}\ \frac{|\mu|}{|\lambda|}\ . \tag{45}$$

Substituting $\varphi(x)$ into $\mathscr{L}$ will allow us to see the physical content of the theory. However, notice that $\mathscr{L}$ is now invariant under the local transformation

$$\varphi(x) \to \varphi'(x) = e^{i\theta(x)}\ \varphi(x); \qquad A_i \to A'_i = A_i + \frac{\partial_i \theta(x)}{q}\ .$$

By choosing the gauge function $\theta(x)$ to be

$$\theta(x) = -\ \frac{|\lambda|}{|\mu|}\ \beta(x) \tag{46}$$

we get

$$\varphi'(x) = \frac{1}{\sqrt{2}}\left(\frac{|\mu|}{|\lambda|} + \alpha(x)\right)\ ; \qquad A'_i(x) = A_i(x) - \frac{|\lambda|}{q|\mu|}\ \partial_i\beta(x) \tag{47}$$

so that the Lagrangian becomes

$$\mathscr{L} = \tfrac{1}{2}\,(\partial_i\alpha\ \partial^i\alpha - \mu^2\alpha^2) + \left(-\tfrac{1}{4}\,F_{ik}F^{ik} + \tfrac{1}{2}\,q^2\,\frac{\mu^2}{\lambda^2}\,A_iA'^i\right) +$$

$$+ A'_iA'^i\cdot\tfrac{1}{2}\,q^2\left(\alpha^2 + 2\,\frac{|\mu|}{|\lambda|}\,\alpha\right) - \frac{1}{8}\,\lambda^2\alpha^4 - \tfrac{1}{2}\,\lambda^2\,\frac{|\mu|}{|\lambda|}\,\alpha^3 + (\text{const}). \tag{48}$$

This describes a system consisting of one massive scalar field $\alpha(x)$ and one massive vector field $A'_i(x)$! Out of the two degrees of freedom of the scalar field denoted by $\alpha(x)$ and $\beta(x)$, one degree of freedom, viz. $\beta(x)$, combines with the original massless gauge field of two degrees of freedom, $A_i(x)$, to produce one massive vector field A'^i. In other words by choosing a vacuum state which breaks the original symmetry of the Lagrangian, we have found a way of making the gauge fields massive. It must be borne in mind that the Lagrangian *always* respects the original symmetry; it is transparent in terms of $[\varphi(x), A_i(x)]$ but hidden in $\alpha(x)$, $A'_i(x)$. Thus we expect the nice features of the theory, such as renormalizability, to survive the algebraic transformations of the fields.

Example 5.4. In the absence of a long-range gauge field, spontaneous symmetry breaking implies the existence of massless states. When a long-range field is present, the situation is entirely different. The physics of this phenomenon, called the 'Higgs mechanism', is well illustrated by the following many-body theory model with the second quantized Hamiltonian,

$$H = \int d^3\mathbf{x}\ \tfrac{1}{2}\ [\pi^2(t, \mathbf{x}) + |\Delta\varphi|^2\,] + \tfrac{1}{2}\int d^3\mathbf{x}\ d^3\mathbf{y}\,\pi(t, \mathbf{x})V(|\mathbf{x} - \mathbf{y}|)\pi(t, \mathbf{y}).$$

This Hamiltonian is invariant under the transformation $\varphi \to \varphi + \lambda$. Clearly, any vacuum expectation value for φ will violate this symmetry.

Quantizing the system in a large box of volume V, we can write the momentum space Hamiltonian,

$$H = \frac{1}{2V} \sum_{\mathbf{k}} \{ |\mathbf{k}|^2 \varphi^*_{\mathbf{k}} \varphi_{\mathbf{k}} + (1 + V_{\mathbf{k}} \pi^*_{\mathbf{k}} \pi_{\mathbf{k}}) \} ,$$

where

$$V_{\mathbf{k}} = \int e^{i\mathbf{k} \cdot \mathbf{x}} V(\mathbf{x}) \, d^3\mathbf{x}.$$

The equations of motion

$$\dot{\pi} = - |\mathbf{k}| \varphi_{\mathbf{k}}; \qquad \dot{\varphi}_{\mathbf{k}} = (1 + V_{\mathbf{k}})\pi_{\mathbf{k}}$$

can be solved immediately to give the dispersion relation

$$\omega^2_{\mathbf{k}} = |\mathbf{k}|(1 + V_{\mathbf{k}}).$$

Suppose that $V_{\mathbf{k}}|\mathbf{k}|^2 \to 0$ as $|\mathbf{k}| \to 0$. In other words $V(\mathbf{x})$ is *not* a long-range (Coulomb like) potential. Then $\omega_{\mathbf{k}} \propto |\mathbf{k}|$ for small $\mathbf{k}$ indicating the dispersion relation of a *massless* mode. On the other hand for long range interactions (like $V(\mathbf{x}) = g^2/|\mathbf{x}|$), $\omega^2_{\mathbf{k}}$ goes to a constant as $\mathbf{k} \to \mathbf{0}$. This is the dispersion relation characteristic of a massive excitation. Thus, long-range interactions are capable of providing mass to originally massless modes. □

Example 5.5. The physical fields can also be identified in our model by the following argument. In the Lagrangian (42) choose a gauge such that $\varphi(x) = \varrho(x)$ is real. Varying $A^i(x)$ and $\varrho(x)$, we get the field equations

$$\partial_i F^{ik} = - 2q^2 \varrho^2 A^k,$$

$$(\partial^i + iqA^i)(\partial_i - iqA_i)\varrho = 2\lambda\varrho(\varphi_0^2 - \varrho^2),$$

(where $\varphi_0 = \langle 0| \varphi |0\rangle$). In the linear approximation, we can put $\varrho(x) = \varphi_0 + \eta(x)$ and retain terms up to first order in η. Then these equations become

$$(\Box + 2q^2\varphi_0^2)A^i = 0,$$

$$(\Box + 4\lambda\varphi_0^2)\eta = 0.$$

These equations show that both A^i and η are massive with masses $\sqrt{2}\, q\varphi_0$ and $2\sqrt{\lambda}\varphi_0$, respectively. □

5.6. SSB with a Nonabelian Gauge Field

We shall now consider the spontaneous symmetry breaking in the presence of a nonabelian gauge symmetry. As in Section 5.2, consider a doublet scalar field described by the Lagrangian

$$L = (\partial_i\varphi^\dagger)(\partial^i\varphi) + \tfrac{1}{2}\mu^2\varphi^\dagger\varphi - \tfrac{1}{2}\lambda^2(\varphi^\dagger\varphi)^2. \tag{49}$$

As we discussed before, this Lagrangian possesses an SU(2) invariance under the global transformation

$$\varphi \to \varphi' = \exp\left(ig\frac{\boldsymbol{\tau}}{2} \cdot \boldsymbol{\alpha}\right)\varphi. \tag{50}$$

We have changed the notation slightly in (50) as compared to Section 5.2. Since τ_A and α^A has three components (i.e. $A = 1, 2, 3$), we have introduced a vector notation in the internal symmetry space and have put $\boldsymbol{\tau} = (\tau^1, \tau^2, \tau^3)$. Besides, we have rescaled the parameters α^A by introducing a $(-g)$ factor in the exponent. We shall follow this vector notation for other three component entities as well. In addition to this invariance the Lagrangian *also* possesses a U(1) invariance under the transformation,

$$\varphi \to \varphi' = \exp(ig'\beta)\,\varphi. \tag{51}$$

(In Section 5.2 we did not bother about this extra symmetry since we were only interested in the nonabelian aspects of the theory.) Thus we can say that the full invariance group of the Lagrangian is SU(2) × U(1). We can make the Lagrangian invariant under the local transformations by introducing a triplet of gauge fields $\mathbf{A}_i$ for the SU(2) part and an abelian gauge field B_i for the U(1) part. The locally invariant Lagrangian will have the form

$$L = (D_i\varphi^\dagger)\,(D^i\varphi) + \tfrac{1}{2}\mu^2\varphi^\dagger\varphi - \tfrac{1}{2}\,\lambda^2(\varphi^\dagger\varphi)^2 - \tfrac{1}{4}\,\mathbf{F}_{ik}\mathbf{F}^{ik} - \tfrac{1}{4}G_{ik}G^{ik}, \tag{52}$$

where

$$D_i\varphi = \left(\partial_i - \frac{ig}{2}\,\tau\cdot\mathbf{A}_i - \frac{ig'}{2}\,B_i\right)\varphi \tag{53}$$

and

$$G_{ik} - \partial_i B_k - \partial_k B_i; \quad \mathbf{F}_{ik} = \partial_i\mathbf{A}_k - \partial_k\mathbf{A}_i + g(\mathbf{A}_k \times \mathbf{A}_i),$$

where the 'cross product' in the internal symmetry space is defined by the usual formula

$$(\mathbf{A}_i \times \mathbf{A}_k)^N = \varepsilon^{NPQ}A_{iP}A_{kQ}.$$

As in the previous case, we shall take the vacuum expectation value for φ to be

$$\langle 0|\,\varphi\,|0\rangle = \begin{pmatrix} 0 \\ f/\sqrt{2} \end{pmatrix}, \quad f = \frac{|\mu|}{|\lambda|}. \tag{54}$$

Let us introduce the physical fields $\theta(x)$ and $\sigma(x)$ by the decomposition

$$\varphi(x) = \exp\left(i\boldsymbol{\tau}\cdot\frac{\boldsymbol{\theta}(x)}{2f}\right)\begin{pmatrix} 0 \\ \dfrac{f}{\sqrt{2}} + \dfrac{\sigma(x)}{\sqrt{2}} \end{pmatrix}. \tag{55}$$

As in the case of (46), an SU(2) transformation with $\alpha(x) = -(1/fg)\theta(x)$ will 'rotate away' the phase part of φ. Substituting the physical fields into the Lagrangian will lead to the following peculiar structure:

$$L = \left\{\tfrac{1}{2}(\partial_i\sigma)\,(\partial^i\sigma) - \tfrac{1}{2}\mu^2\sigma^2\right\} + \frac{g^2f^2}{8}\,(A_i^1A^{i1} + A_i^2A^{i2}) +$$
$$+\frac{1}{8}\,f^2(gA_i^3 - g'B_i)\,(gA^{i3} - g'B^i) + I + K, \tag{56}$$

where I and K denote the interaction terms and the kinetic energy terms for $\mathbf{A}^i$, B^i, respectively. In order to make sense out of the third term, define

$$\begin{aligned} Z_i &= \cos\theta \cdot A_i^3 - \sin\theta \cdot B_i \\ X_i &= \sin\theta \cdot A_i^3 + \cos\theta \cdot B_i \end{aligned} \quad ; \quad \tan\theta = g'/g. \tag{57}$$

Then L can be written as

$$\begin{aligned} L = {} & \tfrac{1}{2}(\partial_i\sigma)(\partial^i\sigma) - \tfrac{1}{2}\mu^2\sigma^2 - \\ & -\tfrac{1}{4}(\partial_i A_k^1 - \partial_k A_i^1)^2 + \frac{1}{8} g^2 f^2 (A_i^1)^2 - \\ & -\tfrac{1}{4}(\partial_i A_k^2 - \partial_k A_i^2)^2 + \frac{1}{8} g^2 f^2 (A_i^2)^2 - \\ & -\tfrac{1}{4}(\partial_i Z_k - \partial_k Z_i)^2 + \frac{1}{8} f^2 (g^2 + g'^2)(Z_i)^2 - \\ & -\tfrac{1}{4}(\partial_i X_k - \partial_k X_i)^2 + \text{Interaction terms.} \end{aligned} \tag{58}$$

We have one massive scalar field $\sigma(x)$ with mass μ; one massive vector field Z_i with mass $\frac{1}{2}f(g^2 + g'^2)^{1/2}$; two massive vector fields with the same mass $\frac{1}{2}fg$; and one massless vector field X_i. We started with three massless $\mathbf{A}_i$ and one massless B_i and four scalars giving a total of 12 degrees of freedom. After the redefinitions, there are three massive vector fields and one massless vector field giving 11 degrees of freedom. The remaining degree of freedom is carried by the scalar $\sigma(x)$.

There are several nontrivial features when a nonabelian symmetry is spontaneously broken. Notice that our Lagrangian possesses the symmetry of a group $G = \mathrm{SU}(2) \times \mathrm{U}(1)$. The individual SU(2) and U(1) transformations commute with each other. This allows us to choose one coupling constant g for the SU(2) sector and another g' for the U(1) sector. People who want to reduce this freedom in the independent choice of coupling constants like to work with a simple group that cannot be separated as direct products.

Gauge invariance of the theory implies masslessness while the spontaneous symmetry breaking gives masses to the gauge fields. In our case, we do have one massless field left. This is due to the fact that only part of the $\mathrm{SU}(2) \times \mathrm{U}(1)$ is 'broken' by our choice of the vacuum expectation value. Consider, for example a situation when the components φ_A ($A = 1, 2, \ldots, r$) of a field transform under the r-dimensional representation of some gauge group G. Let us denote the matrix generators for this representation by $i\upsilon^\alpha_{PQ}$ where υ^α_{PQ} are a set of N antisymmetric $r \times r$ matrices. [We assume that the group G has N generators.] Let the vacuum expectation value of φ_A be

$$\langle 0 |\, \varphi_A \,| 0 \rangle = F_A, \tag{59}$$

where F_A is the minimum point for the potential $V(\varphi)$; i.e. $V'(F_A) = 0$.

Notice that the vectors $v^{\alpha}_{AB}F_B(\alpha = 1, \ldots, N)$ span a subspace of dimension $n \leqslant r - 1$ (since $F_A F_B v_{AB} = 0$). It is possible that our vacuum expectation value still respects a subset of symmetries of G. Since the potential $V(\varphi) \equiv V(\varphi^{\dagger}\varphi)$ is invariant under the infinitesimal transformation

$$\varphi_A \rightarrow \varphi'_A = (\delta_{AB} - \varepsilon^{\alpha} v^{\alpha}_{AB})\varphi_B, \tag{60}$$

we must have

$$\frac{\partial V}{\partial \varphi_A} v^{\alpha}_{AB}\, \varphi_B = 0. \tag{61}$$

Differentiating again and using $(V')_{\varphi=F} = 0$,

$$\left(\frac{\partial^2 V}{\partial \varphi_A\, \partial \varphi_N} \right)_{\varphi=F} v^{\alpha}_{AB}\, F_B = 0. \tag{62}$$

The bracketed term is the 'mass matrix' for the fields and it annihilates all vectors in the n-dimensional subspace $v^{\alpha}F$. Thus there are n zero eigenvalues to the mass matrix – or, in other words, n massless bosons. In more complicated gauge theory models the structure of F_A plays a crucial role as it determines which residual symmetry is still left intact. One often says that a group G is broken 'to' another group G_1 (say) by a particular choice of F_A.[3]

5.7. The Salam–Weinberg Model

The popularity and acceptance of gauge field theories stem largely from the success of the Salam–Weinberg model which provides a unified treatment of electromagnetic and weak interactions. Though we may have reservations about the present concept of 'unification', the model does provide a workable, renormalized theory of weak interactions, the need for which had long been felt.

In this model, we work with the group SU(2) × U(1) and proceed as described in the previous section. We need three fields $\mathbf{A}_i$ to 'gauge' SU(2) and one field B_i for the U(1) part of the group (usually called the 'SU(2) sector' and the 'U(1) sector'). Since the group SU(2) × U(1) has two factors, we can have two coupling constants, say g and g' in the theory. The spontaneous symmetry breaking will give masses to three gauge fields and will leave one as massless. The massless field will be identified with the photon while the massive modes will describe heavy vector particles that mediate the weak interaction.

We have already introduced most of these ideas in the last section. The only extra ingredient we have to put in is the fermionic component of the

[3] Some specific choices for Higgs potential $V(\varphi)$, F_A, and the resulting group G_1 will be discussed in the next section.

theory. There should be four spin-$\frac{1}{2}$, Dirac fields $e(x)$, $\mu(x)$, $\nu_e(x)$, $\nu_\mu(x)$ representing the electron, muon, electron–neutrino, and muon–neutrino (and possibly more if there are more fermions[4]). We have to arrange these fields suitably – i.e. we have to choose the representation for the fields so that their observed features are well described by the theory. Clearly, this part of 'unification' is achieved in collaboration with the experimentalist rather than by any fundamental reasoning. We shall describe here the simplest scheme. Many variants can be constructed without deviating far from the experimental results.

Given a spinor field $\psi(x)$ we define the left-handed (ψ_L), and right-handed (ψ_R) components of the field by

$$\psi_L = \tfrac{1}{2}(1 - \gamma_5)\psi \quad \text{and} \quad \psi_R = \tfrac{1}{2}(1 + \gamma_5)\psi, \tag{63}$$

where γ_5 is the 4×4 matrix

$$\gamma_5 = \begin{pmatrix} 0 & I \\ I & 0 \end{pmatrix}. \tag{64}$$

The fields ψ_R and ψ_L are clearly eigenstates of the γ_5 operator with eigenvalues ± 1. Also, we note that any Dirac spinor ψ can be written as $\psi_L + \psi_R$. The reason for this rather perverse decomposition lies in the fact that weak interactions happen to distinguish (because of parity violation) between the right-handed and left-handed parts.

In the simplest variant of the Salam–Weinberg model, we assume that the multiplets

$$L_e \equiv \begin{bmatrix} \nu_e \\ e_L \end{bmatrix} ; \qquad L_\mu \equiv \begin{bmatrix} \nu_\mu \\ \mu_L \end{bmatrix} \tag{65}$$

transform under the 2×2 representation. On the other hand, $e_R = \frac{1}{2}(1 + \gamma_5)e \equiv R_e$ and $\mu_R \equiv R_\mu$ are assumed to transform as singlets under the group. This constitutes the fermionic part of the theory.

The SU(2) part is 'gauged' by a triplet of gauge fields $\mathbf{A}_i$ with coupling constant g and the U(1) will have the gauge field B_i with a coupling constant which (for the sake of future convenience) we take to be $g'/2$. The symmetry breaking is achieved through the scalar fields called the 'Higgs scalars' contained in a doublet,

$$\varphi = \begin{pmatrix} \varphi_1 \\ \varphi_2 \end{pmatrix}. \tag{66}$$

The 'Higgs potential' for the scalars φ_1, φ_2 is given by

$$V(\varphi^\dagger\varphi) = \mu^2\varphi^\dagger\varphi + \lambda(\varphi^\dagger\varphi)^2. \tag{67}$$

[4] For example, we can include the τ-lepton and its associated neutrino.

Here $\mu^2 < 0$ and the vacuum expectation value is taken to be

$$\langle 0|\ \varphi\ |0\rangle = \frac{1}{\sqrt{2}} \begin{pmatrix} 0 \\ v \end{pmatrix}; \qquad v^2 = -\frac{\mu^2}{\lambda}\ . \tag{68}$$

The total Lagrangian which includes all the above ingredients looks like the following:

$$\begin{aligned} L = &-\tfrac{1}{4}\,\mathbf{A}_{ik}\mathbf{A}^{ik} - \tfrac{1}{4}\,B_{ik}B^{ik} + \left| \left(\partial_i\varphi - i\,\frac{g'}{2}\,B_i\varphi - \frac{ig}{2}\,\boldsymbol{\tau}\cdot\mathbf{A}_i\varphi\right)\right|^2 - V(\varphi^\dagger\varphi) + \\ &+ \left[\bar{R}_e(i\gamma^i\partial_i - g'\not{B})R_e + \bar{L}_e\left(i\gamma^i\partial_i - \frac{g'}{2}\,\not{B} + g\,\frac{\boldsymbol{\tau}}{2}\,\not{\mathbf{A}}\right)L_e - \right. \\ &\left. - \,G_e(\bar{L}_eR_e\varphi + \varphi^\dagger\bar{R}_eL_e)\right] + \\ &+ [\text{Last term with } e \text{ replaced by } \mu]. \end{aligned} \tag{69}$$

The first two terms are the kinetic terms for the gauge fields. The third term is the kinetic energy term for the Higgs field with proper covariant derivative and the fourth term represents the Higgs potential. The rest of the terms represent the fermionic part. (Notice that we have made the right-handed and left-handed parts to have different interactions.) The electronic part and muonic part couple to the scalar field with coupling constants G_e and G_μ.

To decide on the physical fields, we shall write the Higgs doublet in the form

$$\varphi(x) = \begin{pmatrix} 0 \\ \dfrac{v + \varrho}{\sqrt{2}} \end{pmatrix} \exp i\,\frac{\boldsymbol{\tau}}{2}\cdot\boldsymbol{\xi}, \tag{70}$$

where $\varrho(x)$ and $\boldsymbol{\xi}(x)$ are physical fields with zero vacuum expectation values. The complex doublet φ has four degrees of freedom, which is the same as that of $(\varrho, \boldsymbol{\xi})$. We shall absorb the phase by an SU(2) transformation,

$$\varphi \to \varphi' = \begin{pmatrix} 0 \\ \dfrac{v + \varrho}{\sqrt{2}} \end{pmatrix}, \tag{71}$$

$$\frac{\boldsymbol{\tau}}{2}\cdot\mathbf{A}^i \to \frac{\boldsymbol{\tau}}{2}\cdot\mathbf{A}'^i = \mathrm{e}^{-i\boldsymbol{\xi}\cdot\boldsymbol{\tau}/2}\left(\frac{i}{g}\,\partial^i + \frac{\boldsymbol{\tau}}{2}\cdot\mathbf{A}^i\right)\mathrm{e}^{+i\boldsymbol{\xi}\cdot\boldsymbol{\tau}/2} \tag{72}$$

$$\left.\begin{aligned} &L \to L' = \mathrm{e}^{-i\boldsymbol{\xi}\cdot\boldsymbol{\tau}/2}\,L \\ &B' = B\,; \qquad R' = R \end{aligned}\right\}. \tag{73}$$

In terms of these transformed fields, the mass spectrum of the theory is obvious: the Lagrangian is,

$$L = -\tfrac{1}{4}\mathbf{A}_{ik}\cdot\mathbf{A}^{ik} - \tfrac{1}{4}B_{ik}B^{ik} + \Big[\bar{e}_R(i\gamma^i\partial_i - g'\not{B})e_R +$$

$$+ \bar{L}_e\Big(i\gamma^i\partial_i - \frac{g'}{2}\not{B} + g\frac{\tau}{2}\cdot\mathbf{A}\Big)L_e - G_e\frac{v+\varrho}{2}(\bar{e}_Re_L + \bar{e}_Le_R) + (e \leftrightarrow \mu)\Big] +$$

$$+ \tfrac{1}{2}\partial_i\varrho\,\partial^i\varrho + \frac{1}{8}(v+\varrho)^2\Big[|(g'B_i - gA_i^3)|^2 + g^2(A_i^1A_1^i + A_i^2A_2^i\Big] - V. \quad (74)$$

The Higgs field has the mass $\sqrt{|2\mu^2|}$. The electron and muon have acquired the masses $m_e = G_e v/\sqrt{2}$ and $m_\mu = G_\mu v/\sqrt{2}$, respectively; the charged vector field

$$W_i^{\pm} = \frac{1}{\sqrt{2}}(A_i^1 \mp iA_i^2) \quad (75)$$

has mass $m_w = vg/2$. Finally, the quadratic form in A_i^3 and B_i is diagonalized by the choice

$$Z_i \equiv (g^2 + g'^2)^{1/2}(-gA_i^3 + g'B_i),$$
$$A_i \equiv (g^2 + g'^2)^{-1/2}(gB_i + g'A_i^3), \quad (76)$$

leading to a neutral Z boson of mass $M_z = v/2(g^2 + g'^2)^{1/2}$ and to a vector boson of zero mass. We identify the massless vector field A_i with the photon. The two charged fields $W_i^{\pm}$ and the neutral Z_i are the massive vector bosons that mediate the short-range weak interactions. The Fermi coupling constant is given by

$$\frac{G}{\sqrt{2}} = \frac{g^2}{8M_w^2} = \frac{1}{2v^2}\ . \quad (77)$$

The known value of G can be used to put bounds on the masses of vector bosons.

The Salam–Weinberg model provides a renormalizable theory of weak interactions. The weak interactions are mediated by the exchange of vector bosons described by the fields W^+, W^-, and Z. Notice that among these three, the Z particle is neutral, i.e. carries no charge. The effects predicted because of the exchange of the Z boson constituted one of the earliest experimental tests of the Salam–Weinberg theory. The verification of these effects, as well as the discovery of the vector bosons in the laboratory, may be considered a triumph for this model.

At a fundamental level the Salam–Weinberg model still leaves many questions unanswered. Why are weak interactions parity violating? Why does m_μ/m_e have the particular value? Why is there no right-handed neutrino? etc., etc. Only future research can give simple answers to these questions.

Encouraged by the success of this model, particle physicists have considered more general structures which incorporate the strong interactions into the unification scheme. One of the popular models is based on the group SU(5), which is the multiplication group of 5×5 unitary complex matrices of unit determinant. It is easy to see that SU(n) will have $(n^2 - 1)$ generators, so that SU(5) has 24. In the gauge theory based on SU(5) we would require 24 gauge fields.

We can denote these gauge fields by A_i^N with N running over the set (1, 2, 3, . . . , 24). It is more convenient to write these gauge fields as a 5×5 matrix A_{iQ}^P with a constraint Tr $A_i = 0$. This constraint expresses one of the 25 entries in the matrix in terms of the others, leaving 24 independent components. The gauge bosons comprise of eight 'gluons' which are responsible for strong interactions, three vector bosons for the weak interactions, one photon for the electromagnetic interaction and 12 new X bosons. These new X bosons are capable of violating the baryon number conservation.

The symmetry breaking is achieved through a Higgs field which transforms under the adjoint representation of SU(5). Thus we need 24 Higgs components, which are represented as a matrix

$$\Phi_b^a = \sum_{A=1}^{24} (\lambda_A)_b^a \varphi^A \tag{78}$$

(λ_A are the generators of SU(5)). It is usual to take the Higgs potential in the form

$$V(\Phi) = -\frac{\mu^2}{2} \mathrm{Tr}(\Phi^2) + \tfrac{1}{4} a \left[\mathrm{Tr}(\Phi^2)\right]^2 + \tfrac{1}{2} b \,\mathrm{Tr}(\Phi^4) + \tfrac{1}{3} c \,\mathrm{Tr}\, \Phi^3. \tag{79}$$

The symmetry breaking on the minima of this potential and requires detailed examination. It can be shown that for $b < 0$, diag$(v, v, v, v, -4v)$ will minimize V while for $b > 0$ we need diag $(v, v, v, -\frac{3}{2}v, -\frac{3}{2}v)$. In the first case we are still left with a $SU(4) \times U(1)$ symmetry while the second case has a $SU(2) \times SU(3) \times U(1)$ symmetry. The second choice is usually preferred. We shall have occasion to discuss this model later.

We emphasize the fact that the scalar field φ, usually called the 'Higgs field' plays a crucial role in all the above discussions. We have assumed *a priori* a potential for $V(\varphi)$ with minima at nonzero values of φ. Thus, the whole scheme depends critically on the form of $V(\varphi)$. In the tree level of quantum theory, at zero temperature, the form of $V(\varphi)$ can be classically prescribed; this is what we have done in this section. However, two kinds of quantum effects alter the form of $V(\varphi)$. First, closed loops produce quantum corrections to $V(\varphi)$ which can alter the nature of the minima. Second, the potential picks up corrections due to temperature effects when the field system is kept in a heat bath. This feature is an idle curiosity as far as laboratory physics is concerned, but when we consider particle physics in the Early Universe, such effects play an extremely important role. In the next section we shall take up a closely related topic called 'dynamical symmetry breaking'.

5.8. The Coleman–Weinberg Mechanism

In order to produce the symmetry breaking, we had to introduce in our discussion a potential $V(\varphi)$ with nontrivial classical minima. While there is nothing in the theory which forbids this, we must admit certain unnaturalness in this choice. It would have been much better if we could produce a potential $V(\varphi)$ (with nontrivial minima) through some dynamical argument. It was shown by S. Coleman and E. Weinberg that in the renormalized version of 'massless' scalar electrodynamics the scalar field picks up an arbitrary vacuum expectation value $\langle\varphi\rangle$ even though classically $\langle\varphi\rangle$ is zero. We shall now indicate their line of argument. The massless scalar electrodynamics is described by the Lagrangian

$$\mathscr{L} = -\tfrac{1}{4} F^{ik}F_{ik} + [(\partial_i - igA_i)\varphi^\dagger][(\partial + igA^i)\varphi] - \frac{\lambda_0}{6}(\varphi\varphi^\dagger)^2. \tag{80}$$

The first two terms describe a charged massless scalar field interacting with the electromagnetic field. The last term introduces a quartic self-coupling for the φ field with a bare coupling constant λ_0. At the classical level, we clearly have

$$\langle 0|\ \varphi\ |0\rangle = 0 \quad \text{and} \quad \langle 0|\ A_i\ |0\rangle = 0. \tag{81}$$

This is in sharp contrast with the models in the previous section where a $(+\mu^2\varphi^2)$ term produced a minimum for $V(\varphi)$ at $\varphi \neq 0$. Here the potential for the scalar field, in the classical level is just

$$V(\varphi) = \frac{\lambda_0}{6}(\varphi\varphi^\dagger)^2. \tag{82}$$

We now want to consider an effective potential for the scalar field that takes into account the interaction with the A_i. We would like to examine the possibility that the radiative corrections from quantum fluctuations of the vector field produce an effective potential with nonzero minima.

At first sight, the reader may think that this is impossible. The Lagrangian in (80) does not contain any dimensional coupling constants. Thus it may be considered that the effective potential, whatever its form, cannot have a dimensional constant. Since φ has a nontrivial dimension, we cannot see *a priori* where a new scale can enter into the theory.

The answer has to do with the process of renormalization. We have already seen in Chapter 4 that renormalization can introduce an arbitrary scale into the theory. Thus, it is possible for $\langle\varphi\rangle$ to be nonzero, but arbitrary when radiative corrections are taken into account. We shall now indicate the derivation of this result.

By making a gauge transformation, we can completely eliminate the phase part of φ. After absorbing a factor of $2^{-1/2}$, the Lagrangian can be written (in this specific gauge) as

$$L = -\tfrac{1}{4}F^{ik}F_{ik} + \tfrac{1}{2}q^2\varphi^2 A_i A^i + \tfrac{1}{2}\partial_i\varphi\,\partial^i\varphi - \frac{\lambda_0}{4!}\,\varphi^4. \tag{83}$$

The action corresponding to this Lagrangian is

$$S[A^i, \varphi] = \int \left[\tfrac{1}{2}A^i(\Box + q^2\varphi^2)A_i - \tfrac{1}{2}\varphi\Box\varphi - \frac{\lambda_0}{4!}\,\varphi^4\right] \mathrm{d}^4x. \tag{84}$$

The effective action for the scalar field can be obtained by performing the path integral over A_i [compare with our discussion in Section 4.7]

$$\exp iS_{\text{eff}}(\varphi) = \int \exp iS[A_i, \varphi]\,\mathcal{D}A_i. \tag{85}$$

Each component of the vector field will give, on integration the usual determinant factor

$$\begin{aligned}[\det(\Box + q^2\varphi^2)]^{-1/2} &= \exp[-\tfrac{1}{2}\ln\det(\Box + q^2\varphi^2)] \\ &= \exp[-\tfrac{1}{2}\operatorname{Tr}\ln(\Box + q^2\varphi^2)].\end{aligned} \tag{86}$$

Since there are only three independent components, we obtain

$$S_{\text{eff}}[\varphi] = -\int \left(\tfrac{1}{2}\varphi\Box\varphi + \frac{\lambda_0}{4!}\varphi^4\right) \mathrm{d}^4x - \tfrac{3}{2}\operatorname{Tr}\ln(\Box + q^2\varphi^2). \tag{87}$$

By an analysis similar to the one in Section 4.6 we can show that to the lowest order

$$\begin{aligned}V_{\text{eff}}(\varphi) &= \frac{\lambda_0}{4!}\,\varphi^4 + \frac{3}{2}\int \ln(k_E^2 + q^2\varphi^2)\,\frac{\mathrm{d}^4k_E}{(2\pi)^4} \\ &= \varphi^4\left[\frac{\lambda}{4!} + \frac{3q^4}{64\pi^2}\left(-\frac{1}{2} + \ln\frac{q^2\varphi^2}{\mu^2}\right)\right].\end{aligned} \tag{88}$$

Here λ is a running coupling constant at the renormalization scale μ. To find the minimum value of the potential we compute

$$V'_{\text{eff}}(\varphi) = 4\varphi^3\left(\frac{\lambda}{4!} + \frac{3q^4}{64\pi^2}\ln\frac{q^2\varphi^2}{\mu^2}\right), \tag{89}$$

which vanishes for

$$\frac{3q^4}{64\pi^2}\ln\frac{q^2\langle\varphi\rangle^2}{\mu^2} = -\frac{\lambda}{4!}. \tag{90}$$

Eliminating μ for $\langle\varphi\rangle$ we can write the effective potential as

$$V_{\text{eff}}(\varphi) = \frac{3q^4\varphi^4}{64\pi^2}\left(\ln\frac{\varphi^2}{\langle\varphi\rangle^2} - \tfrac{1}{2}\right). \tag{91}$$

As in any renormalized theory, this potential involves an arbitrary scale

$\langle\varphi\rangle$. Note that $V_{\text{eff}}(\varphi)$ has a minimum at $\varphi = \langle\varphi\rangle$, as shown in Figure 5.2. Physical excitations around this minimum give the scalar field the mass

$$m_s^2 = \frac{3q^4}{8\pi^2}\langle\varphi\rangle^2. \tag{92}$$

The vector field acquires the mass

$$m_v^2 = q^2\langle\varphi\rangle^2. \tag{93}$$

In a realistic theory, either of these relationships can be used to eliminate the scale $\langle\varphi\rangle$ in terms of the known physical masses in the theory.

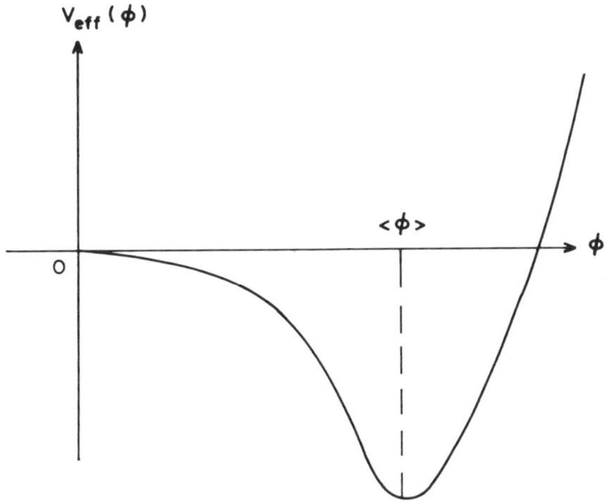

Fig. 5.2. The Coleman–Weinberg effective potential of Equation (91). Note that the curve is very flat near $\varphi = 0$.

This effective potential, called the ‘Coleman–Weinberg potential’, will be used in Chapter 10 to discuss the dynamics of the Early Universe. It clearly illustrates the importance of radiative corrections in the theory.

5.9. The Gauge Field as a Physical System

In our discussions so far, we have always considered gauge fields as part of an interacting system of fields. This is in the spirit of Sections 5.1 and 5.2 where we introduced the gauge fields as subsidiary entities which are required to maintain local gauge invariance. However, it is quite possible to think of the gauge field as a complete physical system by itself. In fact, we consider the free electromagnetic field as a physical system by itself.

In general, a gauge theory based on some Lie group G will be described by a gauge field A_i^M. Here M takes values $1, 2, \ldots, N$, where N is the number of generators for G (which is also sometimes called the dimension of G). The index i denotes the vector nature of A_i^N under space–time Lorentz transformations. The Lagrangian describing the system is taken to be

$$L = -\tfrac{1}{4} F_{ik}^N F_N^{ik} \tag{94}$$

with

$$F_{ik}^N = \partial_i A_k^N - \partial_k A_i^N - g C^{NMP} A_{\mathrm{M}i} A_{PK} \ . \tag{95}$$

Such a theory was first discussed in detail by C. N. Yang and R. L. Mills in 1954. A gauge field based on a nonabelian gauge group G is often called a 'Yang–Mills field'. We have described the formalism for a Yang–Mills field in Section 5.3. We remind the reader that C^{ABC} are the structure constants of the group G and that we do not distinguish between upper and lower indices in the internal symmetry space (i.e. $A_i^N = A_{Ni}$, etc.). The theory described by (94), (95) is invariant under the gauge transformations

$$A_i \to A_i' \equiv U A_i U^{-1} - \frac{i}{g} U \partial_i U^{-1}, \tag{96}$$

with

$$A_i = A_i^N T_N \tag{97}$$

and

$$U = \exp[-iT_A \lambda^A(x)]; \tag{98}$$

where T_A are the generators of the group G. We shall assume that the generator matrices T_A are so normalized as to satisfy

$$\mathrm{Tr}(T_A T_B) = \tfrac{1}{2} \delta_{AB}. \tag{99}$$

(It is shown in Appendix B that this can always be done in the adjoint representation.) Then, we can write (94) as

$$L = -\tfrac{1}{4} F_{ik}^A F_A^{ik} = -\tfrac{1}{2} \mathrm{Tr}(F_{ik} F^{ik}), \tag{100}$$

with

$$F_{ik} \equiv F_{ik}^A T_A. \tag{101}$$

When the gauge group G is U(1), the dimension of the group is 1 and the structure constants vanish. In this case our formalism reduces to ordinary electrodynamics. In general, however, the theory will be much more complicated. The field equations of the theory can be obtained by varying A_i^N in the Lagrangian. This gives

$$\partial_i F^{ik} + ig[A_i, F^{ik}] = 0. \tag{102}$$

From definition (95) it follows that

$$\partial_i \tilde{F}^{ik} + ig[A_i, \tilde{F}^{ik}] = 0, \tag{103}$$

where, the 'dual' of F^{ik} is defined by

$$\tilde{F}^{ik} = \varepsilon^{ik\ell m} F_{\ell m}. \tag{104}$$

Equations (102) and (103) are generalizations of the Maxwell equations of an electromagnetic field. Unlike the equations of electrodynamics, these equations are nonlinear. Thus, we cannot construct new solutions by a linear superposition of known solutions.

It is conventional to define a generalized derivative for gauge fields in the adjoint representation as

$$\nabla_i \equiv \partial_i + ig[A_i, \quad]. \tag{105}$$

Then Equations (102) and (103) can be written in a more concise form as

$$\nabla_i F^{ik} = 0; \qquad \nabla_i \tilde{F}^{ik} = 0. \tag{106}$$

Example 5.6. It is instructive to compare electromagnetism with its nonabelian counterpart. To do this, define the generalized electric and magnetic field matrices

$$\mathbf{E} = \mathbf{E}^A T_A; \qquad \mathbf{B} = \mathbf{B}^A T_A.$$

Note that the bold print here refers to the ordinary vector nature in 3-space (i.e. $\mathbf{E} = (E_x, E_y, E_z)$ each of which is a matrix: $(E_x)_{ij} = E_x^A(\tau_A)_{ij}$). These fields are related to F_{ik}^A by the equations

$$E^\mu = F^{\mu 0}; \qquad B^\mu = -\tfrac{1}{2}\varepsilon^{\mu\alpha\beta} F_{\alpha\beta} \quad (\alpha, \beta, \mu, \nu, = 1, 2, 3).$$

As in the case of electromagnetism, $\mathbf{E}$ and $\mathbf{B}$ are expressible in terms of the potential $A_i = (A^0, -\mathbf{A})$, with

$$E_\mu = \partial_\mu A_0 - \partial_0 A_\mu + ig[A_\mu, A_0],$$

$$B_\mu = \left[(\partial_\alpha A_\beta - \partial_\beta A_\alpha) - ig[A_\alpha, A_\beta]\right] \varepsilon_\mu^{\alpha\beta}.$$

The Lagrangian density can be expressed in terms of $\mathbf{E}$ and $\mathbf{B}$ as

$$\tfrac{1}{4} F_{ik}^A F_A^{ik} = \tfrac{1}{2}(\mathbf{B}_A \cdot \mathbf{B}^A - \mathbf{E}_A \cdot \mathbf{E}^A).$$

The field equations become equations for $\mathbf{E}_A$ and $\mathbf{B}_A$. Of particular interest is the divergence equation

$$\nabla \cdot \mathbf{B}^A = \tfrac{1}{2}\, g C^A{}_{BC}\, \nabla \cdot (\mathbf{A}^B \times \mathbf{A}^C)$$

which shows that, in general, 'magnetic charge' can exist in theory. We shall later construct explicit solutions with a nonzero magnetic charge. □

The quantum version of the above theory holds considerable theoretical interest for the following reasons:

(i) It is believed that strong interaction dynamics are described by a gauge theory based on the SU(3) group. Thus, from the purely pragmatic point of view quantum gauge field theory is required to understand hadron physics. Since such a theory is renormalizable, we can use perturbation expansions and Feynman diagrams to evaluate cross sections for various processes. As we are not interested here in perturbative approaches, we shall not discuss the topic of perturbative quantum chromodynamics.

(ii) The gauge field theory is also of interest because of its similarity with the gravitational field. Pure Yang–Mills theroy and pure gravity have a great deal of structural similarity. (However, gravity is not perturbatively renormalizable; thus the results of perturbative quantum chromodynamics are not of direct relevance to gravity.) By analysing various nonperturbative quantum effects in the Yang–Mills gauge theory, we hope to obtain valuable insights into the structure of quantum gravity.

In the next few sections we shall discuss some simple nonperturbative features of gauge theories. Our aim is not to present a complete discussion but to indicate to the reader how topological effects can be important in a quantum gauge theory.

5.10. The Gauge Field Vacuum and Instantons

In Section 4.6 we discussed the nonperturbative approach in quantum theory. The Euclidean value of the action was shown to be of crucial importance in nonperturbative calculations. It turns out that the Euclidean version of the Yang–Mills theory possesses some interesting surprises. These surprises come in the form of the nontrivial structure of the vacuum of a nonabelian gauge field.

Naïvely speaking, we expect the gauge field vacuum to be a zero field configuration

$$F^A_{ik}(x) = 0, \tag{107}$$

where the x^i now denote the coordinates in Euclidean four-dimensional space. This means that the potential A^A_i must be expressible as a pure gauge:

$$A_i = -\frac{i}{g}\, U\,\partial_i U^{-1}. \tag{108}$$

(We are again using the matrix notation; $A_i \equiv A^A_i \tau_A$, etc.) $U(x)$ in (108) is an element of the gauge group which we shall take to be the SU(2) group. We might think that this potential A_i can be gauge transformed to the trivial $A_i = 0$, and indeed this would have been true for an Abelian gauge field (like the electromagnetic field). But for SU(2) and other gauge groups, it is possible to distribute the $U(x)$ into many 'equivalence classes' such that members of two different classes are *not* connected by a continuous transformation. These equivalence classes are characterized by an integer n called the 'winding number' of the transformation.

When considering the dynamics of the Yang–Mills field it is necessary to take note of this feature. In the path integral formalism, the central object of interest is the Euclidean transition amplitude (cf. Chapter 4)

$$\langle \text{vac}|\text{vac}\rangle = \int \exp\{-S[A_i]\}\ \mathcal{D}A_i, \tag{109}$$

where $S[A_i]$ is the Euclidean value for the Yang–Mills action. Clearly, we are only interested in those configurations of A_i for which S is (i) finite and (ii) minimum. Let us now see what these configurations are.

Notice that, in Euclidean space

$$\text{Tr}\ [(F_{ik} - \tilde{F}_{ik})\ (F^{ik} - \tilde{F}^{ik})] \geq 0, \tag{110}$$

so that

$$\text{Tr}(F_{ik}F^{ik} + \tilde{F}_{ik}\tilde{F}^{ik}) \geq 2\ \text{Tr}(F_{ik}\tilde{F}^{ik}). \tag{111}$$

Since

$$\text{Tr}(F_{ik}F^{ik}) = \text{Tr}(\tilde{F}_{ik}\tilde{F}^{ik}), \tag{112}$$

we get

$$\text{Tr}(F_{ik}F^{ik}) \geq \text{Tr}\ F^{ik}\tilde{F}^{ik}. \tag{113}$$

It is straightforward to show that

$$\tfrac{1}{4}\ \text{Tr}\ \tilde{F}_{ik}F^{ik} = \partial_i X^i, \tag{114}$$

where

$$X^i = \varepsilon^{iabc}\ \text{Tr}\left[\tfrac{1}{2} A_a \partial_b A_c + \frac{i}{3}\ g A_a A_b A_c\right]. \tag{115}$$

Example 5.7. This relation can be obtained by direct index gymnastics. We start with

$$\tfrac{1}{4}\ \text{Tr}\ \tilde{F}_{ik}F^{ik} = \tfrac{1}{2}\varepsilon^{ikmn}\ \text{Tr}[(\partial_i + igA_i)A_k(\partial_m + igA_m)A_n]$$

$$= \tfrac{1}{2}\varepsilon^{ikmn}\ \text{Tr}[(\partial_m A_n)\ (\partial_i A_k) + 2ig(\partial_m A_n)A_iA_k - g^2 A_iA_kA_mA_n]\ .$$

The last term being symmetric in m, n does not contribute. In the other two terms, write

$$(\partial_m A_n)\ (\partial_i A_k) = \partial_m(A_n\partial_i A_k) - A_n(\partial_m\partial_i A_k)$$

Now,

$$\varepsilon^{ikmn}\ \text{Tr}((\partial_m A_n)A_iA_k) = \varepsilon^{ikmn}\ \text{Tr}\{\partial_m(A_nA_iA_k) - A_n(\partial_m A_i)A_k - A_nA_i\partial_m A_k\}$$

$$= \varepsilon^{ikmn}\ \text{Tr}\{\partial_m(A_nA_iA_k)\} - 2\varepsilon^{ikmn}\ \text{Tr}((\partial_m A_n)A_iA_k),$$

so that

$$\varepsilon^{ikmn}\ \text{Tr}((\partial_m A_n)A_iA_k) = \frac{1}{3}\ \varepsilon^{ikmn}\partial_m\{\text{Tr}(A_nA_iA_k)\}\ .$$

Substituting back we get the desired result,

$$\tfrac{1}{4}\,\mathrm{Tr}\,\tilde{F}_{ik}F^{ik} = \varepsilon^{mnik}\partial_m\{\mathrm{Tr}(\tfrac{1}{2}A_n\partial_iA_k + \tfrac{1}{3}\,igA_nA_iA_k)\}. \qquad \square$$

It is usual practice to define a quantity

$$q \equiv \frac{g^2}{16\pi^2}\int \mathrm{d}^4x\;\mathrm{Tr}(\tilde{F}_{ik}F^{ik}), \tag{116}$$

which, because of (114), can be written as

$$q = \frac{g^2}{4\pi^2}\int \mathrm{d}^4x\;\partial_jX^j = \frac{g^2}{4\pi^2}\int_{S^3}\mathrm{d}S_jX^j, \tag{117}$$

where the surface integral is taken over the 3-sphere at the Euclidean infinity. This quantity q is a topological invariant (called the 'topological charge') characterizing the field distribution. From the inequality

$$S = \tfrac{1}{4}\int \mathrm{d}^4x\;\mathrm{Tr}(F_{ik}F^{ik}) \geq \int \mathrm{d}^4x\;\partial_jX^j = \frac{4\pi^2}{g^2}\,q, \tag{118}$$

we see that the minimum value for S is determined by the topological charge q. On the other hand, q is completely determined by the distribution of fields at the Euclidean infinity and is independent of the values at finite X^j. Furthermore, we are only interested in field configurations that produce finite value for the action. This implies that, at the Euclidean infinity

$$\operatorname*{Lim}_{r\to\infty}\; F^{ik} \to \mathrm{O}(r^{-3}) \quad \text{with } r = (x^0)^2 + |\mathbf{x}|^2\;; \tag{119}$$

and hence

$$A_i \to -\;\frac{i}{g}\;U\partial_iU^{-1} \equiv -\;\frac{i}{g}\;\lambda_i\;(\text{say}). \tag{120}$$

For this form of A_i, (115) gives

$$X^i = \frac{1}{6g^2}\;\varepsilon^{iabc}\,\mathrm{Tr}(\lambda_a\lambda_b\lambda_c) \tag{121}$$

and

$$q = \frac{1}{24\pi^2}\int_{S^3}\mathrm{d}S_i\;\varepsilon^{imnp}\,\mathrm{tr}(\lambda_m\lambda_n\lambda_p). \tag{122}$$

This integral can be evaluated conveniently by parametrizing the S^3 by three angles θ_1, θ_2, θ_3. Choosing the associated unit (four) vectors as $\hat{\theta}_1^i$, $\hat{\theta}_2^i$, $\hat{\theta}_3^i$ with

$$\hat{\theta}_A^i\hat{\theta}_{iB} = \delta_{AB}, \tag{123}$$

we get

$$q = \frac{1}{24\pi^2}\int_{S^3}\mathrm{d}S\;\varepsilon^{ABC}\,\mathrm{Tr}(\lambda_A\lambda_B\lambda_C). \tag{124}$$

Here $\lambda_A = U\partial_A U^{-1}$ with $\partial_A = \theta_A^i\, \partial_i$. The differential d$S$ integrates on the 'surface' (i.e. $\mathrm{d}S_i = n_i\, \mathrm{d}S$) of S^3.

Thus the minimum action configurations are completely characterized by the gauge group element $U(x)$ which determines the behaviour of A_i at spatial infinity. We shall now show that q is invariant under a continuous gauge transformation. Consider the infinitesimal change in the group element U

$$U \rightarrow U(1 - i\alpha); \qquad U^{-1} \rightarrow (1 + i\alpha)U^{-1}. \tag{125}$$

Now

$$\delta\lambda^A = \delta(U\partial^A U^{-1}) = (\delta U)\,(\partial^A U^{-1}) + U\partial^A(\delta U^{-1}) = iU(\partial^A\alpha)U^{-1}, \tag{126}$$

so that,

$$\begin{aligned}\delta q &= \frac{1}{8\pi^2}\int \mathrm{d}S\; \varepsilon^{ABC}\; \mathrm{Tr}(\lambda_A\lambda_B\, \delta\lambda_C)\\ &= \frac{i}{8\pi^2}\int \mathrm{d}S\; \varepsilon^{ABC}\; \mathrm{Tr}[(U\partial_A U^{-1})\,(U\partial_B U^{-1})\,(U\partial_C\alpha U^{-1})]\,.\end{aligned} \tag{127}$$

Integrating by parts, we get

$$\begin{aligned}\delta q = \frac{i}{8\pi^2}\Bigg\{&\varepsilon^{ABC}\int \mathrm{d}S\; \partial_C\; \mathrm{Tr}[(\partial_A U^{-1})\,(\partial_B U^{-1})\alpha] - \\ &-\int \mathrm{d}S\; \varepsilon^{ABC}\; \mathrm{Tr}[(\partial_A\partial_B U^{-1})\,(\partial_C U)\alpha + (\partial_A U^{-1})\,(\partial_B\partial_C U^{-1})\alpha]\Bigg\}.\end{aligned} \tag{128}$$

The first term vanishes because S^3 does not have a boundary and the second term vanishes because of the antisymmetry of ε^{ABC}. Thus the value of q does not change under continuous gauge transformations. In other words, the class of all $U(x)$ which are related by continuous gauge transformations lead to the same value for q. For example, the class of all $U(x)$ connected to the identity $U = 1$, gives zero topological charge q, and conforms to our naïve idea of vacuum.

Curiously enough, this is not the whole story. There do exist group elements $U(x)$ which are not connected to the identity element by continuous transformations. For $x \,\varepsilon\, S^3$, the element $U(x)$ may be considered to be a mapping of S^3 to the group manifold SU(2). However, the group manifold of SU(2) itself is the Euclidean unit sphere S^3 (see Section 5.3). Thus each $U(x)$ represents the mapping of S^3 onto S^3. Such mappings fall into what is known as 'homotopy classes' which are labelled by an integer n. This integer denotes the number of times spatial S^3 is covered by the mapping with the group manifold of SU(2).

Those with a healthy suspicion for abstract topics like 'homotopy classes' may be convinced by the following explicit construction. Consider the elements of the group

$$U_n(x) = [P(x)]^n, \quad \mathrm{n} = 0, \pm 1, \pm 2, \ldots \tag{129}$$

where

$$P(x) = \frac{1}{r}\,(x^0 + \mathbf{x}\cdot\tau); \quad (\mathbf{x})^A = x^\mu; \quad (\tau)^A = \tau^A. \tag{130}$$

By substituting into (122), we can directly show that

$$q[U_n] = n. \tag{131}$$

The evaluation is indicated in Example 5.8.

Example 5.8. The integral for q in (124) can be evaluated by making use the symmetries of the situation. First, note that, for U_n given by (129),

$$q[U_n] = nq[U_1].$$

So we only have to calculate

$$q_1 \equiv q[U_1] = \frac{1}{24\pi^2}\int_{S^3} \mathrm{d}S\, \mathrm{e}^{ABC}\, \mathrm{Tr}(\lambda_A\lambda_B\lambda_C),$$

with

$$\lambda^A = P(x)\partial^A P(x)^{-1}.$$

Now the integral for q_1 extends over all x on S^3. But going from one value of x to another, on S^3, is a rotation of S^3 which can always be undone by a continuous (gauge) change of P. Under such a continuous change the integrand changes only by a total divergence which makes no contribution to the integral. Thus q_1 is given by the value of the integrand at any point of S^3 multiplied by the volume of S^3, which is $2\pi^2$. Choosing the point to be (1, 0, 0, 0) in the Euclidean 4-sphere, we get,

$$P\partial_A P^{-1} = 1 \times i\tau_A.$$

Therefore,

$$q_1 = (2\pi^2)\left(-\frac{i}{24\pi^2}\right)\varepsilon^{ABC}\,\mathrm{Tr}(\lambda_A\lambda_B\lambda_C) = 1.$$

Since $q_n = nq$, we get result (131) quoted above. □

We have thus demonstrated two facts:

(i) The topological charge q is invariant under continuous gauge transformations.

(ii) There exist gauge group elements of the form of (129) leading to $q = n$ for $n = 0, \pm 1, \pm 2, \ldots$

It immediately follows that there does not exist a continuous gauge transformation that will take $U = 1$ to, say, $U = P(x)$ of (130). Nevertheless, all these configurations correspond to vanishing F_{ik} and can make rightful claims for being the vacuum. Thus the vacuum of nonabelian gauge theory has a nontrivial topological structure.

We can parametrize these different vacua of the theory by the winding number n. When treated as a quantum ground state, we must allow for tunnelling between these different vacua. In Chapter 3 we discussed the double-well potential which has two classical ground states that are connected by an 'instanton' solution. Similarly, here also we can find instanton solutions to the Euclidean equations of motion that represent tunnelling between different gauge vacua. Such solutions can be obtained in the following way:

Note first that the pure configuration for $q = 1$ is given (using (130), (120)) by

$$A_i = \frac{1}{g} \frac{\tau_{ik} x^k}{r^2}, \tag{132}$$

where,

$$\tau_k \equiv (\tau, i); \qquad \tau_{ik} = \mathrm{i}(\tau_i \tau_k^{\dagger} - \delta_{ik}). \tag{133}$$

Consider now a different configuration given by

$$A_i = \frac{1}{g} \frac{\tau_{ik} x^k}{r^2} f(r^2), \tag{134}$$

where

$$f(0) = 0 \quad \text{and} \quad f(\infty) = 1. \tag{135}$$

Substitution into the vacuum field equations leads to the condition

$$f' - \frac{f(1-f)}{r^2} = 0, \tag{136}$$

which is solved by

$$f(r^2) = \frac{r^2}{\varrho^2 + r^2}, \tag{137}$$

where ϱ is an arbitrary parameter. With this choice for $f(r^2)$ in (134) the field tensor has the form

$$F_{ik} = \tilde{F}_{ik} = \frac{2}{g} \frac{\varrho^2}{(\varrho^2 + r^2)^2} \tau_{ik}. \tag{138}$$

We shall now show that this solution actually represents a tunnelling between two different gauge vacua in Euclidean time. Note that this instanton solution can be written as

$$\mathbf{A}(\mathbf{x}, x^0) = \frac{1}{g} \frac{i[\boldsymbol{\tau}(\boldsymbol{\tau} \cdot \mathbf{x}) - \mathbf{x}] + x^0}{|\mathbf{x}|^2 + (x^0)^2 + \varrho^2}, \tag{139}$$

$$A^0(\mathbf{x}, x^0) = -\frac{1}{g} \frac{\tau \cdot \mathbf{x}}{|\mathbf{x}|^2 + (x^0)^2 + \varrho^2}. \tag{140}$$

We make a transformation to a gauge in which $A^0 = 0$. This can be done through the group element

$$U(\mathbf{x}, x^0) = -\exp\left[\frac{-i\pi\mathbf{x}\cdot\tau}{(x^2+\varrho^2)^{1/2}}\,\varphi(x^2, x^0)\right] \tag{141}$$

with

$$\varphi(x^2, x^0) = \frac{1}{2} + \frac{1}{\pi}\tan^{-1}\frac{x^0}{(x^2+\varrho^2)^{1/2}}\;. \tag{142}$$

In the new gauge, $A^0 = 0$ and $\mathbf{A}$ is given by

$$\mathbf{A}^{(\text{new})}(\mathbf{x}, x^0) = U(\mathbf{x}, x^0)\mathbf{A}^{(\text{old})}U^{-1}(\mathbf{x}, x^0) + \frac{i}{g}\,U(\mathbf{x}, x^0)\nabla\,U^{-1}(\mathbf{x}, x^0). \tag{143}$$

Notice that

$$U(\mathbf{x}, x^0) = \begin{cases} v(\mathbf{x}) & \text{for } x^0 \to +\infty, \\ 1 & \text{for } x^0 \to -\infty, \end{cases}$$

where

$$v(\mathbf{x}) = -\exp\frac{-i\pi\tau\cdot\mathbf{x}}{(x^2+\pi^2)^{1/2}}\;. \tag{144}$$

Thus, our instanton field has the asymptotic behaviour

$$\begin{aligned} \mathbf{A}(\mathbf{x}, x^0) &\to 0 \quad \text{for } x^0 \to -\infty, \\ &\to \frac{i}{g}\,v(\mathbf{x})\nabla\,v^{-1}(\mathbf{x}) \quad \text{for } x^0 \to +\infty. \end{aligned} \tag{145}$$

We can verify that this $v(\mathbf{x})$ is nothing but the $P(x)$ of (130) in the new gauge and thus carries the topological charge $q = 1$. In other words, the instanton solution corresponds to a transition from a vacuum with $q = 0$ at $x^0 = -\infty$ to another vacuum with $q = 1$ at $x^0 \to +\infty$.

Instantons of the Yang–Mills theory thus represent transition amplitudes between classically forbidden vacuum sectors. When physical processes are computed it is essential to take such instanton contributions also into account.

We shall see later that the 'gravitational vacuum' will correspond to the flat spacetime. However, it is possible to have flat space with nontrivial topology, carrying a nonzero topological charge. The instanton solutions discussed above can then be generalized to the gravitational case.

5.11. Solitons – Monopole Solution

A soliton is defined to be a solution to classical equations of motion (in the usual spacetime) which has finite energy and which is, conventionally, static. Since the total energy has to be finite the fields have to fall off sufficiently quickly and can exist in only a limited region of spacetime. This feature, added to the fact that a soliton solution may be static, gives a suggestive interpretation of the solution as a 'particle'. There is, of course, nothing strange in this interpretation. This is rather a moot distinction between a 'particle' and 'field'. In fact, the photon (for example) is nothing but a quantized field configuration. In the case of solitons, we carry out this interpretation at the classical level itself.

In our future work, we shall be interested in one particular soliton solution, usually called the 'magnetic monopole'. The magnetic monopole – in its simplest form – is a solution to classical equations of motion of a system consisting of a *gauge field and a Higgs scalar field*. This solution is everywhere regular, has finite total energy, and asymptotically behaves like the field of a magnetic charge placed at the origin. While its physical features are easy to visualize, the explicit solution, unfortunately, requires a numerical integration of the equations.

Example 5.9. The reader would have noticed that we have now introduced a Higgs scalar field in addition to the gauge field. This is because in 3 + 1 dimensions a pure gauge field cannot have solitonic solution. To see this result consider the energy momentum tensor for the gauge field given by

$$\theta^{ik} = -F_A^{ij} F_j^{Ak} + \tfrac{1}{4}\, g^{ik} F_A^{mn} F_{mn}^A \,.$$

In component form

$$\theta^i{}_i = \tfrac{1}{4}(n-3) F_A^{ij}\, F_{ij}^A \,,$$

$$\theta^{00} = \tfrac{1}{2} F_k^{A0}\, F_A^{ko} + \tfrac{1}{4} F_A^{\mu\nu}\, F_{\mu\nu}^A \,,$$

$$\theta^{0\mu} = F_\nu^{A0}\, F_A^{\nu\mu} \,,$$

where we have assumed that we are working in an $(n + 1)$ dimensional Minkowski spacetime. The requirement that a soliton should be static implies that

$$F_A^{\mu 0} = 0.$$

Now consider the relation

$$\begin{aligned}\partial_\mu(x^\nu \theta^\mu{}_\nu) &= \theta^\mu{}_\mu + x^\alpha \partial_\mu \theta^\mu{}_\alpha \\ &= \theta^\mu{}_\mu - x^\alpha\, \partial_0 \theta^0{}_\alpha = \theta^\mu{}_\mu \,,\end{aligned}$$

where we have used $\partial_i \theta^i{}_k = 0$ and the fact that $\theta^0{}_k$ vanishes for static solutions. We immediately obtain from the above

$$\begin{aligned}0 &= \int \mathrm{d}^n x\ \partial_\mu(x^\nu \theta_\nu^\mu) = \int \mathrm{d}^n x\ \theta_\mu^\mu \\ &= (4-n) \int \mathrm{d}^n x\ F_A^{\mu\nu} F_{\mu\nu}^A \,.\end{aligned}$$

Thus unless $n = 4$ (*five*-dimensional spacetime) one cannot have pure gauge field solitons. □

We consider the system described in Section 5.6 with the Lagrangian

$$L = -\tfrac{1}{4} F_{ik}^A F_A^{ik} + (D_k\varphi)\,(D^k\varphi^\dagger) - V(\varphi^\dagger\varphi) \,. \tag{146}$$

To be explicit, we take the gauge group to be SU(2) with three gauge fields A_k^A. We shall also assume that the scalar field transforms as a triplet. From the

Lagrangian in (146) we can construct the expression for the total energy to be

$$\mathcal{E} = \int \mathrm{d}^3\mathbf{x}\, [\tfrac{1}{2}(\mathbf{B}_A \cdot \mathbf{B}^A + \mathbf{E}_A \cdot \mathbf{E}^A) + \tfrac{1}{2}(D^0\varphi, D^0\varphi) + \tfrac{1}{2}(D^k\varphi, D_k\varphi) + V]\,. \tag{147}$$

The requirement of finite energy leads to the following condition on the asumptotic behaviour of fields:

$$F^{ik}(x) \to O(r^{-2})\,, \qquad A^k(x) \to -\frac{i}{g}\, U\, \partial^k U^{-1} + O(r^{-1}), \tag{148}$$

$$D^k\varphi \to O(r^{-2})\,, \qquad \varphi(x) \to \varrho(x) + O(r^{-2})\,;$$

where we assume that $V(\varphi)$ has a nontrivial minimum for $\varphi(x) = \varrho(x)$. (These conditions contain a large amount of information. Application of some simple results of homotopy theory will now immediately lead to the existence condition and properties of the monopole.) We shall try to construct a nontrivial solution with the following properties; viz. that it is static and it satisfies the above boundary conditions. To this end, consider the following ansatz:

$$A^0_N = 0; \qquad A^\mu_N = \frac{1}{g}\, \varepsilon^{N\mu}{}_\nu\, \frac{x^\mu}{r^2}\, F(r); \quad \mu, \nu = 1, 2, 3,$$

$$\varphi_N\,(x) = \frac{x_N}{r}\, \varrho_0 \eta(r). \tag{149}$$

Quite clearly, we have already chosen a particular gauge such that the minimum of V occurs at $|\varphi|^2 = \varrho_0^2$. The static nature of the solution is seen explicitly in this gauge: A^0_N vanishes and other components are independent of x^0. We further impose the boundary conditions

$$F(0) = \eta(0) = 0; \qquad F(\infty) = \eta(\infty) = 1, \tag{150}$$

so tnat the solution will be regular at origin and will possess the correct asymptotic behaviour. The 'electric' and 'magnetic' fields are given by

$$E^\mu_A = 0; \qquad B^\mu_A = \frac{1}{g}\, \frac{x^A x^\mu}{r^4} F(1-F) + \frac{1}{g}\left(\delta_{A\mu} - \frac{x^A x^\mu}{r^2}\right) \frac{F'}{r}\,. \tag{151}$$

Consider now the expression for the total energy,

$$\mathcal{E} = \int \mathrm{d}^3\mathbf{x}\, \left\{ \frac{1}{g^2}\left(\frac{F'}{r}\right)^2 + \frac{1}{2g^2}\left[\frac{F(1-F)}{r^2}\right]^2 + \varrho_0^2\left[\frac{\eta(1-F)}{r}\right]^2 + \frac{\varrho_0^2}{2}\left[\frac{\eta(1-F)}{r} + \eta' F\right]^2 + V \right\}. \tag{152}$$

We see that all terms are positive definite. However, neither $F = 0$ nor $F = 1$ gives the lowest minimum. From the variational principle it follows that a nontrivial solution exists for F.

Such a solution can be obtained by numerical integration. For a simple quartic potential, the mass of the monopole turns out to be of the order of

$$M = \frac{m_v}{g^2/4\pi} , \tag{153}$$

where m_v is the vector boson mass. This mass is obtained by evaluating the minimum value of $\mathcal{E}$ in (152).

What does the solution mean physically? We know from our previous analysis that the Lagrangian in (146) describes a system containing the following physical fields: (1) one massless vector field; (ii) two massive vector fields; and (iii) one massive scalar field. This can be most easily seen in the gauge in which

$$\varphi(x) = \varrho_0 \begin{pmatrix} 0 \\ 0 \\ 1 \end{pmatrix} \quad \text{as } r \to \infty . \tag{154}$$

However, we were so far using a 'Coulomb' gauge with

$$\varphi(x) = \varrho_0 \begin{pmatrix} x/r \\ y/r \\ z/r \end{pmatrix} \quad \text{as } r \to \infty . \tag{155}$$

We shall first make a gauge transformation from (155) to (154). This can be done by the transformation

$$\varphi \to U\varphi \tag{156}$$

with

$$U = \exp(-i\theta L_2) \exp(-i\varphi L_3) , \tag{157}$$

where (θ, φ) are the polar angles of $\mathbf{r}$ and L_2 and L_3 are the usual rotation generators. (There is a somewhat tricky aspect associated with this transformation; we shall comment on it later near the end of this section.) In this gauge, we have a massless vector field B_{ik} and two massive vector fields G^1_{ik} and G^2_{ik}. Because of the static nature of the solution, only spatial components are nonzero. Explicitly, the fields possess the following asymptotic form:

$$G^1_a = G^2_a \to \frac{e^{-m_v r}}{r} \quad \text{as } r \to \infty ,$$

$$\varphi \to \varrho_0 \begin{pmatrix} 0 \\ 0 \\ 1 \end{pmatrix}. \quad \text{as } r \to \infty ,$$

$$B^\mu \to \frac{1}{g}\frac{x^\mu}{r^3} \quad \text{as } r \to \infty \, .$$

(We have used the 'magnetic component' to represent the field, i.e. we have defined B^μ, G^μ, etc., by the relations $B_{\alpha\beta} = \varepsilon_{\alpha\beta A} B^A$; $G^1_{\alpha\beta} = \varepsilon_{\alpha\beta A} G^{1A}$.) The physics is now clear. The two massive vector fields die out exponentially over a distance scale m_v^{-1}. However, the remaining massless vector field produces a magnetic component. This long-range field has the same form as that of a magnetic monopole with magnetic charge g^{-1} located at the origin.

We have chosen our original Lagrangian to be invariant under the SU(2) group. It is true that SU(2) group *a priori* has nothing to do with electromagnetism. However, SU(2) contains U(1) of electromagnetism as a subgroup. Hence, it may be thought that one of the gauge fields corresponds to the electromagnetic degree of freedom. Obviously this identification has to be done in a gauge invariant manner. For example, we cannot just choose A_i^3 to be the photon field. The following is a useful definition:

$$F_{ik} = \varphi_A \, F^A_{ik} - \frac{1}{g} \, \varepsilon^{ABC} \varphi_A (D_i \varphi_B)\,(D_k \varphi_C).$$

(When φ^A takes the form diag(0, 0, 1), this reduces to $(\partial_i A^3_k - \partial_k A^3_i)$.) However, unlike the standard electromagnetism based on U(1), there is also a magnetic charge:

$$\begin{aligned} \tfrac{1}{2}\, \varepsilon_{ikmn} \partial^k F^{mn} &= \left(\frac{1}{2g}\right) \, \varepsilon_{ikmn} \varepsilon_{ABC} \, \partial^k \varphi^A \partial^m \varphi^B \partial^n \varphi^C \\ &= -\left(\frac{4\pi}{g}\right) J_i^{(\text{magnetic})} \; . \end{aligned}$$

One more point about our solution deserves a special mention. Our basic solution in (149) had the Higgs field φ^A of constant value $(\varphi^A \varphi_A)^{1/2}$ but pointing in different directions in isospace (in fact it was pointing radially outward). After the gauge transformation we have made φ^A point along the same direction everywhere! It is well known that such a transformation is singular and ill-defined in at least one direction. (In our case the transformation is singular on the negative z axis.) Therefore, it is better to use (149) in deciding global questions. In general, the monopole solution will require a specific directional dependence of φ^A. It will not be possible to continuously transform the field to point in the same direction everywhere (see Figure 5.3). (This is the tricky aspect we mentioned earlier about (157).)

We shall have occasion to worry about the existence of magnetic monopoles in Chapter 10. It is probably worthwhile to point out the ingredients that have gone into the monopole. In Example 5.9, we have shown that a pure Yang–Mills field (without the Higgs field) cannot have solitonic solutions of finite energy in three-spatial dimensions. *Thus, it is necessary to have the Higgs field*. It is also essential that the Higgs field possesses a nonzero vacuum

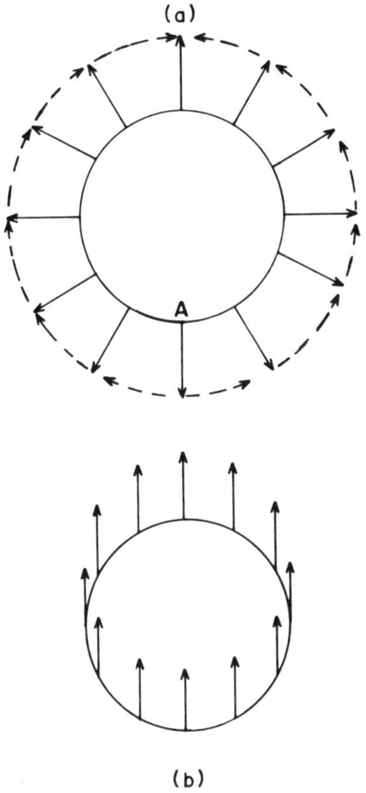

Fig. 5.3. (a) The monopole field configuration which is pointing radially outward everywhere; (b) the monopole field configuration (in a different gauge) which is pointing in the same direction everywhere. Clearly, such a transformation (e.g. one that rotates vectors in (a) to the configuration (b) as indicated by dashed arrows) must be singular somewhere (in the example shown this occurs at A).

expectation value: that is, the minimum of V must occur at nonzero ϱ_0. To see this result, consider the effect of $\varrho_0 \to 0$ in (152). When $\varrho_0 = 0$, $F = 1$ minimizes the energy $\mathcal{E}$. However, $F = 1$ leads to a singularity at the origin and is not a physically acceptable solution. *Thus a magnetic monopole can exist only in the theory with spontaneous symmetry breaking.* Even when the symmetry is broken, if all the vector fields acquire mass, then all field strengths will die down exponentially at large distances. The monopole solution exists in the above example only because of the existence of one vector field which remains massless. We saw in Section 5.6 that this feature is due to the existence of an extra U(1) symmetry in the theory, which stays unbroken. That is the last requirement for the existence of a magnetic monopole, viz. *the breaking of the symmetry should leave one* U(1) *symmetry* intact. We discussed the monopole solution based on the SU(2) gauge group;

similar monopole solutions exist in theories based on larger gauge groups like SU(5), as long as the above criteria are satisfied.

Notes and References

1. The field theory texts referred to in the last chapter discuss many aspects of gauge fields. Of particular relevance is:

   Huang, K.: 1982, *Quarks, Leptons, and Gauge Fields*, World Scientific, chs IV, V, and X.

2. There are many good textbooks in group theory. We mention two useful texts:

   Wyborne, B. G.: 1974, *Classical Groups for Physicists*, Wiley, New York.

   Gilmore, R.: *Lie Groups, Lie Algebras and Some of Their Applications*, 1974, Wiley, New York.

3. A good discussion of grand unification (and group theory to the extent required!) can be found in the review article:

   Langacker, P.: 1981, 'Grand Unified Theories and Proton Decay', *Phys. Repts* **72**, 185.

4. The original reference relevant to Section 5.8 is:

   Coleman, S., and Weinberg, E.: 1973, *Phys. Rev.* **D7**, 1888.

5. A good discussion of topological aspects of gauge theories can be found in:

   Jackiw, R.: 1983, 'Topological Investigations of Quantized Gauge Theories' (Les Houches School on Relativity, Groups and Topology), MIT-CTP #1108.

6. There exists a monograph on the subject of solitons and instantons:

   Rajaraman, R.: 1982, *Solitons and Instantons*, North-Holland, Amsterdam.

Part II
Classical General Relativity

Chapter 6

General Theory of Relativity

6.1. The Need for a General Theory of Relativity

In 1905 Einstein proposed the special theory of relativity, a theory that revolutionized the concepts of space, time, and motion which had been considered well established since the days of Newton. Einstein had been led to special relativity from considerations of symmetry, from the deduction that Maxwell's equations of electromagnetic theory were invariant under Lorentz transformations (and not under Galilean transformations). Observational considerations did not motivate the theory: the null result of the Michelson–Morley experiment followed as a simple deduction from the theory *after* it was formulated.

The important aspect of special relativity was the unification of space and time into a single entity which we shall refer to as 'spacetime'. This contrasts with the Newtonian notions of absolute space and absolute time as two separate entities. The contrast may be seen from a comparison of Lorentz and Galilean transformations.

Let us take rectangular Cartesian coordinates $\mathbf{r} \equiv (x, y, z)$ to specify a point in space and t the time, as used by an observer O. Let another observer O' move with uniform velocity $\mathbf{v}$ with respect to O. Let t', $\mathbf{r}' \equiv (x, y, z)$ denote the time and space coordinates in the rest frame of O' with O, O' coinciding at $t = t' = 0$. According to the Galilean transformations we have

$$t' = t, \qquad \mathbf{r}' = \mathbf{r} - \mathbf{v}t. \tag{1}$$

Over and above these transformations we also have rotations in space which preserve the Euclidean metric

$$\mathrm{d}l^2 = \mathrm{d}x^2 + \mathrm{d}y^2 + \mathrm{d}z^2. \tag{2}$$

These are the basic symmetries of Newtonian spacetime physics.

The orthogonal transformations preserving (2) and the Galilean transformation (1) are replaced in special relativity by the Lorentz transformations. To describe these it is convenient to use the superscript notation for coordinates: $x \equiv x^1$, $y \equiv x^2$, $z \equiv x^3$ and $ct \equiv x^0$, where c = speed of light. We also write the spacetime metric in the form

$$ds^2 = \eta_{ik}\, dx^i\, dx^k, \tag{3}$$

where η_{ik} = diag (+1, −1, −1, −1), i, k = 0, 1, 2, 3 and the summation convention operates. The spacetime with the above metric is named after H. Minkowski (1864–1909) whose work greatly contributed to the understanding of spacetime geometry and its relationship to special relativity. The Lorentz transformation is a general linear transformation of spacetime coordinates:

$$x'^i = L^i_k\, x^k \tag{4}$$

that preserves (3). Hence the coefficients L^i_k satisfy the relation

$$\eta_{ik} L^i_m\, L^k_n = \eta_{mn}. \tag{5}$$

Example 6.1. Corresponding to the situation which led to the Galilean transformation (1) above, the Lorentz transformation is given by

$$t' = \frac{t - \mathbf{r} \cdot \mathbf{v}/c^2}{\sqrt{1-(v^2/c^2)}}, \qquad \mathbf{r}' = \frac{\mathbf{r}-\mathbf{v}t}{\sqrt{1-(v^2/c^2)}}.$$

□

In special relativity (4) describes the general coordinate transformation between observers in uniform relative motion. It is also assumed, however, that these observers are of a special class: they are *inertial* observers; that is, they are not acted on by any external force. According to Newton's first law of motion these observers are *unaccelerated*

It should be emphasized here that in the Newtonian kinematics it is assumed that there is an absolute space relative to which all motions can be measured. Thus velocities as well as accelerations can be measured relative to absolute space. In special relativity there is no absolute space but observers which are unaccelerated are assumed to exist. Such observers are able to measure absolute accelerations in their rest frames; but the notion of absolute velocity does not exist.

Particle mechanics and the Newtonian law of gravitation were known to the nineteenth-century physicists to be invariant under Galilean transformations and spatial rotations. The Maxwell equations, on the other hand, appeared to be invariant under Lorentz transformations. For a consistent blending of mechanics and gravitation with electromagnetism it was necessary either to modify the former and make them Lorentz invariant or to recast the latter to make it invariant under Galilean transformations. In his 1905 paper, Einstein adopted the former course and he was subsequently vindicated by experiments.

Example 6.2. To see the invariance of Maxwell's equations in vacuum

$$\nabla \cdot \mathbf{H} = 0, \quad \nabla \times \mathbf{E} + \frac{1}{c}\frac{\partial \mathbf{H}}{\partial t} = 0, \ \nabla \cdot \mathbf{E} = \mathbf{0}, \ \nabla \times \mathbf{H} - \frac{1}{c}\frac{\partial \mathbf{E}}{\partial t} = 0$$

under the Lorentz transformation of Example 6.1, define

$$(\mathbf{E}')_\perp = \frac{\left(\mathbf{E} + \dfrac{\mathbf{v} \times \mathbf{H}}{c}\right)_\perp}{\sqrt{1 - \dfrac{v^2}{c^2}}} \quad (\mathbf{H}')_\perp = \frac{\left(\mathbf{H} - \dfrac{\mathbf{v} \times \mathbf{E}}{c}\right)_\perp}{\sqrt{1 - \dfrac{v^2}{c^2}}}; \mathbf{E}'_\parallel = \mathbf{E}_\parallel \ \ \mathbf{B}'_\parallel = \mathbf{B}_\parallel .$$

Here $(\mathbf{E})_\perp$ and $(\mathbf{E})_\parallel$ denote the components of **E**, perpendicular and parallel to **v**, respectively. It can be verified that **E**′ and **H**′ satisfy the same Maxwell equations in the rest frame of O'. This notation is, however, cumbersome: the invariance of Maxwell's equations becomes easily manifest in the four-dimensional notation used in Section 6.7.

While it was comparatively straightforward (by hindsight of course!) to make particle mechanics Lorentz invariant, there were problems with gravitation. Let us briefly get a flavour of the difficulty.

The Newtonian inverse square law of gravitation was an 'action at a distance' law. Moreover, it presumed that the gravitational effect is transmitted across any spatial distance instantaneously. Special relativity and Lorentz invariance required, on the other hand, that the interaction should propagate with the speed of light. This difference is illustrated in Figure 6.1.

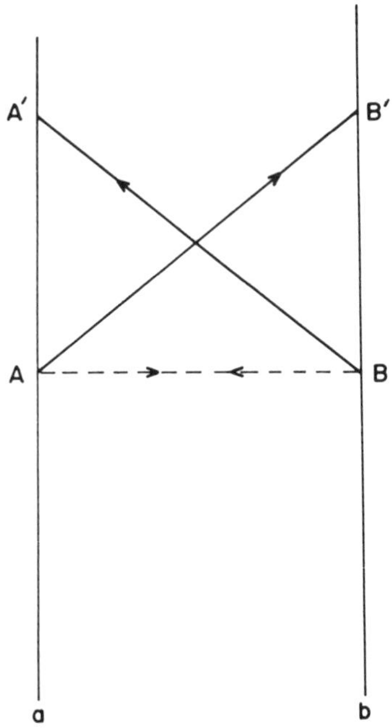

Fig. 6.1. The spacetime diagram shows two worldlines of particles a and b. The Newtonian instantaneous interaction propagates along the dashed line AB. According to special relativity, the interaction from A travels to worldline b with the speed of light, reaching it at the worldpoint B'. Similarly the interaction B reaches the worldline a at worldpoint A'.

By the early twentieth century the success of electromagnetic theory had established the 'field' as the popular mode of description of a physical interaction. Thus, instead of using the mysterious notion of action at a distance we could use the gravitational *field* to describe Newton's law. The work of Laplace and Poisson in the early nineteenth century had already shown how to do this.

In the field theoretic picture we have a gravitational potential φ which is related to the gravitational field $\mathbf{F}$ by the relation

$$\mathbf{F} = \nabla \varphi. \tag{6}$$

The source equation for $\mathbf{F}$ is then given by Gauss's theorem

$$\nabla \cdot \mathbf{F} = -4\pi G \varrho, \tag{7}$$

Combining (6) and (7) we get

$$\nabla^2 \varphi = -4\pi G \varrho. \tag{8}$$

The source ϱ is, of course, the matter density.

The difficulty encountered when modifying (8) to ensure Lorentz invariance is as follows. First, on the left-hand side we must change the Laplacian operator ∇^2 to the wave operator $\Box$:

$$\nabla^2 \varphi \rightarrow \nabla^2 \varphi - \frac{1}{c^2} \frac{\partial^2 \varphi}{\partial t^2} \equiv - \Box \varphi \tag{9}$$

What about the right-hand side? The mass energy equivalence established by special relativity requires us to include on the right-hand side not only the matter density ϱ but also the radiation density u divided by c^2. But relativistic mechanics also tells us that $u + \varrho c^2$ is not a scalar quantity but is the (time, time) component of the energy momentum tensor T_{ik}, $i, k = 0, 1, 2, 3$. In other words, (8) is replaced by

$$\Box \varphi = \frac{4\pi G}{c^2} T_{00}. \tag{10}$$

Equation (10) above is not the end of the trail! For, by itself (10) is not Lorentz invariant. Rather, the right-hand side suggests that it is the (0, 0) component of a tensor equation

$$\Box \varphi_{ik} = \frac{4\pi G}{c^2} T_{ik} \tag{11a}$$

with $\varphi_{00} = \varphi$. Thus from a scalar theory we have been led to a tensor theory in which the relativistic analogue of the Newtonian gravitational potential appears to be the (time, time) component of a tensor potential.

Have we reached the desired end at last? Not yet, for we have neglected another aspect of the mass–energy equivalence. Every physical field has dynamical degrees of freedom which imply that the field has energy and momentum. Our tensor field is no exception. If the right-hand side of (11a) is supposed to include contributions from all forms of matter and radiation, it must also include the energy momentum tensor of the field generated by φ_{ik}.

The precise form of this tensor need not concern us for the present, but it is expected to include terms quadratic in $\varphi_{ik,\,l} \equiv \partial\varphi_{ik}/\partial x^l$. Hence, the source equation (11a) for the potential is nonlinear and includes itself as a source. This nonlinearity demonstrates why the problem of generalizing Newtonian gravitation into a form consistent with special relativity is nontrivial. We do not end up with a linear theory like Maxwell's.

This inherent nonlinearity also shows up even if we try to preserve the scalar character of φ by modifying (10) to the form

$$\Box\,\varphi = \frac{4\pi G}{c^2}\,T_i^i, \tag{11b}$$

wherein the right-hand side is scalar. T_i^i includes stresses as well as the energy density of matter and radiation. As in the tensor case, we have to include the contributions from the gravitational field also to the right-hand side.

So far we have considered the problem from one end – that of gravitation. The problem is no less difficult if we examine the consistency of special relativity in the presence of the phenomenon of gravitation. Consider the very basis of special relativity: the inertial observers. Where can we find them in the real world? Nowhere! There is no way we can isolate a spacetime region to make it free from gravitational force. In this case also we notice a difference between gravitation and electromagnetism. It is possible to design a chamber in which the observer feels *no* electromagnetic force, but we cannot design a chamber in which the observer is free from gravitational effects in a permanent and exact way.[1]

It follows, therefore, that with gravitation around, special relativity cannot even get off the first base. We have no real observers who can be identified as inertial. So, how do we introduce Lorentz transformations in real life?

The problems of coexistence of gravitation and special relativity are thus nontrivial and cannot be solved in a piecemeal fashion. That is why the solution offered by Einstein, in the form of a general theory of relativity, was a *tour de force* of intellectual achievement. In the remainder of this chapter we shall lead up to this remarkable theory which took Einstein a decade to formulate and perfect.

6.2. Curved Spacetime

The significant feature of general relativity that set it apart from any other physical theory was the idea that gravitation manifests itself through the

[1]What about the weightless astronaut in a satellite round the Earth? We shall consider this apparent exception to the above rule later in Section 6.7.

curvature of spacetime. In putting forth this idea Einstein was guided by the property of gravitation we noted in the last section, viz. its permanence. If gravitational fields are present in any region of space they cannot be removed. Indeed, it can be argued that gravitation is as much intrinsic to the region as are space and time. But how can this idea be expressed mathematically?

It is here that Einstein found applications for a concept that had hitherto remained buried in abstract mathematics. Einstein argued that the presence of gravity manifests itself through the spacetime having a non-Euclidean (or, more precisely, a non-Minkowskian) geometry. To understand this idea further let us first consider some examples of *non-Euclidean* geometries.

Any geometrical system that is not built on the same set of axioms as Euclid's may be designated 'non-Euclidean'. The logical completeness and practical applicability of Euclid's geometry initially created the erroneous impression that it was the only possible geometry. Looking at the history of geometry we find that one of Euclid's basic postulates played the role of a red herring. This was the famous *parallel* postulate which is illustrated in Figure 6.2(a). This postulate states that, given a straight line *l* and a point *P* outside it, we can draw one and only one straight line through *P* parallel to *l*. Such a line is drawn in Figure 6.2(a).

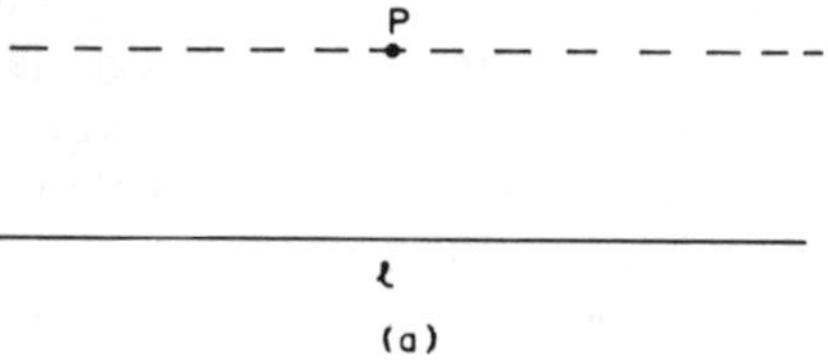

Fig. 6.2(a). In Euclid's geometry it is *assumed* that given a straight line *l* and a point *P* outside it, one and only one line *l'* can be drawn through *P* parallel to *l*.

Is it really an assumption or can it be proved on the basis of other assumptions or axioms? If the latter alternative is correct, we do not need this statement as an extra assumption. For several centuries mathematicians regarded the parallel postulate in this light and a number of proofs of its validity were also given. All such proofs were subsequently shown to be false and finally it dawned on some mathematicians of the nineteenth century that the parallel postulate was unprovable, that it was indeed an assumption needed to complete the Euclidean system.

This being the case, is it possible to replace the parallel postulate by some alternative postulate? Figures 6.2(b) and (c) show two logical alternatives: (b) through *P no* straight line can be drawn parallel to ℓ or (c) through *P* more than one line can be drawn parallel to ℓ. The work of great geometors like Bolyai (1802–1860), Lobatchevsky (1793–1856), Gauss (1777–1855) and Riemann (1826–1866) was to demonstrate the existence of non-Euclidean geometrical systems based on (b) or (c).

In Figure 6.2(b) we see the example of a non-Euclidean geometry of type (b). This is the geometry on the surface of a Euclidean sphere. The 'straight lines' on a sphere, being the lines of shortest distance, are arcs of the great circles. All straight lines are closed and any two straight lines intersect. Naturally we cannot draw parallel lines on this surface. By contrast, the geometry on a saddle-shaped surface will be found to be of type (c) [see Figure 6.2(c)].

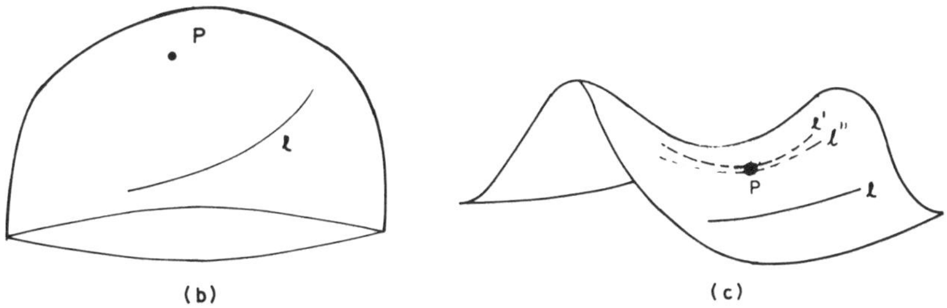

Fig. 6.2(b). On the surface of a sphere no line can be drawn through P parallel to l. (c) On a saddle surface lines l', l'' drawn through P do not meet the line l.

Theorems of Euclidean geometry do not hold for figures drawn on such surfaces. Thus a triangle on the spherical surface will have its three angles adding up to a total invariably exceeding π. On a saddle-shaped surface the corresponding sum is less than π. For reasons which will become clearer in later sections, it is customary to refer to such geometries as curved space geometries in contrast to the flat space Euclidean geometry.

Example 6.3. In the triangle ABC drawn on the sphere of Figure 6.3 the three angles add up to a sum exceeding π. Spherical geometry shows that the area of the triangle is given by $(A+B+C-\pi)a^2$, where a is the radius of the sphere. (It is easy to prove this result by completing all the great circles forming the sides of $\triangle ABC$). □

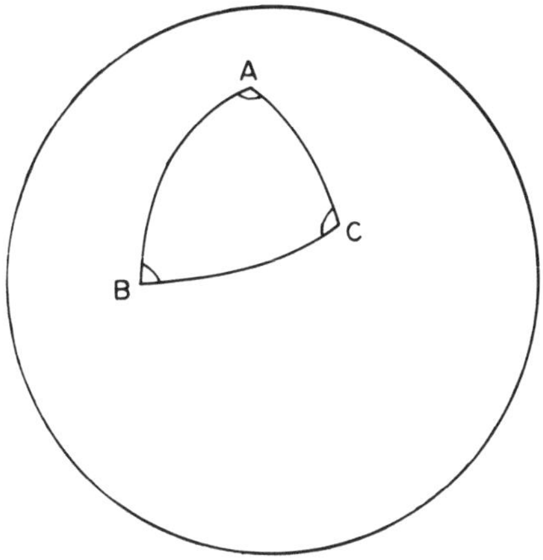

Fig. 6.3. The area of the spherical triangle is given by $(A + B + C - \pi)a^2$, where a is the radius of the sphere.

Let us now see how these ideas can be extended from pure space to a union of space and time, for in doing so we shall catch a glimpse of what general relativity is all about.

Consider the dynamical problem of a cannonball shot out of a cannon, neglecting air resistance. According to Newton's first law of motion, the cannonball would have continued to move in a straight direction with uniform speed *if* there were no forces acting on it. The dashed line in the three-dimensional spacetime plot of Figure 6.4 illustrates this imaginary worldline. The real worldline of the cannonball is the continuous curve whose projection on the spatial plane is the familiar parabola. According to Newton's second law, this parabolic trajectory results because the force of gravity acts on the ball pulling it towards the Earth's surface. How would Einstein interpret this situation?

According to Einstein's general relativity, the Earth's gravitational field produces a non-Euclidean geometry around it. Or, to put it differently, all gravitational phenomena in the vicinity of the Earth have to be viewed in this

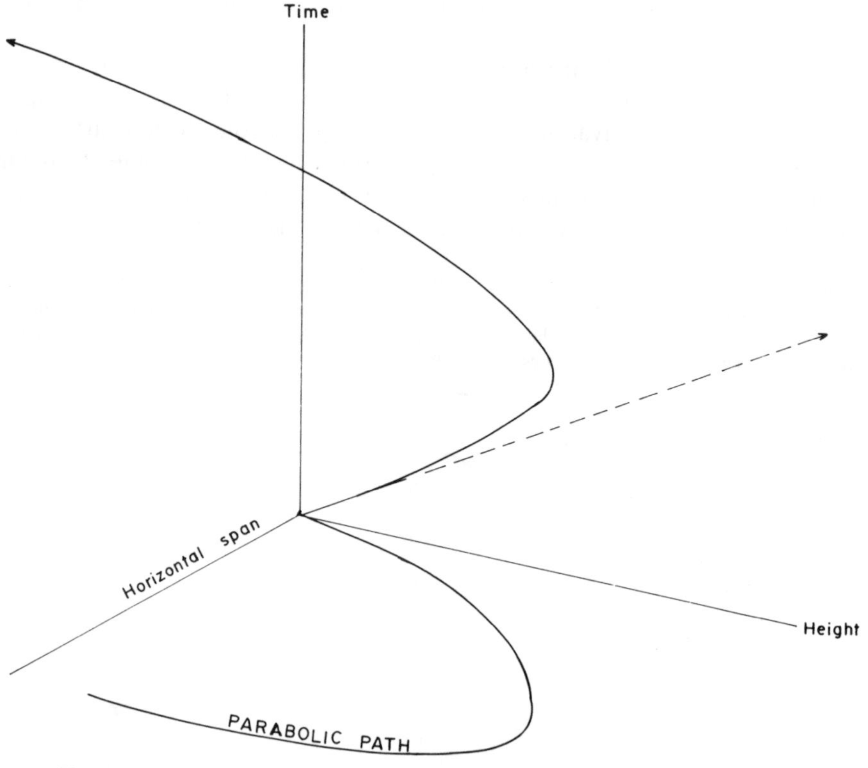

Fig. 6.4. The solid curve is the worldline of the cannonball whose projection is the parabolic path traced out in space by the ball. The dashed line is the imaginary worldline which the cannonball would have followed in the absence of gravity.

non-Euclidean geometry of spacetime. Thus the dashed worldline has no significance since it refers to a gravity-free environment which clearly cannot be achieved in practice. On the other hand, having taken gravity into account through its manifestation as non-Euclidean geometry, we can assert that the cannonball is moving in a curved spacetime but under *no* force. Thus it moves according to the first law of motion 'in a straight line and with uniform speed'.

To one who is brought up on Euclidean–Minkowskian flat geometry, it becomes difficult to accept that the curved worldline of Figure 6.4 is indeed one describing uniform motion in a straight line. The result is nevertheless true because the 'straightness' of a line and the 'uniformity' of speed are to be judged with reference to the new set of rules which characterize the geometry of the curved spacetime around the Earth.

In the Newtonian framework the laws of motion and the law of gravitation together solved the problem of bodies moving under the gravitational attraction of one another. In general relativity we need instead the following two items of information:

i How is the non-Euclidean geometry of spacetime determined by the presence in it of sources of matter and energy?

ii How is uniform motion in a straight line defined in a given non-Euclidean geometry?

Further, since a physicist is also concerned with other interactions besides gravitation, he needs to know how to state and handle the various laws of physics in a curved spacetime. For example, all the topics we covered in Part I will have to be described in the curved spacetime of general relativity.

In the following sections we shall develop mathematical machinery which will help us answer these questions.

6.3. Vectors and Tensors

Vectors and tensors play an important role in describing physics in the flat spacetime of special relativity. When the physicist insists on Lorentz *invariance* of physical laws, in practical terms he needs a way of writing them that *ensures* that invariance. The quantities which help him express 'invariant truths' are, of course, scalars, vectors, and tensors.[2] In expressing physics in a curved spacetime we need to develop analogous concepts in a more general way. Although we used vectors and tensors freely in Part I, we did so in the flat Minkowski background. These concepts need to be generalized now.

To begin with, in special relativity, the transformation under which we require quantities to be invariant is the Lorentz transformation given by (4). A scalar function $\varphi(x^i)$ of coordinates remains unchanged by (4):

$$\varphi'(x'^i) = \varphi(x^i), \tag{12}$$

[2] and spinors, if we wish to introduce the quantum mechanical notion of spin.

while a vector function $V^k(x^i)$ transforms as

$$V'^k(x'^i) = L^k_m V^m(x^i). \tag{13}$$

In general relativity we shall be concerned with general coordinate transformations. Thus we shall assume that x'^i are C^2 functions of x^i and that inverse functions x^i of x'^i exist over the region of spacetime under consideration. (It is possible to give more formal and mathematically more rigorous definitions of the various quantities that we encounter here and elsewhere. The purist may find them in the books cited at the end of the chapter. We shall adopt an intuitive approach which would appeal to a physicist but not to a mathematician.) We shall require that our physically significant quantities should not depend on any special choice of coordinates and, as such, they should be expressible by scalars, vectors, tensors, etc.

Thus, a scalar function $\varphi(x^i)$ continues to satisfy (12) although we now extend our study to general coordinate transformations. The generalization of (13) is correspondingly given by

$$V'^k(x'^i) = \frac{\partial x'^k}{\partial x^m} V^m(x^i). \tag{14}$$

This relation reduces to (13) for the Lorentz transformation given by (4). Vectors which transform according to the rule given by (14) are called 'contravariant' vectors. The transformation coefficients $\partial x'^k/\partial x^m$ are, in general, functions of position.

We also define another set of vectors, called 'covariant' vectors, by the transformation rule

$$V'_k(x'^i) = \frac{\partial x^m}{\partial x'^k} V_m(x^i). \tag{15}$$

Note that the transformation matrices of the two types of vectors are reciprocals of each other:

$$\frac{\partial x'^k}{\partial x^m} \frac{\partial x^n}{\partial x'^k} = \delta^n_m , \tag{16}$$

where δ^n_m is the Kronecker delta.

Example 6.4. If $x^i(\lambda)$ denote the parametric functions of a curve, then the tangent vector

$$u^i \equiv \frac{dx^i}{d\lambda}$$

transforms as a contravariant vector. Likewise, if we denote by the function $\varphi(x^i)$ = const, a series of 3-surfaces, then their normals

$$n_i \equiv \frac{\partial \varphi}{\partial x^i}$$

transform as covariant vectors. The proofs of these statements are straightforward. □

These definitions can be extended to include, *tensors*, which can be contravariant (all upper indices), covariant (all lower indices) or mixed. The transformation rule for a contravariant tensor of rank n is illustrated by the following example:

$$T'^{i_1 \dots i_n} = \frac{\partial x'^{i_1}}{\partial x^{k_1}} \dots \frac{\partial x'^{i_n}}{\partial x^{k_n}} T^{k_1 \dots k_n} . \tag{17}$$

For each contravariant index i_n a transformation matrix $\partial x'^{i_n}/\partial x^{k_n}$ appears. Similarly, for a typical covariant index i_n, coefficient $\partial x^{k_n}/\partial x'^{i_n}$ appears in the transformation formula. For example, it can be easily verified that δ^m_n is a mixed tensor.

In due course we shall encounter tensors with special symmetrics. Tensors can be symmetric or antisymmetric with respect to the interchange of any two of their indices. Similarly, a tensor like δ^i_k may be isotropic, i.e. have the same components in all coordinate frames.

Example 6.5. It is easily seen that the property of symmetry or antisymmetry is coordinate independent. Suppose, for example, that $T^{ik} = T^{ki}$ in one coordinate frame. Then

$$\begin{aligned} T'^{mn} &= \frac{\partial x'^m}{\partial x^i} \frac{\partial x'^n}{\partial x^k} T^{ik} = \frac{\partial x'^m}{\partial x^i} \frac{\partial x'^n}{\partial x^k} T^{ki} \\ &= \frac{\partial x'^n}{\partial x^k} \frac{\partial x'^m}{\partial x^i} T^{ki} = T'^{nm}. \end{aligned}$$

□

Note that the vectors and tensors defined above are functions of position in spacetime, i.e. they are vector *fields* and tensor *fields*. Tensors of the same rank can be added or subtracted at the *same* point to give a tensor of the same rank, but the operation of addition or subtraction cannot be carried out at *different* spacetime points because, as pointed out earlier, the transformation coefficients $\partial x^i/\partial x'^k$ vary from point to point in general.

The product of two tensors is a tensor, again when defined at the same spacetime point. Thus, if A_i and B_{jk} are multiplied, the resulting product is a third-rank tensor $A_i B_{jk}$.

Although product with another tensor, in general, increases the rank of a tensor, there is a special situation when the rank of the product is lowered. Thus, A_i has rank 1 and B^{jk} rank 2, giving rank 3 for the product $A_i B^{jk}$. If, however, we put $j = i$, the product $A_i B^{ik}$ transforms as a vector with free index k. Such an operation is called *contraction*.

We end this section by stating another important result concerning products of tensors, known as the quotient law. Suppose a relation of the following kind holds in all coordinate frames

$$P = QR,$$

where P and R are arbitrary tensors of any rank. Then Q is also a tensor. The proof follows the same lines as given below for the simple example

$$P_i = Q_{ik}R^k.$$

Writing the above relation in primed coordinates we get

$$P'_i = Q'_{ik}R'^k \Rightarrow \frac{\partial x^m}{\partial x'^i} P_m = Q'_{ik} \frac{\partial x'^k}{\partial x^n} R^n$$

Multiply by $\partial x'^i/\partial x^\ell$ and use (16) to get

$$P_\ell = \frac{\partial x'^k}{\partial x^n} \frac{\partial x'^i}{\partial x^\ell} Q'_{ik}R^n = Q_{\ell n}R^n.$$

Since P_ℓ and R^n are arbitrary we get the required result:

$$Q_{\ell n} = \frac{\partial x'^k}{\partial x^n} \frac{\partial x'^i}{\partial x^\ell} Q'_{ik}.$$

6.4. Metric and Geodesics

So far we have not imposed any geometrical structure on spacetime. We do so now by first defining the spacetime metric.

Consider two points P and Q with coordinates $\{x^i\}$ and $\{x^i + \mathrm{d}x^i\}$. Their coordinate difference is expressed in differential form to indicate that P and Q are 'near' each other. However, as mentioned earlier, only coordinate-independent concepts are expected to have physical significance. In the present context the physical quantity we are looking for is the invariant separation distance between P and Q, which we denote by $\mathrm{d}s$.

How should $\mathrm{d}s$ be related to $\mathrm{d}x^i$? If we are guided by (2) or (3) we should look for a quadratic relationship like

$$\mathrm{d}s^2 = g_{ik}\, \mathrm{d}x^i\, \mathrm{d}x^k, \tag{18}$$

where g_{ik} may be functions of x^k. For $\mathrm{d}s^2$ to be invariant we expect (18) to hold in all coordinate systems. Let a changeover to coordinates x'^i change g_{ik} to g'_{ik}. Then the equality

$$g'_{ik}\, \mathrm{d}x'^i\, \mathrm{d}x'^k = g_{ik}\, \mathrm{d}x^i\, \mathrm{d}x^k$$

immediately gives us the transformation law for g_{ik}, viz.

$$g'_{ik} = \frac{\partial x^l}{\partial x'^i} \frac{\partial x^m}{\partial x'^k} g_{lm}. \tag{19}$$

Thus g_{ik} must transform as a second-rank covariant tensor. It is called the 'metric tensor' and (18) is called the 'metric' or the 'line element' of the spacetime. We shall assume that $\det\|g_{ik}\| \neq 0$.

We assume, without loss of generality, that g_{ik} is a symmetric tensor. We can find new coordinates which will *locally* diagonalize the metric (18). That is, given any spacetime point P we can find coordinates T, X, Y, Z such that *at that point* the metric becomes

$$\mathrm{d}s^2 = c^2\,\mathrm{d}T^2 - \mathrm{d}X^2 - \mathrm{d}Y^2 - \mathrm{d}Z^2,$$

the same as in Minkowski spacetime.

Later we shall find that we can achieve even closer similarity to the Minkowski line element. We shall show that we can use (19) to have, at P,

$$g'_{ik}(P) = \eta_{ik}; \qquad g'_{ik,\ell}(P) = 0.$$

For the time being, however, we simply note the indefinite nature of the metric (18). This enables us to identify spacelike, timelike and null separations in the same way as in the special relativity, i.e. by

$$\mathrm{d}s^2 < 0, \quad \mathrm{d}s^2 > 0, \quad \mathrm{d}s^2 = 0,$$

respectively.

The metric tensor is used for an important operation known as raising and lowering the suffix of a tensor or a vector. Thus for a contravariant vector V^i define a covariant associate by

$$V_i = g_{ik}V^k. \tag{20}$$

Since the matrix $\|g_{ik}\|$ has an inverse matrix denoted by $\|g^{ik}\|$, we may invert (20) to write

$$V^k = g^{ik}V_i. \tag{21}$$

It is customary to use the same symbol for vectors and tensors connected by relations like (20) or (21).

Example 6.6. The reciprocal g^{ik} transforms as a contravariant tensor. To see this consider the identity

$$g'^{ik}g'_{km} = \delta^i_m$$

and use (19) to get

$$g'^{ik}\,\frac{\partial x^\ell}{\partial x'^k}\,\frac{\partial x^n}{\partial x'^m}\,g_{n\ell} = \delta^i_m\,.$$

Multiply both sides by $(\partial x^p/\partial x'^i)\,(\partial x'^m/\partial x^q)$ to get

$$\left(g'^{ik}\,\frac{\partial x^p}{\partial x'^i}\,\frac{\partial x^l}{\partial x'^k}\right)\left(\frac{\partial x^n}{\partial x'^m}\,\frac{\partial x'^m}{\partial x^q}\right)g_{nl} = \delta^i_m\,\frac{\partial x^p}{\partial x'^i}\,\frac{\partial x'^m}{\partial x^q}\,.$$

Simplify using (16) to get

$$\left(g'^{ik}\,\frac{\partial x^p}{\partial x'^i}\,\frac{\partial x^l}{\partial x'^k}\right)g_{ql} = \delta^p_q\,.$$

Clearly, by definition the expression in brackets must be g^{pl}. Hence, etc., etc. □

Another important operation, involving contraction with g_{ik}, gives the trace of a tensor. Thus, for a second-rank tensor T^{ik} we define

$$T = g_{ik}T^{ik} \tag{22}$$

as the trace of the tensor. It is easy to verify that $g_{ik}g^{ik} = 4$. For a vector V^i we define the magnitude V by

$$V^2 = V_i V^i = g_{ik}V^i V^k. \tag{23}$$

The vector is spacelike, timelike, or null, according as $V^2 <, >, = 0$.

We shall frequently encounter the quantity

$$g = \det \|g_{ik}\| \,. \tag{24}$$

Note that g is *not* a scalar. In fact, under a coordinate transformation, g transforms as follows:

$$g \to g' = gJ^2, \tag{25}$$

where

$$J = \frac{\partial(x^i)}{\partial(x'^k)} \tag{26}$$

is the Jacobian of the transformation.

Example 6.7. Consider the totally antisymmetric symbol

$$\varepsilon_{iklm} = \begin{cases} 1 & \text{if } (i, k, l, m) \text{ is an even permutation of } (0, 1, 2, 3) \\ -1 & \text{if } (i, k, l, m) \text{ is an odd permutation of } (0, 1, 2, 3) \\ 0 & \text{otherwise.} \end{cases}$$

Then we show that

$$e_{iklm} \equiv (-g)^{1/2}\varepsilon_{iklm}$$

transforms as a tensor. For, from (25)

$$\begin{aligned} \varepsilon_{iklm} \frac{\partial x^i}{\partial x'^p} \frac{\partial x^k}{\partial x'^q} \frac{\partial x^l}{\partial x'^r} \frac{\partial x^m}{\partial x'^s} (-g)^{1/2} &= J\varepsilon_{pqrs}(-g)^{1/2} \\ &= (-g')^{1/2}\varepsilon_{pqrs} \\ &= e'_{pqrs} \,. \end{aligned}$$

Notice that e_{iklm} is really a 'pseudotensor' since it changes signs under reflections. It is called the *Levi–Civita tensor.* □

6.4.1. *Geodesics*

The metric tells us how to compute the distances between points along specific curves. Thus, given two points P_1 and P_2 with coordinates $x^i_{(1)}$ and $x^i_{(2)}$ connected by a curve Γ, we define the 'length' of the curve Γ by the integral

$$S_{21} = \int_{P_1}^{P_2} \mathrm{d}s_\Gamma \,, \tag{27}$$

where $\mathrm{d}s_\Gamma$ is the metric distance computed along Γ. To perform the integral (27) we note that Γ may be parametrized by a parameter λ so that a typical point P on Γ has coordinates $x^i(\lambda)$ determined as functions of λ, and

$$x^i_{(1)} = x^i(0), \quad x^i_{(2)} = x^i(1). \tag{28}$$

We then have

$$S_{21} = \int_0^1 \left\{ g_{ik} \dot{x}^i \dot{x}^k \right\}^{1/2} \mathrm{d}\lambda, \tag{29}$$

where $\dot{x}^i \equiv \mathrm{d}x^i/\mathrm{d}\lambda$.

Can we now generalize the Euclidean notion of a straight line as a curve of shortest distance? Before we do so we must take note of certain complications which arise not because of curvature of space but because of our use of an indefinite metric. The expression for $\mathrm{d}s$ may not always be real or pure imaginary; it may even be zero for null displacements. Thus, the notion of shortest distance needs to be examined with care.

For a timelike curve S_{12} is real and positive, for a spacelike curve S_{12} is pure imaginary, while for a null curve S_{12} is zero. In any of these cases we may look for stationary values of S_{12}. A curve along which S_{12} is stationary is called a *geodesic*. A geodesic may be spacelike, timelike, or null (see Figure 6.5).

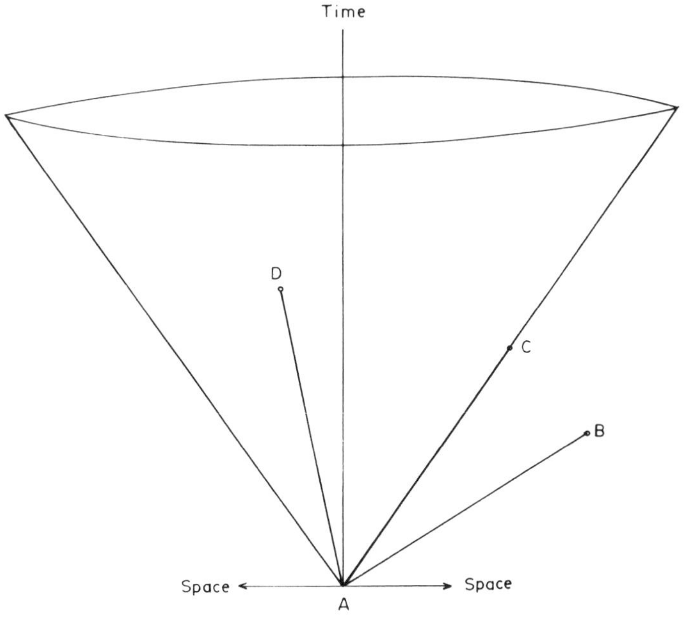

Fig. 6.5. The lines AB, AC, and AD are, respectively, spacelike, null, and timelike geodesics in Minkowski spacetime.

Writing

$$\varphi^2 = g_{ik}\dot{x}^i\dot{x}^k, \tag{30}$$

the stationary value of φ is given by the Euler–Lagrange equations

$$\frac{d}{d\lambda}\left(\frac{\partial\varphi}{\partial\dot{x}^i}\right) - \frac{\partial\varphi}{\partial x^i} = 0. \tag{31}$$

Since φ does not depend explicitly on λ, Equations (31) have a first integral

$$\dot{x}^i\frac{\partial\varphi}{\partial\dot{x}^i} - \varphi = \text{const},$$

that is,

$$g_{ik}\,\dot{x}^i\,\dot{x}^k = \text{const}. \tag{32}$$

Thus if we choose a parameter λ such that $ds/d\lambda$ = const, our geodesic equation takes a simpler form. This can always be achieved for spacelike or timelike geodesics but not for null geodesics. In the former case Equations (31) reduce to

$$\frac{d}{d\lambda}\left(g_{ik}\frac{dx^k}{d\lambda}\right) - \frac{1}{2}\frac{\partial g_{mn}}{\partial x^i}\frac{dx^m}{d\lambda}\frac{dx^n}{d\lambda} = 0. \tag{33}$$

The parameter $\lambda \propto s$ is called an 'affine parameter'.

Example 6.8. Consider the two-dimensional space on the surface of a unit sphere. Ignoring the minus sign in the metric we may write the line element as

$$ds^2 = d\theta^2 + \sin^2\theta\, d\varphi^2.$$

Take $\theta = x^1$, $\varphi = x^2$. Then (33) gives two equations (with $\lambda \equiv s$)

$$\frac{d^2\theta}{ds^2} - \sin\theta\cos\theta\left(\frac{d\varphi}{ds}\right)^2 = 0, \qquad \frac{d}{ds}\left(\sin^2\theta\,\frac{d\varphi}{ds}\right) = 0.$$

Integrating the second equation we get

$$\frac{d\varphi}{ds} = k\,\text{cosec}^2\,\theta, \quad \text{k} = \text{const}.$$

The first integral (32) gives

$$\sin^2\theta\left(\frac{d\theta}{ds}\right)^2 + k^2 = \sin^2\theta.$$

Thus the (θ, φ) equation for the geodesic is given by

$$\text{cosec}^2\,\theta\left(\frac{d\theta}{d\varphi}\right)^2 + 1 = k^{-2}\sin^2\theta.$$

Write $\mu = \cot\theta$ and $a^2 = k^{-2} - 1$. Then the above differential equation integrates to $\mu = a\sin(\varphi - \varphi_0)$, where φ_0 is a constant. This curve can be easily identified as a great circle. □

For null geodesics, the procedure outlined above breaks down because $\varphi^2 = 0$, $\mathrm{d}s = 0$. Nevertheless, (33) can be obtained even for null geodesics by stationarizing the integral

$$\int_0^1 \varphi^2 \, \mathrm{d}\lambda.$$

In this case s cannot be used as affine parameter. The parameter λ in (33) is then determined by

$$g_{ik} \dot{x}^i \dot{x}^k = 0. \tag{34}$$

Later we shall recast the geodesic equation in a form that is more commonly known in literature.

6.5. Parallel Transport

The second geometrical property that we need to impose on the curved spacetime is the notion of parallel transport. The need for this property is felt for two reasons, one physical and the other mathematical. Let us consider the mathematical reasons first.

The Euclidean notion of a straight line contains two important criteria for determining whether a curve is a straight line. We have already examined one criterion in the previous section, that of shortest distance. The notion of shortest distance is replaced by that of stationary distance in curved spacetime. The second property is contained in the adjective 'straight'. A straight line does not change its direction from point to point: if we let a point P run along the straight line, the tangent to the line at P does not change in direction. How do we describe the change in the direction of a vector as it moves along a curve?

The physical reason relates to the expression of physical laws by differential equations of scalars, vectors, or tensors, with respect to space and time variables. Do spacetime derivatives of vectors transform as tensors? An affirmative answer is essential if we want to give a generally covariant form to these equations. The answer to the above question is, however, negative as can be verified below.

Define $V'^k{}_{,l} \equiv \partial V'^k/\partial x'^l$ and differentiate the vector relation (14) to get

$$V'^k{}_{,l} = \frac{\partial x^n}{\partial x'^l} \frac{\partial}{\partial x^n} \left\{ \frac{\partial x'^k}{\partial x^m} V^m \right\}$$

$$= \frac{\partial x^n}{\partial x'^l} \frac{\partial x'^k}{\partial x^m} V^m{}_{,n} + \frac{\partial^2 x'^k}{\partial x^n \, \partial x^m} \frac{\partial x^n}{\partial x'^l} V^m. \tag{35}$$

For $V^m{}_{,n}$ to transform as a tensor the second term on the right-hand side should not appear. This anomalous term gives the clue to why $V^m{}_{,n}$ is not a tensor.

By definition $V^m{}_{,n}$ is given by

$$\lim_{\delta x^n \to 0} \frac{V^m(x^k + \delta x^k) - V^m(x^k)}{\delta x^n}. \tag{36}$$

The numerator in the above expression takes the difference of two vectors at neighbouring points. As we saw earlier in Section 6.3, this subtraction will not give a vector. Hence the ratio whose limit is $V^m{}_{,n}$ is not itself a tensor.

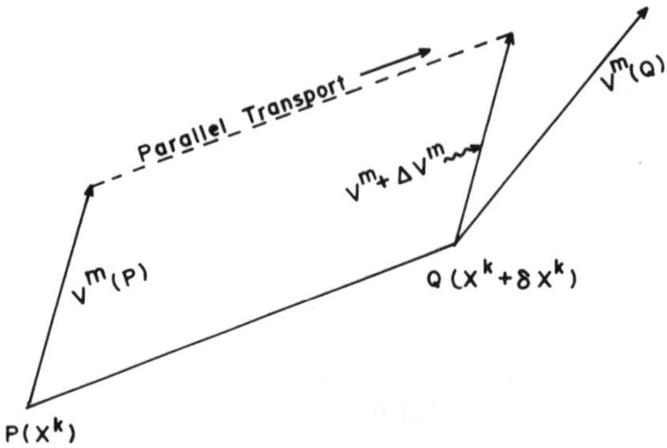

Fig. 6.6. The parallel transport of a vector from P to Q.

Figure 6.6 shows what needs to be done to ensure that we do get a tensor and at the same time obtain a derivative of V^m. P is the point $\{x^k\}$ and Q the point $\{x^k + \delta x^k\}$. At Q we are given the value of the vector field V^m. To find the difference between this value and the value of V^m at P we should first transport V^m parallel to itself from P to Q, ensuring that its value is not changed.

Since the general coordinate system is curvilinear, we do not expect the components of our parallel-transported vector to be the same at P and Q. In general, we expect the change in V^m to scale with V^m as well as with δx^n, the shift in coordinates. Let us write the change as

$$\Delta V^m = -\Gamma^m_{pn} V^p \, \delta x^n. \tag{37}$$

The quantities Γ^m_{pn} are functions of x^k and are known as the *affine connection.* Since they were explicitly considered first by Christoffel in 1869, they are known as 'Christoffel symbols' or '3-index symbols'.

The above operation is called the *parallel transport* of a contravariant vector. Note that there is nothing in the geometry so far developed to tell us what the Γ^i_{kl} should be. Rather, our choice of the Γ's will determine the nature of spacetime geometry. Later we shall give a prescription that will determine the Christoffel symbols completely in terms of the metric tensor and its derivatives. First, we shall outline a few general properties of these quantities.

To begin with, we note that a scalar quantity is not expected to change by parallel transport. Hence,

$$\Delta\varphi = 0. \tag{38}$$

Next, by writing $\varphi = V^i B_i$, where B_i is an arbitrary covariant vector, and using (37) we get

$$\Delta B_i = \Gamma^p_{in} B_p \, \delta x^n. \tag{39}$$

Similarly, assuming that the change in a tensor T_{ik} is like that of the product $V_i V_k$, we get

$$\Delta T_{ik} = (\Gamma^p_{in} T_{pk} + \Gamma^p_{kn} T_{ip}) \, \delta x^n. \tag{40}$$

Using these definitions we can now define the derivative of a vector in a covariant fashion. To do so we modify the numerator in (36) by replacing $V^m(x^k)$ by $V^m + \Delta V^m$, and then taking the limit. To distinguish it from the ordinary derivative $V^m{}_{,n}$ we write the covariant derivative as $V^m{}_{;n}$. Thus

$$V^m{}_{;n} = V^m{}_{,n} + \Gamma^m_{pn} V^p. \tag{41}$$

Similarly, the covariant derivative of a covariant vector is given by

$$B_{m;n} = B_{m,n} - \Gamma^p_{mn} B_p. \tag{42}$$

How do the Γ's transform under a general coordinate transformation? Knowing that $V^m{}_{;n}$ does transform as a tensor we can use (35) and write

$$V'^k{}_{;l} = \frac{\partial x^n}{\partial x'^l} \frac{\partial x'^k}{\partial x^m} V^m{}_{;n}$$

to deduce that

$$\Gamma'^k_{il} = \frac{\partial x'^k}{\partial x^m} \frac{\partial x^p}{\partial x'^i} \frac{\partial x^n}{\partial x'^l} \Gamma^m_{pn} + \frac{\partial^2 x'^m}{\partial x^i \, \partial x^l} \frac{\partial x^k}{\partial x'^m} \;. \tag{43}$$

Example 6.9. Let us prove (43) with the help of (42). In primed coordinates

$$B_m = \frac{\partial x'^l}{\partial x^m} B'_l \;,$$

and hence

$$B_{m,n} = \frac{\partial^2 x'^l}{\partial x^n \, \partial x^m} B'_l + \frac{\partial x'^l}{\partial x^m} \frac{\partial x'^k}{\partial x^n} B'_{l,k} \,.$$

Therefore, from (42)

$$B_{m;n} = \frac{\partial^2 x'^l}{\partial x^n \, \partial x^m} B'_l + \frac{\partial x'^l}{\partial x^m} \frac{\partial x'^k}{\partial x^n} B'_{l,k} - \Gamma^p_{mn} \frac{\partial x'^l}{\partial x^p} B'_l \,.$$

However, because $B_{m;n}$ transforms as a tensor,

$$B_{m;n} = \frac{\partial x'^p}{\partial x^m} \frac{\partial x'^k}{\partial x^n} B'_{p;k}$$

$$= \frac{\partial x'^p}{\partial x^m} \frac{\partial x'^k}{\partial x^n} B'_{p,k} - \frac{\partial x'^p}{\partial x^m} \frac{\partial x'^k}{\partial x^n} \Gamma'^l_{pk} B'_l.$$

Equate the two expressions for $B_{m;n}$ and use the fact that B'_l is arbitrary to get

$$\Gamma^p_{mn} \frac{\partial x'^l}{\partial x^p} = \frac{\partial^2 x'^l}{\partial x^n \partial x^m} + \frac{\partial x'^p}{\partial x^m} \frac{\partial x'^k}{\partial x^n} \Gamma'^l_{pk} .$$

Multiply by $\partial x^q/\partial x'^l$ and the result follows. □

Thus the Γ's do not transform as a tensor but their antisymmetric part

$$\Omega^p_{mn} = \tfrac{1}{2} (\Gamma^p_{mn} - \Gamma^p_{nm}) \tag{44}$$

does transform as a tensor. As such, Ω^p_{mn} should carry some fundamental significance for spacetime geometry. There are non-Euclidean geometries in which $\Omega^p_{mn} \neq 0$ but Einstein himself preferred the simpler version with

$$\Omega^p_{mn} = 0, \qquad \Gamma^p_{mn} = \Gamma^p_{nm}. \tag{45}$$

Spacetimes with nonzero Ω^p_{mn} are said to have torsion.

Example 6.10. Suppose we have two affine connections Γ^i_{kl} and $\bar{\Gamma}^i_{kl}$ defined on the same spacetime and the same coordinate system. Then

$$Q^i_{kl} = \bar{\Gamma}^i_{kl} - \Gamma^i_{kl}$$

transforms as a tensor. This is because the 'inhomogeneous' term present in (43) is eliminated in the transformation law of Q^i_{kl}. □

In addition, Riemann's prescription for determining Γ^p_{mn} was adopted by Einstein. The resulting geometry is called 'Riemannian geometry'. In this geometry Γ^p_{mn} are so chosen that

$$g_{ik;l} \equiv 0. \tag{46}$$

These are 40 equations for the 40 unknown Γ^p_{mn} satisfying the symmetry condition (45). Their solution gives

$$\Gamma^p_{mn} = \tfrac{1}{2} g^{pq} \left[g_{qm,n} + g_{qn,m} - g_{mn,q} \right] . \tag{47}$$

Example 6.11. Using (45) and (46) we can solve for Γ^p_{mn} as follows. Writing (46) out fully we get

$$\Gamma_{k/il} + \Gamma_{i/kl} = g_{ik,l}$$

where $\Gamma_{k/il} \equiv g_{km}\Gamma^m_{il}$ are symmetric in (i, l). By a cyclic interchange of (i, k, l) we get two more relations:

$$\Gamma_{l/ki} + \Gamma_{k/li} = g_{kl,i}, \quad \Gamma_{i/lk} + \Gamma_{l/ik} = g_{il,k}.$$

Subtract the first relation from the sum of the above two. Since $\Gamma_{k/il} = \Gamma_{k/li}$, $\Gamma_{i/kl} = \Gamma_{i/lk}$ we get

$$2\Gamma_{l/ik} = g_{kl,i} + g_{li,k} - g_{ik,l}.$$

From this (47) follows. □

The following are identities involving the covariant derivatives and the Christoffel symbols:

$$\text{(i)} \quad \Gamma^i_{li} \equiv \frac{\partial}{\partial x^l}(\ln\sqrt{-g}) \tag{48}$$

$$\text{(ii)} \quad \Gamma^m_{np}\, g^{np} \equiv -\frac{1}{\sqrt{-g}}\frac{\partial}{\partial x^p}(\sqrt{-g}\; g^{pm}) \tag{49}$$

$$\text{(iii)} \quad V^k{}_{;k} \equiv \frac{1}{\sqrt{-g}}\frac{\partial}{\partial x^k}(\sqrt{-g}\; V^k) \tag{50}$$

$$\text{(iv)} \quad F^{ik}_{;k} = \frac{1}{\sqrt{-g}}\frac{\partial}{\partial x^i}(\sqrt{-g}\; F^{ik}) \quad (\text{for } F^{ik} = -F^{ki}). \tag{51}$$

The first two identities can be proved with the help of the relation

$$\mathrm{d}g = g g^{ik}\, \mathrm{d}g_{ik}, \tag{52}$$

which follows from the properties of determinants.

Example 6.12. To prove (48), for example, use (47):

$$\Gamma^i_{li} = \tfrac{1}{2} g^{ik}\,[g_{kl,i} + g_{ki,l} - g_{li,k}]\,.$$

By symmetry the first term cancels the last, leaving

$$\tfrac{1}{2} g^{ik} g_{ik,l} \equiv \frac{1}{2g}\; g_{,l} = (\ln\sqrt{-g}\,)_{,l}.$$

To prove (49) take the differential of the relation

$$g_{ik} g^{kl} = \delta^l_i$$

to get

$$g_{ik,m}\, g^{kl} = -\, g_{ik}\, g^{kl}_{,m}\;.$$

This relation, together with (52), gives (49). □

We end this section with the answer to the mathematical query raised in the beginning. Let us determine the condition on the curve $x^i \equiv x^i(\lambda)$ to be 'straight'.

The tangent to the curve at a point λ is given by the vector

$$u^i = \frac{\mathrm{d}x^i}{\mathrm{d}\lambda} \; . \tag{53}$$

The change in u^i under parallel transport along the curve to a neighbouring point specified by the parameter $\lambda + \delta\lambda$ is given by

$$\Delta u^i = -\Gamma^i_{kl} u^k u^l \, \delta\lambda.$$

This must be subtracted from the positional change δu^i in going from λ to $\lambda + \delta\lambda$ to get the real change:

$$\delta u^i - \Delta u^i = (u^i_{,k} u^k + \Gamma^i_{kl} \, u^k u^l) \, \delta\lambda.$$

For 'straightness' this must vanish. Thus, we get the equations of the straight line as

$$\frac{\mathrm{d}u^i}{\mathrm{d}\lambda} + \Gamma^i_{kl} u^k u^l = 0. \tag{54}$$

It is easily verified that for the Riemannian affine connection, Equation (54) is the same as the Equation (33) for a geodesic. Thus, in Riemannian geometry the 'shortest distance' and 'straightness' are properties of the same class of curves. We shall henceforth use (54) for describing a geodesic. For a non-null geodesic λ may be replaced by s.

Next we shall turn to the most crucial geometrical property of spacetime, namely its curvature, as the key to the phenomenon of gravitation is contained in this property.

6.6. The Curvature Tensor

It may be naïvely supposed that in going from metric (3) to metric (18) wherein g_{ik} are functions of coordinates (and are not constant) we have made a transition from a flat to a curved spacetime. This supposition is not correct. Consider two examples of metrics of type (18). In the first example we have

$$\mathrm{d}s^2 = c^2 \, \mathrm{d}t^2 - \mathrm{d}r^2 - r^2(\mathrm{d}\theta^2 + \sin^2\theta \, \mathrm{d}\varphi^2). \tag{55}$$

For $x^0 = t$, $x^1 = r$, $x^2 = \theta$, $x^3 = \varphi$, g_{22} and g_{33} are evidently not constant. However, a coordinate transformation can be found which changes (55) to form (3). Hence, a variable metric by itself does not guarantee a spacetime nontrivially different from Minkowski spacetime.

In the second example we have

$$\mathrm{d}s^2 = c^2 \, \mathrm{d}t^2 - \frac{\mathrm{d}r^2}{1 - r^2} - r^2(\mathrm{d}\theta^2 + \sin^2\theta \, \mathrm{d}\varphi^2). \tag{56}$$

In this case it is *not* possible to find a coordinate transformation that will change (56) to (3). We thus have a spacetime which is nontrivially different from Minkowski spacetime.

But how do we decide, without having to try out various coordinate transformations, which spacetimes are different from (3)? The answer to this question is provided by a fourth-rank tensor called the 'curvature tensor'. Denoted by R_{iklm}, this tensor vanishes for (3). The vanishing of this tensor implies that the spacetime is 'flat'. If this tensor does not vanish for a given spacetime, we say that it is 'curved'.

Before we give the prescription for computing R_{iklm} we try to undertand its geometrical significance. We shall do it in two different ways.

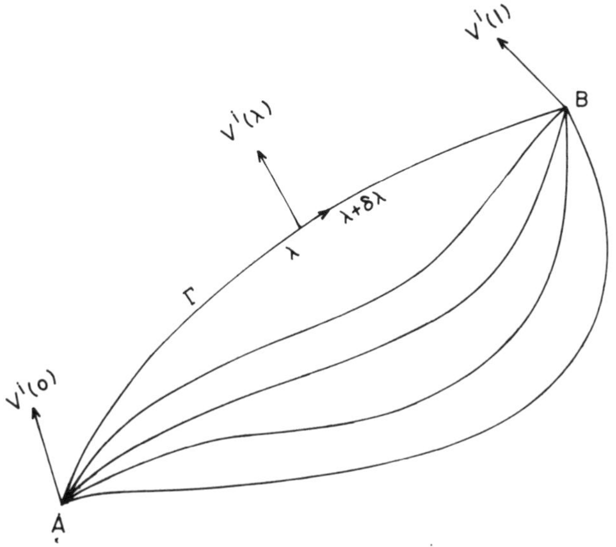

Fig. 6.7. The parallel transport of a vector V^i from A to B will depend on the curve Γ connecting A to B, unless the spacetime is 'flat'.

In Figure 6.7 we see various curves connecting two points A and B. Let us select a particular curve and label it by Γ. Suppose we have a vector V^i defined at A, which we shift to B by parallel transport along Γ. Let $x^i(\lambda)$ denote a typical point on Γ with $\{x^i(0)\} \equiv A$ and $\{x^i(1)\} \equiv B$.

At an intermediate point λ the vector in question may be denoted by $V^i(\lambda)$. When $\lambda \rightarrow \lambda + \delta\lambda$, we have

$$\begin{aligned} V^i(\lambda + \delta\lambda) &= -\,\Gamma^i_{kl} V^k(\lambda)\, \delta x^l \\ &= -\,\Gamma^i_{kl} V^k(\lambda) \dot{x}^l(\lambda)\, \delta\lambda. \end{aligned}$$

Thus $V^i(\lambda)$ satisfies the differential equation

$$\frac{\mathrm{d}V^i}{\mathrm{d}\lambda} = -\,\Gamma^i_{kl} V^k \dot{x}^l(\lambda). \tag{57}$$

For a given curve Γ, (57) has a unique solution for $V^i(1)$ at B. However, are we sure that the answer would have been the same had we chosen some other curve Γ' from A to B? The affirmative answer depends on our being able to solve the differential equations

$$\frac{\partial V^i}{\partial x^l} = -\Gamma^i_{kl} V^k . \tag{58}$$

The solvability of these equations is not guaranteed for arbitrary Γ^i_{kl}. To find the necessary condition on Γ^i_{kl} we differentiate (58) and use it again to get

$$\frac{\partial^2 V^i}{\partial x^m \partial x^l} = -\frac{\partial}{\partial x^m}(\Gamma^i_{kl} V^k)$$

$$= -\frac{\partial \Gamma^i_{kl}}{\partial x^m} V^k + \Gamma^i_{kl}\, \Gamma^k_{mn}\, V^n .$$

An interchange of indices m and l gives

$$\frac{\partial^2 V^i}{\partial x^l \partial x^m} = -\frac{\partial \Gamma^i_{km}}{\partial x^l} V^k + \Gamma^i_{km}\, \Gamma^k_{ln} V^n .$$

Since $V^i{}_{,lm} = V^i{}_{,ml}$, we get from the above relations the condition

$$R^i{}_{nml} V^n = 0, \tag{59}$$

where

$$R^i{}_{nml} = \frac{\partial \Gamma^i_{nl}}{\partial x^m} - \frac{\partial \Gamma^i_{nm}}{\partial x^l} - \Gamma^i_{kl}\, \Gamma^k_{mn} + \Gamma^i_{km}\, \Gamma^k_{ln} . \tag{60}$$

Since V^n is arbitrary, our condition (59) reduces to

$$R^i{}_{nml} = 0. \tag{61}$$

If, therefore, the above condition is not satisfied we cannot generate a vector field by parallely transporting V^i from A to arbitrary spacetime points. What if the condition is satisfied? A more sophisticated argument than that given above is needed to show that if (61) holds then it is immaterial what curve we use to transport V^i from A to B; we shall always get the same answer.

In Figure 6.8 we see an example of the former kind. Parallel transport of a vector along a closed curve on the surface of a sphere leads to a change in the direction of the vector from its initial direction. We shall use this property as an indicator of whether a given spacetime has curvature.

Before we consider another manifestation of the curvature property we note that the quantity $R^i{}_{nml}$ is indeed a tensor. To see this most directly, first prove the identity involving second covariant derivatives of an arbitrary vector:

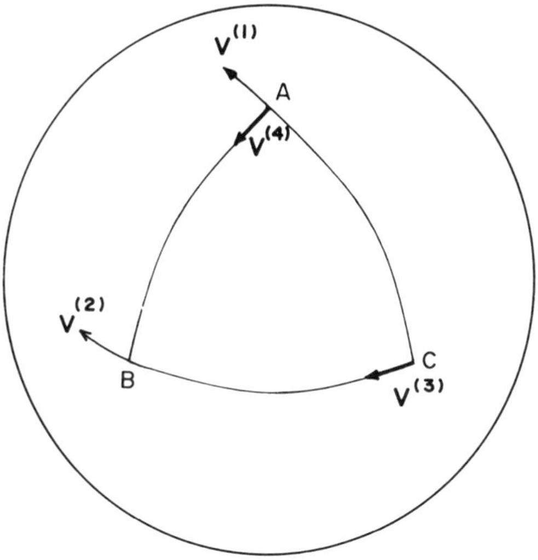

Fig. 6.8. The vector **V** is parallely transported along the spherical triangle *ABC*, with three right angles at *A*, *B*, and *C*. The superscript on **V** denotes the stage during the transport. The vector has turned through a right angle between stages 1 and 4.

$$V^i_{;lm} - V^i_{;ml} \equiv R^i_{nml} V^n. \tag{62}$$

Then use the quotient law to prove that R^i_{nml} *is a tensor.*

Example 6.13. The proof of (62) is straightforward. We have

$$V^i_{;l} = V^i_{,l} + \Gamma^i_{nl} V^n$$

$$V^i_{;lm} = V^i_{,lm} + \Gamma^i_{nl,m} V^n + \Gamma^i_{nl} V^n_{,m} + \Gamma^i_{mn} V^n_{;l} - \Gamma^n_{lm} V^i_{;n}$$

$$= V^i_{,lm} + \Gamma^i_{nl} V^n_{,m} + \Gamma^i_{mn} V^n_{,l} - \Gamma^n_{lm} V^i_{,n}$$

$$+ \left\{ \Gamma^i_{nl,m} + \Gamma^i_{mk}\Gamma^k_{nl} - \Gamma^k_{lm}\Gamma^i_{nk} \right\} V^n.$$

$V^i_{;ml}$ can be similarly written down by interchanging (l, m).
Hence,

$$V^i_{;lm} - V^i_{;ml} = \{ \Gamma^i_{nl,m} - \Gamma^i_{nm,l} + \Gamma^i_{mk}\Gamma^k_{nl} - \Gamma^i_{lk}\Gamma^k_{nm} \} V^n.$$

Since the left-hand side is a tensor and the V^n on the right-hand side is a vector, the 'quotient' in { } is a tensor. □

The second demonstration of spacetime curvature is illustrated in Figure 6.9. In (a) we have two straight lines emerging from a point O on a plane. Notice that the relative separation between the line increases at a uniform rate as we move away from O. In (b) we have two meridional great circles from the polar point O. These are the geodesics on the spherical surface.

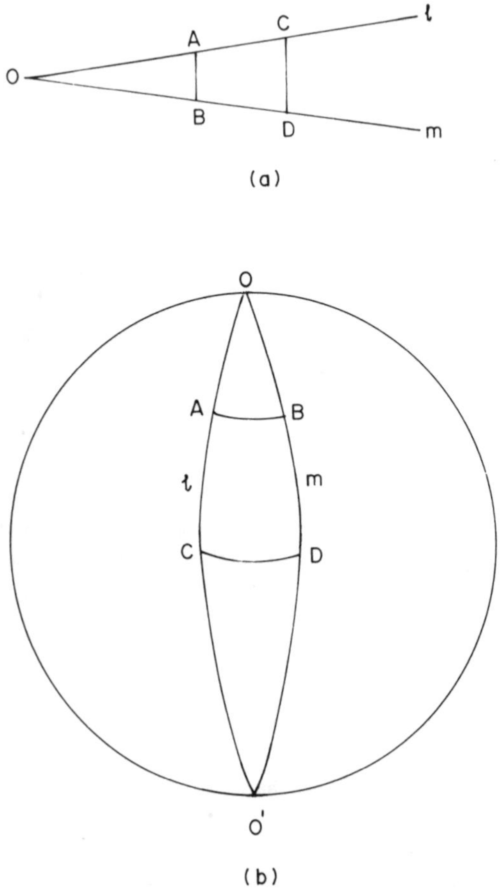

Fig. 6.9. In (a) we see two straight lines l and m diverging from point O, on a plane. Their separation increases uniformly. Thus $AB/CD = OA/OC$, indicating that as we move away from O the separation between l and m increases in proportion to the distance from O. In (b), l and m are arcs of great circles on a sphere. Their separation does not increase uniformly. It reaches a maximum (CD) on the equator with respect to O and then declines to zero at the antipode O'.

These geodesics diverge from each other at a rate that slows down to a halt, and eventually becomes negative.

This nonuniformity of the rate at which neighbouring geodesics separate distinguishes a curved spacetime from a flat one. We shall now work out this result quantitatively. To this end let $x^i(\lambda, \mu)$ denote a bundle of geodesics diverging from a point O. A fixed μ identifies a unique geodesic in the bundle on which λ is the affine parameter measured from O while for a fixed λ, μ identifies points of neighbouring geodesics all located at the same λ distance from O. We define two vectors u^i and v^i by

$$u^i = \left.\frac{\partial x^i}{\partial \lambda}\right|_\mu \quad \text{(tangent vector)},$$

$$v^i = \left.\frac{\partial x^i}{\partial \mu}\right|_\lambda \quad \text{(deviation vector)}. \tag{63}$$

We first note that

$$u^i_{;k}\, v^k = \left(\frac{\partial u^i}{\partial x^k} + \Gamma^i_{lk} u^l\right) v^k = \frac{\partial^2 x^i}{\partial \mu\, \partial \lambda} + \Gamma^i_{lk} u^l v^k$$

$$= \frac{\partial^2 x^i}{\partial \lambda\, \partial \mu} + \Gamma^i_{lk} u^l\, v^k = v^i_{;k}\, u^k. \tag{64}$$

Hence we get

$$(v^i_{;k}\, u^k)_{;l} u^l = (u^i_{;k}\, v^k)_{;l} u^l$$

$$= u^i_{;kl}\, v^k\, u^l + u^i_{;k}\, u^l\, v^k_{;l}$$

$$= (u^i_{;lk} + R^i_{\ mlk}\, u^m)\, v^k u^l + u^i_{;k}\, u^k_{;l}\, v^l$$

$$= (u^i_{;l}\, u^l)_{;k}\, v^k + R^i_{\ mlk}\, u^m v^k u^l,$$

where we have used (64). Since we are talking about geodesics $u^i_{;l} u^l = 0$. Hence, the above may be written as

$$\frac{\delta^2 v^i}{\delta \lambda^2} = R^i_{\ mlk} u^m u^l v^k, \tag{65}$$

where $\delta/\delta\lambda$ denotes differentiation along a geodesic μ = const.

Thus for $R^i_{\ mlk} = 0$, v^i increases uniformly with λ as in Figure 6.9(a). The general case of nonzero $R^i_{\ mlk}$ shows that the separation between neighbouring geodesics changes nonuniformly with the common affine parameter. Equation (65) is called the 'equation of geodetic deviation'.

Thus we again encounter the tensor $R^i_{\ klm}$ which was first proposed by Riemann in 1854 in his dissertation on geometry. It is sometimes called the "Riemann tensor' or simply the 'Riemann–Christoffel tensor' or simply the 'curvature tensor'. We shall use the last name more often. The curvature tensor has certain symmetry properties which we shall discuss in Section 6.8.

6.7. Physics in Curved Spacetime

Having outlined some of the basic features of Riemannian geometry we now turn to some physical inputs. We begin with a discussion of the so-called *principle of equivalence*.

In Section 6.5 we noted that the Christoffel symbols are not tensors. This property raises the following question: "Is it possible to choose a coordinate system in which all Γ's vanish at a given point?" The answer is 'Yes' and the transformation law (43) suggests how to go about it.

First choose near the given point P a coordinate system $\{x^i\}$ such that

$$x^i_p = 0\,, \qquad g_{ik}(P) = \eta_{ik}.$$

This is always possible. Suppose we have $\Gamma^m_{pn} \neq 0$ at P. Then choose a new coordinate system $\{x'^i\}$ in the neighbourhood of P with the following transformation law:

$$x'^i = x^i - \tfrac{1}{2}(\Gamma^i_{kl})_P x^k x^l. \tag{66}$$

Then at P we have

$$\frac{\partial x'^i}{\partial x^k} = \delta^i_k, \qquad \frac{\partial^2 x'^i}{\partial x^k\,\partial x^l} = -\,(\Gamma^i_{kl})_P.$$

From (43) it then follows that $\Gamma'^k_{il} = 0$ at P.

The vanishing of the Γ's at P ensures that $g_{ik,\,l}$ also vanish at P. Thus the metric *appears to be* that of special relativity not only at P but also in a small neighbourhood of P. The words 'appears to be' are important, for in reality the spacetime at P is *not flat*. Because the derivatives of Γ^i_{kl} are not zero at P, the curvature tensor is nonzero.

In fact, it can be shown that the vanishing of R_{iklm} in a small domain N_p around P ensures the existence of coordinates in N_p which give the metric (3) throughout N_p. If the curvature tensor is nonvanishing at P, the utmost simplification that can be achieved is the vanishing of the Γ's at P. This is like the 'flat-Earth' approximation which is valid over any small area on the Earth's surface.

Such a coordinate system is called 'locally inertial'. Bearing in mind our goal of identifying gravitation with curved spacetime, we can say that an observer using the locally inertial frame has the momentary and local illusion that he is under the influence of no forces. For, in his local spacetime neighbourhood the metric appears to him to be that of special relativity. However, as the man on Earth discovers the limits of his flat-Earth approximation by looking over large distances, so would a locally inertial observer find that his simulation of a truly inertial observer is valid over a small region of spacetime in his vicinity.

In formulating the general theory of relativity Einstein made considerable use of the notion of locally inertial observers. His celebrated example of a man in a freely falling lift describes this very notion. A man in such a lift feels a temporary relief from the Earth's gravitational force. The astronaut in a terrestrial satellite feels weightless for the same reason. His weightlessness is only local.

These examples of weightlessness do not give prescriptions for eliminating the Earth's gravitational field entirely. Nor do two freely falling observers fail to notice the effect of Earth's gravity if they observe it closely enough. For, such observers follow timelike geodesics and we have seen that geodetic deviation reveals the existence of the curvature tensor.

Example 6.14. Consider two observers freely falling on the Earth under Newtonian gravitation. To fix ideas, suppose that both the observers start from rest at $t = 0$ from positions $\mathbf{r} = \mathbf{a}$ and $\mathbf{r} = \mathbf{a} + \Delta_0$ measured from the centre of the Earth. The trajectory of the first observer is given by

$$\mathbf{r} = \mathbf{a}\, f(t),$$

where, from Newtonian physics, we get

$$\ddot{f} = -\frac{GM}{a^3 f^2}$$

with M = mass of the Earth. Writing $\tau = (2GM/a^3)^{1/2} t$ we get

$$2\frac{d^2 f}{d\tau^2} = -\frac{1}{f^2},$$

which has the integral

$$\left(\frac{df}{d\tau}\right)^2 = \frac{1-f}{f},$$

since at $\tau = 0$, $(t = 0)$, $f = 1$, $\dot{f} = 0$. Let $f = F(\tau)$ denote the solution of this equation with the given initial conditions. Then

$$\mathbf{r} = \mathbf{a}F\left(\sqrt{\frac{2GM}{a^3}}\, t\right).$$

Similarly, for the second observer we get

$$\mathbf{r} + \Delta = (\mathbf{a} + \Delta_0)\, F\left(\sqrt{\frac{2GM}{|\mathbf{a} + \Delta_0|^3}}\, t\right).$$

Notice that the relative separation Δ between the observers at any given time t varies with t. Hence, even in a freely falling lift the observers can in principle discover the effect of Earth's gravity. For $|\Delta_0| \ll a$ we get

$$\Delta(t) \approx \Delta_0 F - \frac{3}{2}\tau\, \frac{\mathbf{a}\cdot\Delta_0}{a^2}\sqrt{\frac{1-F}{F}}\, \mathbf{a}\,. \qquad \square$$

Nevertheless, the notion that a suitably accelerated frame can simulate or neutralize a gravitational field, albeit locally, is called the 'principle of equivalence' in its *weak* form. In a refined form this principle is the same as Galileo's law of falling bodies. And like Galileo's law and its quantification by Newton's laws of motion and gravitation, the weak principle of equivalence implies equality of inertial and gravitational masses of any particle.

There is, however, a *strong* version of the principle of equivalence that tells us how to write the equations of any physical interaction in a generally covariant form. The prescription is as follows. Use special relativity to write down the equations in the locally inertial frame. Then change all derivatives from ordinary derivatives to covariant ones. We illustrate this with the example of electromagnetic theory.

In flat spacetime we write the electromagnetic field tensor as the skew derivative of the potential:

$$F_{ik} = A_{k,i} - A_{i,k}. \tag{67}$$

The electric field is identified with the components (F_{01}, F_{02}, F_{03}) and the magnetic field with (F_{23}, F_{31}, F_{12}). In the covariant language of curved spacetime (67) becomes

$$F_{ik} = A_{k;i} - A_{i;k}. \tag{68}$$

Because of symmetries of the Christoffel symbols, (68) reduces to (67), however. The gauge condition changes as follows:

$$A^i_{,i} = 0 \rightarrow A^i_{;i} = 0. \tag{69}$$

The Maxwell source equations change likewise:

$$F^{ik}_{,k} = j^i \rightarrow F^{ik}_{;k} = j^i. \tag{70}$$

The Lorentz force equation is altered to

$$m\frac{\mathrm{d}^2x^i}{\mathrm{d}s^2} = eF^i_k\frac{\mathrm{d}x^k}{\mathrm{d}s} \rightarrow m\left(\frac{\mathrm{d}^2x^i}{\mathrm{d}s^2} + \Gamma^i_{kl}\frac{\mathrm{d}x^k}{\mathrm{d}s}\frac{\mathrm{d}x^l}{\mathrm{d}s}\right) = eF^i_k\frac{\mathrm{d}x^k}{\mathrm{d}s} \tag{71}$$

Notice that the above rule allows us to write in an unambiguous way the laws of physics in curved spacetime. Since we are going to identify spacetime curvature with gravitation, the strong principle of equivalence tells us how the laws of physics operate in the presence of gravitation. Such a link of a physical interaction with gravitation is called *minimal* coupling. We are now able to use it to rewrite the formal expressions of Part I in a given curved spacetime.

We end this section by describing a novel aspect introduced into physics by minimal coupling. This concerns the derivation of the energy momentum tensor of an interaction from an action principle. We state the problem in general terms.

Let φ be a field (scalar, vector, or a tensor) which is introduced through a Lagrangian density

$$\mathscr{L} = \mathscr{L}\{\varphi, \varphi_{,i}\}. \tag{72}$$

In flat spacetime the action is written as

$$S = \int_{\mathcal{V}} \mathscr{L}\,\mathrm{d}^4x, \tag{73}$$

where $\mathcal{V}$ is a four-dimensional region. By the strong principle of equivalence we change L by replacing $\varphi_{,i}$ by $\varphi_{;i}$. In writing the action we have to replace the 4-volume element d^4x by $\sqrt{-g}\,\mathrm{d}^4x$, thus

$$S = \int \mathscr{L}\{\varphi, \varphi_{;i}\}\sqrt{-g}\,\mathrm{d}^4x. \tag{74}$$

(For, under a change of coordinates $\sqrt{-g}\, \mathrm{d}^4x$ behaves like a scalar, as can be seen by using (25) and (26).)

To obtain the field equations from S we use the variational principle. Thus, formally, the field equations are given by

$$\frac{\delta S}{\delta \varphi} = 0. \tag{75}$$

If the field interacts with some forms of matter, the latter may appear in (75) as source terms.

Notice, however, that S now contains an additional piece of the unknown, the spacetime geometry. The metric tensor and its derivatives which appear through the Γ's could also be regarded as independent functional variables. What will we get by the variation $g_{ik} \to g_{ik} + \delta g_{ik}$? We may define the answer formally as

$$\delta S = -\int \tfrac{1}{2} T^{ik}\, \delta g_{ik} \sqrt{-g}\, \mathrm{d}^4x. \tag{76}$$

For the electromagnetic field we have

$$S = -\ \tfrac{1}{4} \int F_{ik} F^{ik} \sqrt{-g}\, \mathrm{d}^4x \tag{77}$$

and hence

$$T^{ik} = -\ [F^i_l F^{kl} - \tfrac{1}{4} F_{mn} F^{mn} g^{ik}]. \tag{78}$$

T^{ik} is, of course, the energy momentum tensor of the electromagnetic field. From this example we may use (76) to define the energy momentum tensor of the φ field. Notice that since g_{ik} is symmetric, T^{ik} is also symmetric.

Example 6.15. Let us work out the energy tensor for a system of particles, using the above technique. Suppose we have a system of particles $a, b, \ldots$ of masses $m_a, m_b, \ldots$ and coordinates $x^i_a, x^i_b, \ldots$ in $\mathcal{V}$. Using $s_a, s_b, \ldots$ to denote the proper times of these particles we have the action describing them as

$$S_m = -\sum_a \int m_a\, \mathrm{d}s_a.$$

Under a variation of the metric tensor

$$\mathrm{d}s^2 = g_{ik}\, \mathrm{d}x^i\, \mathrm{d}x^k \to 2\mathrm{d}s\, \delta\, (\mathrm{d}s) = \delta g_{ik}\, \mathrm{d}x^i\, \mathrm{d}x^k,$$

that is,

$$\delta S_m = -\sum_a \int \tfrac{1}{2} m_a \frac{\mathrm{d}x^i_a}{\mathrm{d}s_a} \frac{\mathrm{d}x^k_a}{\mathrm{d}s_a}\, \delta g_{ik}\, \mathrm{d}s_a.$$

To be precise, the variation of δg_{ik} effective for particle a is only from that thin tube of $\mathcal{V}$ through which the particle passes. We may express this by writing

$$\underset{(\mathrm{m})}{T^{ik}}(x) = \sum_a \int m_a \frac{\mathrm{d}x^i_a}{\mathrm{d}s_a} \frac{\mathrm{d}x^k_a}{\mathrm{d}s_a} \frac{\delta_4(x - x_a)}{\sqrt{-g(x)}}\, \mathrm{d}s_a.$$

where the underscript (m) denotes matter and the delta function contributes only if the field point x falls on any particular worldline. □

Example 6.16. The above general expression for a system of particles can be simplified in certain typical situations. We shall continue to assume $c = 1$.

In the 'dust' approximation, all particles move essentially parallel to each other with no relative motion. Thus we can represent the flow of dust by a single velocity vector $v^k = dx^k_a/ds_a$ for all a. Writing

$$\varrho(x) = \sum_a \int m_a \frac{\delta_4(x-x_a)}{\sqrt{-g(x)}} \, ds_a$$

for the 'density' of dust we get

$$\underset{(m)}{T^{ik}}(x) = \varrho v^i v^k .$$

In the 'fluid' approximation we assume that although there is an overall flow vector v^k for the fluid, the individual particles have small random motions about it which give rise to pressures. Thus, we write

$$\frac{dx^k_a}{ds_a} = v^k + w^k_a ,$$

where w^k_a denotes the random motion part, and assume that $|w^k_a| \ll 1$ as well as $v_k w^k_a = 0$. Then

$$m_a \frac{dx^i_a}{ds_a} \frac{dx^k_a}{ds_a} = m_a(v^i + w^i_a)(v^k + w^k_a) .$$

We continue to define density here as we did earlier for dust. The isotropic pressure in the fluid is defined by p, where

$$p(v^i v^k - g^{ik}) = \sum_a \int m_a \frac{\delta_4(x - x_a)}{\sqrt{-g(x)}} w^i_a w^k_a \, ds_a .$$

The coefficient of p arises because we have assumed that the vectors of random motion are orthogonal to the flow vector v^k. It is also assumed that linear terms in random motions average out to zero. Thus, we get

$$\underset{(m)}{T^{ik}} = (p + \varrho)\, v^i v^k - p g^{ik} .$$

Finally, when we are dealing with a relativistic gas, where particles with speeds comparable to the speed of light are moving randomly, the form for $\underset{(m)}{T^{ik}}$ becomes

$$\underset{(m)}{T^{ik}} = \tfrac{4}{3} \varrho v^i v^k - \tfrac{1}{3} \varrho g^{ik} .$$

Note that in this case the trace of $\underset{(m)}{T^{ik}}$ vanishes. □

This method of obtaining T^{ik} may be contrasted with the method of flat space field theories (cf. Chapter 4) which arrive at T^{ik} in an *ad hoc* manner and which still leave its final form undetermined to within a tensor of zero divergence. The above method gives a unique and symmetric tensor for T^{ik}.

The fact that S is independent of coordinates enables us to deduce a conservation law, as would be expected by Noether's theorem.

Consider an infinitesimal change of coordinates:

$$x^i = x'^i + \xi^i(x'). \tag{79}$$

Then g_{ik} changes to

$$g'_{ik} = (\delta^l_i + \xi^l_{,i})\,(\delta^m_k + \xi^m_{,k})\, g_{lm}$$
$$\cong g_{ik} + \xi_{i,k} + \xi_{k,i}\,;$$

that is,

$$\delta g_{ik} = g'_{ik} - g_{ik} = \xi_{i,k} + \xi_{k,i}\,. \tag{80}$$

Here we have neglected quantities of second or higher order in ξ_i and its derivatives.

Since S does not change by (79) we get from (76)

$$\int_{\mathcal{V}} T^{ik}\,(\xi_{i,k} + \xi_{k,i})\,\sqrt{-g}\;\; \mathrm{d}^4x = 0,$$

that is,

$$\int_{\mathcal{V}} T^{ik}\xi_{i,k}\,\sqrt{-g}\;\; \mathrm{d}^4x = 0;$$

for $T^{ik} = T^{ki}$.

Choose $\mathcal{V}$ to be an infinitesimal neighbourhood around a point P and use locally inertial coordinates at P. Then the above relation becomes

$$\int_{\mathcal{V}} T^{ik}\xi_i n_k\; \mathrm{d}^3x - \int_{\mathcal{V}} T^{ik}{}_{,k}\xi_i\; \mathrm{d}^4x = 0,$$

where we have used Green's theorem and $\partial\mathcal{V}$ denotes the boundary surface of $\mathcal{V}$. Assuming that ξ_i are arbitrary but vanish on $\partial\mathcal{V}$, we get from the above

$$T^{ik}{}_{,k} = 0.$$

In generally covariant form, this equation becomes

$$T^{ik}{}_{;k} = 0. \tag{81}$$

This is the law of conservation of energy in curved spacetime.

Example 6.17. In the above calculation we have used Green's theorem. In a covariant form this theorem may be stated as follows:

$$\int_{\mathcal{V}} A^i{}_{;i}\,\sqrt{-g}\;\; \mathrm{d}^4x = \int_{\partial\mathcal{V}} A^i n_i\,\sqrt{-g}\;\; \mathrm{d}^3x,$$

where A^i is a contravariant vector field. To prove this use (50) and write the left-hand side as

$$\int_v \frac{\partial}{\partial x^i} (\sqrt{-g}\; A^i)\, d^4x = \int_{\partial v} \sqrt{-g}\; A^i n_i\, d^3x.$$

□

6.8. Einstein's Field Equations

We now come to the central issue of general relativity, the quantification of Einstein's notion that non-Euclidean geometry means gravitation. To this end we recall from Section 6.1 that the energy momentum tensor T_{ik} should serve as the 'source' of gravitation, if we want to generalize the Newtonian–Poissonian equation. However, we do not yet have any tensor potentials. What we need is a second-rank tensor that will serve as the potential tensor but which will also have a geometrical significance.

A clue to this problem is provided by Einstein's example of the falling lift described in Section 6.7. There the gravitational field of the Earth was seen to be cancelled, albeit locally, by the acceleration of the freely falling lift. We shall now consider the reverse situation: in a flat Minkowski spacetime we shall place a uniformly accelerated observer and then try to construct the line element in his rest frame. If we can cancel gravity by a local acceleration then this example will tell us whether uniform acceleration can generate a gravity-like situation.

First consider an inertial rectangular coordinate system F_1: $Oxyz$ in Minkowski spacetime and let the observer A start at rest from O and move along Ox with uniform acceleration g. Then special relativity tells us that if A measures his proper time t' from the instant he was at O at $t = 0$, then he is at x at time t where

$$x = \frac{c^2}{g}\left(\cosh\frac{gt'}{c} - 1\right), \qquad t = \frac{c}{g}\sinh\frac{gt'}{c}\,, \tag{82}$$

and A has velocity u in the x direction, given by

$$u = \frac{gt}{\sqrt{1 + (g^2t^2/c^2)}} = c\tanh\frac{gt'}{c}\,. \tag{83}$$

Now introduce another inertial frame F_2: $O'\ X'\ Y'\ Z'$ whose origin instantaneously coincides with A, has speed u in the direction Ox and has $O'\ X'$ lying along Ox while Oy, Oz are parallel to $O'\ Y'$, $O'\ Z'$. The Lorentz transformation relates the coordinates of a typical event:

$$x = \frac{c^2}{g}\left(\cosh\frac{gt'}{c} - 1\right) + \frac{X' + uT'}{\sqrt{1 - (u^2/c^2)}}\,, \quad y = Y',\ z = Z', \tag{84}$$

$$t = \frac{c}{g}\sinh\frac{gt'}{c} + \frac{T' + (uX'/c^2)}{\sqrt{1 - (u^2/c^2)}}\,. \tag{85}$$

Here we have measured T' from the instant O' coincides with A.

From these two inertial frames we now generate a third noninertial frame F' comoving with A with Cartesian axes parallel to Ox, Oy, Oz. In this frame the coordinates of the above event are (x', y', z', t') of which the spatial coordinates are identified with the coordinates (X', Y', Z') measured relative to F_2 when its origin O' happens to coincide with A. From (84) and (85) these values are obtained by setting $X' = x'$, $Y' = y'$, $Z' = z'$ and $T' = 0$. We then get

$$x = \frac{c^2}{g}\left(\cosh\frac{gt'}{c} - 1\right) + x'\cosh\frac{gt'}{c}, \qquad y = y', \quad z = z', \tag{86}$$

$$t = \frac{c}{g}\sinh\frac{gt'}{c} + \frac{x'}{c}\sinh\frac{gt'}{c}. \tag{87}$$

It is now a simple matter to construct the line element for the coordinate system (x', y', z', t'). A simple calculus gives

$$ds^2 = c^2\left(1 + \frac{gx'}{c^2}\right)^2 dt'^2 - dx'^2 - dy'^2 - dz'^2. \tag{88}$$

For $|gx'/c^2| \ll 1$ we may approximate g_{00} in (88) by

$$g_{00} \cong c^2(1 + h_{00}), \quad h_{00} = \frac{2gx'}{c^2}. \tag{89}$$

If we invoke an artificial gravity with potential φ to compensate for this acceleration, we may write

$$\varphi' \approx -gx', \quad h_{00} = -\frac{2\varphi'}{c^2}. \tag{90}$$

In other words, g_{00} appears to have some relationship with the Newtonian potential, at least when $|h_{00}| \ll 1$.

At first sight we may therefore conclude that it is the metric tensor that plays the role of the tensor potential. However, the covariant wave operator applied on g_{ik} gives

$$\Box g_{ik} \equiv g^{lm}g_{ik;lm} \equiv 0 \tag{91}$$

identically, because of (46). We therefore need something else.

The only other geometrical tensor we have so far is the curvature tensor R_{iklm}, a tensor of fourth rank. Can we get a second-rank tensor out of this? Before we answer this question we first consider the symmetry properties of this tensor.

From (62) we note that

$$R_{iklm} = -R_{ikml} \tag{92}$$

There are, however, more symmetries in this tensor. To discover them it is convenient to go to the locally inertial frame. In this frame all Γ's vanish and

$g_{ik} = \eta_{ik}$ at the point in question. A little calculation will show that in this frame

$$R_{iklm} = g_{il,km} - g_{im,kl} + g_{km,il} - g_{kl,im}, \tag{93}$$

From this expression it is easy to deduce that, apart from (92), the following symmetries also hold:

$$R_{iklm} = -R_{kilm} = R_{lmik}, \tag{94}$$

$$R_{iklm} + R_{imkl} + R_{ilmk} = 0. \tag{95}$$

Because of these properties the number of independent components of R_{iklm} is considerably smaller than the total ($4^4 = 256$) number of all its components. To determine the former, note that because of antisymmetry the first two indices generate only six combinations, as do the last two components. Thus the total number of independent components cannot exceed $6 \times 6 = 36$. However, since the first two and last two components can be exchanged without affecting the tensor, the number of independent components is really not more than $36 - ((6 \times 5)/2) = 21$. Condition (95) imposes one more linear constraint on the components, so that the effective number of independent components is finally reduced to 20.

In a symmetric energy momentum tensor there are ten independent components. Correspondingly we need a symmetric geometrical tensor of second rank with ten independent components. Can such a tensor be constructed from R_{iklm}, say, by the process of contraction? It is easy to see that there is essentially one second-rank tensor that can be so constructed. This is called the 'Ricci tensor' and is defined by

$$R_{kl} = g^{im} R_{iklm}. \tag{96}$$

It is easy to verify that this tensor is symmetric; for,

$$R_{lk} = g^{im} R_{ilkm} = g^{im} R_{kmil} = g^{im} R_{mkli} = R_{kl}. \tag{97}$$

If we contract further we get a scalar

$$R = g^{lk} R_{lk}. \tag{98}$$

This is called the 'scalar curvature'. Using (96) and (98) we define another symmetric second-rank tensor:

$$G_{lk} = R_{lk} - \tfrac{1}{2} g_{lk} R. \tag{99}$$

This tensor is known as the 'Einstein tensor'. It was this tensor that Einstein finally selected as the required second-rank tensor for his field equations.

The hueristic argument used by Einstein to arrive at the field equations for general relativity can now be followed. First notice that both R_{ik} and G_{ik} contain spacetime derivatives of g_{ik} up to second order. We shall later see that in the linearized approximation of Einstein's equations we do get something like the wave equation. Further, the tensor G_{ik} satisfies the so-called 'Bianchi identity'

$$G^{ik}{}_{;k} \equiv 0. \tag{100}$$

This can be proved from the more general form of the identity (also proved by Bianchi)

$$R_{iklm;n} + R_{iknl;m} + R_{ikmn;l} = 0. \tag{101}$$

Example 6.18. To prove (101) use the locally inertial frame with R_{iklm} given by (93). The result follows immediately. For (100) multiply (101) by g^{im} and use (96). We get

$$R_{kl;n} + g^{im}R_{kiln;m} - R_{kn;l} = 0.$$

Now multiply by g^{kn} to get

$$R^n{}_{l;n} + R^m{}_{l;m} - R_{;l} = 0,$$

that is,

$$2(R^n{}_l - \tfrac{1}{2}\delta^n{}_l R)_{;n} = 0.$$ □

Hence, if write

$$G_{ik} = -\varkappa T_{ik}, \quad \varkappa = \text{const}, \tag{102}$$

we have, by virtue of (100), the result

$$T^{ik}{}_{;k} \equiv 0. \tag{103}$$

Thus, while we could have used R_{ik} in place of G_{ik} so far as the criteria of symmetry, rank, and spacetime derivatives of the metric tensor are concerned, the Bianchi identity decides in favour of G_{ik}. By choosing G_{ik} we automatically ensure that any tensor appearing on the right-hand side of (102) will have zero divergence.

Although (102) contains ten partial differential equations for the ten different g_{ik}, the identity (100) imposes four constraint equations, thus reducing the number of independent equations to six. Does his mean that the system as a whole is undetermined? Not so! For we have to remember that the equations are covariant under coordinate transformations $x^i \to x'^i$. Thus, if $g_{ik}(x^l)$ is a solution, then so is $g'_{ik}(x'^l)$, where

$$g'_{ik} = \frac{\partial x^l}{\partial x'^i}\,\frac{\partial x^m}{\partial x'^k}\,g_{lm}.$$

Although the four arbitrary functions $x'^i(x^k)$ remain undetermined, their indeterminacy does not imply an indeterminacy of physical information. Notice also that (103) was derived in Section 6.7 from precisely these considerations of invariance.

Equations (102) are nonlinear and fulfil our expectation of Section 6.1 that the gravitational field equations should be nonlinear. It remains for us to know the magnitude of the coupling constant $\varkappa$, which we shall determine in the following section.

To end the present section we show that the field equations (102) can also be deduced from an action principle. This derivation was given by D. Hilbert in 1915 almost immediately after Einstein proposed his equations. We recall from (73) and (76) that T^{ik} can be derived from a variation of the metric tensor. By taking a variation of the identity

$$g^{ik}g_{kl} = \delta^i_l,$$

we deduce that

$$\delta g^{ik} = -g^{im}g^{kn}\,\delta g_{mn}. \tag{104}$$

Hence (76) becomes

$$\delta S = -\frac{1}{2}\int_{\mathcal{V}} T_{ik}\sqrt{-g}\;\;\delta g^{ik}\,\mathrm{d}^4x. \tag{105}$$

Suppose we now add to S an additional term given by

$$S_g = \frac{1}{2\varkappa c}\int_{\mathcal{V}} R\sqrt{-g}\;\;\mathrm{d}^4x. \tag{106}$$

Then, provided we can prove that

$$\delta S_g = \frac{1}{2\varkappa c}\int_{\mathcal{V}} G_{ik}\sqrt{-g}\;\;\delta g^{ik}\,\mathrm{d}^4x, \tag{107}$$

we have arrived at the field equations (102) from the variational equations

$$\frac{\delta S}{\delta g^{ik}} = 0. \tag{108}$$

with

$$S = S_g + \int_{\mathcal{V}} L\sqrt{-g}\;\;\mathrm{d}^4x. \tag{109}$$

Hilbert's method therefore consists of proving (107) which demands a straightforward but tedious application of the results from Riemannian geometry we have outlined in previous sections. Whereas Einstein's approach is heuristic and intuitive, Hilbert's method is aesthetically more pleasing since it appeals to a general principle. Indeed, when we discuss quantum gravity in Part IV, we shall find Hilbert's method a very useful starting point.

To prove (107) proceed as follows. Using (52) and (104) we get

$$\delta\sqrt{-g} = -\tfrac{1}{2}\sqrt{-g}\;g_{ik}\,\delta g^{ik},$$

and hence

$$\delta(R\sqrt{-g}\,) = \delta(R_{ik}g^{ik}\sqrt{-g}\,)$$
$$= (R_{ik} - \tfrac{1}{2}\, g_{ik}R)\sqrt{-g}\;\delta g^{ik} + \delta R_{ik}g^{ik}\sqrt{-g}\,.$$

Thus our object is achieved if we can show that

$$\int \delta R_{ik}g^{ik}\sqrt{-g}\;\; \mathrm{d}^4x = 0. \tag{110}$$

To prove this result choose a locally inertial frame at a typical spacetime point x^i. At this point we find from (60) and (97) that

$$\delta R_{nm}\, g^{nm} = \left(\frac{\partial \delta\Gamma^i_{nm}}{\partial x^l} - \frac{\partial \delta\Gamma^i_{nl}}{\partial x^m}\right)\, \delta^l_i g^{nm} = \frac{\partial w^l}{\partial x^l}\,,$$

where

$$w^l = \delta\Gamma^i_{nm}\, g^{nm}\delta^l_i - \delta\Gamma^i_{nm}\,\delta^m_i g^{nl}. \tag{111}$$

Notice that in the locally inertial frame the derivative of g_{ik} vanishes at x^i. We also note that although the Γ's do not transform as a tensor the δΓ's do (see Example 6.10). We can therefore consider w^l as a vector in *all* coordinate frames and write

$$\delta R_{nm}\, g^{nm} = w^l{}_{;l}$$

Then

$$\int_{\mathcal{V}} \delta R_{ik}\, g^{ik}\sqrt{-g}\;\mathrm{d}^4x = \int_{\mathcal{V}} w^l{}_{;l}\sqrt{-g}\;\mathrm{d}^4x = \int_{\partial\mathcal{V}} w^l n_l\;\mathrm{d}^3x, \tag{112}$$

where we have used (50) and the Green's theorem (Example 6.17).

The result (110) is therefore proved for all metric variations whose derivatives also vanish on $\partial\mathcal{V}$. Thus the δΓ's vanish on $\partial\mathcal{V}$ and so w^l vanishes also. The vanishing of δg_{ik} on $\partial\mathcal{V}$ is consistent with normal variational principles. We are, however, asked to assume here that $\delta g_{ik,l}$ also vanish on $\partial\mathcal{V}$. This requirement is unusual, but necessary here since the action integrand R includes second derivatives of g_{ik}. Normally the field Lagrangian contains spacetime derivatives of first order only. We shall touch upon this curious aspect of the Hilbert action principle when we discuss path integration for quantum gravity in Part IV.

It is possible to retain only first derivatives of $g_{ik,l}$ in the Hilbert action by modifying Equation (106) to

$$\tilde{S}_g = \frac{1}{2\varkappa c}\int g^{nm}(\Gamma^i_{km}\,\Gamma^k_{in} - \Gamma^i_{ki}\,\Gamma^k_{mn})\sqrt{-g}\;\mathrm{d}^4x. \tag{113}$$

The variation $g_{ik} \to g_{ik} + \delta g_{ik}$ gives the same Einstein tensor as before. However, there is one fundamental defect in $\tilde{S}_g$: it is not a scalar action. The integrand can be made to vanish at any chosen point by using the locally inertial frame. Nevertheless, $\bar{S}_g$ has played a useful role in the development of general relativity. We shall refer to it again in Section 7.4.

6.9. The Newtonian Approximation

The Einstein field equations (102) at first sight do not have any structural similarity with the equations of Newtonian theory. Yet such a similarity does become apparent in the case of weak gravitation fields. We now investigate this problem and determine the value of the coupling constant $\varkappa$ in terms of the Newtonian gravitational constant G and the velocity of light c.

The weak field approximation consists of the following assumptions:

(a) The curvature of spacetime is small so that it is possible to write the metric tensor in the form

$$g_{ik} = \eta_{ik} + h_{ik}\,, \quad |h_{ik}| \ll 1. \tag{114}$$

(b) The time variations of the gravitational field are small and can be ignored in comparison with space variations. This means that

$$|h_{ik,0}| \ll |h_{ik,\mu}|\ . \tag{115}$$

(c) The particle velocities are nonrelativistic so that in the spacetime metric we may effectively set all h_{ik} except h_{00} equal to zero.

Consider first what happens to timelike geodesics. According to Einstein these describe particles moving under no forces while according to Newton they are trajectories of particles moving under the gravitational field. Let φ be the gravitational potential in the Newtonian picture so that the equation of motion becomes

$$\frac{\mathrm{d}^2x^\mu}{\mathrm{d}t^2} = \frac{\partial\varphi}{\partial x^\mu}\ . \tag{116}$$

The geodesic equation, on the other hand, is given by (54), with the affine parameter $\lambda = s$. In the nonrelativistic approximation ((c) above), $s \approx ct$ and we may consider only the spacelike components $i = \mu$ of (54). Now

$$\Gamma^\mu_{ik} = \tfrac{1}{2} g^{\mu\nu}[g_{\nu i,k} + g_{\nu k,i} - g_{ik,\nu}]$$

is nonzero only for $i = k = 0$ for which case

$$\Gamma^\mu_{00} = +\tfrac{1}{2}\, g_{00,\mu}.$$

Hence the geodesic equation becomes

$$\frac{1}{c^2}\cdot\frac{\mathrm{d}^2x^\mu}{\mathrm{d}t^2} + \tfrac{1}{2}\, h_{00,\mu} = 0. \tag{117}$$

A comparison with (116) gives $h_{00,\mu} = -2\varphi_{,\mu}$ which imply

$$h_{00} = -\frac{2\varphi}{c^2} \, . \tag{118}$$

Our example of the accelerated observer had already given us a hint of this result. We have absorbed the arbitrary constant of integration in the definition of the Newtonian potential φ. Thus we get

$$g_{00} = 1 - \frac{2\varphi}{c^2} \, . \tag{119}$$

We next consider the Einstein field equations. By defining $T = T_{ik} \, g^{ik}$ we rewrite (102) in the form

$$R_{ik} = -\varkappa \, [\, T_{ik} - \tfrac{1}{2} \, g_{ik} \, T \,] . \tag{120}$$

We consider this equation in the above approximate situation. First we note that

$$R_{ik} \cong \tfrac{1}{2} \, \eta^{lm} h_{ik,lm} + \tfrac{1}{2} \, \{ h_{,ik} - h^l_{l,ik} - h^l_{k,il} \}$$

and hence because of assumptions (a) – (c) we get the only nonzero component of R_{ik} as

$$R_{00} \cong -\tfrac{1}{2} \nabla^2 h_{00} . \tag{121}$$

Similarly for slow-moving matter we may use the 'dust' approximation, i.e. treat it as pressure-free fluid. Taking ϱ to be the density of dust, we have in this approximation

$$-\varkappa \, [\, T_{00} - \tfrac{1}{2} \, g_{00} T] \cong -\tfrac{1}{2} \, \varkappa \varrho c^2 . \tag{122}$$

Using (118) – (122) we get the following equation

$$\frac{1}{c^2} \, \nabla^2 \varphi = -\tfrac{1}{2} \, \varkappa \varrho c^2 .$$

If we want to establish complete equivalence to the Newtonian theory in this limit, the above equation must be identical to the Poisson equation (8). This requirement determines $\varkappa$ as

$$\varkappa = \frac{8\pi G}{c^4} \, . \tag{123}$$

Therefore, we have the Einstein equations in the final form as

$$R_{ik} - \tfrac{1}{2} \, g_{ik} \, R = - \frac{8\pi G}{c^4} \, T_{ik} . \tag{124}$$

Example 6.19. Let us consider the cannonball problem of Section 6.2 in Einstein's theory in its Newtonian approximation. With $c = 1$ the line element is given by

$$ds^2 = \left(1 - \frac{2GM}{r}\right) dt^2 - (dx^2 + dy^2 + dz^2),$$

where M = mass of the Earth and r = distance of the cannonball from the centre of the Earth.
In the 'flat earth' approximation we shall write $r = R_0 + z$, where R_0 = Earth's radius and z = height of the ball above the Earth's surface (x and y measure the horizontal displacements from the point of firing). Then

$$\frac{2GM}{r} \approx \frac{2GM}{R_0}\left(1 + \frac{z}{R_0}\right)^{-1} \approx \frac{2GM}{R_0} - 2gz,$$

where $g = GM/R_0^2$ is the acceleration due to gravity. Thus we have

$$g_{00} \approx 1 - \frac{2GM}{R_0} + 2gz, \qquad g^{00} \approx 1 + \frac{2GM}{R_0} - 2gz.$$

The other components of g_{ik} are simply η_{ik}. The geodetic equations (33) therefore become (with $\lambda \cong t$)

$$\frac{d^2x}{dt^2} = 0, \qquad \frac{d^2y}{dt^2} = 0, \qquad \frac{d^2z}{dt^2} + g = 0.$$

These are the Newtonian equations of motion of the cannonball. □

6.10. The Λ Term

These gravitational equations could incorporate another term without altering the heuristic derivation given by Einstein. The modified equations become

$$R_{ik} - \tfrac{1}{2} g_{ik} R + \Lambda g_{ik} = -\frac{8\pi G}{c^4} T_{ik}. \tag{125}$$

The constant Λ has to be small enough not to interfere with the satisfactory working of these equations in the Newtonian approximation. In this approximation the Λ term describes a force of repulsion of magnitude Λr between any two particles of matter separated by a distance r. The force may therefore become important on the cosmological scale, although its magnitude on the terrestrial or solar system scale may be negligible.

Example 6.20. Let us compare the Λ term with Newtonian gravity on the scale of the solar system.
At a distance r from the Sun the acceleration due to its gravity is GM/r^2, while that due to the Λ force of repulsion is Λr. Thus we need

$$\Lambda r \ll \frac{GM}{r^2}, \qquad \text{that is,} \quad \Lambda \ll \frac{GM}{r^3}.$$

For the inequality to hold on the solar surface we substitute $M \approx 2 \times 10^{33}$ g and $r = 7 \times 10^{10}$ cm in the above to get

$$\Lambda \ll 4 \times 10^{-7}\ \mathrm{s}^{-2}.$$

We shall see in Chapter 8 that the cosmological models require $\Lambda \sim 10^{-35}\ \mathrm{s}^{-2}$. □

In the Hilbert derivation, the above equations are obtained if the R in (106) is replaced by $R - 2\Lambda$. If Λ is treated as an undetermined multiplier, then the Hilbert problem may be interpreted as stationarizing (109) while preserving the 4-volume of υ. We shall discuss the Λ term in the cosmological applications of general relativity.

6.11. Conformal Transformations

Consider the transformation

$$g^*_{ik} = \Omega^2 g_{ik}, \tag{126}$$

where Ω is a well-behaved (C^2) function of spacetime coordinates. Such a transformation is called 'conformal transformation' because it preserves the ratios of lengths and the angle between any two infinitesimal line segments at any given point. We discuss some properties of this transformation because it will play a major role in our later work.

First, note that (126) preserves the meaning of the words 'spacelike', 'timelike', and 'null'. The Christoffel symbols change from Γ^i_{kl} to

$$\Gamma^{*i}_{kl} = \Gamma^i_{kl} + \frac{1}{\Omega}\left\{\Omega_k\delta^i_l + \Omega_l\delta^i_k - \Omega^i g_{kl}\right\}, \tag{127}$$

and hence timelike and spacelike geodesics do not remain geodesics under this transformation. Null geodesics are, however, still null geodesics. This result is expressed by the statement that *the global light cone structure is unchanged under conformal transformations*.

Quantities, equations, etc., which are unaltered in form by a conformal transformation are called 'conformally invariant'. It is easy to verify that R, R_{ik}, R_{iklm} are *not* conformally invariant, but the tensor

$$C^h{}_{ijk} \equiv R^h{}_{ijk} + \tfrac{1}{2}\left[\delta^h_j R_{ik} - \delta^h_k R_{ij} + g_{ik}R^h_j - g_{ij}R^h{}_k\right] + \frac{1}{6}R\left[\delta^h_k g_{ij} - \delta^h_j g_{ik}\right] \tag{128}$$

is conformally invariant. It is called 'Weyl's conformal curvature tensor', or simply, the Weyl tensor'. A spacetime that is conformal to the flat spacetime is called 'conformally flat'. For such a spacetime the Weyl tensor vanishes.

It is also easy to verify that under (126) Maxwell's equations remain unchanged. While the scalar wave operator $\Box\varphi$ is messed up by (126) the combination $(\Box + \frac{1}{6}R)\varphi$ transforms to

$$(\Box + \tfrac{1}{6}R)^*\varphi^* = \frac{1}{\Omega^3}(\Box + \tfrac{1}{6}R)\varphi, \tag{129}$$

with $\varphi^* = \varphi/\Omega$. In dealing with scalar wave equations the combination $\Box + \frac{1}{6}R$. will therefore occur frequently.

This concludes our present discussion whose motivation was to lead to the formulation of a formal theory of gravity. In this approach we have been somewhat old fashioned in writing the various geometrical properties in the suffix notation of vectors and tensors. This is necessary once we decide to use coordinates: it must, however, be remembered that coordinates are means to an end and that in the last analysis the geometrical or physical reality must be appreciated. Although the general covariance guarantees this, often the coordinate dependent approach may pose artificial difficulties. The intrinsic approach, using differential forms, is preferred by many workers for its direct relevance to physically or geometrically meaningful quantities. This approach is, however, more difficult to understand than the coordinate dependent approach used here. The essential results of this chapter are derived in Appendix C, using the technique of differential forms. The reader is recommended to read it only after he has familiarized himself with this chapter.

Notes and References

1. A discussion of differential geometry with relevance to general relativity is found in the classic book:

   Eisenhart, L. P.: 1926, *Riemannian Geometry*, Princeton, New Jersey.

2. A modern mathematical treatment of differential geometry and general relativity is given by:

   Hawking, S. W., and Ellis, G. F. R.: 1973, *Large Scale Structure of Spacetime*, Cambridge.

3. Advanced treatments of general relativity are given in:

   Misner, C. W., Thorne, K. S., and Wheeler, J. A.: 1973, *Gravitation*, Freeman.
   Weinberg, S.: 1972, *Gravitation and Cosmology*, Wiley, New York.

4. A detailed treatment of general relativity at the level of this book will be found in:

   Narlikar, J. V.: 1978, *General Relativity and Cosmology*, Macmillan, New York.

Chapter 7

Gravitating Massive Objects

7.1. The Schwarzschild Solution

The general theory of relativity developed in the previous chapter is usually applied in two different contexts. The first of these is concerned with the gravitational effects near (and within) a massive object. The idea here is that the presence of a massive object in a bounded region of space generates non-Euclidean effects of spacetime geometry which die out as we go away from the object. Thus, far away from the object the spacetime is that of special relativity. Such solutions of Einstein's equations are said to have boundary conditions implying an *asymptotically flat* spacetime. We shall discuss such solutions in this chapter.

The second context in which Einstein's equations have found applications is in cosmology. The astronomical observations tell us that, as far as we can see, the universe is filled with matter and radiation. Hence, asymptotic flatness cannot be the correct concept in describing the real universe. The cosmological applications of relativity will be discussed in the following chapter.

We take up the simplest of the local problems, first solved by K. Schwarzschild in 1916: "What is the geometry outside a spherical distribution of matter of gravitational mass M?"

First we note that far away from the mass the Newtonian approximation holds, and hence

$$g_{00} \sim 1 - \frac{2GM}{c^2 r}\,, \tag{1}$$

since the gravitational potential at a distance r from the centre of M in the Newtonian theory is given by $\varphi = GM/r$. As $r \to \infty$, $g_{00} \to 1$ as required by the condition of asymptotic flatness. The solution of Einstein's equations that we shall look for must possess the above property.

Although the Einstein equations are ten in number and are nonlinear, our job is simplified because of the inherent spherical symmetry of the problem. Taking the centre of the object as the origin we can use the spherical polar coordinates (r, θ, φ) to label a point in space. We now write down a line element which is most general under the assumption of spherical symmetry:

$$ds^2 = A(r, t)\, dt^2 + 2H(r, t)\, dt\, dr + B(r, t)\, dr^2 + \\ + F(r, t)\, (d\theta^2 + \sin^2\theta\, d\varphi^2), \tag{2}$$

where A, H, B, and F are arbitrary functions of spacetime. Appendix D describes how spacetime symmetries are expressed quantitatively. The proof of (2) is based on those ideas and for further details we refer the reader to a specialized text on general relativity.

A coordinate transformation

$$r = r', \qquad t = k(r', t') \tag{3}$$

can, however, be found that will diagonalize the line element. A straightforward calculation shows that the condition for this is

$$A\, \frac{\partial k}{\partial r'} + H = 0. \tag{4}$$

Hence, without loss of generality we can set $H = 0$ in (2). To ensure proper signature we shall express (2) in a somewhat different form:

$$ds^2 = c^2\, e^{\sigma}\, dt^2 - e^{\omega}\, dr^2 - e^{\mu}\, (d\theta^2 + \sin^2\theta\, d\varphi^2), \tag{5}$$

where σ, ω, and μ are real functions of r and t.

A new way of writing (5) is to use coordinates (r', t') such that

$$r' = e^{\mu/2} \tag{6}$$

with t' chosen to diagonalize the new line element. This leads to an equation similar to (4). Again we drop primes and rewrite (5) in the form

$$ds^2 = c^2\, e^{\nu}\, dt^2 - e^{\lambda}\, dr^2 - r^2\, (d\theta^2 + \sin^2\theta\, d\varphi^2). \tag{7}$$

The coordinate r has now a special significance: the area of the sphere r = constant is given by the Euclidean expression $4\pi r^2$. These coordinates (t, r, θ, φ) are called 'Schwarzschild coordinates'. Let us suppose that the surface of the object is given by $r = r_0$. Using these coordinates and the metric (7) we solve Einstein's equations

$$R_{ik} - \tfrac{1}{2} g_{ik}\, R = -\, \frac{8\pi G}{c^4}\, T_{ik}. \tag{8}$$

The right-hand side is assumed to be zero outside the object, $r > r_0$.

The metric tensor has the following nonzero components

$$g_{00} = e^{\nu}, \quad g_{11} = -e^{\lambda}, \quad g_{22} = -r^2, \quad g_{33} = -r^2 \sin^2\theta. \tag{9}$$

Using formula (7) we find the nonzero components of Γ^i_{kl} to be

$$\Gamma^0_{00} = \tfrac{1}{2}\dot{\nu},\ \Gamma^1_{00} = \tfrac{1}{2} e^{\nu-\lambda}\, \nu',\ \Gamma^1_{11} = \tfrac{1}{2}\lambda',\ \Gamma^0_{11} = \tfrac{1}{2} e^{\lambda-\nu}\, \dot{\lambda},$$

$$\Gamma^0_{01} = \tfrac{1}{2}\nu', \ \Gamma^1_{01} = \tfrac{1}{2}\dot{\lambda}, \ \Gamma^2_{12} = \Gamma^3_{13} = \frac{1}{r},$$

$$\Gamma^1_{22} = -r\,\mathrm{e}^{-\lambda}, \qquad \Gamma^1_{33} = -r\sin^2\theta\,\mathrm{e}^{-\lambda}, \qquad \Gamma^2_{23} = -\sin\theta\cos\theta,$$

$$(\ ') \equiv \partial/\partial r, \qquad (\ \dot{}\ ) \equiv \partial/\partial t. \tag{10}$$

Next we use the expression for R_{ik} and R as defined in Chapter 6 to compute the left-hand side of (8). A straightforward but tedious calculation gives the following independent equations:

$$\mathrm{e}^{-\lambda}\left(\frac{1}{r^2} - \frac{\lambda'}{r}\right) - \frac{1}{r^2} = -\frac{8\pi G}{c^4}\,T^0_0, \tag{11}$$

$$\mathrm{e}^{-\lambda}\left(\frac{\nu'}{r^2} - \frac{1}{r^2}\right) - \frac{1}{r^2} = -\frac{8\pi G}{c^4}\,T^1_1, \tag{12}$$

$$\mathrm{e}^{-\lambda}\,\frac{\dot{\lambda}}{r} = -\frac{8\pi G}{c^4}\,T^1_0, \tag{13}$$

$$\tfrac{1}{2}\mathrm{e}^{-\lambda}\left(\nu'' + \tfrac{1}{2}\nu'^2 + \frac{\nu' - \lambda'}{r} - \tfrac{1}{2}\nu'\lambda'\right) - \tfrac{1}{2}\mathrm{e}^{-\nu}(\ddot{\lambda} + \tfrac{1}{2}\dot{\lambda}^2 - \tfrac{1}{2}\dot{\nu}\dot{\lambda})$$

$$= -\frac{8\pi G}{c^4}\,T^2_2 = \frac{8\pi G}{c^4}\,T^3_3. \tag{14}$$

The first of these equations can be integrated directly to give

$$\mathrm{e}^{-\lambda} = 1 - \frac{2Gm(r, t)}{c^2 r}, \tag{15}$$

where

$$m(r, t) = \int_0^r 4\pi r_1^2\,T^0_0\,\mathrm{d}r_1. \tag{16}$$

Note that because $T^0_0 = 0$ for $r > r_0$, we have $m(r, t)$ independent of r for $r > 0$. Further, because $T^1_0 = 0$ for $r > r_0$, $\dot{\lambda} = 0$ and hence $m(r, t)$ is independent of t. Thus, for $r > r_0$,

$$m(r, t) = M = \text{const.} \tag{17}$$

For $r > r_0$, the equations $T^0_0 = T^1_1 = 0$ give, from the difference of (11) and (12), the simple relation

$$\nu' + \lambda' = 0$$

which integrates to

$$\nu + \lambda = f(t),$$

where $f(t)$ is an arbitrary function of t.

However, an arbitrary additive function of t in ν can be eliminated by a simple time transformation. Thus we set $f(t) = 0$ without loss of generality. Hence, for $r > r_0$ we get

$$e^{\nu} = e^{-\lambda} = 1 - \frac{2GM}{c^2 r} \,. \tag{18}$$

By a comparison with the asymptotic requirement (1) we see that the arbitrary constant M can be identified with the gravitational mass of the object.

The line element for $r > r_0$ is thus given by

$$ds^2 = c^2\, dt^2 \left(1 - \frac{2GM}{c^2 r}\right) - \frac{dr^2}{1 - (2GM/c^2 r)} - r^2(d\theta^2 + \sin^2\theta\, d\varphi^2). \tag{19}$$

This is called the 'Schwarzschild line element'.

Formula (19) gives the answer we were looking for: the geometry of spacetime outside a massive object. The solution is exact and hence valid for all r. The surprising result that emerges is that nowhere did we make an assumption of a static system, but we have ended with a time-independent metric. In other words, even if the object were collapsing, expanding, or pulsating in a spherically symmetric manner, the outside spacetime will be static. This result was first noticed by G. D. Birkhoff and is known as 'Birkhoff's theorem'.

The metric seems to exhibit peculiarities at

$$r = r_s = \frac{2GM}{c^2} \,. \tag{20}$$

At this radius, known as the 'Schwarzschild radius', g_{00} vanishes while g_{11} diverges. We shall discuss the physical aspects of the Schwarzschild radius later. We note that r_s has any real physical status only provided $r_s > r_0$, since otherwise the solution (19) is not valid. Further, even if $r_0 \leq r_s$, the spacetime curvature at $r = r_0$ is finite; that is, the scalars like

$$R, \quad R_{ik}R^{ik}, \quad R_{iklm}R^{iklm}, \quad R_{iklm}R^{lmpq}R_{pq}{}^{ik}, \ldots,$$

are either zero or bounded at $r = r_s$.

Example 7.1. Let us compute the quantity $R_{iklm}R^{iklm}$ for the Schwarzschild solution and then evaluate it at $r = r_s$.

Using the nonzero Christoffel symbols from (10) we find that the only nonzero components of R_{iklm} are (with c = 1):

$$R_{0101} = -\tfrac{1}{2}e^{\nu}(\nu'' + \nu'^2), \quad R_{0202} = -\tfrac{1}{2}r\, e^{2\nu}\,\nu', \quad R_{0303} = -\tfrac{1}{2}r\, e^{2\nu}\,\nu' \sin^2\theta,$$

$$R_{1212} = \tfrac{1}{2}r\nu', \quad R_{1313} = \tfrac{1}{2}r\nu' \sin^2\theta, \quad R_{2323} = -r^2\sin^2\theta(1 - e^{\nu}),$$

Hence

$$R_{iklm}R^{iklm} = 4\left[(R_{0101})^2 + (R_{0202})^2 \cdot \frac{1}{r^4}\, e^{-2\nu} + \right.$$

$$+ (R_{0303})^2 \frac{1}{r^4\sin^4\theta}\, e^{-2\nu} + (R_{1212})^2 \frac{e^{2\nu}}{r^4} +$$

$$+ \quad (R_{1313})^2 \frac{e^{2\nu}}{r^4 \sin^4\theta} + (R_{2323})^2 \frac{1}{r^8 \sin^4\theta} \Bigg] = \frac{48G^2M^2}{r^6} .$$

At $r = 2GM$ this invariant has the finite value $3/4G^4M^4$. □

Example 7.2. Let us work out the radial tidal force on a body under radial free fall towards $r = 0$ in the Schwarzschild's metric.

First note that under radial free fall we have, for $c = 1$,

$$u^0 = \frac{dt}{ds} = \gamma\left(1 - \frac{2GM}{r}\right)^{-1}, \qquad u^1 = \frac{dr}{ds} = -\sqrt{\frac{2GM}{r} + \gamma^2 - 1},$$

where γ = const. [This is a special case of (35) with $h = 0$.] To compute the radial tidal force we take the separation vector to be $v^i = (0, v, 0, 0)$. The tidal term is given by the right-hand side of (6.65):

$$f^i = R^i{}_{mlk} u^m u^l v^k .$$

From Example 7.1 we can compute the nonzero components of f^i. These are

$$f^0 = R^0{}_{101} u^1 u^0 v = \frac{2GM}{r^3} u^0 v\, e^{-\nu} u^1 ,$$

$$f^1 = \frac{2GM}{r^3} u^0 v\, e^{\nu} u^0 = \frac{2GM}{r^3} u^0 v \gamma .$$

We have to find the force as 'felt' by the body and so we transform f^i and v^i to the rest frame of the body. Consider f^i first. In the rest frame[1] it must have the form $F^i = (0, F, 0, 0)$. Since $f_i f^i = F^i F_i$ we get

$$F^2 = -\left(\frac{2GM}{r^3} u^0 v\right)^2 [e^{-\nu}(u^1)^2 - e^{+\nu}(u^0)^2] ,$$

that is,

$$F = \frac{2GM}{r^3} u^0 v .$$

Let v^i become $V^i = (b, a, 0, 0)$ in the rest frame of the body. Here a corresponds to the linear extent of the body in the radial direction. Since $v^i f_i = V^i F_i$ we have

$$aF = v f^1 e^{-\nu}, \qquad v = a \frac{F}{f^1} e^{\nu} = a\gamma^{-1} e^{\nu} = \frac{a}{u^0} .$$

Hence, $F = 2GMa/r^3$. □

7.2. Experimental Tests of General Relativity

Having formulated the general theory of relativity and used it to solve a concrete problem, we are now in a position to confront the theory with some

[1] This is a locally inertial frame with one Cartesian axis in the radial direction and the other two in the local directions of increase of θ and φ.

experimental tests. In fact we have already taken care that the theory agrees with the Newtonian theory in the area of weak gravitational fields where the latter theory has already been tested and found satisfactory. We should therefore look for such experiments wherein the differences in the predictions of the Newtonian theory and general relativity are tested. Only if the theory succeeds in these tests shall we have confidence that it can be taken seriously enough for extrapolations into the uncharted territories.

We briefly enumerate four tests in this section. Three of them, dealing with the passage of light in curved spacetime, have been conducted with the latest technology available. The fourth test concerns motions of planets round the Sun and requires careful comparison with astronomical data collected over centuries. (For a thorough discussion of these tests see the texts listed at the end of the chapter.)

For all these tests we shall require the Schwarzschild solution (19).

7.2.1. *Gravitational Redshift*

Consider monochromatic light waves leaving the surface of a massive body and reaching a remote observer A. Figure 7.1 shows a light ray leaving a surface point B and reaching A, its trajectory being determined by a null geodesic from B to A. Let the coordinates of the point of emission be (t_B, r_0, θ_B, φ_B) and those of the point of reception, (t_A, r_A, θ_A, φ_A). The null geodesic

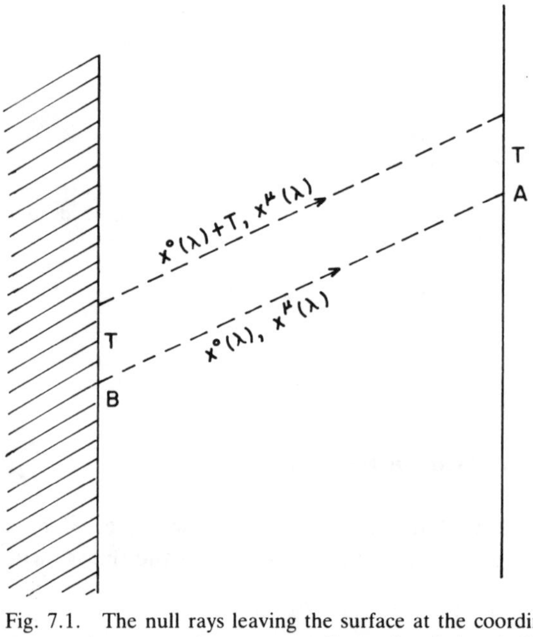

Fig. 7.1. The null rays leaving the surface at the coordinate time interval T reach the remote observer at the same coordinate time interval. The corresponding proper time intervals at these locations are, however, different.

may be described by $x^i(\lambda)$, with $\lambda = 0$ the point of emission and $\lambda = 1$ the point of reception. Thus,

$$x^0(0) = t_B, \qquad x^0(1) = t_A. \tag{21}$$

Let us identify t_B and t_A as the instants of emission and reception of a wavecrest. Let the next wavecrest leave B at $t_B + T$. When will it reach A? Because the line element is static, the answer to the question is easy. If $x^i(\lambda) \equiv \{x^0(\lambda), x^\mu(\lambda)\}$ is a solution of the null geodesic equations, then so is $\{x^0(\lambda) + T, x^\mu(\lambda)\}$. Thus the point of reception of the next wavecrest at A is $t_A + T$.

Since both A and B are at rest in this coordinate system, their proper time intervals τ_A and τ_B are related to their coordinate time interval T by the ratio of $\mathrm{d}s$ to $\mathrm{d}t$ evaluated according to (19) at constant, r, θ, φ at the locations of A and B, respectively. Thus,

$$\tau_A = \sqrt{1 - \frac{2GM}{c^2 r_0}}\; T, \qquad \tau_B = \sqrt{1 - \frac{2GM}{c^2 r_s}}\; T. \tag{22}$$

The wavelengths of light measured by A and B are, respectively, $\lambda_A = c\tau_A$, $\lambda_B = c\tau_B$ so that

$$1 + z \equiv \frac{\lambda_A}{\lambda_B} = \frac{\sqrt{1 - \dfrac{2GM}{c^2 r_A}}}{\sqrt{1 - \dfrac{2GM}{c^2 r_0}}}\;. \tag{23}$$

The quantity z measures the fractional *increase* in wavelength suffered by light in going from a stronger gravitational field at B to a weaker field at A. Since $r_A \gg r_s$ we may write z as

$$z \cong \left(1 - \frac{2GM}{c^2 r_0}\right)^{-1/2} - 1. \tag{24}$$

This is the formula for gravitational *redshift*; red, because the visible spectrum of the body shifts towards the red end.

This formula has been verified for $r_0 \gg r_s$ when it becomes

$$z \cong \frac{GM}{c^2 r_0}\;. \tag{25}$$

In white dwarf stars Sirius B and 40 Eridani B the measured z values are, respectively, 3×10^{-4} and 7×10^{-5}.

It is possible to check formula (23) in the laboratory by observing the increase in the frequency of a gamma ray photon emitted by an excited nucleus of iron (Fe^{57*}) and absorbed by a detector iron nucleus in the ground state (Fe^{57}). In falling through a height of 22.5 m (in the 1960 experiment conducted by R. V. Pound and G. A. Rebka) the photon increased in energy and hence in frequency by a fraction 2.44×10^{-15}. The absorption can be achieved by resonance after the detector has been matched to receive the photon of increased frequency by being given a suitable Doppler blueshift with respect to the source.

Notice that we are using (23) in the reverse sense. The detector is in the stronger gravitational field and hence observes a gravitational *blueshift* from the source.

Example 7.3. Consider a photon falling through a height H on to the Earth's surface. The gravitational potential due to mass M of the Earth at a distance r from its centre is given by GM/r. Thus the relativistic formula (23) gives

$$\frac{\nu_B}{\nu_A} = \frac{\lambda_A}{\lambda_B} = \frac{\sqrt{1 - \dfrac{2GM}{c^2 r_A}}}{\sqrt{1 - \dfrac{2GM}{c^2 r_s}}} \approx 1 - \frac{GM}{c^2 r_A} + \frac{GM}{c^2 r_B},$$

where $r_A = R$, the radius of the Earth, and $r_B = R + H$. The increase in frequency of the photon is therefore given by

$$\nu_A - \nu_B \cong \nu_B \frac{GM}{c^2 R^2} H = \frac{gH}{c^2} \nu_B ,$$

where g = acceleration due to gravity on the Earth. The fractional increase of frequency for a free fall through H metres is therefore $\sim 10^{-16} H$. To achieve resonance for absorption of the blueshifted photon the detector has to be given a velocity $\sim 3 \times 10^{-8} H \text{ ms}^{-1}$ towards the source. □

7.2.2. *Bending of Light*

Just as in the example of the cannonball discussed in Section 6.1, light photons are also bent by the gravitational pull of a massive object. In Figure 7.2 the situation is illustrated for light rays from a star reaching the terrestrial observer. The rays get bent if they happen to pass close to the surface of the Sun, with the result that the star's direction changes.

The mathematical derivation of the bending formula for the situation of Figure 7.2 is briefly as follows:

Equation (6.54) for null geodesics, when written out for the line element (19), gives in the case of $x^2 = \theta$,

$$\frac{d^2\theta}{d\lambda^2} + \frac{2}{r}\frac{d\theta}{d\lambda}\frac{dr}{d\lambda} - \sin\theta\cos\theta\left(\frac{d\varphi}{d\lambda}\right)^2 = 0. \tag{26}$$

This has a solution $\theta = \pi/2$. Without loss of generality we may choose $\theta = \pi/2$ for our light ray since in choosing our coordinate system we are free

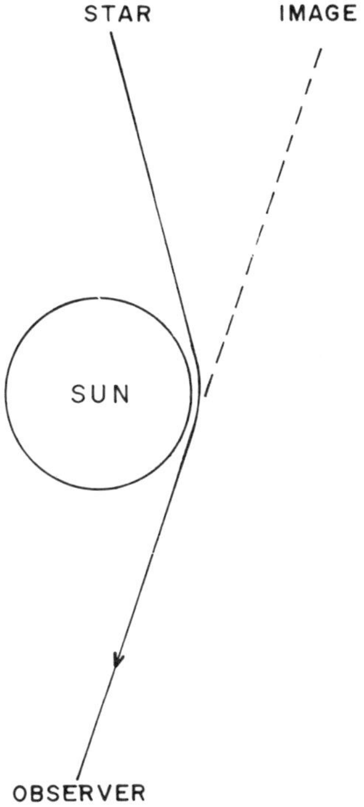

Fig. 7.2. The light ray from the star to the observer is 'bent' in the neighbourhood of the Sun with the result that the observer sees the star's image in the direction of the dashed line.

to select any direction as the polar axis. This step simplifies other components of (6.54). The x^0 and x^3 equations integrate to

$$c\left(1-\frac{2GM}{c^2r}\right)\frac{\mathrm{d}t}{\mathrm{d}\lambda}=\gamma,\quad r^2\,\frac{\mathrm{d}\varphi}{\mathrm{d}\lambda}=h,\quad (\gamma,\ h\ \text{constants}). \tag{27}$$

We could also write down the $x^1 = r$ equation. However, it is simpler to use the first integral (6.52):

$$c^2\left(1-\frac{2GM}{c^2r}\right)\left(\frac{\mathrm{d}t}{\mathrm{d}\lambda}\right)^2-\left(1-\frac{2GM}{c^2r}\right)^{-1}\left(\frac{\mathrm{d}r}{\mathrm{d}\lambda}\right)^2-r^2\left(\frac{\mathrm{d}\varphi}{\mathrm{d}\lambda}\right)^2=0. \tag{28}$$

Write $u = 2GM/c^2r$ and eliminate λ and t between (27) and (28) to get

$$\left(\frac{\mathrm{d}u}{\mathrm{d}\varphi}\right)^2+u^2=k^2+u^3\ , \tag{29}$$

where $k = \gamma r_s/h$. Differentiate (29) with respect to φ to get rid of the arbitrary constant k to get

$$\frac{d^2u}{d\varphi^2} + u = \frac{3u^2}{2} . \tag{30}$$

We normally expect u to be small. Neglecting the right-hand side gives us the trivial Euclidean straight line $u \cong u_0$,

$$u_0 = \frac{r_s}{r_0} \cos \varphi. \tag{31}$$

This is the dashed line of Figure 7.2. Substitute this value of u_0 on the right-hand side and solve for the next approximation $u \cong u_0 + u_1$:

$$\frac{d^2u_1}{d\varphi^2} + u_1 = \frac{3r_s^2}{2r_0^2} \cos^2 \varphi. \tag{32}$$

Thus, to the next order in r_s/r_0 we get

$$u \cong u_0 + u_1 = \frac{r_s}{r_0} \cos \varphi + \frac{r_s^2}{2r_0^2} (2-\cos^2 \varphi). \tag{33}$$

Since the star and the observer are far away from the Sun, we need the asymptotes of the above curve to work out the bending angle in Figure 7.2. Anticipating that the bending angle is small we write the asymptotes as

$$\varphi \cong \pm \left\{ \frac{\pi}{2} + \eta \right\}, \quad \eta \ll 1.$$

Substitute this value of φ in the right-hand side of (33) and set $u = 0$ to solve for η. We get

$$\eta \cong \frac{r_s}{r_0} .$$

The net bending is therefore

$$\Delta\varphi \equiv 2\eta = \frac{4GM}{c^2 r_0} . \tag{34}$$

For the Sun this works out at $\simeq$ 1.75 arcsec. This small shift in the apparent direction of a star has to be measured during a total solar eclipse. (In any other situation the star cannot be picked up against the bright solar light.) The observations taken during several solar eclipses in many different locations seem to support the general relativistic prediction. However, it is hard to disentangle the real gravitational effect from the bending effects produced by refraction in the solar corona.

The difficulties of observing a star occulted by the solar disc and the refraction effects are lessened considerably if the observations are made with radio or microwaves. Such measurements could be made during the mid-1970s for microwaves from the quasar 3C 279 as it passed behind the solar disc. Because the Sun is a weak emitter of microwaves, the observations do not have to be made during an eclipse. An unambiguous result confirming formula (34) within ~1% error bars has been claimed by two independent groups of observers.

Incidentally, although Newton did not assume light to be subject to gravitational force we can adapt Newtonian gravitation to argue that light is made of particles which are subject to the inverse square law of gravitation. Calculation based on this assumption gives half the relativistic bending and seems to be ruled out by the above observations.

Example 7.4. To work out the Newtonian bending formula, assume that light photons describe hyperbolic trajectories with asymptotic speed equal to c. The hyperbolic orbit may be described by radial-polar coordinates centred on the Sun (whose mass is assumed to be M):

$$\frac{l}{r} = e \cos \theta + 1$$

where l = semilatus rectum and e = eccentricity of the orbit. The asymptotes are in the directions given by $\theta = \pm\cos^{-1}(-1/e)$. Thus we have to find e for an orbit whose closest approach to the Sun is at $r = r_0$. For $e \gg 1$ the bending angle is $\approx 2/e$.

Properties of conics tells us that $l = r_0(e + 1)$. The radial and transverse velocities are given by

$$\dot{r} = \frac{le \sin \theta}{(e \cos \theta + 1)^2}\dot{\theta}, \qquad r\dot{\theta} = \frac{l\dot{\theta}}{e \cos \theta + 1}.$$

Since the transverse force is zero, $r^2\dot{\theta}$ is constant. Thus

$$r^2\dot{\theta} = \frac{l^2\dot{\theta}}{(e \cos \theta + 1)^2} = h = \text{const.}$$

The radial equation of motion gives, with the help of the above relation,

$$-\frac{GM}{r^2} = \ddot{r} - r\dot{\theta}^2 = \frac{\mathrm{d}}{\mathrm{d}t}\left(\frac{he}{l}\sin\theta\right) - \frac{h^2(e\cos\theta + 1)^3}{l^3}$$

$$= \frac{h^2 e \cos\theta}{l^3}(e\cos\theta + 1)^2 - \frac{h^2(e\cos\theta + 1)^3}{l^3}$$

$$= -\frac{h^2}{lr^2}.$$

Thus, we have $h^2 = GMl$. The value of h can also be determined by the asymptotic condition $h = cp$, where p is the perpendicular from the focus (the Sun) on the initial asymptote. Thus

$$h = \frac{cl}{\sqrt{e^2 - 1}} \Rightarrow (e^2 - 1) = \frac{c^2l^2}{h^2} = \frac{c^2 l}{GM}.$$

Remembering that $l = r_0(e + 1)$ we get

$$(e - 1) = \frac{c^2 r_0}{GM}, \qquad \text{i.e.} \quad e \approx \frac{c^2 r_0}{GM}.$$

Thus the bending angle is $2/e \approx 2GM/c^2 r_0$. □

7.2.3. *The Delay in Radar Echos*

In this test non-Euclidean effects appear in time measurements of radar signals bounced off a planet or a satellite. Such a signal travels along a null geodesic. If it happens to pass though the strong gravitational field of a massive object, the arrival of the bounced signal is delayed by a time interval of the order of GM/r_0c^3.

Example 7.5. Consider Equations (27) and (28). Eliminate λ and φ betweem them to get

$$\left(\frac{dr}{dt}\right)^2 = c^2\left(1-\frac{2GM}{c^2r}\right)^2 - \frac{h^2c^2}{r^2\gamma^2}\cdot\left(1-\frac{2GM}{c^2r}\right)^3$$

$$= c^2\left(1-\frac{2GM}{c^2r}\right)^2\left[1-\frac{r_0^2}{r^2}\left(1-\frac{2GM}{c^2r}\right)\left(1-\frac{2GM}{c^2r_0}\right)\right],$$

where we have chosen h/γ such that dr/dt vanishes at $r = r_0$.

We now make the approximation that r, $r_0 \gg r_s$, and expand the binomials accordingly. Hence,

$$\left(\frac{dr}{dt}\right) \cong c\left(1-\frac{2GM}{c^2r}\right)\left[1-\frac{r_0^2}{r^2}\left(1-\frac{2GM}{c^2r}+\frac{2GM}{c^2r_0}\right)\right]^{1/2}$$

$$\cong c^2\left(1-\frac{r_0^2}{r^2}\right)^{1/2}\left[1-\frac{r_sr_0}{2r(r+r_0)}-\frac{r_s}{r}\right]$$

$$\cong c^2\left(1-\frac{r_0^2}{r^2}\right)\left[1-\eta(r)\right], \quad \text{say,}$$

where $\eta(r)$ denotes the general relativistic effect. The time delay between $r = r_0$ and $r = r_1$ is thus given by

$$\Delta(r_1) = \frac{1}{c}\int_{r_0}^{r_1}\frac{r\,dr}{\sqrt{r^2-r_0^2}}\left\{\frac{1}{1-\eta(r)}-1\right\} \cong \frac{1}{c}\int_{r_0}^{r_1}\frac{r\eta(r)\,dr}{\sqrt{r^2-r_0^2}}.$$

This integral has to be evaluated numerically. □

Such signals were sent from the Earth and bounced off spacecrafts Mariner 6 and Mariner 7 using S-band radio waves, when these spacecrafts were on the opposite side of the Sun. A time delay of the order of 200 μs was observed and it confirmed the theoretical prediction within 3% accuracy.

7.2.4. *The Advance of the Perihelion of Mercury*

Given that the mass of the Sun is considerably larger than the masses of the planets in the solar system, we may assume the planets to be test particles following geodesics in the Schwarzschild spacetime around the Sun.

The relevant equations in this case are, of course, those for a timelike geodesic. The θ-component again allows us to assume $\theta = \pi/2$ for the orbital plane. As in (27) we get the first integrals

$$c\left(1 - \frac{2GM}{c^2 r}\right)\frac{\mathrm{d}t}{\mathrm{d}s} = \gamma, \qquad r^2\frac{\mathrm{d}\varphi}{\mathrm{d}s} = h, \tag{35}$$

where γ and h are constants. Similarly, the line element (19) gives the first integral in the form

$$\left(1 - \frac{2GM}{c^2 r}\right) c^2 \left(\frac{\mathrm{d}t}{\mathrm{d}s}\right)^2 - \left(1 - \frac{2GM}{c^2 r}\right)^{-1}\left(\frac{\mathrm{d}r}{\mathrm{d}s}\right)^2 - r^2\left(\frac{\mathrm{d}\varphi}{\mathrm{d}s}\right)^2 = 1. \tag{36}$$

Again we define $u = 2GM/c^2 r$ and use (35) to write (36) in the form

$$\left(\frac{\mathrm{d}u}{\mathrm{d}\varphi}\right)^2 + u^2 = \frac{(\gamma^2 - 1)}{h^2} r_s^2 + \frac{r_s^2}{h^2} u + u^3.$$

Differentiate with respect to φ to arrive at the following equation:

$$\frac{\mathrm{d}^2 u}{\mathrm{d}\varphi^2} + u = \frac{r_s^2}{2h^2} + \frac{3}{2} u^2. \tag{37}$$

In the Newtonian theory the second term on the right-hand side is absent and we get the familiar ellipse

$$u_0 = \frac{1}{l_1}(1 + e\cos\theta), \quad l_1 = \frac{2h^2}{r_s^2}$$

as the solution. To obtain an approximate solution of (37) we write $u = u_0 + u_1$, where

$$\frac{\mathrm{d}^2 u_1}{\mathrm{d}\varphi^2} + u_1 \cong \frac{3}{2} u_0^2 = \frac{3}{2l_1^2}(1 + 2e\cos\varphi + e^2\cos^2\varphi).$$

Because of the cos φ term on the right-hand side u_1 has a secular behaviour. This part of u_1 will be detected in long-term observations. Hence we confine ourselves to the secular part only, which is

$$u_{1s} \cong \frac{3e}{2l_1^2}\varphi\sin\varphi.$$

Hence,

$$u \cong u + u_{1s} \cong \frac{1}{l_1}\left[1 + e\cos\varphi + \frac{3e}{2l_1}\varphi\sin\varphi\right]$$

$$\cong \frac{1}{l_1}[1 + e\cos(\varphi - \varphi_0)], \tag{39}$$

where

$$\varphi_0 \simeq \frac{3}{2l_1}\,\varphi\,. \tag{40}$$

φ_0 gives the angular coordinate of the perihelion of the orbit. Over one planetary period, φ_0 changes by $3\pi/l$. Hence, the average precession rate of the perihelion is

$$n = \frac{3\pi}{l_1 T} = \frac{6\pi GM}{c^2 l T}\,, \qquad l = \frac{2GM}{c^2}\, l_1, \tag{41}$$

T being the period of the planet and $2l$ its latus rectum (cf. Figure 7.3).

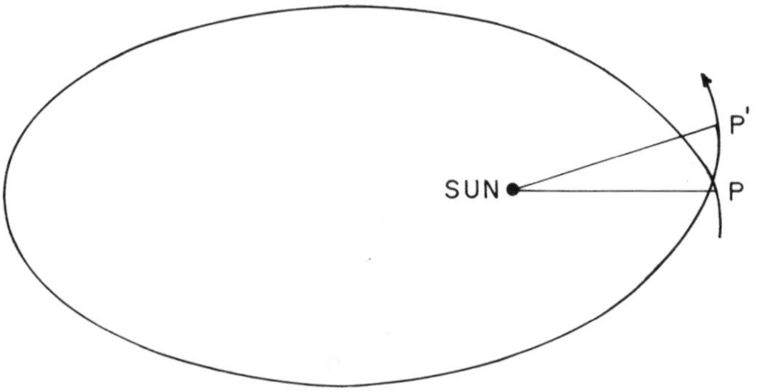

Fig. 7.3. A schematic diagram showing how the direction of the perihelion of Mercury shifts relative to the Sun from P to P' as the planet completes one orbit.

This effect is largest for the planet Mercury which has the least values of l and T. Even so, the effect predicted is small:

$$n = 43.03 \text{ arcsec century}^{-1}. \tag{42}$$

The actual observed precession rate with respect to the local inertial frame centred on the Sun is ~ 575.73 arcsec cy^{-1} of which ~ 532 arcsec cy^{-1} is explained by the perturbative effects of other planets on Mercury's orbit. The residual to be explained is ~ 43.73 arcsec cy^{-1}, which agrees very well with (42) within observational errors. This test therefore adds another feather in the cap of general relativity.

We have completed the list of direct tests of relativity. The reader will not have failed to notice that in all the tests the theory has been tested in weak gravitational fields. To test it fully we need to have situations of strong gravitational fields accessible to experimentation. So far this has not been possible, nor does it seem possible in the foreseeable future. So any discussion of strong-field scenarios in astrophysics must rest on the implicit belief that we are using the correct theory, although it has been tested and found satisfactory only under limited weak-field conditions.

7.3. Gravitational Radiation

Before proceeding to strong-field scenarios we shall deal with another weak-field situation which, at least theoretically, suggests that time dependent gravitational disturbances travel with the speed of light. We investigate below the phenomenon of gravitational radiation. To this end we go back to Section 6.9 and of the three assumptions made herein retain only assumption (a).

With the metric tensor given by (6.114) we linearize Einstein's equations. First note that

$$R_{iklm} \cong \tfrac{1}{2}\,\{h_{im,kl} + h_{kl,im} - h_{km,il} - h_{il,km}\}. \tag{43}$$

Before we proceed to calculate R_{ik} and R we make a coordinate transformation which is analogous to the gauge transformation in Maxwell's theory.

Define the gravitational potentials by

$$\psi_i^k = h_i^k - \tfrac{1}{2}\,\delta_i^k h, \qquad h = h^l{}_l. \tag{44}$$

Suppose we make a change of coordinates

$$x^i = x'^i + \xi^i, \tag{45}$$

where ξ^i are quantities of first-order smallness. Then from (6.80)

$$h'_{ik} = h_{ik} + \left(\frac{\partial \xi_i}{\partial x^k} + \frac{\partial \xi_k}{\partial x^i}\right), \qquad h' = h + 2\,\frac{\partial \xi^k}{\partial x^k}. \tag{46}$$

Thus,

$$\begin{aligned} \psi'^k{}_{i,k} &\equiv h'^k{}_{i,k} - \tfrac{1}{2}\,h'{}_{,i} = \psi_i^k{}_{,k} + (\xi_i{}^{,k} + \xi^k{}_{,i})_{,k} - \xi^k{}_{,ki} \\ &= \psi^k{}_{i,k} + \Box\xi_i = 0 \end{aligned} \tag{47}$$

if we choose ξ_i suitably.

Thus, without loss of generality we may assume that

$$\psi^k_{i,k} = 0. \tag{48}$$

Just as the gauge condition $A^i{}_{,i} = 0$ in electromagnetic theory still leaves some freedom for further manoeuvre, so in the present case, a coordinate transformation

$$x'^i = x^i + \xi^i, \quad \Box\xi^i = 0 \tag{48}$$

still ensures (48) in primed coordinates.

.With (48) holding we get, from (43),

$$R_{ik} \cong \tfrac{1}{2}\,\Box h_{ik}, \qquad R \cong \tfrac{1}{2}\,\Box h, \qquad G_{ik} \cong \tfrac{1}{2}\,\Box\psi_{ik}. \tag{49}$$

Thus Einstein's equations become

$$\Box\psi_{ik} = -\,\frac{16\pi G}{c^4}\,T_{ik}\,. \tag{50}$$

This is a linearized tensor wave equation. By analogy with electromagnetism we can consider a spin-2 particle, the 'graviton' as the carrier of gravitational radiation. The wave equation (50) can be solved for given sources.

Birkhoff's theorem tells us that a spherically symmetric source does not radiate, since the spacetime outside it is static. This is analogous to the electromagnetic case wherein also a spherically symmetric system does not radiate. The primary source for electromagnetic radiation is an accelerated electric charge which generates an electric dipole with nonzero second time derivative. For (50) the primary source is the *third* time derivative of the mass *quadrupole* moment. Denoting the quadrupole moment of a source by I, we get the rate at which energy is radiated to be

$$\mathcal{E} = \frac{G}{5c^5} \, |\dddot{I}_{\mu\nu}|^2 . \tag{51}$$

The following examples illustrate some features of gravitational radiation. It is clear from them that only huge astronomical events can generate gravitational waves in sufficient quantity to enable their detection in the laboratory.

Example 7.6. We consider a laboratory source of gravitational radiation made of a uniform cyclinder of length L and a circular cross section of radius R. The density of the cyclinder is ϱ (say) so that its mass is

$$M = \pi R^2 L \varrho .$$

This cyclinder is suspended horizontally from the midpoint of its length and is made to turn round the vertical axis with angular velocity ω. For this cylinder the quadrupole moment tensor with respect to Cartesian axes x_μ fixed in space is

$$I_{\mu\nu} = \frac{ML^2}{12} \begin{bmatrix} \cos^2\theta - \frac{1}{3} & \cos\theta\sin\theta & 0 \\ \cos\theta\sin\theta & \cos^2\theta - \frac{1}{3} & 0 \\ 0 & 0 & -\frac{1}{3} \end{bmatrix}$$

where the 3-axis is vertical and the cyclinder is instantaneously making an angle θ with the x_1 axis. Then

$$\dddot{I}_{\mu\nu} = \frac{ML^2\omega^3}{3} \begin{bmatrix} \sin^2 2\theta & -\cos^2 2\theta & 0 \\ \cos^2 2\theta & \sin^2 2\theta & 0 \\ 0 & 0 & 0 \end{bmatrix} ; \quad |\dddot{I}_{\mu\nu}|^2 = \frac{2M^2L^4\omega^6}{9} .$$

Hence, $\mathcal{E} = 2GM^2L^4\omega^6/45c^5$.

For rapid rotation ω is limited above by the fact that the material has a finite tensile strength $\bar{t}$. Thus

$$\varrho l^2 \omega^2 < 8\bar{t} .$$

For iron, $\varrho = 7.8$ g cm^{-3}, $\bar{t} = 3 \times 10^9$ dyne cm^{-2} and $\omega \lesssim 28$ rad s^{-1}. For $R = 1$ m, $l = 20$ m, $M \cong 4.9 \times 10^8$ g and at the limiting value of ω, we find $\mathcal{E} = 2.2 \times 10^{-22}$erg s^{-1}. □

Example 7.7. As an astronomical source of gravity waves consider a binary star system of masses m_1 and m_2 in circular orbit.

For this two-body system take the centre of mass at origin. Then if $\mathbf{r}^{(1)}$ and $\mathbf{r}^{(2)}$ are the position vectors of m_1 and m_2 we can write

$$\frac{\mathbf{r}^{(1)}}{m_2} = -\frac{\mathbf{r}^{(2)}}{m_1} = \frac{r}{m_1 + m_2}[\cos\theta \sin\theta, 0]$$

where the x_3 axis is $\perp$ the plane of the orbit and θ is the angle made by the line joining and the stars with the x_1 axis at time t. Then $\dot{\theta} = \omega$ where from Newtonian physics (which holds in the weak-field limit) we get

$$\omega^2 r^3 = G(m_1 + m_2).$$

For this system we have

$$I_{\mu\nu} = \frac{r m_1 m_2}{(m_1 + m_2)}\begin{bmatrix} \cos^2\theta - \frac{1}{3} & \cos\theta\sin\theta & 0 \\ \cos\theta\sin\theta & \cos^2\theta - \frac{1}{3} & 0 \\ 0 & 0 & -\frac{1}{3} \end{bmatrix}.$$

As in the previous example we get

$$|\dddot{I}_{\mu\nu}|^2 = \frac{32 m_1^2 m_2^2}{(m_1 + m_2)^2} r^4 \omega^6 = \frac{32 m_1^2 m_2^2 G^{4/3}}{(m_1 + m_2)^{2/3}} \omega^{10/3}.$$

Thus the energy is lost at a rate

$$\frac{\mathrm{d}E}{\mathrm{d}t} = -\frac{32}{5}\frac{G^{7/3}}{c^5}\frac{m_1^2 m_2^2}{(m_1 + m_2)^{2/3}}\omega^{10/3}.$$

The binary system has a total energy equal to

$$E = -\frac{G m_1 m_2}{2r} = -\frac{G^{2/3} m_1 m_2}{2(m_1 + m_1)^{1/3}}\omega^{2/3}$$

The rate of loss of energy may be related to the rate of change of period $T = 2\pi/\omega$. We get

$$\dot{T} = -\frac{2\pi}{\omega^2}\dot{\omega} = \frac{6\pi}{\omega^{5/3}}\frac{(m_1 + m_2)^{1/3}}{g^{2/3} m_1 m_2}\dot{E}$$

$$= -\frac{192\pi}{5}\frac{G^{5/3}}{c^5}\frac{m_1 m_2}{(m_1 + m_2)^{1/3}}\left(\frac{2\pi}{T}\right)^{5/3}.$$

Thus the period of the orbit gradually decreases and the orbit shrinks. □

How can we design equipment to detect gravitational waves? A changing gravitational field generates a time-dependent curvature tensor. Equation (6.65) of geodetic deviation helps detect such a tensor; for it generates a force between two neighbouring particles of matter. This force strains elastic matter and the resulting stresses can be detected by electrical methods if the material of the detector is a piezo-electric cystal.

In the early 1970s J. Weber's detectors set up in Maryland and Chicago showed coincident events which he interpreted as due to gravitational waves. Subsequent studies by other groups have, however, failed to detect similar waves, and the general consensus is that such waves, if present, are still to be detected. Detectors much more sensitive than Weber's pioneering instrument are under construction.

Astronomical events that can generate large quantities of gravitational waves are supernova explosions. Close binaries are also sources of such waves. In such a system the waves are continually emitted and the stars gradually get closer to each other (cf. Example 7.7). In this process the period of the binary is reduced. Such a reduction was noticed for the binary pulsar PSR 1913 + 16 and it is believed to be an indirect proof of the existence of gravitational radiation.

Before we move to another topic we remind the reader that all the above calculations are for the linearized theory and not for the full nonlinear theory. In the latter theory much more care is needed to disentangle the concept of radiation from the spacetime geometry of the background. We shall not touch upon those complexities here.

7.4. Geometrodynamics

This word was coined by John Wheeler to describe the general situation wherein changes in spacetime are related to the dynamic changes in its matter/radiation content according to general relativity, in analogy with Maxwell–Lorentz electrodynamics. There are, however, more conceptual problems with geometrodynamics than with electrodynamics, because in the former the 'field' variables g_{ik} also describe the spacetime metric, whereas in the latter the background spacetime remains unchanged as fields propagate through it. This is in fact the same difficulty that we mentioned in the context of gravitational radiation. Nevertheless, the following formalism (which we shall need later while discussing quantum gravity) may be used to describe geometrodynamics.

Like electrodynamics, we now try to correctly formulate the general relativistic problem implied by the question: 'How does the spacetime geometry change around changing sources of gravity?' As discussed in Section 6.8 we have the correct number of differential equations for the unknown quantities in g_{ik} and T_{ik}. However, the formulation of the initial value problem poses difficulties that are nontrivial.

First note that the Einstein equations are of second order in g_{ik} and so we should expect to specify $g_{ik}(0, x^{\mu})$ and $\dot{g}_{ik}(0, x^{\mu})$ as the initial data. However, on examination of the field equations we find that the second timer derivatives of the metric tensor appear only through the component $R_{\mu 00 \nu}$ and these are confined to $\ddot{g}_{\mu\nu}$ only. Thus for $i, k = 1, 2, 3$ we are allowed to specify $g_{ik}(0, x^{\mu})$ and $\dot{g}_{ik}(0, x^{\mu})$. The $\ddot{g}_{\mu\nu}$ terms appear through $R^{\mu}{}_{\nu}$ while the $R^{0}{}_{0}$ and $R^{0}{}_{\nu}$ terms contain only first time derivatives. Therefore, the $\binom{0}{0}$ and $\binom{0}{\nu}$ equations act as constraints on the initial data.

To simplify the overall problem we choose a coordinate system such that (with $c = 1$)

$$ds^2 = dt^2 + g_{\mu\nu}\, dx^\mu\, dx^\nu, \tag{52}$$

and thus assume the spacetime to be foliated by spacelike hypersurfaces Σ given by t = const. Without affecting the structure of the line element (52) we can make coordinate transformations $x^\mu \to x'^\mu$ which change $g_{\mu\nu}$ *to* $g'_{\mu\nu}$, say. Thus the real degrees of freedom of $g_{\mu\nu}$ are only three. Of these three degrees of freedom, one could be assigned to specify how the hypersurfaces (Σ) are embedded in the spacetime while the other two degrees of freedom tell us about the intrinsic geometry of Σ. We shall now discuss how this is done.

In the frame of reference giving (52) we denote by n^i the unit vector field normal to the hypersurfaces $\{\Sigma\}$ and introduced the quantity

$$K_{il} \equiv n_{i;l} \equiv n_{i,l} - \Gamma^p_{il}\, n_p. \tag{53}$$

K_{il} is called the 'extrinsic curvature tensor', and its trace K the extrinsic curvature of Σ. Since

$$n^i = n_i = (1, 0, 0, 0), \tag{54a}$$

we get the only nonzero components of K_{il} as the spacelike ones (i, l = 1, 2, 3). Further,

$$\Gamma^0_{\mu\nu} = -\tfrac{1}{2}\dot{g}_{\mu\nu} = -K_{\mu\nu}, \qquad \Gamma^\mu_{0\nu} = K^\mu{}_\nu. \tag{54b}$$

We now consider the gravitational action involving first derivatives only (cf. Equation (6.113)). In the above notation it takes the form

$$\tilde{S} = \int \sqrt{-g}\ {}^4\mathscr{L}\, d^4x \equiv \int_{t_1}^{t_2} \int \sqrt{-g}\ \{{}^3\mathscr{L} - (\mathrm{Tr}\, K)^2 + \mathrm{Tr}\, K^2\}\, d^3\mathbf{x}\, dt, \tag{55}$$

where

$${}^3\mathscr{L} = (\Gamma^\alpha_{\beta\gamma}\Gamma^\gamma_{\alpha\sigma} - \Gamma^\alpha_{\beta\sigma}\Gamma^\mu_{\alpha\mu})g^{\beta\sigma} \tag{56}$$

is the three-dimensional Lagrangian exactly analogous to the four-dimensional Lagrangian we started with. Since the scalar curvature ${}^3R$ for the 3-geometry on Σ differs from ${}^3\mathscr{L}$ by an unimportant 3-divergence, we can write

$$S = \int_{t_1}^{t_2} \int \sqrt{-{}^3g}\,\{{}^3R - (\mathrm{Tr}\, K)^2 + \mathrm{Tr}(K^2)\}\, d^3\mathbf{x}\, dt. \tag{57}$$

This action is quadratic in $\dot{g}_{\mu\nu}$ and can serve as the variational principle for geometrodynamics. However, there are problems still! We have to consider what happened to the $\binom{0}{\mu}$ and $\binom{0}{0}$ parts of the field equations.

First define the conjugate momenta by

$$\pi^{\mu\nu} \equiv \frac{\delta S}{\delta \dot{g}_{\mu\nu}} = \sqrt{-{}^3g}\,\{g^{\mu\nu}(\mathrm{Tr}\, K) - K^{\mu\nu}\} \tag{58}$$

and write the action in a modified form as

$$S_{\text{modified}} = \int_{t_1}^{t_2}\int \{\pi^{\mu\nu}\dot{g}_{\mu\nu} - NX^0 + N_\alpha X^\alpha\}\,\mathrm{d}^3\mathbf{x}\,\mathrm{d}t,$$

$$X^0 \equiv \sqrt{-{}^3g}\,\{{}^3R + (\mathrm{Tr}\,K)^2 - \mathrm{Tr}\,K^2\},$$

$$X^\alpha \equiv -2\sqrt{-{}^3g}\,(K^{\mu\alpha} - g^{\mu\alpha}K)_{;\mu}, \tag{59}$$

where N and N_α are Lagrange multipliers and $X^0 = 0$ $X^\alpha = 0$ are constraints imposed on the variation of S as given by (57). Thus the $g_{\alpha\beta}$ and N, N_α are all freely varied in (59). A little calculation will show that this problem is now no different from the variational problem of Hilbert for the line element

$$\mathrm{d}s^2 = (N^2 - N_\alpha N^\alpha)\,\mathrm{d}t^2 + 2N_\alpha\,\mathrm{d}x^\alpha\,\mathrm{d}t + g_{\mu\nu}\,\mathrm{d}x^\mu\,\mathrm{d}x^\nu. \tag{60}$$

In the example from electrodynamics discussed below the constraint equation is seen to be a consequence of the gauge invariance. Likewise the constraints $X^0 = 0$, $X^\alpha = 0$ are the consequences here of the general coordinate invariance.

Example 7.8. Let us consider the initial value problem in electrodynamics. The electromagnetic action (6.77) can be written in the form

$$S = \frac{1}{8}\int_{t_1}^{t_2}\int(\mathbf{E}^2 - \mathbf{B}^2)\,\mathrm{d}^3\mathbf{x}\,\mathrm{d}t$$

in Minkowski spacetime. The electromagnetic fields $\mathbf{E}$, $\mathbf{B}$ are derivable from potentials $\mathbf{A}$, φ:

$$\mathbf{B} = \nabla\times\mathbf{A},\quad \mathbf{E} = -\nabla\varphi - \frac{\partial\mathbf{A}}{\partial t}.$$

However, we can use gauge transformations

$$\mathbf{A}\to\mathbf{A} + \nabla\chi,\qquad \varphi\to\varphi - \frac{\partial\chi}{\partial t}$$

to choose $\varphi = 0$. Then the above action becomes

$$S = \frac{1}{8}\int_{t_1}^{t_2}\int[\dot{\mathbf{A}}^2 - (\nabla\times\mathbf{A})^2]\,\mathrm{d}^3\mathbf{x}\,\mathrm{d}t.$$

Now vary $\mathbf{A}$ to get from $\delta S = 0$ the equation

$$\frac{\partial\mathbf{E}}{\partial t} = \nabla\times\mathbf{B}.$$

The definitions of $\mathbf{B}$ and $\mathbf{E}$ in terms of $\mathbf{A}$ have already guaranteed the two Maxwell equations

$$\nabla\cdot\mathbf{B} = 0,\qquad \nabla\times\mathbf{E} = -\frac{\partial\mathbf{B}}{\partial t}.$$

Notice, however, that the fourth Maxwell equation $\nabla\cdot\mathbf{E} = 0$ has not yet been obtained! This is a constraint on Maxwell's equations which, if satisfied initially, is then satisfied always since

$$\frac{\partial}{\partial t}\nabla\cdot\mathbf{E} = \nabla\cdot\frac{\partial\mathbf{E}}{\partial t} = \nabla\cdot(\nabla\times\mathbf{B}) \equiv 0.$$

To introduce this constraint we use the method of Lagrange multipliers and write

$$S_{\text{modified}} = \frac{1}{8}\int_{t_1}^{t_2}\int [\dot{\mathbf{A}}^2 - (\nabla\times\mathbf{A})^2 + \varphi(\nabla\cdot\mathbf{E})]\,\mathrm{d}^3\mathbf{x}\,\mathrm{d}t,$$

where φ is the Lagrange multiplier. In this way we have 'recovered' the scalar potential φ that had been 'gauged away' earlier.

If we had not known beforehand this extra condition $\nabla\cdot\mathbf{E} = 0$, we could have guessed it by using gauge invariance. For, when performing the variation $S \to S + \delta S$ we have ignored the surface terms at $t = t_1 = t_2$:

$$\delta S = \int \mathbf{E}\cdot\delta\mathbf{A}\,\mathrm{d}^3\mathbf{x}.$$

Suppose we require δS to vanish for all variations of $\mathbf{A}$ of the kind

$$\delta\mathbf{A} = \nabla\delta\chi$$

Then $\delta S = 0$ for arbitrary $\delta\chi$ gives us $\nabla\cdot\mathbf{E} = 0$. □

Considerable discussion has gone in the literature as to how to specify the correct initial values on a spacelike hypersurface. The natural prescription of the 3-geometry ${}^3G$ on Σ in the form of $g_{\mu\nu}$ decides not only the initial geometry but also how the spacetime is foliated. The alternative suggested by Isenberg and Wheeler in 1979 is, however, free from the ambiguities that beset the above prescription because of the constraint conditions. This alternative requires the specification of Tr K and the conformal part of ${}^3G$. We shall return to this discussion in Part IV. Our purpose in presenting it here was to make the reader aware of the difficulties present in the formulation of the initial value problem in general relativity.

7.5. Gravitational Collapse

We next consider a particularly simple example of geometrodynamics: the problem of gravitational collapse of a dustball. The problem has relevance to astrophysics in the following way.

In a typical star, equilibrium is achieved between the contracting force of gravity and the counteracting outward force of thermal pressures. The pressures are maintained for a long time in the life of the star by the thermonuclear processes in its core. These processes are ultimately exhausted. A massive star ($M \gtrsim 6$ solar masses), however, explodes and becomes a supernova, leaving behind a hot, dense core. This core cannot call upon thermonuclear processes to generate pressures since the nuclear fuel is by now exhausted.

There is, however, another type of pressure, that due to degenerate matter in the form of neutrons, that a star can call upon to oppose its gravitational contraction. These pressures also have their limitations: they can support a star of mass not exceeding three solar masses. (Some calculations put this limit at two-thirds this value.) What happens to a core which is more massive than this limit? It begins to contract and as it does so the imbalance between the internal pressures and gravity grows and the contraction becomes catastrophic. This is known as *gravitational collapse*. Since dust as a pressure-free fluid gives a good approximation to this final state, the problem considered below has physical relevance.

We shall consider spherically symmetric collapse and use the line element (5) to describe the spacetime geometry within the dustball. The outside solution is, of course, given by the Schwarzschild line element (*vide* Birkhoff's theorem). We shall use comoving coordinates (r, θ, φ) to label a typical dust particle and write the energy momentum tensor as

$$T^{ik} = \varrho v^i v^k, \qquad v^i = (e^{-\sigma/2}, 0, 0, 0). \tag{61}$$

For convenience we have taken $c = 1$. The conservation law,

$$T^{ik}{}_{;k} = 0, \tag{62}$$

gives $\sigma^1 \equiv \partial\sigma/\partial r = 0$ so that σ is a function of t only. Such a function can be trivially absorbed in a time transformation. Hence we shall set $\sigma = 0$.

Example 7.9. Let us consider the line element (5) and assume that the matter has the form of a fluid of pressure p and density ϱ. If (5) refers to comoving coordinates, we have $T^i_k = \mathrm{diag}(\varrho, -p, -p, -p)$. Consider $T^i_{k;i} = 0$ for $k = 0$ and $k = 1$:

$$\begin{aligned} 0 = T^i_{0;i} &\equiv T^i_{0,i} + \Gamma^i_{li} T^l_0 - \Gamma^m_{0i} T^i_m \\ &= \dot{\varrho} + \Gamma^i_{0i}\varrho - \Gamma^0_{00}\varrho + (\Gamma^1_{01} + \Gamma^2_{02} + \Gamma^3_{03})p \\ &= \dot{\varrho} + (\dot{\mu} + \tfrac{1}{2}\dot{\omega})(p + \varrho), \\ 0 = T^i_{1;i} &= T^i_{1,i} + \Gamma^i_{li} T^l_1 - \Gamma^i_{1i} T^i_m \\ &= -p' - \tfrac{1}{2}(p + \varrho)\sigma'. \end{aligned}$$

If p is a function of ϱ we get

$$\sigma = -\int \frac{2\mathrm{d}p}{p + \varrho} + \text{Function of } t,$$

$$\mu + \tfrac{1}{2}\omega = -\int \frac{\mathrm{d}\varrho}{p + \varrho} + \text{Function of } r.$$

Hence, for a dustball ($p = 0$) we can effectively set $\sigma = 0$. Also, we get

$$\varrho = (\text{Function of } r) \times e^{-(\mu + \omega/2)}. \qquad \square$$

We next set up the Einstein equations by following the procedure adopted in Section 7.1. We shall not go through the details of computations of the Γ^i_{kl}

and R_{ik} but use the final answers. First we find that the (1, 0) component of Einstein's equations gives

$$2\dot{\mu}' + \dot{\mu}\mu' - \dot{\omega}\mu' = 0, \tag{63}$$

which integrates to

$$e^{\omega} = \frac{e^{\mu}\,\mu'^2}{4(1+f)}, \tag{64}$$

where $f(r)$ is an arbitrary function of r.

Next we consider the (1, 1) equation which takes the form

$$\tfrac{1}{2}\mu'^2\,e^{-\omega} - (\ddot{\mu} + \tfrac{3}{4}\dot{\mu}^2) - e^{-\mu} = 0. \tag{65}$$

We substitute from (64) for e^{ω} and then integrate (65) to get

$$\dot{\mu}^2 = 4f(r)\,e^{-\mu} + 4F(r)\,e^{-3\mu/2}, \tag{66}$$

where $F(r)$ is another function of the radial coordinate r.

Finally we consider the (0, 0) equation which gives

$$\mu'' + \frac{3}{4}\mu'^2 - \tfrac{1}{2}\mu'\omega' - \tfrac{1}{2}e^{\omega}(\dot{\omega}\dot{\mu} + \tfrac{1}{2}\dot{\mu}^2) - e^{-\mu} = 8\pi G\varrho. \tag{67}$$

To solve this, differentiate (66) and (64) with respect to r and also (64) with respect to t to eliminate ω' and $\dot{\omega}$. A little manipulation gives

$$F'(r) = 4\pi G\varrho\,e^{3\mu/2}\,\mu'. \tag{68}$$

(Compare this result with that obtained in Example 7.9.)

We now consider the initial conditions. Suppose at $t = 0$ the ball was at rest so that $\dot{\omega} = 0$, $\dot{\mu} = 0$ at $t = 0$. Let $\varrho = \varrho_0(r)$ be a prescribed function of r initially. Remembering that we can relabel the r-coordinate provided that we preserve its monotonic nature, suppose that we choose r-coordinate such that

$$e^{\mu(r,0)} = r^2. \tag{69}$$

Thus initially, r has the same meaning that is given to the Schwarzschild coordinate. Then (68) at $t = 0$ gives

$$F'(r) = 8\pi G\varrho_0 r^2,$$

that is,

$$F(r) = 8\pi G\int_0^r \varrho_0(r_1)r_1^2\,dr_1. \tag{70}$$

(We have set the arbitrary constant in the above integration to zero. Why?) Further, since $\dot{\mu} = 0$ at t = 0, (66) gives

$$f(r) = -\frac{1}{r}F(r). \tag{71}$$

The arbitrary functions are thus fully determined.

Equation (66) can then be integrated. It is convenient to write

$$e^{\mu/2} = R(r, t). \tag{72}$$

Then (66) becomes

$$\frac{\dot{R}^2}{R^2} = F(r)\left\{\frac{1}{R^3} - \frac{1}{rR^2}\right\}$$

that is,

$$\dot{R}^2 = \frac{F(r)}{r}\left\{\frac{r}{R} - 1\right\}. \tag{73}$$

The reader can complete the solution in the general case. We discuss the special case of uniform initial density ϱ_0. Then (70) and (71) become

$$F(r) = \frac{8\pi G}{3}\varrho_0 r^3, \qquad f(r) = -\frac{8\pi G\varrho_0}{3}r^2. \tag{74}$$

Write $R = rS(t)$. Then (73) becomes

$$\dot{S}^2 = \frac{8\pi G\varrho_0}{3}\left(\frac{1}{S} - 1\right). \tag{75}$$

The line element for the internal solution then takes the form

$$ds^2 = dt^2 - S^2(t)\left[\frac{dr^2}{1 - \alpha r^2} + r^2(d\theta^2 + \sin^2\theta\, d\varphi^2)\right], \tag{76}$$

where $\alpha = 8\pi G\varrho_0/3$.

In this solution the dustball collapses while keeping uniform density at all stages. The 'final' state is of $S = 0$ when $\varrho \to \infty$. The general case also has this final state of infinite density, although the collapsing object is not uniform.

Example 7.10. Let us discuss the more general case of (73). We write $R = r\cos^2\psi$ so that this equation becomes

$$\dot{\psi} = -\sqrt{\frac{F}{4r^3}}\sec^2\psi.$$

which integrates to

$$\sqrt{\frac{F}{r^3}}\,[t - g(r)] = \psi + \sin\psi\cos\psi.$$

Now $R \to 0$ when $\psi \to \pi/2$, i.e. when

$$t \to t_0(r) = g(r) + \frac{\pi}{2}\sqrt{\frac{r^3}{F(r)}}.$$

Thus the spherical surface of comoving radius r shrinks to a point at $t = t_0(r)$. Unlike the homogeneous situation the entire body does not collapse at the same instant.

Also, from (64) we have

$$e^{\omega} = \frac{R'^2}{1+f} = \frac{R'^2}{1 - F/r}.$$

At $\psi = \pi/2$,

$$R' = -2r[\cos\psi \sin\psi\psi']_{\pi/2}.$$

However, from the above we get

$$(1 + \cos 2\psi)\psi' = t\alpha(r) + \beta(r),$$

where $\alpha(r)$ and $\beta(r)$ are related to F and g. Hence, at $\psi = \pi/2$, $\beta' \to \infty$ so fast that $R' \to \infty$. Thus radial displacements diverge while transverse displacements shrink. The overall effect is, however, to shrink the proper volume to zero. □

For the comoving observer the collapse time-scale in the homogeneous case is easily seen to be

$$T = \frac{\pi}{2\sqrt{\alpha}}. \tag{77}$$

The final state has divergent components of R_{iklm} and the various invariants in Section 7.1 are infinite. This is known as the state of *spacetime singularity*. Since the spacetime geometry breaks down it is not possible to continue the solution beyond the singular epoch $t = T$.

Let us consider outward light signals sent by an observer B on the surface of the collapsing body to an external observer A. For simplicity, assume that A is at rest in the Schwarzschild reference frame which holds outside the massive body. To avoid confusion we shall use R instead of r as the Schwarzschild radial coordinate and suppose that $R = R_A$ for A. Further, suppose that A is located in the radially outward direction from B.

The Schwarzschild radial coordinate of B is given by

$$R_B = r_0 S(t), \tag{78}$$

since we expect the angular part of the Schwarzschild line element

$$ds^2 = \left(1 - \frac{2GM}{R}\right) dT^2 - \frac{dR^2}{1 - (2GM/R)} - R^2(d\theta^2 + \sin^2\theta\, d\varphi^2) \tag{79}$$

to be continuous with (5) on the surface of the body. We have also changed the time coordinate here from t to T since t is used for the comoving system.

The constant M can be related to the parameters r_0, ϱ_0 of the dustball as follows. At the initial instant $t = 0$ the r coordinate had the meaning of the Schwarzschild coordinate. Thus we can directly compare (76) with (79) for $R = r = r_0$. Since $S(0) = 1$, we get from a comparison of coefficients of dr^2 and dR^2

$$1 - \alpha r_0^2 = 1 - \frac{2GM}{r_0},$$

that is,

$$M = \frac{\alpha r_0^3}{2G} = \frac{4\pi}{3}\varrho_0 r_0^3. \tag{80}$$

Let the light ray leave B at $T = T_B$ to reach A at $T = T_A$. Then the equations of radial null geodesic give us

$$T_A - T_B = \int_{R_B}^{R_A} \frac{\mathrm{d}R}{1 - (2GM/R)}. \tag{81}$$

The above integral diverges as $R_B \to 2GM = r_s$ (cf. Section 7.1). This means that the outside observer cannot receive any signals from B after B has crossed into the Schwarzschild sphere. Likewise, A remains causally disconnected from any subsequent developments in the collapsing body.

This stage is identified with the formation of the 'event horizon'. The object itself is said to have become a black hole after the event horizon is formed. The horizon in the above case is the surface $R = 2GM$. It is not difficult to show that in the above case the light waves from B to A arrive at A with increasing redshift which diverges as B approaches the horizon. Because of the redshift the luminosity of the object drops sharply until at $R = 2GM$ the object becomes absolutely invisible, thus justifying the name 'black hole'.

Example 7.11. Let us evaluate the function $T(t)$ on the boundary of the freely collapsing dustball.

At $r = r_0$, the line element (76) gives $\mathrm{d}s = \mathrm{d}t$. From (79) for the same observer B we have

$$\mathrm{d}s^2 = \left(1 - \frac{2GM}{R}\right)\left(\frac{\partial T}{\partial t}\right)^2_{r_0} \mathrm{d}t^2 - \left(1 - \frac{2GM}{R}\right)^{-1}\left(\frac{\partial R}{\partial t}\right)^2_{r_0} \mathrm{d}t^2.$$

Equating $\mathrm{d}s$ to $\mathrm{d}t$ and using (78) we get

$$\left(\frac{\partial T}{\partial t}\right)^2_{r_0} = \frac{1}{1 - (2GM/R)}\left\{1 + \frac{r_0^2 \dot{S}^2(t)}{1 - (2GM/R)}\right\}.$$

This relation determines $T(t)$ at $r = r_0$. □

Example 7.12. Suppose B sends the crest of a light wave of wavelength λ_B at time t_B which A receives at time T_B. The next crest will leave B at $t_B + \Delta t_B$ where $\Delta t_B = \lambda_B$. Let it be received by A at $T_A + \Delta T_A$. Ignoring the term $2GM/R_A$ at A we conclude that the light wave arrives at A with a wavelength $\lambda_A = (\Delta T_A)$. Thus the spectral shift z is given by

$$1 + z = \frac{\lambda_A}{\lambda_B} = \frac{\Delta T_A}{\Delta t_B}.$$

In Example 7.11 above we have worked out $\Delta T_B/\Delta t_B$. So we need to know $\Delta T_A/\Delta T_B$.

From (81) we have

$$T_A - T_B = R_A - R_B + 2GM \ln \frac{R_A - 2GM}{R_B - 2GM}.$$

As the body shrinks, the change in R_B during the proper time interval Δt_B is given by $\Delta R_B = -r_0 \dot{S}(t) \Delta t_B$. Therefore, from the above relation we get

$$\Delta T_A - \Delta T_B = \frac{r_0 \dot{S}(t) \Delta t_B}{1 - (2GM/R)}$$

The net redshift follows, with the help of Example 7.11,

$$1 + z = \frac{\Delta T_A}{\Delta t_B} = \frac{\Delta T_B}{\Delta t_B} + \frac{r_0 \dot{S}(t)}{1 - (2GM/R)}$$

$$= \frac{1}{1 - (2GM/R)} \left\{ \left(1 - \frac{2GM}{R} + r_0^2 \dot{S}^2(t) \right)^{1/2} + r_0 \dot{S}(t) \right\}.$$

Notice that the redshift observed by A is larger than in the case if the object were static (*vide* Equation (23)). This is because we have here the combination of gravitational redshift and Doppler redshift, the latter arising from the recession of B from A. □

Example 7.13. Suppose that instead of collapse the dustball were exploding from a singularity. This time-reversed version of the collapse problem is very similar to the Big Bang cosmology to be discussed in the next chapter, and is called a 'white hole'. The spectral shift formula now changes from that derived the Example 7.12 above by replacing $\dot{S}(t)$ by $-\dot{S}(t)$.

We therefore get

$$1 + z = \frac{1}{1 - (2GM/R)} \left\{ \left(1 - \frac{2GM}{R} + r_0^2 \dot{S}^2 \right)^{1/2} - r_0 \dot{S} \right\}$$

$$= \left\{ \left(1 - \frac{2GM}{R} + r_0^2 \dot{S}^2 \right)^{1/2} + r_0 \dot{S} \right\}^{-1}.$$

First we notice that the spectral shift does *not* diverge at $R = 2GM$, as it did in the collapse situation. Moreover, it is possible to have $z < 0$ especially in the early stages when $\dot{S}\,(>0)$ was large.

From (75) we get

$$r_0^2 \dot{S}^2 = \frac{8\pi G \varrho_0}{3} r_0^2 \left(\frac{1}{S} - 1 \right) = \frac{8\pi G \varrho_0 r_0^3}{3} \left(\frac{1}{R} - \frac{1}{r_0} \right).$$

But with (80), we have

$$1 - \frac{2GM}{R} + r_0^2 \dot{S}^2 = 1 - \alpha r_0^2.$$

Thus the spectral shift formula for white holes is

$$1 + z = \frac{1}{\sqrt{1 - \alpha r_0^2} + r_0 \dot{S}},$$

and blueshifts are possible for

$$r_0 \dot{S} > 1 - \sqrt{1 - \alpha r_0^2}.$$

□

7.5.1. *The Kruskal Diagram*

It is clear from the above discussion that the Schwarzschild coordinates are not suitable for describing the passage of null rays across the event horizon $R = 2GM$. In 1960, M. D. Kruskal and G. Szekeres independently proposed a new coordinate system to describe the empty Schwarzschild spacetime outside the collapsing spherical ball.

The new coordinates retain the angular variables θ and φ but change R and T in (79) to (u, v) given by the following set of relations:

(I) $R \geqslant 2GM, \quad u \geqslant 0$

$$u = \Phi \cosh\left(\frac{T}{4GM}\right), \quad v = \Phi \sinh\left(\frac{T}{4GM}\right),$$

$$\Phi = \left(\frac{R}{2GM} - 1\right)^{1/2} \exp\left(\frac{R}{4GM}\right). \tag{82a}$$

(II) $R \geqslant 2GM, \quad u \leqslant 0, \quad \Phi$ as in (I):

$$u = -\Phi \cosh\left(\frac{T}{4GM}\right), \quad v = -\Phi \sinh\left(\frac{T}{4GM}\right), \tag{82b}$$

(III) $R \leqslant 2GM, \quad v \geqslant 0$

$$u = \Phi \sinh\left(\frac{T}{4GM}\right), \quad v = \Phi \cosh\left(\frac{T}{4GM}\right),$$

$$\Phi = \left(1 - \frac{R}{2GM}\right)^{1/2} \exp\left(\frac{R}{4GM}\right). \tag{82c}$$

(IV) $R \leqslant 2GM, v \leqslant 0, \Phi$ as in (III):

$$u = -\Phi \sinh\left(\frac{T}{4GM}\right), \quad v = -\Phi \cosh\left(\frac{T}{4GM}\right), \tag{82d}$$

The line element in these coordinates is given by

$$\mathrm{d}s^2 = \frac{32G^3M^3}{R} \exp\left(-\frac{R}{2GM}\right) [\mathrm{d}v^2 - \mathrm{d}u^2] - R^2(\mathrm{d}\theta^2 + \sin^2\theta\, \mathrm{d}\varphi^2), \tag{83}$$

where

$$\left(\frac{R}{2GM} - 1\right) \exp \frac{R}{2GM} = u^2 - v^2. \tag{84}$$

Figure 7.4 shows the (u, v) plane and the propagation of radial null rays within it. The radial null rays have the simple equations

$$u \pm v = \text{const.} \tag{85}$$

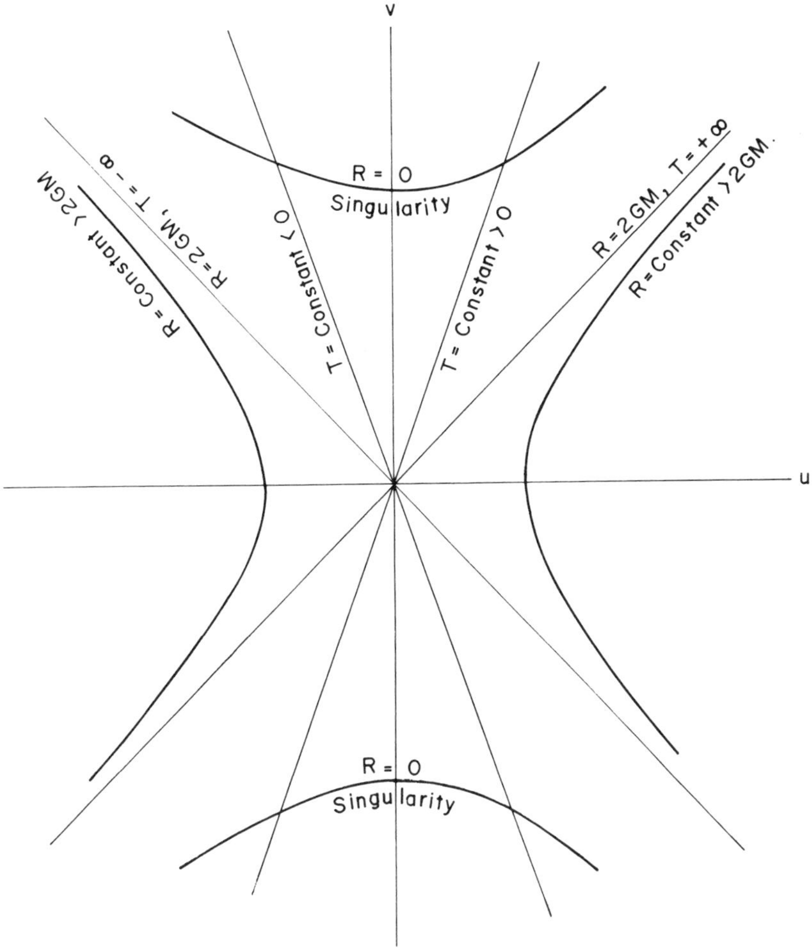

Fig. 7.4. The Kruskal diagram (not drawn to scale).

The T = constant lines are radial while the R = constant curves the hyperbolae. This diagram is commonly known as the 'Kruskal diagram'.

From (83) we see that dv^2 always has a positive coefficient, implying, thereby that v is globally timelike. (This cannot be said of the T-coordinate which becomes spacelike and R timelike for $R < 2GM$.) We shall refer to this property in Chapter 9.

7.6. Black Holes

The notion of a black hold developed above for spherically symmetric collapse can also be extended qualitatively to nonspherical collapse. Considerable scientific literature and folklore have evolved on black holes over the last fifteen years or so. Below we present highlights of some of these results.

First, what can we say about the collapse of a body of arbitrary shape? No exact solution is available in general relativity to answer this question. Various things could happen; for example, the body might fragment, or it may radiate gravitational waves while changing its shape. In the latter event the body is expected to become a black hole; that is, it will be enveloped by an event horizon. However, unlike the initial irregular shape of the body the event horizon is highly regular. This conjecture is based on *Price's theorem* which actually deals with small deviations from spherical symmetry. Price's theorem is as follows.

Suppose a collapsing body generates a disturbance which is characterized by a zero rest mass field of integral spin s. (For electrical disturbances $s = 1$ and for gravitational disturbances $s = 2$, for example.) Suppose the field $\Phi^{(s)}$ is expanded in spherical harmonics as follows:

$$\Phi^{(s)} = \sum_n A_n^{(s)} S_n(\theta, \varphi). \tag{86}$$

The functions A_n depend on the radial coordinate r and the time coordinate t. The field is believed to be small enough not to exert any back reaction on the metric which, to this order of approximation, is that of Schwarzschild in the external spacetime.

If we now study the behaviour of $\Phi^{(s)}$ as the body collapses it is found that, as $r \to 2GM/c^2$, all spherical harmonics $n > s$ tend to zero. They are radiated away during collapse. Only those harmonics for which $n < s$ survive. Thus for gravitational radiation the surviving moments are $n = 0$ (mass) and $n = 1$ (angular momentum). For a body containing electric charges and currents, only the $n = 0$ moment (electric charge) survives. This is the essence of Price's theorem.

On this basis we can say that the only surviving item of information after a black hole is formed are mass, electric charge, and angular momentum – if the collapsing body is subject to the two basic interactions of gravity and electromagnetic theory. This final state of a black hole is known as an exact solution of Einstein's equations. We shall discuss its properties shortly.

We remind the reader that the above conclusion, leading to the formation of a black hole, is a *conjecture, not a rigorous theorem*, if we depart from Price's simplifying assumptions. The details of gravitational collapse and its final outcome in the general case cannot be deduced with the present techniques of applied mathematics. Meanwhile in black-hole folklore this conjecture is stated in the words: "A black hole has no hair." The solution for the most general final state of a charged massive rotating black hole was obtained by E. T. Newman and his colleagues, as a generalization of the solution obtained in 1963 by R. Kerr for a rotating *uncharged* black hole. The former solution can be described by the following line element ($c = 1$):

$$ds^2 = \frac{\Delta}{\varrho^2}(dt - a\sin^2\theta\, d\varphi)^2 - \frac{\sin^2\theta}{\varrho^2}[(r^2 + a^2)\, d\varphi - a\, dt]^2 -$$

$$- \frac{\varrho^2}{\Delta}dr^2 - \varrho^2 d\theta^2, \tag{87}$$

with

$$\left.\begin{aligned} &M = \text{Mass} \\ &a = \text{Angular momentum per unit mass} \\ &Q = \text{Electric charge} \end{aligned}\right\} \tag{88}$$

of the black hole and

$$\Delta = r^2 + a^2 + GQ^2 - 2GMr, \tag{89}$$

$$\varrho = r^2 + a^2\cos^2\theta. \tag{90}$$

The *Kerr solution* is given by setting $Q = 0$ in (87). If we put $a = 0$, $Q = 0$ we recover the Schwarzschild solution, while for $a = 0$, $Q \neq 0$ we get the metric obtained in 1916–18 by H. Reissner and G. Nördström, for a massive spherical charge:

$$ds^2 = e^{\nu}\, dt^2 - e^{-\nu}\, dr^2 - r^2(d\theta^2 + \sin^2\theta\, d\varphi^2),$$

$$e^{\nu} = 1 - \frac{2GM}{r} + \frac{GQ^2}{r^2}. \tag{91}$$

We shall discuss the Kerr black hole in some detail since it is the most likely of all types to arise in astrophysical situations. We assume that the black hole is rotating with respect to the rest frame of distant stars. Consider an outside observer at (r, θ, φ). Suppose this observer wishes to move in such a way that he sees the distant parts of the universe nonrotating. To do so the observer must remain at constant (r, θ, φ) by exerting a force that will counteract the rotational effect of the black hole.

From (87) we get for $dr = 0$, $d\theta = 0$.

$$ds^2 = \frac{\Delta}{\varrho^2}(dt - a\sin^2\theta\, d\varphi)^2 - \frac{\sin^2\theta}{\varrho^2}[(r^2 + a^2)d\varphi - a\, dt\,]^2.$$

We see from above that ds^2 must be positive in order that the observer's motion is timelike. For $d\varphi = 0$ we get

$$(r^2 - 2GMr + a^2) - a^2\sin^2\theta > 0,$$

that is,

$$r > Gm + \sqrt{G^2M^2 - a^2\cos^2\theta} \equiv r(\theta). \tag{92}$$

This means that for $r > r(\theta)$ the observer can achieve his above objective; for $r < r(\theta)$ he cannot. From the above expression for ds^2 we find that $|d\varphi| > 0$ for $ds^2 > 0$ at $r < r(\theta)$. This limiting surface $r = r(\theta)$ is called the 'static limit' (see Figure 7.5).

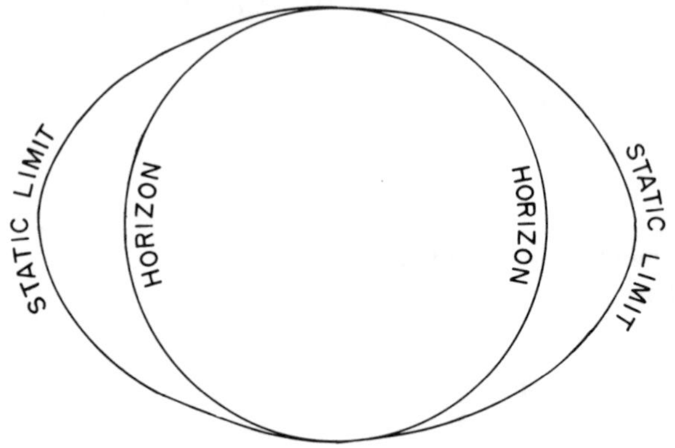

Fig. 7.5. A schematic diagram showing the meridional section of a Kerr black hole. The region between the static limit and the horizon is called the ergosphere.

The second important surface is the event horizon at $r = r_+$, where r_+ is the larger root of $\Delta = 0$. That is,

$$r_+ = GM + \sqrt{G^2M^2 - a^2}. \tag{93}$$

No light signal can reach an outside observer from $r < r_+$.

The region of spacetime between $r = r_+$ and $r = r(\theta)$ is called the 'ergosphere'. It is so called because a particle moving in it is made to rotate by the black hole along its axis and is thereby given kinetic energy. This store of energy can be tapped in principle, and hence energy extraction from a rotating black hole in feasible.

Example 7.14. The following dynamical problem will illustrate how energy can be extracted from the ergosphere of a Kerr black hole. Consider a particle of unit rest mass fired into the black hole (we have set $G = 1$ in what follows).

Let the particle move in the equatorial plane of the black hole. Its geodesic equation satisfies the first integrals of constant energy E and angular momentum H (see Appendix D). From these we get

$$\frac{\mathrm{d}t}{\mathrm{d}s} = \frac{E(r^3 + ra^2 + 2Ma^2) - 2MaH}{r(r^2 - 2Mr + a^2)}, \quad \frac{\mathrm{d}\varphi}{\mathrm{d}s} = \frac{H(r - 2M) + 2ME}{r(r^2 - 2Mr + a^2)}.$$

Since $\mathrm{d}\theta/\mathrm{d}s = 0$ we also know $\mathrm{d}r/\mathrm{d}s$. We assume $E > 0$, $H > 0$.

Consider the momentum balance just outside the horizon ($r = \eta + r_+$, $\eta \ll r_+$) in a break-up of the particle into two pieces of rest masses m_1, m_2 and with constants of motion (E_1, H_1), (E_2, H_2), respectively. Then we get

$$E = m_1E_1 + m_2E_2, \quad H = m_1H_1 + m_2H_2, \quad \frac{\mathrm{d}r}{\mathrm{d}s} = m_1\frac{\mathrm{d}r_1}{\mathrm{d}s} + m_2\frac{\mathrm{d}r_2}{\mathrm{d}s}.$$

Near the horizon we have, for $E < aH/2Mr_+$,

$$\frac{dr}{ds} = \pm \frac{2M}{r_+}\left(E - \frac{aH}{2Mr_+}\right) = \frac{2M}{r_+}\left(E - \frac{aH}{2Mr_+}\right).$$

Since r is decreasing we take the plus sign. For the two pieces take $H_1 = 0$, $dr_2/ds = 0$. Then for $E_1 < 0$ we can take

$$\left(\frac{dr_1}{ds}\right) = \frac{2M}{r_+} E_1, \qquad E_2 = \frac{aH_2}{2Mr_+} \Rightarrow m_2E_2 = \frac{aH}{2Mr_+}.$$

The first part falls into the black hole while the second emerges with energy m_2E_2. Then the net energy gain is

$$\Delta E = m_2E_2 - E = \frac{aH}{2Mr_+} - E > 0.$$

The method first suggested by Roger Penrose will not work for $a = 0$.

7.6.1. *The Laws of Black-Hole Physics*

The above example of a rotating black hole as a storehouse of energy is a special case of physical behaviour of black holes in general. This physical behaviour is quantified by the four laws of black-hole physics that are remarkably analogous to the four laws of classical thermodynamics. We shall state these laws and illustrate them with the help of the Kerr black hole.

The *first law* of black-hole physics states that in all processes involving black holes the energy, momentum, angular momentum, spin, electric charge, etc., are conserved. This law states nothing more than the fact that black holes, as part of any physical process, are subject to the various conservation laws of physics.

The *second law* of black-hole physics states that in any physical process the sum of the surface areas of all participating black holes can never decrease.

The second law brings in the important concept of the *area* of a black hole. The area can be given a meaning under the assumption of *stationarity*. That is, it is *assumed* that there exists a time coordinate t such that the spacetime including the black hole is invariant under a time translation $t \to t + \varepsilon$ for $|\varepsilon| \neq 0$ (see Appendix D for a discussion of this idea in general). The Schwarzschild and Kerr metrics satisfy this condition since their metric tensors are manifestly time-independent. Let us consider the Kerr solution. Set $t = \text{constant}$ and consider its section with the horizon given by $r = r_+ = \text{constant}$. On this section the line element is given by

$$ds^2 = -\varrho^2\, d\theta^2 - \frac{\sin^2\theta}{\varrho^2}\,(r_+^2 + a^2)^2\, d\varphi^2. \tag{94}$$

The proper area of this section of the horizon is defined as the area of the black hole. From (94) we see that this area is

$$\begin{aligned} A &= \int_{\theta=0}^{\pi}\int_{\varphi=0}^{2\pi} \varrho\, d\theta \times \frac{\sin\theta}{\varrho}\,(a^2 + r_+^2)\, d\varphi \\ &= 4\pi(a^2 + r_+^2). \end{aligned} \tag{95}$$

Consider now a Kerr black hole with mass M, total angular momentum $H \equiv Ma$, and area A. We have from our previous formulae

$$A = 8\pi GMr_+ = 8\pi GM\left[GM + \sqrt{G^2M^2 - \frac{H^2}{M^2}}\right]. \tag{96}$$

Let us consider a change in the black hole as a result of some physical interaction. By the second law A cannot decrease, i.e. a change in A must satisfy the condition

$$\delta A > 0. \tag{97}$$

We shall suppose that $\delta A = 0$ in the 'best possible' situation, a situation which can be compared with reversible processes in thermodynamics in which the change of entropy S is zero. Let $M \to M + \delta M$ and $H \to H + \delta H$. Then from (96) we get for $\delta A = 0$.

$$\delta M = \frac{a\,\delta H}{a^2 + r_+^2}.$$

Thus, if we wish to extract energy from a black hole we must have $\delta M > 0$ and, hence, $\delta H < 0$. The ergosphere can be used to achieve this. An explicit process first discussed by Penrose in shown in Figure 7.6.

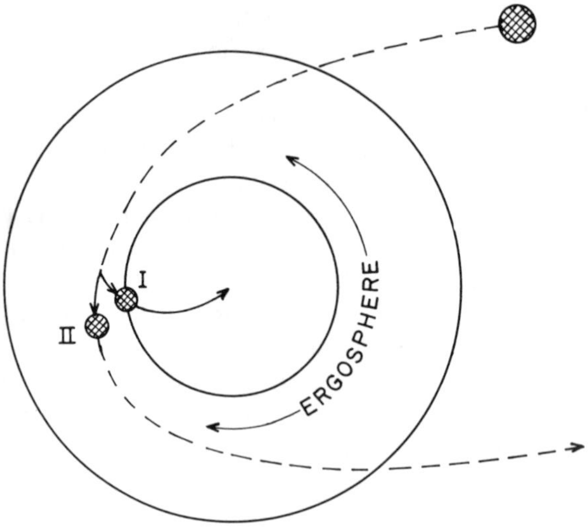

Fig. 7.6. In the Penrose process a piece of matter sent into the ergosphere of a Kerr black hole breaks into two pieces. Part I falls into the black hole while Part II escapes with greater energy than the parent piece. See Example 7.16.

Evidently this process will no longer work when the angular momentum of the black hole is reduced to zero. Thus the Schwarzschild black hole represents the final state of the energy extraction process. It has an irreducible mass which can be found easily from (96). If this mass is M_0, the area of the final Schwarzschild black hole must be the same as A, the initial area of the Kerr black hole when energy extraction began:

$$16\pi G^2 M_0^2 = 8\pi GM\left[GM + \sqrt{G^2M^2 \frac{H^2}{M^2}}\right]. \tag{98}$$

The Kerr black hole is said to be 'extreme' when

$$H = GM^2. \tag{99}$$

The horizon of such a black hole has zero surface area. In fact, for $H > aGM$ the horizon disappears and the external observer can 'see' the singularity at $r = 0$. Such a singularity is called a 'naked singularity'.

Black-hole physicists conjecture that naked singularity does not occur in nature. The "proof" of the second law of black-hole physics assumes that naked singularity does not exist. Can condition (99) be achieved in a Kerr black hole? The answer is in the negative and is expressed by the third law of black-hole physics.

The *third law* of black-hole physics states that by no finite series of processes can we make the *surface gravity* of a black hole zero.

To understand this law and the concept of surface gravity we have to consider the fourth law, which is in fact the *zeroth law* of black-hole physics. This states that in a stationary axisymmetric black hole in an asymptotically flat spacetime the surface gravity is constant over the horizon.

As the term implies, surface gravity $\varkappa$ is a parameter which measures how rapidly a freely falling particle falls on the black hole. Of course, in the curved spacetime of general relativity a freely falling particle describes a timelike geodesic and is technically unaccelerated. However, in a stationary spacetime the special time coordinate t gives us a special clock with which to measure speeds and accelerations, although it is not the proper time of any observer. Surface gravity is defined as the acceleration measured with this coordinate, on the horizon.

Example 7.15. Consider the particle held at rest at r = constant just outside the horizon of a Schwarzschild black hole. The velocity vector of the particle is

$$v^i \equiv (v, 0, 0, 0), \quad v^i v_i = 1 \Rightarrow v^2 \, e^{\nu} = 1,$$

that is,

$$v = e^{-\nu/2},$$

where $e^{\nu} = 1 - 2GM/r$ $(c = 1)$.

Let us compute the acceleration f^i of this particle. We have

$$f^i = v^k v^i{}_{;k} = \left(\frac{\partial v^i}{\partial \xi^k} + \Gamma^i_{kl}\, v^l\right) v^k.$$

Since $v^k \neq 0$ only for $k = 0$ and v^i is independent of time, we get

$$f^i = \Gamma^i_{00}\, v^2 = e^{-\nu}\, \Gamma^i_{00}.$$

Thus, only $i = 1$ gives the nonzero component of acceleration. This is

$$f^1 = e^{-\nu}\, \Gamma^1_{00} = \tfrac{1}{2} e^{\nu}\, \nu' = \frac{GM}{r^2}\,.$$

If the acceleration were measured with respect to the coordinate time t we would have got $f^1\, e^{\nu/2} = F^1$ (say). The magnitude of this acceleraiton is then F, where

$$|F|^2 = e^{-\nu}\, |F^1|^2 = |f^1|^2, \quad \text{i.e. } |F| = GM/r^2$$

This acceleration is identified with $\varkappa$, the surface gravity. □

For the Kerr black hole the surface gravity is given by the following expression which vanishes in the 'extreme' case:

$$\varkappa = \frac{\sqrt{G^2M^2 - a^2}}{a^2 + r_+^2}\,. \tag{100}$$

Using this we may write, for the Kerr black hole, the identity

$$\delta M = \varkappa\, \frac{\delta A}{8\pi G} + \frac{a\, \delta H}{a^2 + r_+^2}\,. \tag{101}$$

This is analogous to the thermodynamic relation

$$\delta u = T\, \mathrm{d}s - p\, \mathrm{d}V, \tag{102}$$

with $\varkappa$ playing the role of temperature and A the role of entropy. The reason for labelling the various laws of black-hole physics in this particular numerical sequence is now obvious to a student of thermodynamics.

7.6.2. *Black-Hole Radiation*

The analogy

Surface gravity	~	Temperature
Area	~	Entropy

between black-hole parameters and thermodynamic quantities goes deeper than what we have so far discussed. It was shown by Stephen Hawking in 1974 that within a certain multiple the surface gravity does define the temperature of the black hole and, likewise, area is a multiple of the black hole's entropy. Comparing (101) with (102) we find that if we write

$$\varkappa = \alpha T, \quad \alpha = \text{const.} \tag{103}$$

then

$$A = \frac{8\pi G}{\alpha}\, S. \tag{104}$$

What is α?

The input missing in the present discussion so far is that of quantum theory, and this is what Hawking used to determine α. We shall discuss the Hawking process in Chapter 9 where we shall show it as the outcome of field quantization in the curved spacetime of the black hole. Here we derive the coefficient α on dimensional grounds, within an unknown numerical constant.

First we restore c to its rightful place in (101) and introduce $\hbar$, the only constant that quantum theory could contribute. Since in absolute units kT denotes energy, with k = Boltzmann's constant, we have to find a constant made of c, $\hbar$, and G that will convert acceleration $\varkappa$ to energy. Thus,

$$\frac{\text{Energy}}{\text{Acceleration}} \sim \text{Mass} \times \text{Distance} \sim \frac{\hbar}{c}$$

giving

$$qkT = \frac{\hbar}{c}\,\varkappa,$$

where q is a pure number; i.e.,

$$\alpha = q\,\frac{c}{\hbar}\,k. \tag{105}$$

A body with temperature T and area A will radiate at the rate

$$L = \tfrac{1}{4} acT^4 A, \tag{106}$$

where a (the radiation constant) is given by

$$a = \frac{\pi^2 k^4}{15 h^3 c^3}\,. \tag{107}$$

Thus we arrive at the surprising conclusion that a black hole radiates! For the Schwarzschild black hole

$$\varkappa = \frac{c^4}{4GM}\,, \qquad A = \frac{16\pi G^2 M^2}{c^4}\,, \tag{108}$$

and hence (104) – (108) together give

$$L = \frac{\pi^3 \hbar c^6}{960 q^4 G^2 M^2}\,. \tag{109}$$

Radiating at this luminosity, the black hole will be evaporated away in a time

$$t = \frac{320 q^4 G^2 M^3}{\pi^3 \hbar c^4}\,. \tag{110}$$

These formulae presuppose a thermal black-body spectrum for the radiation. A justification of these assumptions will be given when we discuss the Hawking process.

Example 7.16. To get an idea of the magnitude of the Hawking effect we shall express (109) first for M in grammes and then for M in units of $M_\odot$, the mass of the Sun. Thus we have

$$\begin{aligned} t &\cong 4 \times 10^{-29} M_g^3 q^4 \text{ seconds} \\ &\cong 3 \times 10^{71} \left(\frac{M}{M_\odot}\right)^3 q^4 \text{ seconds.} \end{aligned}$$

For $M_g \sim 10^{15}$ g, the time scale $t \sim$ age of the Big Bang universe as discussed in Chapter 8. This led to the speculation that 'mini' black holes, if formed shortly after the Big Bang, would survive to this day provided that their masses exceed 10^{15} g. The second equality shows that the characteristic time of evaporation of black holes of stellar mass is far too long (compared to the 'age' of the universe) for the process of black-hole radiation to be of relevance to astrophysics. □

7.6.3. *Detection of Black Holes*

Figure 7.7 illustrates an astrophysical scenario for the detection of black holes. Note first that the Hawking process is ineffective for stellar mass black holes (Example 7.16). Hence, for their detection we have to rely on the gravitational influence that these objects exert on the surroundings.

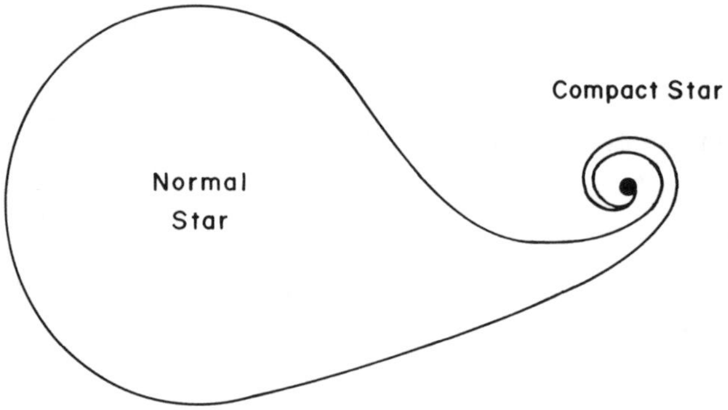

Fig. 7.7. The double-star scenario common to many X-ray sources. The compact member may be a black hole.

The double-star scenario of Figure 7.7 uses the result that if one member of a binary system is a black hole, the other member will be seen to have elliptical motion in space, apparently without a companion. Although in principle this fact should be sufficient to pick out binaries containing a black hole, in practice a search based on this criterion is not easy. The detection is, however, made easier by the phenomenon of accretion. In Figure 7.7 the black hole draws matter from the companion and this matter circulates round the black hole before ultimately falling into it. This matter forms a disc which radiates, mainly in soft X-rays.

The accretion process can work for both neutron stars and black holes as the compact members of binaries. In the 1970s the detection of several X-ray sources in binary systems lent credence to the above picture. In most cases the orbital parameters suggest the compact companion to be a neutron star. In Cygnus X-1, however, the mass estimate for the compact member is $\sim 8\ M_\odot$, far in excess of the 2–3 $M_\odot$ limit for neutron stars. It is suspected, therefore, that the object in question is a black hole.

Black holes on more massive scale with masses of the order of $\sim 10^3\ M_\odot$ in globular clusters and of the order $\sim 10^5$–$10^9\ M_\odot$ in galactic nuclei, quasars, etc, have been proposed as sources of energy. The main purpose of such black holes is to draw matter from the surroundings, impart kinetic energy to it at the expense of the (black hole's) gravitational energy and thus to pave the way for subsequent conversion of kinetic energy into energy of radiation. All such theories or scenarios assume a (not proven) high degree of efficiency of conversion of energy from one form to another, and their viability is judged today largely by subjective criteria.

This completes our discussion of gravitational fields associated with local sources wherein the spacetime is highly non-Euclidean near the sources but is asymptotically flat. In the next chapter we take up the discussion of cosmology which simplifies the local geometry by the assumption of uniformity, but gives up the assumption of asymptotic flatness.

Notes and References

1. A discussion of spacetime symmetrics at various levels can be found in:

Eisenhart, L. P.: 1927, *Riemannian Geometry*, Princeton, New Jersey.
Misner, C. W., Thorne, K. S., and Wheeler, J. A.: 1973, *Gravitation*, Freeman.
Weinberg, S.: 1972, *Gravitation and Cosmology*, Wiley, New York, ch. 13.
Narlikar, J. V.: 1978, *General Relativity and Cosmology*, Macmillan, New York.

2. Experimental tests of general relativity are discussed in Misner *et al.* (1973), Pt IX; Weinberg (1972), ch. 8; and more recently in Narlikar (1978), ch. 9 – see Note 1 above.

3. A discussion of gravitational waves will be found in the volume:

Trautmann, A., Pirani, F. A. E., and Bondi, H. (eds): 1965, *Brandeis Summer Institute in Physics*, Vol. I: *Lectures in General Relativity*, Prentice-Hall.

4. For a detailed discussion of geometrodynamics see Misner *et al.* (1973), ch. 21 (see Note 1, above), and

Isenberg, J., and Wheeler, J. A.: 1979, in *Relativity, Quanta and Cosmology* (eds M. Pantaleo and F. de Finis), Johnson, New York, p. 267.

5. Gravitational collapse is discussed in detail from different angles in Misner *et al.* (1973) and Narlikar (1978) – see Note 1 above.

6. For a detailed discussion of properties of black holes, etc., see:

Hawking, S. W., and Ellis, G. F. R.: 1973, *Large Scale Structure of Spacetime*, Cambridge.

See also Misner *et al.* (1973) and Narlikar (1978) – see Note 1 above.

Chapter 8

Relativistic Cosmology

8.1. Cosmological Symmetries

From isolated sources of gravity we now turn our attention to the largest system imaginable, viz. the universe. So far we have built a picture of the universe on the basis of astronomical observations which cover very distant regions, of the order of billions of light years (1 light year = distance travelled by light in one year $\cong 9.460 \times 10^{17}$ cm). This picture suggests that although local inhomogeneities like stars, galaxies, and clusters of galaxies exist, the universe on the large scale is homogeneous. Table 8.1 will help us to visualize the various distance scales involved in astronomy.

TABLE 8.1 Astronomical distance scales

Object	Size or distance (cm)
Sun's diameter	1.4×10^{11}
Typical interstellar distance	3×10^{18}
Diameter of the galaxy	10^{23}
Typical intergalactic separation in a cluster	10^{24}
Typical cluster size	10^{25}–10^{26}
The Hubble radius of the universe	10^{28}

Even if we take the cluster size as a typical inhomogeneity, the overall Hubble radius of the universe is ~100 times larger. Hence, on this scale we may ignore all inhomogeneities smaller than galaxies or even clusters of galaxies. In the first approximation the cosmologist considers the universe as a distribution of point particles, each particle being as large as a galaxy when examined locally.

A distribution of matter could consist of randomly moving particles with nonuniform densities. Fortunately for the cosmologist, the observed motions and distributions of galaxies show a great deal of symmetry. In 1929 E. P. Hubble was the first to discover that galaxies are moving away from one another in a systematic manner. Hubble's observations revealed that the redshift of a typical galaxy increases with faintness. As a first approximation, if we assume that all galaxies are intrinsically of equal brightness, then the above discovery amounts to the rule

$$z = \frac{c}{H} D, \tag{1}$$

where z is the redshift, D the distance of the galaxy, and H is a constant, known as *Hubble's constant*. This linear redshift–distance law is often called 'Hubble's law'.

Hubble's law gives us the important result that the large-scale motion of the galaxies is not random but is systematic: for if we interpret z as due to a radial motion of the galaxy away from us, then we find that all galaxies are moving away from us with radial speeds proportional to distance. This result does not place us in any 'special' position. The linear law guarantees that we would have seen the same radial motion from any other galaxy as the vantage point. Indeed, probes of different parts of the universe show an overall homogeneous and isotropic distribution of galaxies.

Hubble's observations were anticipated in the early 1920s by mathematical models constructed by A. Friedmann who, in 1922, obtained a global solution of Einstein's equations in which the separation between galaxies increased with time systematically. We shall first consider the Friedmann models which still provide the basic framework for modern cosmology.

It is convenient to begin with the so-called 'Weyl's postulate', based on the work of the mathematician Hermann Weyl in the early 1920s. This postulate states an assumption about the large-scale structure of the universe in the following way: "The worldlines of galaxies form a bundle of nonintersecting geodesics $\{\Gamma\}$ orthogonal to a family of spacelike hypersurfaces $\{\Sigma\}$."

We may assume that, according to Weyl's postulate, through each spacetime point P there passes a unique geodesic Γ_P of the above bundle $\{\Gamma\}$. Let x^μ label the three space coordinates of P which are the same for *all* points on this geodesic.

Similarly, the time coordinate t of P may be taken to label the spacelike hypersurface Σ_P of the above family, orthogonal to the geodetic bundle, which passes through P. All points on this hypersurface will have the same time coordinate (see Figure 8.1). We shall take $c = 1$ in this chapter.

It is clear that the above definitions simplify the spacetime metric to the form

$$ds^2 = g_{00}\, dt^2 + g_{\mu\nu} dx^\mu\, dx^\nu, \tag{2}$$

where g_{00}, $g_{\mu\nu}$ are functions of x^μ and t. If we use the assumption that the lines x^μ = constant are geodesics, we get the result from (6.54):

$$\Gamma^\mu_{00} = 0, \qquad \Gamma_{\mu|00} = 0, \qquad \frac{\partial g_{00}}{\partial x^\mu} = 0,$$

i.e. g_{00} depends on t only. Clearly we can redefine a new coordinate t' for which

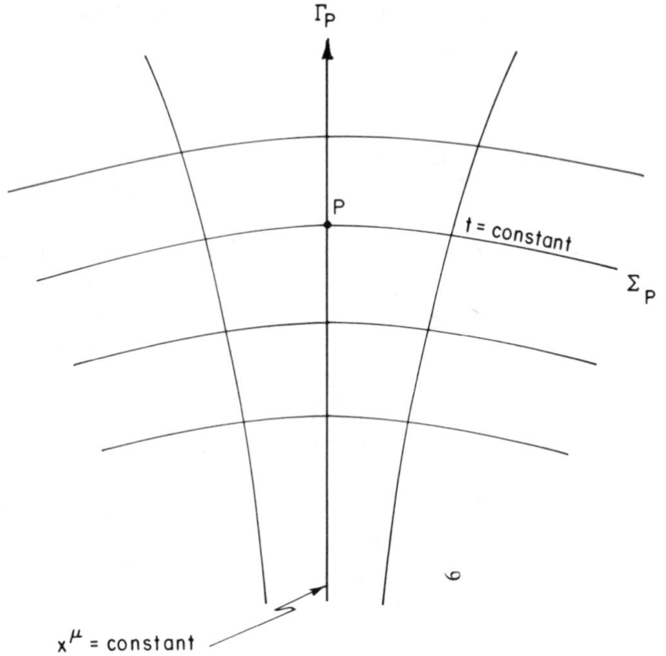

Fig. 8.1. The worldlines of galaxies form a geodesic bundle which allows the fixation of a unique coordinate system for the cosmological spacetime.

$$dt' = \sqrt{g_{00}}\,dt, \qquad g'_{00} = 1.$$

Thus we take $g_{00} = 1$ without loss of generality. The time t is then the proper time of any observer at rest on any of the galaxies. In view of the fact that all observers with their world points on the same hypersurface Σ record the same time t, we shall call t the 'cosmic time'. Whenever we refer to any epoch of the universe our understanding will be that the above time coordinate is used.

Example 8.1. Let us suppose that we have worldlines $\{\Gamma\}$ as given in Weyl's postulate but that they are *not* orthogonal to the surfaces $\{\Sigma\}$. How far is the line element

$$ds^2 = g_{00}\,dt^2 + 2g_{0\mu}\,dt\,dx^\mu + g_{\mu\nu}\,dx^\mu\,dx^\nu$$

simplified? Since the velocity vector of matter is given by $u^i = (f, 0, 0, 0)$ and since $u^i u_i = 1$, we get $f = (g_{00})^{-1/2}$.

The geodesic equations are then

$$\frac{df}{ds} + \Gamma^0_{00} f^2 = 0, \qquad \Gamma^\mu_{00} f^2 = 0.$$

Since $df/ds = f(\partial f/\partial t) = \frac{1}{2}(\partial f^2/\partial t)$ we have from the first of the above equations

$$\Gamma^0_{00} = -\frac{1}{2f^2}\frac{\partial f^2}{\partial t} = \frac{1}{2g_{00}}\frac{\partial g_{00}}{\partial t},$$

since $f^2 = g_{00}^{-1}$. Hence, we have

$$\Gamma_{\mu|00} = g_{\mu 0}\,\Gamma^0_{00} = \frac{g_{\mu 0}}{2g_{00}}\,\frac{\partial g_{00}}{\partial t}\,.$$

But we also have

$$\Gamma_{\mu|00} = \frac{\partial g_{\mu 0}}{\partial t} - \frac{1}{2}\,\frac{\partial g_{00}}{\partial x^{\mu}}\,.$$

Thus, equating the two relations gives

$$2g_{00}\,\frac{\partial g_{\mu 0}}{\partial t} - g_{\mu 0}\frac{\partial g_{00}}{\partial t} - g_{00}\frac{\partial g_{00}}{\partial x^{\mu}} = 0,$$

i.e.

$$g_{0\mu} = (g_{00})^{1/2}\,\frac{\partial \varphi}{\partial x^{\mu}}\,, \qquad \varphi = \int g_{00}^{1/2}\,\mathrm{d}t.$$

If we assume that the family $\{\Sigma\}$ is orthogonal to $\{\Gamma\}$ then $g_{\mu 0} = 0$ and we recover our earlier result. If we assume that $g_{\mu 0} \neq 0$ but t still measures the proper time of a typical observer with $x^{\mu} =$ const, then g_{00} can be made equal to unity and the above result becomes

$$\frac{\partial g_{\mu 0}}{\partial t} = 0.$$

We shall refer to this result later. □

We now use the symmetries of homogeneity and isotropy to simplify the line element

$$\mathrm{d}s^2 = \mathrm{d}t^2 + g_{\mu\nu}\,\mathrm{d}x^{\mu}\,\mathrm{d}x^{\nu} \tag{3}$$

further. These symmetries will be dignified by the name 'cosmological principle'.

The 'weak cosmological principle' states that the hypersurfaces Σ are homogeneous. The 'strong cosmological principle' states that the hypersurfaces Σ are isotropic from all world points on it. In Appendix D the mathematical description of these properties is given, and a reference is made to the theorem that isotropy at all points of Σ implies homogeneity and that Σ is then a *maximally symmetric* subspace of the spacetime. The line element (2) then becomes further simplified to

$$\mathrm{d}s^2 = \mathrm{d}t^2 - Q^2(t)\left\{\frac{\mathrm{d}r^2}{1-kr^2} + r^2(\mathrm{d}\theta^2 + \sin^2\theta\, d\varphi^2)\right\}\,, \tag{4}$$

where r, θ, φ are the new constant labels of Γ. The parameter k takes the values 0, 1, or -1. For $k = \pm 1$ we can look upon Σ as the 'surface' of a hypersphere ($k = +1$) or a hyperpseudosphere ($k = -1$). In the former case Σ is compact and the model universe with $k = +1$ is called 'closed'. By contrast Σ is noncompact for $k = 0, -1$, and the corresponding model universes are called 'open'. $Q(t)$ is an arbitrary function of t.

Example 8.2. Consider a hypersphere of radius Q, whose surface is given by the Cartesian coordinate equation

$$X_1^2 + X_2^2 + X_3^2 + X_4^2 = Q^2.$$

Define (r, θ, φ) coordinates intrinsic to the surface by

$$X_1 = Qr \sin\theta \sin\varphi, \qquad X_2 = Qr \sin\theta \cos\varphi,$$

$$X_3 = Qr \cos\theta, \qquad X_4 = \pm Q\sqrt{1 - r^2}.$$

Simple differentiation will show that the Euclidean metric in the 4-space of (X_1, X_2, X_3, X_4) induces on the above surface the metric

$$\mathrm{d}X_1^2 + \mathrm{d}X_2^2 + \mathrm{d}X_3^2 + \mathrm{d}X_4^2 = Q^2\left[\frac{\mathrm{d}r^2}{1 - r^2} + r^2(\mathrm{d}\theta^2 + \sin^2\theta\,\mathrm{d}\varphi^2)\right].$$

Notice that if we are to cover the complete surface, then besides $0 < \varphi < 2\pi$, $0 < \theta < \pi$, we need r to cover the range $0 < r < 1$ twice. For $X_4 > 0$ we get the 'upper hemihypersphere' only. A space in which the points $(X_1, X_2, X_3, \pm X_4)$ are identified is called the 'elliptical space'. The space which covers the entire surface is called the 'spherical space'.

Similarly, the $k = -1$ case in (4) will be obtained by considering the pseudo-Euclidean metric

$$\mathrm{d}X_4^2 - \mathrm{d}X_1^2 - \mathrm{d}X_2^2 - \mathrm{d}X_3^2$$

intrinsic to the surface of the hyperpseudosphere

$$X_1^2 + X_2^2 + X_3^2 - X_4^2 = Q^2. \qquad \square$$

Group theoretic arguments of the type discussed in Appendix D were first used by H. P. Robertson and A. G. Walker independently to arrive at the line element (4). Therefore, the above line element is often referred to as the Robertson–Walker line element. We shall call it the RW line element. In the following section we shall derive the solutions of Einstein's equations for the homogeneous isotropic models. In Section 8.6 we shall discuss the homogeneous (but *not* isotropic) models.

8.2. The Friedmann Models

We shall not go through all the details of computing the Einstein tensor for the RW line element. We simply state the result that the only nonzero components out of the ten are:

$$R_0^0 - \tfrac{1}{2}R \equiv -3\,\frac{\dot{Q}^2 + k}{Q^2} = -8\pi G\, T_0^0,$$

$$R_1^1 - \tfrac{1}{2}R = R_2^2 - \tfrac{1}{2}R = R_3^3 - \tfrac{1}{2}R$$

$$\equiv -\left(2\frac{\ddot{Q}}{Q} + \frac{\dot{Q}^2 + k}{Q^2}\right) = -8\pi G T_1^1. \qquad (5)$$

To complete the Einstein equations we need the right-hand side, i.e. the components of the energy tensor T^i_k. The present observations of the universe show that the density of matter in visible form, such as galaxies, is $\sim 3 \times 10^{-31}$ g cm^{-3} whereas the energy density of radiation is $\lesssim 10^{-12}$ erg cm^{-3} (i.e. $\lesssim 10^{-33}$ g cm^{-3} in mass density equivalent). Thus the dominant contribution to T^i_k comes at present from matter. Moreover, this matter has negligible pressures because the random motions of galaxies are less then $\sim 10^3$ km s^{-1} $\ll$ speed of light. Therefore the dust approximation (see Example 6.16) is valid at least at the present epoch and in the immediate past and future. So we write

$$T_{ik} = \varrho u_i u_k, \tag{6}$$

where ϱ is the density of matter and u_i is the flow vector of matter. By Weyl's postulate we get

$$u_i = (1, 0, 0, 0). \tag{7}$$

Thus the field equations becomes

$$3 \frac{\dot{Q} + k}{Q^2} = 8\pi G\varrho, \tag{8}$$

$$2 \frac{\ddot{Q}}{Q} + \frac{\dot{Q}^2 + k}{Q^2} = 0. \tag{9}$$

Multiply (8) by Q^3 and differentiate with respect to t to get from (9)

$$\frac{\mathrm{d}}{\mathrm{d}t}(\varrho Q^3) = 0, \qquad \varrho Q^3 = A \text{ (const)}. \tag{10}$$

Hence (8) becomes

$$3 \frac{\dot{Q}^2 + k}{Q^2} = \frac{8\pi GA}{Q^3} . \tag{11}$$

The unknown function $Q(t)$ can now be determined by solving this first-order differential equation for $k = 0, \pm 1$. Before going ahead we first pause and take note of history.

In 1917 when Einstein first tackled the problem of cosmology with his field equations he assumed the universe to be static (Hubble's discovery of the expanding universe came twelve years later). We see that (8) and (9) do not admit a static solution; but if we use the Λ term, the modified field equations become

$$3 \frac{\dot{Q} + k}{Q^2} - \Lambda = 8\pi G\varrho \tag{12}$$

$$2 \frac{\ddot{Q}}{Q} + \frac{\dot{Q}^2 + k}{Q^2} - \Lambda = 0. \tag{13}$$

Suppose we want a static solution with $Q = Q_0$, $\varrho = \varrho_0$ (both Q_0, ϱ_0 constants). Then (12) and (13) give

$$-\Lambda + \frac{3k}{Q_0^2} = 8\pi G\varrho_0, \qquad \Lambda = \frac{k}{Q_0^2}.$$

Thus, we need $k = +1$ and

$$Q_0 = \frac{1}{\sqrt{4\pi G\varrho_0}}, \qquad \Lambda = 4\pi G\varrho_0. \tag{14}$$

Hence Einstein's model universe was necessarily closed. Einstein found this conclusion particularly satisfactory since a closed space explicitly demonstrated the effect of matter on spacetime geometry. In fact he felt that his model would uniquely follow from his theory of relativity.

However, shortly afterwards W. de Sitter found another solution of Equations (12) and (13) with totally different characteristics. He solved these equations for $\varrho = 0$, $k = 0$ to get, for the line element,

$$\mathrm{d}s^2 = \mathrm{d}t^2 - \mathrm{e}^{2Ht}\left[\mathrm{d}r^2 + r^2(\mathrm{d}\theta^2 + \sin^2\theta\,\mathrm{d}\varphi^2)\right], \tag{15}$$

where

$$\Lambda = 3H^2. \tag{16}$$

De Sitter's universe is empty, but it expands continuously; that is, the proper distance between two test particles with fixed values of (r, θ, φ) increases exponentially with time. De Sitter's solution not only demolished the uniqueness of Einstein's solution but also showed that the presence of matter was not necessary to produce spacetime curvature.

The real universe, however, turned out to share some features of both the above solutions: it could be approximated by a model of uniform density like Einstein's model and it showed the property of expansion as in de Sitter's model. Both these features were captured by A. Friedmann in his model of 1922 seven years prior to Hubble's discovery of the expansion of the universe. Friedmann solved Equations (8) and (9) (without the Λ term).

To study the Friedmann models, we solve Equation (11) in the three cases of $k = 0, \pm 1$. In all the cases we take $Q = Q_0$ at the present epoch $t = t_0$.

(i) For $k = 0$, (11) is easily solved to give

$$Q = \left(\frac{3H_0 t}{2}\right)^{2/3} Q_0, \qquad \varrho_0 = \frac{3H_0^2}{8\pi G}, \tag{17}$$

where

$$H_0 = \left.\frac{\dot{Q}}{Q}\right|_{t=t_0} = \frac{2}{3t_0} \tag{18}$$

is Hubble's constant at the present epoch. We shall see later why the Hubble constant has this meaning. The $k = 0$ model is often called the *Einstein–de*

Sitter model since it was jointly proposed by Einstein and de Sitter in 1932.
(ii) For $k = 1$, (11) is solved by using a parameter ψ:

$$Q = Q_1 \sin^2 \frac{\psi}{2} ,$$

$$t = \tfrac{1}{2} Q_1(\psi - \sin \psi),$$

$$Q_1 = \frac{8\pi G A}{3} . \tag{19}$$

It is convenient to express the answer in terms of H_0 and another parameter q_0 defined by

$$q_0 H_0^2 = - \left. \frac{\ddot{Q}}{Q} \right|_{t=t_0} . \tag{20}$$

q_0 is called the 'deceleration parameter'. Using the result that $Q = Q_0$ at $t = t_0$ we get, from (8) and (9),

$$H_0^2 + \frac{1}{Q_0^2} = \frac{8\pi G \varrho_0}{3}, \quad \frac{1}{Q_0^2} = (2q_0 - 1)H_0^2. \tag{21}$$

Thus $q_0 > \frac{1}{2}$ for $k = 1$ and

$$\varrho_0 = \frac{3q_0 H_0^2}{4\pi G}, \qquad Q_0 = \frac{1}{H_0\sqrt{2q_0 - 1}} . \tag{22}$$

Since from (10), $A = Q_0^3 \varrho_0$, (21) and (22) give

$$Q_1 = \frac{8\pi G}{3} \{(2q_0 - 1)H_0^2\}^{-3/2} \cdot \frac{3q_0 H_0^2}{4\pi G}$$

$$= \frac{2q_0}{(2q_0 - 1)^{3/2}} \cdot \frac{1}{H_0} . \tag{23}$$

The 'present' value of ψ is given by

$$\psi_0 = 2 \sin^{-1}\sqrt{\frac{Q_0}{Q_1}} = 2 \sin^{-1} \left(\frac{2q_0 - 1}{2q_0}\right)^{1/2} . \tag{24}$$

(iii) For $k = -1$ we similarly get $q_0 < \frac{1}{2}$ and

$$Q = Q_1 \sinh^2 \frac{\psi}{2}$$

$$t = \tfrac{1}{2} Q_1(\sinh \psi - \psi), \tag{25}$$

$$\left.\begin{aligned} Q_1 &= \frac{2q_0}{(1-2q_0)^{3/2}} \cdot \frac{1}{H_0}, \\ Q_0 &= \frac{1}{(1-2q_0)^{1/2}} \frac{1}{H}, \end{aligned}\right\} \tag{26}$$

$$\varrho_0 = \frac{3q_0H_0^2}{4\pi G}, \tag{27}$$

and

$$\psi_0 = 2\sinh^{-1}\left(\frac{1-2q_0}{2q_0}\right)^{1/2}. \tag{28}$$

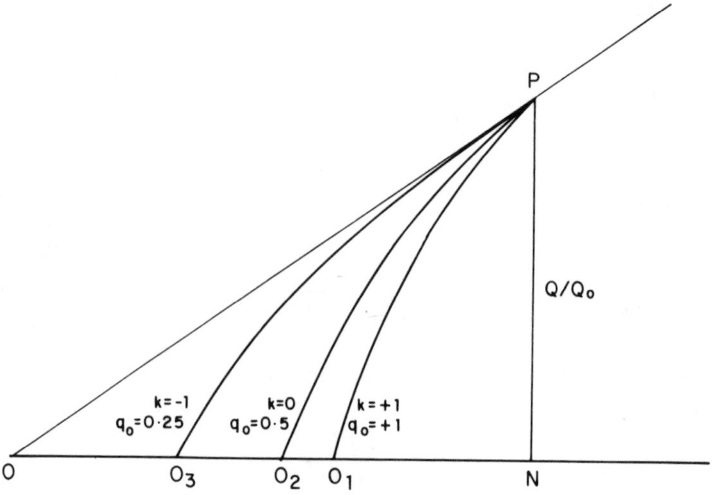

Fig. 8.2. Typical curves showing how Q/Q_0 changes with cosmic time for different Friedmann models.

The three types of models are drawn together in Figure 8.2 where Q/Q_0 is plotted against t. We have assumed that we know H_0 and have shown the models for a range of values of q_0. The time origin is shifted by t_0 so that all curves touch at the point P with ordinate $PN = 1$. The common tangent to all curves meets the time axis at point O. Clearly $ON = H_0^{-1}$, and these curves hit the time axis ($Q = 0$) at points O_1, O_2, . . . lying to the *right* of O.

Example 8.3. In later work we shall need the RW models in another coordinate system, which gives (4) as

$$ds^2 = L^2(\tau)\left[d\tau^2 - \left\{\frac{dr^2}{1-kr^2} + r^2(d\theta^2 + \sin^2\theta\, d\varphi^2)\right\}\right].$$

Clearly the connection between (Q, t) and (L, τ) is given by

$$dt = L(\tau)\, d\tau, \qquad Q(t) = L(\tau).$$

The three Friedmann models have the following values for $L(\tau)$:

$$k = 0: \qquad \tau = \frac{2}{H_0 Q_0}\left(\frac{3H_0 t}{2}\right)^{1/3}, \qquad L(\tau) = Q_0^3\left(\frac{H_0\tau}{2}\right)^2,$$

$$k = 1: \qquad \tau = \psi, \qquad L(\tau) = Q_1 \sin^2\frac{\tau}{2},$$

$$k = -1: \qquad \tau = \psi, \qquad L(\tau) = Q_1 \sinh^2\frac{\tau}{2}.$$

□

The universe had $Q = 0$ at some epoch in the past when the density and spacetime curvature were infinite. We encountered a similar situation in the collapse of the dust ball in Chapter 7. In fact the explicit solution worked out there is the time-reversed version of the $k = 1$ model discussed here. The spacetime singularity occurred in the Friedmann models *in the past* when Q was zero (in the dust ball it occurs in the future).

(There is, however, one Friedmann model which does not have singularity at $Q = 0$. This is the model with $k = 0$, $\varrho = 0$ which has $Q \propto t$. It can be shown that this spacetime is flat. Sometimes this model is referred to as the 'Milne model' because E. A. Milne had arrived at this model from another angle, from his theory of *kinematic relativity*.)

The epoch of $Q = 0$ is called the *Big Bang* epoch. The history of the universe cannot be traced prior to this epoch. It is usually argued that the universe was 'created' at this epoch, presumably in a violent event. General relativity permits us to study the behaviour of the universe after this epoch. We may accordingly call the time interval that has elapsed since this epoch to the present, 'the age' of the universe. This age can be computed easily from the above formulae. Figure 8.2, however, shows that the age cannot exceed H_0^{-1}.

8.2.1. *Radiation Models*

While deriving the cosmological equations (8) and (9) we assumed the dust approximation on the basis of the present state of the universe. Was this approximation valid in the past, right up to the Big Bang epoch $Q = 0$? To find the answer, suppose we include the energy tensor of pure radiation also on the right-hand side of Einstein's equations:

$$R^i_k - \tfrac{1}{2}\,\delta^i_k R = -8\pi G\left\{\underset{(m)}{T^i_k} + \underset{(r)}{T^i_k}\right\}, \tag{29}$$

where the underscripts (m) and (r) denote matter and radiation, respectively. The divergence of the right-hand side gives

$$\underset{(m)}{T}{}^{i}_{k;i} = 0, \quad \underset{(r)}{T}{}^{i}_{k;i} = 0, \tag{30}$$

on the assumption that matter and radiation are decoupled from each other. Using this form of $\underset{(r)}{T}{}^{i}_{k}$ derived in Chapter 6 (Example 6.16) we get, from (30),

$$U = \frac{B}{Q^4}, \quad B = \text{const} > 0. \tag{31}$$

A comparison with (10) shows that, for sufficiently small Q, radiation must dominate over matter.

Accordingly, for small Q Equations (8) and (9) are replaced by

$$\left.\begin{aligned} 3\,\frac{\dot{Q}^2 + k}{Q^2} &= \frac{8\pi GB}{Q^4}, \\ 2\,\frac{\ddot{Q}}{Q} + \frac{\dot{Q}^2 + k}{Q^2} &= -\frac{8\pi GB}{3Q^4}. \end{aligned}\right\} \tag{32}$$

We anticipate that close to $Q = 0$ the curvature term k/Q^2 is unimportant. This assumption has deep significance which we will discuss in Section 8.5. Taking it as correct, we see that (31) and (32) have the solution

$$Q(t) = \beta t^{1/2}, \quad \beta = \text{const}. \tag{33}$$

Thus, our assumption that Q became zero at some time in the past is correct. The rate of expansion is, however, different in the radiation-dominated era than in the matter-dominated one.

Example 8.4. Let us estimate the relative importance of the curvature term which we neglected in obtaining solution (33).

From (32) we see that the curvature term is of magnitude $1/Q^2$ while the term retained is $\dot{Q}^2/Q^2$. Thus we have

$$\frac{\text{Curvature term neglected}}{\text{Term retained}} \cong \frac{1}{\dot{Q}^2} \approx \frac{4t}{\beta^2}.$$

For sufficiently small t, therefore, the approximation is justified. In Section 8.5 this issue will be discussed further. □

8.2.2. Λ *Cosmologies*

We referred earlier in Equations (12) and (13) to the possibility of obtaining a static universe if the Λ term were used. Although the static model is now a historical curiosity, we can get a wider range of models of the expanding universe if we use (12) and (13) instead of (8) and (9). The empty de Sitter model was a special case of this kind.

Abbé Lemaitré and A. S. Eddington in the 1930s proposed that these Λ-cosmologies provided more interesting history of the universe than the simpler Friedmann cosmologies. For example, by choosing Λ slightly higher than the critical value needed for the static model, it is possible to have a model exploding into existence at $t = 0$, then slowing down in expansion to a long quasistationary phase, over $0 < t_1 < t < t_1 + T$ say, and then expanding again. The period T could be made arbitrarily long by choosing Λ close enough to the critical value.

8.3. Observational Cosmology

The Friedmann solutions, like any other models built out of physical theories, can be subjected to observational tests. We shall discuss them briefly here.

8.3.1. *Redshift*

Since the expansion of the universe was deduced from Hubble's observations, let us see how Hubble's law is deducible from Friedmann models. First, consider light propagation between a galaxy G_1 at $(r_1, \theta_1, \varphi_1)$ and the observer at $r = 0$. (Since the universe is homogeneous we may take $r = 0$ at any galaxy.) Suppose a light ray was emitted at $t = t_1$ and received by the observer at $t = t_0$.

Equations (6.54) of a null geodesic, when written out in full, tell us that $\theta = \theta_1$, $\varphi = \varphi_1$ are the integrals for the θ and φ components. Setting $ds = 0$ in (4) gives

$$\frac{dt}{Q(t)} = \pm \frac{dr}{\sqrt{1 - kr^2}}. \tag{34}$$

with the given boundary conditions we get

$$\int_{t_1}^{t_0} \frac{dt}{Q(t)} = \int_0^{r_1} \frac{dr}{\sqrt{1 - kr^2}}. \tag{35}$$

Example 8.5. What is the affine parameter for the above null geodesic? This is determined from the t-component of the geodesic equation:

$$\frac{d^2t}{d\lambda^2} + \Gamma^0_{00}\left(\frac{dt}{d\lambda}\right)^2 + 2\Gamma^0_{01}\left(\frac{dt}{d\lambda}\right)\left(\frac{dr}{d\lambda}\right) + \Gamma^0_{11}\left(\frac{dr}{d\lambda}\right)^2 = 0.$$

We have $\Gamma^0_{00} = 0$, $\Gamma^0_{01} = 0$, $\Gamma^0_{11} = -\frac{1}{2}\dot{g}_{11}$, so that the above equation becomes

$$\frac{d^2t}{d\lambda^2} + \frac{Q\dot{Q}}{1 - kr^2}\left(\frac{dr}{d\lambda}\right)^2 = 0.$$

Use (34) to change it to

$$\frac{d^2t}{d\lambda^2} + \frac{\dot{Q}}{Q}\left(\frac{dt}{d\lambda}\right)^2 = 0$$

which has the first integral

$$Q\,\frac{dt}{d\lambda} = \text{const.}$$

Thus, $\lambda \propto \int Q\,dt$. □

Suppose light waves of wavelength λ_0 in the rest frame of the galaxy G_1 are being emitted by G_1. Let the crest of one wave leave at t_1. The crest of the next wave will then leave at $t_1 + \lambda_0$. Let this crest arrive at $r = 0$ at the time $t_0 + \lambda$. Then, since the right-hand side of Equation (35) for the propagation of this crest will be the same as that for the earlier crest, we get

$$\int_{t_1}^{t_0} \frac{dt}{Q(t)} = \int_{t_1 + \lambda_0}^{t_0 + \lambda} \frac{dt}{Q(t)}\ .$$

For λ_0 and λ small compared to the characteristic time of variation of $Q(t)$, the above relation gives

$$\frac{\lambda}{Q(t_0)} - \frac{\lambda_0}{Q(t_1)} = 0,$$

that is,

$$1 + z \equiv \frac{\lambda}{\lambda_0} = \frac{Q(t_0)}{Q(t_1)}\ . \qquad (36)$$

Thus the redshift z, which measures the fractional increase in wavelength from emission to reception, equals the fractional increase in the scale factor between the time of emission and the time of reception. In a universe that has been continuously expanding from $Q = 0$, the spectra of all galaxies will exhibit redshift $z > 0$. Redshift arising from the expansion of the universe is often referred to as 'cosmological redshift'.

8.3.2. *Hubble's Law*

Our next step is to assign a meaning to distance of a remote galaxy. We shall use the astronomer's method of measuring distance, which is crudely as follows. Suppose an astronomical object has luminosity L, i.e. it emits energy L per unit time. Let the observer hold a detector in such a way that it measures the rate of flux of energy F crossing unit area normal to the line of sight to the source. Then in Euclidean geometry the distance D of the object is related to F and L by the relation

$$F = \frac{L}{4\pi D^2}\ . \qquad (37)$$

This relation is based on the assumption that the object radiates isotropically. This distance D is called the 'luminosity distance' of the object.

We shall adopt (37) to estimate D from F and L in Friedmann models. Let us go back to our earlier example of a galaxy G_1 as the source of light. Consider a small time interval Δ_1 at the epoch t_1 at G_1 during which it emits an energy $L\Delta_1$ isotropically. At t_0 this energy is distributed uniformly across a sphere of coordinate radius r_1 centred on G_1. From the RW line element we get the area of this sphere as

$$4\pi r_1^2 Q^2(t_0).$$

Thus the value of F appears to be

$$\frac{L}{4\pi r_1^2 Q^2(t_0)} \cdot$$

Not quite! First note that because of redshift and relation (36) the energy $L\Delta_1$ is received by O not in the time interval Δ_1 but in the interval $\Delta_1(1 + z)$. Moreover, each light proton reaches O with a frequency reduced by the factor $(1 + z)$. Since the energy of a photon is proportional to its frequency, the reduction of energy per photon is by the same factor. Therefore, the net reduction in the value of F is by a factor $(1 + z)^2$ so that

$$F = \frac{L}{4\pi r_1^2 Q_0^2(t)(1 + z)^2} \tag{38}$$

and the luminosity distance is

$$D(z) = r_1 Q_0(t)(1 + z). \tag{39}$$

Example 8.6. Suppose that instead of receiving all the photons from the galaxy G_1, the observer at $r = 0$ has a measuring equipment that receives radiation over a frequency range $[\nu_1, \nu_2]$. Let $S(\nu)\,d\nu$ denote the flux received by him in the bandwidth $[\nu, \nu + d\nu]$ lying within the above range of frequencies. Because of redshift, the photons arriving in this bandwidth must have started within the bandwidth

$$(1 + z) \times [\nu, \nu + d\nu].$$

If the energy radiated by the source in the bandwidth $(\nu, \nu + d\nu)$ is $J(\nu)\,d\nu$, then from the above argument and from (38) we get

$$S(\nu) = \frac{J(\nu \cdot \overline{1 + z})}{4\pi D^2} \cdot (1 + z).$$

$S(\nu)$ is called the 'flux density' at ν. □

We now compute $D(z)$ for Friedmann models. The case $k = 0$ is simple. Let us take the more difficult case of $k = 1$. Using (19) and (24) we solve for (35) to get

$$\int_0^{r_1} \frac{\mathrm{d}r}{\sqrt{1-r^2}} = \sin^{-1} r_1,$$

$$\int_{t_1}^{t_0} \frac{\mathrm{d}t}{Q(t)} = \frac{1}{2}\int_{\psi_1}^{\psi_0} \frac{(1-\cos\psi)\,\mathrm{d}\psi}{\sin^2(\psi/2)} = (\psi_0 - \psi_1),$$

that is,

$$r_1 = \sin(\psi_0 - \psi_1). \tag{40}$$

Next, from (24) we get

$$\sin\frac{\psi_0}{2} = \left(\frac{2q_0-1}{2q_0}\right)^{1/2}, \qquad \cos\psi_0 = \frac{1-q_0}{q_0}, \qquad \sin\psi_0 = \frac{\sqrt{2q_0-1}}{q_0}. \tag{41}$$

(36) gives the relation

$$\sin^2\frac{\psi_1}{2} = (1+z)^{-1}, \qquad \sin^2\frac{\psi_0}{2} = \frac{2q_0-1}{2q_0(1+z)} \tag{42a}$$

so that

$$\cos\psi_1 = \frac{1+q_0z-q_0}{q_0(1+z)}, \qquad \sin\psi_1 = \frac{\sqrt{(2q_0-1)(1+2q_0z)}}{q_0(1+z)}. \tag{42b}$$

Hence (40) gives

$$r_1 = \frac{\sqrt{2q_0-1}}{q_0}\cdot\frac{(1+q_0z-q_0)}{q_0(1+z)} - \frac{(1-q_0)\sqrt{(2q_0-1)(1+2q_0z)}}{q_0(1+z)}$$

$$= \frac{\sqrt{2q_0-1}}{q_0^2(1+z)}\left[q_0z + (1-q_0)\left\{1-\sqrt{1+2q_0z}\right\}\right]. \tag{43a}$$

We now use (22) for Q_0 and the above value of r_1 in relation (39) to arrive at the required answer

$$D(z) = \frac{1}{H_0}\cdot\frac{q_0z + (1-q_0)\{1-\sqrt{1+2q_0z}\}}{q_0^2} \tag{43b}$$

Calculations for the $q_0 < \frac{1}{2}$ case proceed in the same way and lead to the same final expression (43b) for $D(z)$. It can be verified that for the Einstein–de Sitter model expression (43b) holds in the limit $q_0 \to \frac{1}{2}$. Thus a single analytical expression for $D(z)$ covers all three types of Friedmann models.

This is the redshift–distance relation in Friedmann models, and illustrated in Figure 8.3 for a range of parametric values of q_0. For small z all curves approach the straight line

$$D = \frac{1}{H_0}z, \tag{44}$$

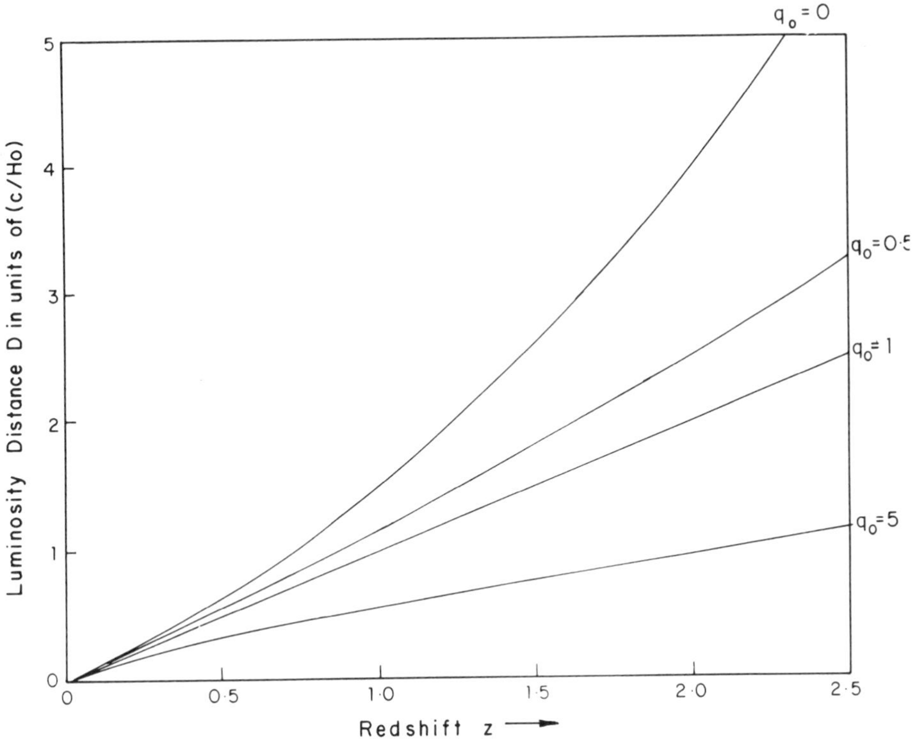

Fig. 8.3. Luminosity distance plotted as a function of redshift for different Friedmann models.

which was the linear law first discovered by Hubble. The Hubble constant H_0 can thus be measured from surveys of nearby galaxies.

The reader is referred to texts at the end of this chapter for a discussion of the observational difficulties of measuring H_0. To express the present range of uncertainty in H_0 its value is often quoted as

$$H_0 = 100 h_0 \text{ km s}^{-1} \text{ Mpc}^{-1}, \tag{45}$$

where the parameter h_0 lies between 0.5 and 1 according to most observers. Hubble's original estimate (which included many unknown systematic errors) was as high as $h_0 = 5.3$. The unit Mpc (≡ megaparsec) for distance is commonly used in extragalactic astronomy:

$$1 \text{ Mpc} = 3.0856 \times 10^{24} \text{ cm} \cong 3.26 \times 10^6 \text{ light years}. \tag{46a}$$

The Hubble constant can also be expressed in units of reciprocal time:

$$H_0^{-1} \cong 9.8 \times 10^9 \, h_0^{-1} \text{ yr}. \tag{46b}$$

Relation (43b) can, in principle, be tested at large redshifts ($z \sim 1$) and the parameter q_0 determined. But the many observational uncertainties make this test operationally inconclusive.

8.3.3. *Density of the Universe*

Another way of determining q_0 is to measure the mean matter density in the universe. The Einstein–de Sitter model has $q_0 = \frac{1}{2}$. The corresponding density is called the closure density. From (17) its value is given by

$$\varrho_c = 2 \times 10^{-29}\, h_0^2 \text{ g cm}^{-3}. \tag{47}$$

$\varrho > \varrho_c \Rightarrow q_0 > \frac{1}{2}$ and $k = 1$ and we get a closed universe. Likewise, the universe is open for $\varrho < \varrho_c$. If we rely on the luminous matter alone to give us an estimate of ϱ, we get $q_0 \lesssim 0.1$ and the universe is open. However, it is suspected that there is considerable matter in the universe in dark form. Guesses about its form include low-mass stars, black holes, massive neutrinos, monopoles, etc. Thus the issue whether the universe is open ($q_0 < \frac{1}{2}$) or closed ($q_0 > \frac{1}{2}$) is still unsettled.

Cosmological literature includes another parameter called Ω_0 which is defined by

$$\Omega_0 = \frac{8\pi G\varrho}{3H_0^2}. \tag{48}$$

Thus, for the Friedmann models, $\Omega_0 = 2q_0$.

Example 8.7. The relation $\Omega_0 = 2q_0$ is changed in Λ-cosmologies. Consider (12) and (13) at the present epoch. We have

$$H_0^2 + \frac{k}{Q_0^2} - \frac{1}{3}\Lambda = \Omega_0 H_0^2$$

$$-2q_0 H_0^2 + H_0^2 + \frac{k}{Q_0^2} - \Lambda = 0.$$

Taking a difference of these two relations gives

$$\Omega_0 = 2q_0 + \frac{2\Lambda}{3H_0^2}.$$

Thus, independent measurements of Ω_0 and q_0 can tell us if $\Lambda \neq 0$.

8.3.4. *Number Counts*

Another way of identifying non-Euclidean geometrical effects is to test the volume–distance relationship. In a Euclidean sphere the volume (V) radius (D) relation is

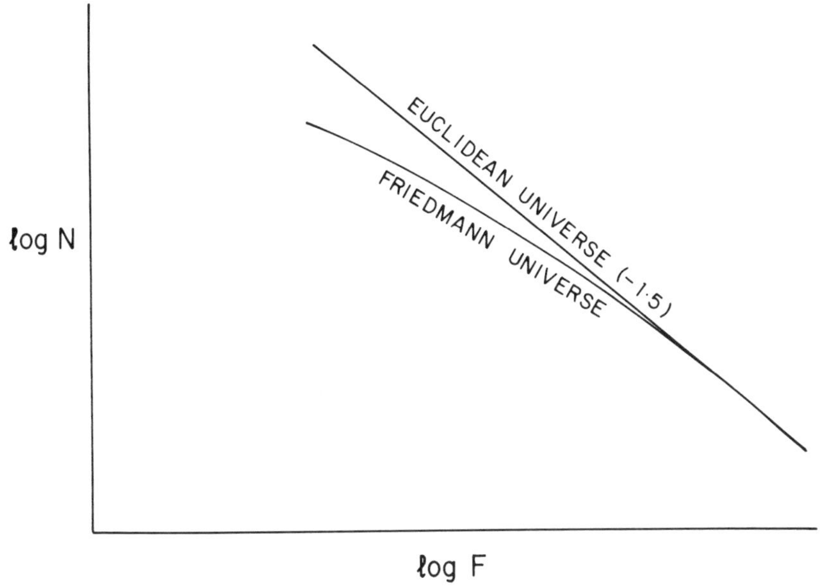

Fig. 8.4. Schematic plot of log N against log F as in the Euclidean universe and in a typical Friedmann universe.

$$V = \frac{4\pi}{3} D^3. \tag{49}$$

The astronomer's way of testing this relation is to count the number N of (supposedly) identical objects brighter than a specified limit of flux F. If the number density of such objects is n and if each has luminosity L, then

$$N = Vn, \qquad L = 4\pi D^2 F. \tag{50}$$

Relations (49) and (50) give the result

$$\frac{\mathrm{d} \log N}{\mathrm{d} \log F} = -1.5. \tag{51}$$

We have already seen how the L, D, F relation is modified in Friedmann cosmology. To see how the N, V, n relation is modified we assume that n = constant for a comoving coordinate volume. Then the number of objects within a coordinate distance r_1 of the origin is

$$N = 4\pi n \int_0^{r_1} \frac{r^2\, \mathrm{d}r}{\sqrt{1 - kr^2}}. \tag{52}$$

This relation can be converted into an $N(z)$ relation, while (38) gives us the $F(z)$ relation. With z as a parameter we can plot log N against log F. Figure 8.4 shows how such curves vary.

Again, starting at low z with the slope given by (51) the different q_0 curves diverge from the Euclidean curve with

$$\left| \frac{\mathrm{d} \log N}{\mathrm{d} \log F} \right| < 1.5.$$

Example 8.8. To compute the N–F relation in the Einstein–de Sitter model ($q_0 = \frac{1}{2}$) we have from (52) and (43b)

$$N(z) = \frac{4\pi n}{3} r_1^3, \qquad D(z) = \frac{2}{H_0} \{z + 1 - \sqrt{1 + z}\}.$$

The relation between r_1 and z is obtained as follows. From (35) and (17) we get

$$r_1 = \left(\frac{3H_0}{2}\right)^{-2/3} (t_0^{1/3} - t_1^{1/3}) = t_0 \left(1 - \frac{1}{\sqrt{1 + z}}\right).$$

Hence

$$\log N = 3 \log \left(1 - \frac{1}{\sqrt{1 + z}}\right) + \text{const}$$

$$\log F = -2 \log (1 + z - \sqrt{1 + z}) + \text{const.}$$

$$\frac{\mathrm{d} \log N}{\mathrm{d} \log F} = -\frac{3}{2} \{2 \sqrt{1 + z} - 1\}^{-1}.$$

Thus the magnitude of the slope gets smaller and smaller than 1.5 as z increases. For $z = 1.25$ the slope predicted is -0.75. □

Although in principle this test can distinguish between the various Friedmann models it is ineffective in practice as an indicator of q_0. The reason is that it has not been possible to ensure the control conditions n = constant, L = constant in the real universe. Thus counts of galaxies, radio sources and quasars have been tried without being successful in determining q_0.

8.3.5. *The Variation of Angular Sizes*

Consider a population of spherical objects of radius a distributed all over a Euclidean universe. From a distance D a typical object will subtend an angle

$$\theta = \frac{2a}{D} \quad \text{for } a \ll D \tag{53}$$

at the observer. Thus, remote objects 'look' smaller because the angles subtended by them at the observer are smaller.

How does this result get modified in the Friedmann universes? Suppose we consider the galaxy G_1 to have radius a and let us consider the plane section φ = constant of this galaxy. The angle subtended by the diameter of G_1 in this plane at the observer is θ, where, by the RW line element we have

$$2a = S(t_1) r_1 \theta = D\theta(1 + z)^{-2},$$

with D, the luminosity distance. Thus,

$$\theta = \frac{2a}{D} (1 + z)^2. \qquad (54)$$

Notice that θ, according to (54), need not be a monotonic function of D since, as D increases, $(1 + z)^2$ also increases. For the Einstein–de Sitter model ($q_0 = \frac{1}{2}$)

$$D = \frac{2}{H_0} \{(1 + z) - \sqrt{(1 + z)}\} \qquad (55)$$

and simple calculation will show that θ has a *minimum* for $z = 1.25$.

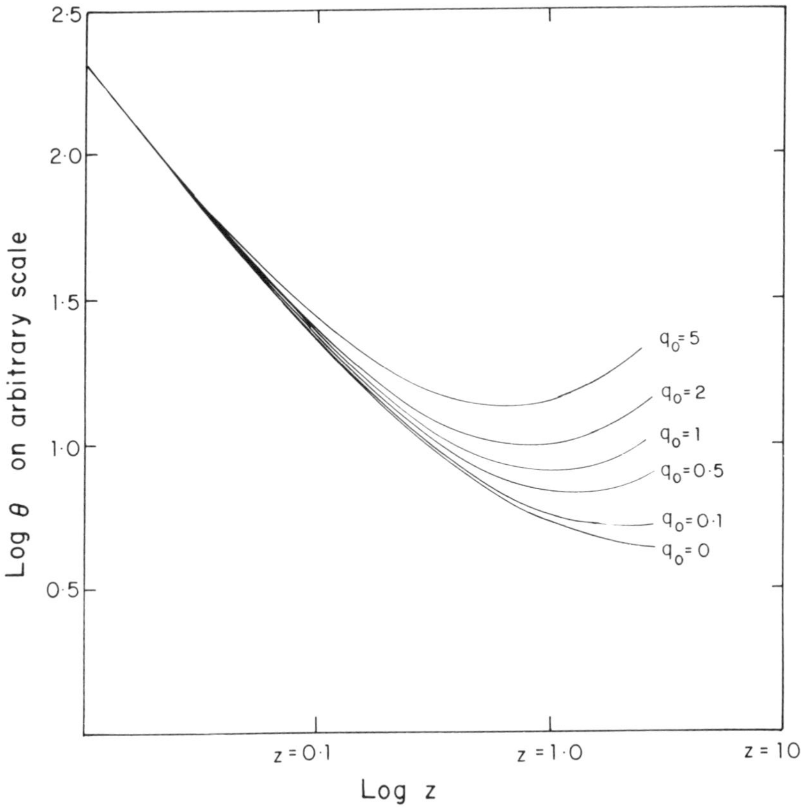

Fig. 8.5. The variation of angular size θ with redshift z in different Friedmann models. The plots are on logarithmic scales for θ and z.

Figure 8.5 shows that qualitatively all Friedmann models display this behaviour.

Unfortunately, as in the previous examples, a cosmological test to decide the value of q_0 based on this result has so far proved ineffective. The measurements of angular sizes of galaxies, quasars, and radio sources have told us more about how these objects vary in their physical properties than about the large-scale geometry of the universe.

8.4. The Early Universe

We have so far emphasized the geometrical aspects of cosmology and viewed the Friedmann models as solutions of Einstein's equations. We next outline briefly the physical aspects of cosmology which are concerned with the following questions:

(i) How and when did galaxy formation take place in a homogeneous (and isotropic) universe?

(ii) How, when, and where were the nuclei of the various chemical elements formed?

(iii) How and when did the various particles like protons, electrons, etc., form?

The answers to these questions require us to consider the early history of the universe, although the relevant epochs are, of course, different for the three phenomena. Also, the sophistication of approach and the measure of success achieved vary in the three cases.

Of the three problems listed above, the second is best understood, and we shall consider that problem first.

8.4.1. *The Origin of Atomic Nuclei*

George Gamow and his younger colleagues R. C. Herman and R. Alpher were pioneers in the calculations of nucleosynthesis of the early universe. It was Gamow's idea in the mid-1940s that in the very early universe the particles (i.e. electrons, protons, neutrons, etc.) and radiation were in thermodynamic equilibrium at high temperatures and that conditions were suitable for synthesis of baryons to form light and heavy nuclei. This idea has subsequently been investigated with the latest inputs from basic theoretical physics using fast computers, and the conclusion is that Gamow was right but only partially. Primordial nucleosynthesis can deliver only the light nuclei with atomic numbers not exceeding 5. The work of Geoffrey and Margaret Burbidge, William Fowler, and Fred Hoyle demonstrated that all heavier nuclei can be made in stars. Below we discuss the highlights of the primordial nucleosynthesis theory.

In the typical early universe scenario considered by Gamow, the matter in the universe – made of electrons, protons, neutrons, neutrinos, photons, etc. – behaved like an ideal gas; that is, these particles moved freely between collisions although the collisions happened frequently enough to establish a thermodynamic equilibrium for all participating systems.

Assuming that various particles were in thermal equilibrium in the early universe, their distribution functions were given (with $c = 1$) by

$$n(P)\,\mathrm{d}P = \frac{g}{2\pi^2\hbar^3} P^2 \left[\exp\left(\frac{E-\mu}{kT}\right) \pm 1\right]^{-1} \mathrm{d}P, \tag{56}$$

where P = momentum, E = energy, and m = the mass of a typical particle, so that

$$E^2 = P^2 + m^2; \tag{57}$$

g denotes the multiplicity of spin states of the particle and the '+' or '−' sign in the denominator applies according to whether the particle obeys Fermi–Dirac or Bose–Einstein statistics. The temperature T of the population may be compared with the 'rest-mass temperature' for the particle

$$T_m = \frac{m}{k}\ . \tag{58}$$

For $T \gg T_m$, (56) can be simplified under the relativistic approximation while for $T < T_m$ the approximation is nonrelativistic. For the time being we shall neglect the chemical potentials μ.

In the relativistic limit the number density N, pressure p, and energy density ε of the particle population are given by

$$N = \frac{1.202}{\pi^2}\left(\frac{kT}{\hbar}\right)^3 g \text{ (bosons)}, \qquad N = \frac{3}{4} \times \frac{1.202}{\pi^2}\left(\frac{kT}{\hbar}\right)^3 g \text{ (fermions)},$$

$$\varepsilon = 3p = \frac{\pi^2(kT)^4}{30\hbar^3} g \text{ (bosons)}, \quad \varepsilon = 3p = \frac{7}{8} \times \frac{\pi^2(kT)^4}{30\hbar^3}\ g \text{ (fermions)}. \tag{59}$$

Thus the energy density of a cosmological mixture containing g_b boson spin states and g_f fermion spin states is given by ε (as given above for bosons), with

$$g = g_b + \frac{7}{8}\ g_f. \tag{60}$$

Example 8.9. For electrons, $g = 2$ and we get the average interparticle distance for the electron population as

$$\langle r \rangle \sim N^{1/3} \sim 0.6 \left(\frac{kT}{\hbar c}\right),$$

where we have restored c in (59). The typical electrostatic potential energy is therefore

$$\frac{e^2}{\langle r \rangle} \sim \frac{5}{3}\ \frac{e^2}{\hbar c}\ (kT).$$

Since kT is of the order of an electron's kinetic energy we find that kinetic energy $\gg$ potential energy, thus justifying the assumption that between collisions the electrons move as free particles.

How frequent are the electron–electron collisions? The Thomson scattering formula gives the collision rate as

$$\Gamma_c \cong N\sigma c \sim 1.5\left(\frac{e^2}{m_e c^2}\right)^2\left(\frac{kT}{\hbar c}\right)^3 c.$$

This rate has to be compared with the rate of expansion of the universe given by (61) below

$$H(t) = \frac{1}{2t} \cong \sqrt{12\pi G a T^2}.$$

Notice that $\Gamma_c \propto T^3$ while $H(t) \propto T^2$. Thus, a sufficiently early $\Gamma_c > H$ and collisions could be considered as sufficiently frequent. By evaluating $H(t)$ and Γ_c we get $\Gamma_c \gg H(t)$ at the time of nucleosynthesis, when $T \gtrsim 10^8$ K.

A similar calculation of neutrino shows that their collisions with leptons (via the electro-weak interaction) became less frequent compared to the expansion of the universe at the much higher temperature of $T \gtrsim 10^{10}$ K. Thus, neutrinos decoupled from the rest of the particles when the universe cooled below $\sim 10^{10}$ K. □

Let us consider the situation when the universe had a population of electrons (e^-), positions (e^+) neutrinos (ν), antineutrinos ($\bar{\nu}$), and photons (γ). For this mixture $g = 9/2$ and the field equations (32) become (neglecting the curvature term)

$$\frac{\dot{Q}^2}{Q^2} = 12\pi G a T^4. \tag{61}$$

However, for $p = \varepsilon/3$ we get from $T^i{}_{k;i} = 0$ result (31), i.e.

$$\varepsilon \propto \frac{1}{Q^4}. \tag{62}$$

From (60), (61), and (62) we get

$$Q \propto t^{1/2}, \quad T = (48\pi G a)^{-1/4}\, t^{-1/2}. \tag{63}$$

The time–temperature relationship may be expressed in numerical terms as

$$T = 1.04 \times 10^{10}\, t^{-1/2}_{\text{seconds}}\ \text{K}. \tag{64}$$

At $t = 1$ s, the cosmological mixture has an appreciable number of pairs $e^{\pm}$. However, as the universe expands the temperature drops and the pairs annihilate to produce γ's. The result is that the photon temperature rises appreciably. The neutrinos ν, $\bar{\nu}$ are at this stage essentially decoupled from other particles since their interaction rate is small compared to the rate of expansion of the universe (see Example 8.9). Hence, their temperature is essentially unaffected by pair annihilation. Thus, we find that after the (e^-, e^+) annihilation is complete the photon temperature T_γ is higher than the neutrino temperature T_ν by a factor $\sim$1.4.

As the universe cools further through the temperature range $\sim 5 \times 10^9$ to $\sim 10^8$ K all the primordial nucleosynthesis takes place. Since the participating nuclei are all nonrelativistic we have to consider their distribution functions in the approximation $T < T_m$. At this stage we cannot ignore the chemical potentials which were neglected in (56). Let us consider the formation of the deuterium (d) nucleus which contains one neutron (n) and one proton (p). Let B_d denote the binding energy of d, so that

$$m_d = m_n + m_p - B_d. \tag{65}$$

In the reaction

$$n + p \rightarrow d + \gamma \tag{66}$$

the chemical potential is conserved; we have therefore

$$\mu_n + \mu_p = \mu_d + \mu_\gamma = \mu_d \tag{67}$$

since $\mu_\gamma = 0$. We shall assume that there are N nucleons per unit volume in the cosmological mixture, some of them bound in deuterium, the others free.

Let X_n, X_p, and X_d denote the mass fractions of the total mixture while N_n, N_p, and N_d denote the number densities of free neutrons, free protons, and deuterium nuclei at temperature T. Then we have

$$N_n = 2 \left(\frac{m_n kT}{2\pi\hbar^2}\right)^{3/2} \exp\left(\frac{\mu_n - m_n}{kT}\right),$$

$$N_p = 2 \left(\frac{m_p kT}{2\pi\hbar^2}\right)^{3/2} \exp\left(\frac{\mu_p - m_p}{kT}\right), \tag{68}$$

$$N_d = 2 \left(\frac{m_d kT}{2\pi\hbar^2}\right)^{3/2} \exp\left(\frac{\mu_d - m_d}{kT}\right), \tag{69}$$

where except in the exponentials we may put $m_d = 2m_n = 2m_p$ and write

$$X_n = \frac{N_n}{N}, \quad X_p = \frac{N_p}{N}, \quad X_d = \frac{2N_d}{N}. \tag{69}$$

From (65)–(69) we get

$$X_d = \frac{3}{\sqrt{2}} X_p X_n \xi \exp\left(\frac{B_d}{kT}\right), \tag{70}$$

where

$$\xi = 2N \left(\frac{m_p kT}{2\pi\hbar^2}\right)^{-3/2}. \tag{71}$$

The above relation illustrates how the primordial nucleosynthesis works. For $B_d = 2.22$ MeV we get the result that for $T \lesssim 8 \times 10^8$ K the exponential factor in (71) becomes large enough to give an appreciable number of deuterium nuclei. (That is, the nuclear binding is able to trap and hold the colliding nucleons.)

The process, of course, does not stop here. Heavier nuclei like ^{3}H, ^{3}He, ^{4}He, etc. are formed by further capture of n, p and d. In the end, of course, the bulk of synthesized nuclei are in the ^{4}He form which is the most stable.

The process does not continue further because nuclei like Li, Be, and B are not stable and break back to ^{4}He. The next stable nucleus is ^{12}C, but to synthesize ^{12}C we need *three* helium nuclei together, a process requiring very high temperatures ($\sim 10^{10}$ K) that are no longer possible in the universe. This type of synthesis can, however, take place in stellar interiors.

The ultimate mass of the universe, after the formation of ^{4}He, is mainly composed of free protons and helium nuclei. The mass fraction Y of helium is therefore given essentially by the neutron to proton ratio (n/p) before nucleosynthesis began:

$$Y = \frac{2(n/p)}{(n/p) + 1} . \tag{72}$$

Calculations show that $Y \sim 0.25$.

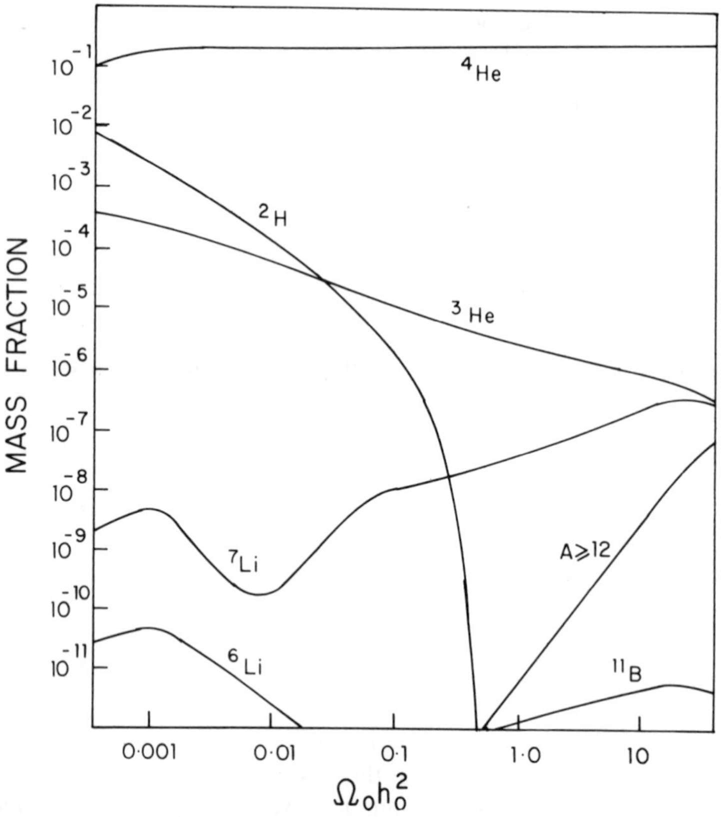

Fig. 8.6. The primordial abundances of light nuclei as functions of the parameter $\Omega_0 h_0^2$.

Figure 8.6 shows the results of recent calculations of primordial nucleosynthesis, in the form of the mass fractions of primordial nuclei in different Friedmann models. Notice that the deuterium fraction X_d is most sensitive to

Ω_0 – it falls rapidly for $\Omega_0 h_0^2 > 0.14$. The present observational estimates of X_d suggest that for $h_0 = 1$, $\Omega_0 \lesssim 0.12$ and hence $q_0 \lesssim 0.06$. Although this calculation favours open models, the case is not yet settled because nonbaryonic matter would not affect X_d and such matter could make up the excess (hidden) matter needed to close the universe. The helium fraction Y is broadly in agreement with observations.

The most telling evidence in favour of the early, hot universe is in the form of relic radiation. The radiation temperature T_γ falls off as $1/Q$ and Gamow and his colleagues had predicted a present-day background of black-body radiation whose temperature they estimated to be ~5 K. Such a background was first discovered in 1965 by A. A. Penzias and R. Wilson. Its spectrum, which is peaked in the microwave region, is shown in Figure 8.7. The effective temperature of this radiation is ~3 K.

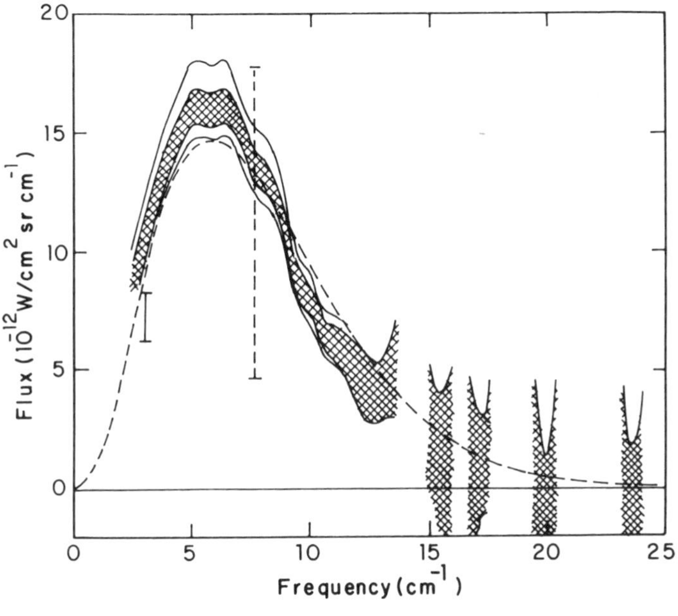

Fig. 8.7. The high-frequency end of the microwave background measurements (shown by hatched regions). The dashed curve is the best-fit Planckian curve for the data with the temperature ~2.9 K.

8.4.2. *The Origin of Galaxies*

The events covered in Section 8.4.1 last up to ~3 minutes from the origin of the universe. At that stage the cosmic brew comprised light nuclei, free protons and electrons, and neutrinos. As the mixture cooled further, a stage would come when the electrostatic binding between electrons and protons becomes strong enough to bring them together to form neutral hydrogen atoms. Saha's ionization equation, analogous to (70), will then determine the degree of ionization of the mixture. Calculations show that most of the ions disappear when the temperature drops to ~3000 K.

Example 8.10. We use Saha's ionization equation to estimate the temperature at which the electrons combined with ions to form neutral H atoms in the early universe.

The equation is

$$\frac{N_e^2}{N_{\mathrm{H}}} = \left(\frac{m_e kT}{2\pi\hbar^2}\right)^{3/2} \exp\left(-\frac{B}{kT}\right).$$

Here N_e = number density of electrons (= number density of ions, i.e. free protons), N_{H} = number density of neutral H atoms, and B = binding energy of electrons in the H-atom = 13.59 eV.

Write N_B = total number density of baryons, so that $N_B = N_{\mathrm{H}} + N_e$. Let $x = N_e/N_B$. Then the above equation becomes

$$\frac{x^2}{1-x} = \frac{1}{N_B}\left(\frac{m_e kT}{2\pi\hbar^2}\right)^{3/2} \exp\left(-\frac{B}{kT}\right)$$

At present the number density of baryons is

$$(N_B)_0 \cong 1.2 \times 10^{-5}(\Omega_0 h_0^2).$$

Since the present radiation temperature is $T_0 \cong 3$ K and $N_B \propto T^3$, we can express N_B in the above Saha equation as a function of T. Thus the equation can be solved for x as a function of T for various values of T. For example, for $\Omega_0 h_0^2 = 0.1$, $x = 0.003$ at $T = 3000$ K. □

From the present value of the radiation temperature of ~3 K we see that this epoch occurred at a redshift of ~10^3. Because of the disappearance of free electrons there is almost no scattering of radiation and the universe becomes optically thin. The black-body spectrum established by this time is simply cooled adiabatically. The epoch when the last scattering took place – the epoch of redshift ~10^3 – is called the 'recombination epoch'.

It is believed that galaxies form out of the inhomogeneities present in the universe around this epoch. The inhomogeneities have to grow with time in order to become galaxies. If a region of space has a sufficient concentration of matter, its stronger gravity, compared to its surroundings, enables it to grow. This qualitative idea was first developed by Sir James Jeans at the turn of the century. His calculations were in the background of a static medium, but can be extended to work in the expanding universe. In particular, his notion of a critical minimum mass for growth has survived unchanged.

This mass is called the 'Jeans mass', denoted by M_J: its physical significance can be understood as follows. Suppose a given inhomogeneity has average density ϱ and radius R so that its mass is

$$M = \frac{4\pi}{3}\varrho R^3. \tag{73}$$

The gravitational energy of this mass is

$$\mathcal{E}_G \sim \frac{GM^2}{R} \sim \frac{16\pi^2}{9} \varrho^2 R^5 G. \tag{74}$$

We compare it with the thermal energy of the sphere. If c_s is the sound speed of the medium, the thermal energy is given by

$$\mathcal{E}_{\mathrm{Th}} \sim \frac{4\pi}{3} \varrho c_s^2 R^3. \tag{75}$$

For the inhomogeneity to grow by compression its gravitational energy must be large compared to its internal thermal energy. The condition $\mathcal{E}_G \gg \mathcal{E}_{\mathrm{Th}}$ gives

$$R^2 \gg \frac{3c_s^2}{4\pi G\varrho} \equiv R_{\mathrm{J}}^2, \tag{76}$$

where R_{J} is the *Jeans length*. The Jeans mass is correspondingly given by

$$M_{\mathrm{J}} = \frac{4\pi}{3} \varrho R_{\mathrm{J}}^3, \tag{77}$$

and the condition for growth is $M \gg M_{\mathrm{J}}$.

Example 8.11. Let us estimate the Jeans mass at the recombination epoch. The present density of matter in the universe is, from (47),

$$\varrho_0 \cong 2 \times 10^{-29} h_0^2 \Omega_0 \text{ g cm}^{-3}.$$

At a redshift of 10^3 the density was

$$\varrho \sim 10^9 \varrho_0 \sim 2 \times 10^{-20} h_0^2 \Omega_0 \text{ g cm}^{-3}.$$

For a mono-atomic gas at 3000 K, the sound speed is

$$c_s^2 = \frac{5}{3}\frac{p}{\varrho} \sim \frac{5}{3} \mathcal{R} T \sim 4 \times 10^{11} \text{ cm}^2 \text{ s}^{-2},$$

where $\mathcal{R}$ is the gas constant. Thus from (76) and (77) we get

$$R_{\mathrm{J}} \cong 8.5 \times 10^{18} (\Omega_0 h_0^2)^{-1/2} \text{ cm},$$

$$M_{\mathrm{J}} \cong 2.5 \times 10^4 (h_0^2 \Omega_0)^{-1/2} M_\odot.$$

For $h_0^2 \Omega_0 = 0.01$,

$$M_{\mathrm{J}} \cong 2.5 \times 10^5 M_\odot.$$ □

At the recombination epoch, $M_{\mathrm{J}} \simeq 10^5$–$10^6 M_\odot$ and this value illustrates the inadequacy of the Jeans theory as we see that the characteristic mass of growth-prone inhomogeneities exceeds $10^6 M_\odot$ but we do not see why the typical galactic masses are in the range 10^{10}–$10^{12} M_\odot$. The various scenarios for galaxy formation have been discussed extensively but none of them is found sufficiently satisfactory in the sense of uniquely predicting the masses and other physical features of galaxies.

One observational difficulty is related to the microwave background. The background is expected to show imprints of galactic inhomogeneities in the form of fluctuations of its temperature on the scale of 1–10 arc min. The expected fluctuations are

$$\frac{\Delta T}{T} \geq 10^{-4}. \tag{78}$$

Present observations do not reveal fluctuations of this order or even lower. Thus, the expected demonstration that galaxies were formed after the recombination epoch has not yet occurred.

8.4.3. *GUTs and Cosmology*

The attempts to unify the strong interaction with the electro-weak interaction to produce a grand unified theory (GUT) have found echos in cosmology for the following reasons.

First, on the cosmological side the basic mystery that has to be explained is the ratio of photons to baryons, currently estimated at

$$\begin{aligned} \frac{N_\gamma}{N_B} &= \frac{2.404}{\pi^2}\left(\frac{kT_0}{ch}\right)^3 \cdot \frac{8\pi G}{3H_0^2\Omega_0 m_p} \\ &\cong 4.57 \times 10^7(\Omega_0 h_0^2)^{-1}\left(\frac{T_0}{3}\right)^3 \end{aligned} \tag{79}$$

where T_0 is the present black-body temperature of microwave background. Since baryons are believed to be indestructible (except in circumstances to be considered below), and photons are produced essentially from annihilations of matter and antimatter, the above ratio has been frozen in from the early hot phase of the universe. The question is, 'How was this ratio arrived at in the early universe?'

Of course, the above argument presupposes that there is such a thing as a baryon number; that is, that the universe has an excess of matter over antimatter. Why and how was this asymmetry introduced? Clearly, we need a recipe from particle physicists which supplies scenarios for generating baryons while at the same time, destroying the symmetry between baryons and antibaryons. GUTs supply such scenarios.

The gain from particle physics to cosmology is reciprocated by the circumstance that the early universe provides the only viable testing ground for the predictions of GUTs. There is no unique grand unified theory but it is generally felt that the electro-weak and strong interactions become comparable in strength and unifiable at energy exceeding $\sim 10^{14}$ GeV. No man-made accelerator is going to generate particles of such high energy (the present level achieved is only $\sim 10^3$ GeV) and hence GUTs are unverifiable in the terrestrial laboratory. This is where the early universe becomes indispensable to particle physicists.

Suppose we express the time–temperature relationship of (64) as an energy–time relationship, with kT as energy. Generalizing (64) to any mixture of relativistic particles we get

$$t_s = 2.4g^{-1/2}E_{\text{MeV}}^{-2}. \tag{80}$$

Here E_{MeV} expresses the energy kT in units of mega-electron volts.

Taking $E_{\text{MeV}} = 10^{17}$ to correspond to the energy at which GUTs become operative, and setting $g \cong 100$ as a characteristic number of spin-degrees of freedom, we get from (80) $t_s \sim 2.4 \times 10^{-35}$. That is, when the universe was younger than $\sim 2.4 \times 10^{-35}$ s, the grand unified theory was fully operative. It is around this time that we may expect baryons to be formed.

Nevertheless, the recipe for producing a baryon excess needs more ingredients than this. First, we need an asymmetry in the behaviour of the mediating bosons (often called the X-bosons) towards matter and antimatter. Second, we need conditions departing from thermodynamic equilibrium in order to ensure that reactions which produce a net N_B are not countered by reverse reactions that destroy N_B. Such recipes, accompanied by ingenious cooking conditions, have been proposed to explain N_B/N_γ.

The role of the vacuum is nontrivial in gauge theories such as are used in the grand unification programme. Phase transitions can occur in which the physical system jumps from a false vacuum to the true vacuum of less energy. We have discussed this behaviour in Part I and shall return to it in Part III. In general relativity such transitions have a dynamical effect on spacetime geometry. As we shall see later (cf. Chapter 10) the above vacuum effect is equivalent to introducing a Λ term in the Einstein field equations and so we get the de Sitter line element (15) for the vacuum state. The value of Λ invoked by these considerations is, however, much larger than those considered by de Sitter or Einstein. The sudden exponential growth in the scale factor Q is called *inflation*.

Finally, we mention the limit to which the considerations of the early universe can be pushed back in time. The limit arises when classical gravity is no longer reliable and we must resort to quantum gravity. The limit in time is the so-called Planck time

$$t_P = \sqrt{\frac{G\hbar}{c^5}} \sim 5 \times 10^{-44}\text{ s}. \tag{81}$$

For $t < t_P$, quantum cosmology (to be discussed in Part IV) generates new models of the early universe.

Example 8.12. Consider the scale factor $Q \propto t^{1/2}$ in the early universe. The characteristic time scale associated with the universe at epoch t is $\sim H(t)^{-1} \equiv Q/\dot{Q} = 2t$. The corresponding distance scale is $2ct$ and the volume of a Euclidean sphere of this radius is $V = 32\pi c^3 t^3/3$. From Einstein's equations, the energy density in the universe at this epoch is

$$\varepsilon = \frac{3c^2 H^2(t)}{8\pi G} = \frac{3c^2}{32\pi G t^2} .$$

Therefore, the total energy in the above cosmological volume is given by

$$E = V\varepsilon = \frac{c^5 t}{G} .$$

The laws of quantum mechanics become relevant to cosmology at this epoch if

$$(2t) \times (E) \lesssim \hbar,$$

that is,

$$t \lesssim \sqrt{\frac{G\hbar}{2c^5}} = \frac{t_P}{\sqrt{2}} .$$

8.5. The Problems of Singularity, Horizon, and Flatness

We now identify three problems of standard cosmology, which are of a fundamental nature.

8.5.1. *Spacetime Singularity*

The instant when the scale factor Q became zero in the Friedmann models is the instant of spacetime singularity. Calculations show that invariants like R, $R_{ik}R^{ik}$, . . . diverge at $Q = 0$. We have already remarked that it is not possible to continue a meaningful discussion of cosmology over a time interval which contains this singular epoch.

It may be that such a singular epoch tells us that the event at $Q = 0$ was so profound that it lies beyond the scope of physics. The identification of this event with the 'creation' of the universe and its elevation to a level beyond the reach of physics appeals to many cosmologists and physicists. There is a class of physicists, however, which regards this singularity not as a metaphysical concept but as a drawback or deficiency of classical general relativity. We share this point of view and suggest that although the general theory has been well tested in the weak field limit (cf. Section 7.2) its viability in situations of strong gravity remains unproven. The appearance of singularity at $Q = 0$ is a manifestation of the inability of the classical general relativity to handle strong gravity situations adequately. Also, even if we exclude from physics the $Q = 0$ singularity, other singularities remain, such as the one discussed in Section 7.5 as the endpoint of gravitational collapse.

In the early period of Friedmann cosmology it was believed that the $Q = 0$ singularity is an artifact of the high degree of symmetry assumed for the cosmological spacetime. It was felt that if the symmetries were removed the singularity would vanish. This expectation was demolished by the singularity theorems proved in the 1960s, which showed that spacetime singularity is a characteristic of all general relativistic solutions whether they describe gravitational collapse or the expanding cosmological models. Thus, spacetime singularity is an inevitable feature of relativistic cosmology.

8.5.2. *Particle Horizons*

Consider the past light cone from the spacetime point $r = 0$, $t > 0$. Suppose that this intersects the singular epoch $t = 0$ in the sphere of coordinate radius r_H. Then

$$\int_0^{r_H} \frac{dr}{\sqrt{1 - kr^2}} = \int_0^t \frac{dt_1}{Q(t_1)} . \tag{82}$$

Example 8.13. Consider the Friedmann models with $k = +1$. From (40) and (41) applied to (82) we get the present particle horizon radius as

$$r_H = \sin \psi_0 = \frac{\sqrt{2q_0 - 1}}{q_0}$$

Similarly, for $k = -1$ we have

$$r_H = \frac{\sqrt{1 - 2q_0}}{q_0} .$$

(For the $k = 0$ case the r coordinate can be scaled arbitrarily, but $Q_0 r_H$ remains invariant under scaling.) □

Example 8.14. We mention here the exception to the rule given by the Milne model. For this model (82) takes the form

$$r_H = \lim_{\varepsilon \to 0} \sinh \ln \frac{t}{\varepsilon} \to \infty .$$ □

Except for the flat spacetime case of $S \propto t$, $k = -1$, the integral in t is convergent and tends to zero as $t \to 0$. Thus (82) defines a function $r_H(t)$ such that at epoch t all particles with $r \leq r_H(t)$ can be seen by the observer at $r = 0$, while those with $r > r_H(t)$ cannot be seen. This sphere of coordinate radius $r_H(t)$ is the particle horizon of the observer at $r = 0$ (see Figure 8.8).

The particle horizon is thus the physical barrier beyond which no particle can influence the observer at $r = 0$. Since, in Friedmann cosmologies, $r_H(t) \to 0$ as $t \to 0$, the spheres of influence around each point shrink to that point. Although we expect that the observed homogeneity of the universe was due to physical mixing, we find that in actuality this mixing could not have taken place sufficiently early in the universe as regions of influence between remote parts of the universe were nonoverlapping in the past. The Big Bang

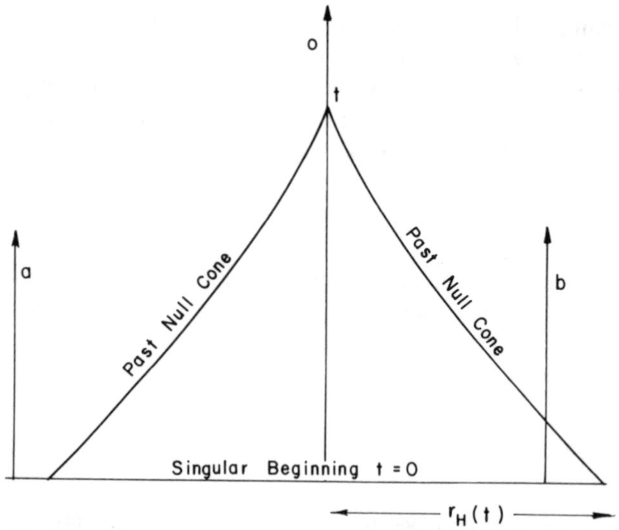

Fig. 8.8. The past null one of the observer O terminates at the singular epoch. There it has a radial coordinate extent of $r_H(t)$. Particles like a which lie beyond this distance are not seen by O at time t. Particles like b which lie within this distance are visible to O.

theory therefore forces us to *assume* highly symmetrical (homogeneous) initial conditions.

Example 8.15. The horizon problem was realized when it was found that the microwave background had a high degree of isotropy. How did one part of the universe 'know' the value of radiation density in another remote part in, say, the direction that was diametrically opposite to us?

At redshift $z \simeq 10^3$, the $k = +1$ model gives, from (42b) and Example 8.13.

$$r_H = \sin \psi_1 = \frac{\sqrt{(2q_0 - 1)(1 + 2q_0 z)}}{q_0(1 + z)} .$$

To get the proper distance corresponding to r_H at the present epoch we have to multiply r_H by Q_0 as given by (22). Thus we get the characteristic diameter of a typical horizon-bound sphere as

$$d_H = 2Q_0\, r_H = \frac{2c}{H_0} \frac{\sqrt{1 + 2q_0 z}}{1 + z} .$$

The typical angle subtended by the region at our position is, for $z \gg 1$,

$$\theta_H \approx \frac{H_0 d_H}{c} \sim \frac{2\sqrt{1 + 2q_0 z}}{1 + z} \sim \sqrt{\frac{8q_0}{z}} \sim 5° \sqrt{q_0} .$$

For the exact derivation of the value of θ_H see Narlikar (1983) in Note 1 at the end of this chapter. □

8.5.3. *Flatness*

We have defined the parameter Ω_0 by (48) at the present epoch. Let us find the value of the same parameter at an earlier hot epoch when the temperature was T. A simple calculation (see Example 8.16) gives

$$\Omega(T) = 1 + 4 \times 10^{-15}\, T_{\text{MeV}}^{-2}(\Omega_0 - 1). \tag{83}$$

Example 8.16. In the early hot epoch, Equations (32) give

$$H^2 + \frac{k}{Q^2} = \frac{8\pi G u}{3}.$$

For $k = 0$ (i.e. for the closure parameter $\Omega = 1$) we have

$$u = \frac{3H^2}{8\pi G} \equiv u_c.$$

Thus, we write $u = \Omega u_c$ and get, in general,

$$\Omega - 1 = \frac{k}{Q^2 H^2} = \frac{k}{\dot{Q}^2}.$$

The right-hand side denotes the 'curvature term' whose magnitude is expected to be small. How small it is can be estimated as follows.

We have, in the early universe, $Q \propto t^{1/2}$. Let $Q = \alpha t_s^{1/2}$, where α = constant. Then $\dot{Q} = \alpha t_s^{-1/2}/2$. Since $T \propto Q^{-1}$ we can estimate $\dot{Q}$ from the present values Q_0 and T_0. For $k = 1$ we have, from (22), $Q_0^{-1} = H_0\sqrt{\Omega_0 - 1}$. Hence,

$$\alpha t^{1/2} T = \frac{T_0}{H_0\sqrt{\Omega_0 - 1}} = (\Omega_0 - 1)^{-1/2} \times 2.5 \times 10^{18} \text{ cm MeV};$$

for, $T_0 = 3\text{ K} \simeq 2.5 \times 10^{-10}$ MeV, $H_0 = 100$ km s^{-1} Mpc^{-1}. However, from (80) we get

$$t_s = 2.4 g^{-1/2}\, T_{\text{MeV}}^{-2}.$$

From these two relations, and after restoring c and ignoring g, we obtain

$$\Omega(T) - 1 \cong 4 \times 10^{-15}\, T_{\text{MeV}}^{-2}(\Omega_0 - 1).$$ □

Let us denote by Δ the difference between $\Omega(T)$ and unity, the value for the flat Friedmann model. Then (83) gives

$$\Delta(T) \cong 4 \times 10^{-15}\, T_{\text{MeV}}^{-2}\Delta_0 = \eta\Delta_0, \text{ say} \tag{84}$$

The present range of Δ_0 is between ~5 and −1, say. If we assume that the decision about the present value of Ω_0 was made very early in the universe, then the permissible range at that time was smaller than the present range of Δ_0 of ~6 by the factor η.

Since Ω_0 and Δ_0 depend on the matter density their present values were determined when the baryon number was fixed. This takes us back to the GUT era of temperature ~10^{14} GeV, which corresponds to $T_{\text{MeV}} \cong 10^{17}$. This gives the permitted range of $\Delta(T)$ in (84) as ~(4–20) × 10^{-49}. In other words, in order to arrive at the present range of Ω_0, the universe had to be very finely tuned near the 'flat value' $\Omega(T) = 1$ when baryon excess was generated.

This somewhat restrictive nature of the early universe situation was first pointed by Dicke and Peebles (who evaluated the range of $\Delta(T)$ at the *later* epoch of nucleosynthesis) but was highlighted in the context of GUTs by A. Guth in 1981. Guth's resolution of this problem through the 'inflationary universe' will be discussed in Chapter 10, and our resolution through quantum cosmology in Part IV.

8.6. Anisotropic Cosmologies

For the time being we stay within the classical framework of general relativity and seek a resolution of some of the above problems through the generalization of the cosmological models. Accordingly, we now assume that they are homogeneous but not isotropic.

Thus we go back to the line element (3) and generalize it to the form

$$ds^2 = dt^2 + 2g_{\mu 0}\, dt\, dx^\mu + g_{\mu\nu}\, dx^\mu\, dx^\nu \tag{85}$$

wherein the worldlines of galaxies are still given by x^μ = constant and t still measures the cosmic time but the surfaces Σ are no longer orthogonal to worldline Γ. Further, we assume $\{\Gamma\}$ to be geodesics, which imply that $g_{\mu 0}$ are functions of x^μ only (see Example 8.1). What is the significance of these modifications?

First we note that $u^i = (1, 0, 0, 0)$ is the velocity field of galaxies and it has a nonvanishing curl:

$$\omega_{ik} = \tfrac{1}{2}(u_{k;i} - u_{i;k}) = \tfrac{1}{2}(g_{k0,i} - g_{i0,k}),$$

i.e. ω_{ik} has only spacelike components defined by

$$\omega_{\mu\nu} = \tfrac{1}{2}(g_{\mu 0,\nu} - g_{\nu 0,\mu}). \tag{86}$$

$\omega_{\mu\nu}$ is the spin-tensor of cosmic fluid. To an observer using a locally inertial frame, the galaxies with worldlines Γ will appear to be going round him. It was K. Gödel who in 1949 first drew attention to such spinning models.

Observations do not show any evidence for spin of the universe in this way. This was known for a long time. In the 1890s, Ernst Mach noted the fact that the background of distant stars is nonrotating relative to the local inertial frame. He read into this result some long-range interaction that relates the inertia of a particle to the background of distant matter. Although Mach did not give details of such an interaction, this concept – dignified by the name 'Mach's principle' – has prompted several theoreticians to construct a Machian theory of cosmology. See Narlikar (1983) in Note 1 at the end of this chapter for details of some such theories.

Let us return to spinning models, which were studied by relativists in the 1950s for the possibility that spin may generate centrifugal force and thereby prevent the spacetime singularity. The reason for this expectation was an equation, derived by A. K. Raychaudhuri in 1955, which we shall discuss briefly below.

Corresponding to (86), define shear by

$$\sigma_{ik} = \tfrac{1}{2}(u_{i;k} + u_{k;i}) - \tfrac{1}{3}(g_{ik} - u_i u_k)u^l{}_{;l}. \tag{87}$$

The reason for the $\frac{1}{3}g_{ik}$ term is that, like rotations, shear also turns out to be three-dimensional in character. Raychaudhuri's equation follows from the $\binom{0}{0}$ component of the Einstein equations:

$$\frac{\ddot{Q}}{Q} = \tfrac{1}{3}\,(2\omega^2 - \sigma^2 - 4\pi G\varepsilon), \tag{88}$$

where $Q = (-g)^{1/6}$ and

$$2\omega^2 = -\omega_{ik}\omega^{ik}, \qquad \sigma^2 = \sigma_{ik}\sigma^{ik}. \tag{89}$$

ε is the energy density of matter. Thus the ω^2 term suggests that, if it dominates at small Q, we could have a 'bounce' with $Q > 0$, $\dot{Q} = 0$, $\ddot{Q} > 0$.

Searches for spinning and bouncing solutions of Einstein's solution proved fūtile and the immediate reason was that, although rotation may lead to bounce, the spin term comes with the wrong sign. Also, while shear without spin ($\sigma^2 > 0$, $\omega = 0$) is possible, spin without shear ($\sigma = 0$, $\omega^2 > 0$) is not. At a deeper level, we now see that because of the singularity theorems mentioned in Section 8.5, bouncing models are impossible.

In the remaining part of this section we shall set $\sigma^2 > 0$, but $\omega = 0$.

Because of the homogeneity of the 3-space $\{\Sigma\}$, we can express the line element as

$$ds^2 = dt^2 + g_{\alpha\beta}\omega^\alpha\omega^\beta, \tag{90}$$

where ω^α are a set of three 1-forms (see Appendix C) and $g_{\alpha\beta}$ depend on t only. The homogeneity can be discussed by means of the techniques of Appendix D and we are able to write the commutation relations

$$[\omega^\alpha, \omega^\beta] = C^{\alpha\beta}{}_\gamma\omega^\gamma, \tag{91}$$

where the C's are the structure constants of the Lie group of motions and satisfy the Jacobi identity. Thus we can write

$$C^{\alpha\beta}{}_\gamma = \varepsilon^{\alpha\beta\sigma}\,\{n_{\sigma\gamma} + \varepsilon_{\sigma\gamma\mu}a^\mu\}$$

$$n_{\sigma\alpha} = n_{\alpha\sigma}. \tag{92}$$

By diagonalizing $n_{\alpha\sigma}$ as diag(n_1, n_2, n_3,) we find that the Jacobi identity gives $n_{\alpha\sigma}\,a^\sigma = 0$ and, hence, if $a^\sigma = (a, 0, 0)$, n_1 or $a = 0$. Thus all 3-parameter groups of motion can be characterized by n_1 or a, n_2 and n_3. There are nine groups, and the corresponding cosmological models are labelled Bianchi types I to IX since Bianchi was the first to classify the spaces in this way. The models with $a = 0$ are called type A models and are simpler to analyze. In type B we have $n_1 = 0$, $a \neq 0$ and there exist two cases, type VI_h $(h > 0)$ and VII_h $(h < 0)$ which are parametrized by another parameter h associated with scale transformations.

For future use, in Part IV we shall express the metric tensor in the form

$$-g_{\mu\nu} = \mathrm{e}^{2\lambda(t)}\, \mathrm{e}^{-2\beta_{\mu\nu}(t)}, \tag{93}$$

where

$$(\beta_{\mu\nu}) = \mathrm{e}^{-\psi k_3}\, \mathrm{e}^{-\theta k_1}\, \mathrm{e}^{-\theta k_3}\, \beta_d\, \mathrm{e}^{\theta k_3}\, \mathrm{e}^{\theta k_1}\, \mathrm{e}^{\psi k_3} \tag{94}$$

and

$$\beta_d = \mathrm{diag}\left[\beta_1, -\tfrac{1}{2}\beta_1 + \frac{\sqrt{3}}{2}\beta_2, -\tfrac{1}{2}\beta_1 - \frac{\sqrt{3}}{2}\beta_2\right], \tag{95}$$

$$k_3 = \begin{bmatrix} 0 & 1 & 0 \\ 1 & 0 & 0 \\ 0 & 0 & 0 \end{bmatrix}, \qquad k_1 = \begin{bmatrix} 0 & 0 & 0 \\ 0 & 0 & 1 \\ 0 & -1 & 0 \end{bmatrix}. \tag{96}$$

The five functions $\psi(t)$, $\theta(t)$, $\beta_1(t)$, $\beta_2(t)$, and $\lambda(t)$ parametrize the metric.

Consider the simpler case of diagonal Bianchi models with $\theta = 0$, $\psi = 0$. The dynamics of the models can then be derived from the action

$$S = \frac{1}{16\pi G} \int L\, \mathrm{d}t \tag{97}$$

with

$$L = -\mathrm{e}^{-3\lambda}\,[6\dot{\lambda}^2 - \frac{3}{2}(\dot{\beta}_1^2 + \dot{\beta}_1^2)] + \mathrm{e}^{3\lambda} R^*. \tag{98}$$

The form of R^* is given by

$$R^* = \tfrac{1}{2}\mathrm{e}^{-2\lambda} f(\beta_1, \beta_2) \quad \text{for type A,}$$

$$R^* = C_1\, \mathrm{e}^{-2\lambda} \exp\{-C_2\beta_1 - C_3\beta_2\} \quad \text{for type B.} \tag{99}$$

The functional form of f and the constants C_1, C_2, C_3 depend on the type of Bianchi model under consideration. We shall return to these models in Part IV when considering quantum cosmology.

We end by describing two simple anisotropic models, one of empty spacetime, the other of dust universe.

8.6.1. *The Kanser Model*

This describes empty spacetime with the line element

$$\mathrm{d}s^2 = \mathrm{d}t^2 - t^{2p_1}(\mathrm{d}x^1)^2 - t^{2p_2}(\mathrm{d}x^2)^2 - t^{2p_3}(\mathrm{d}x^3)^2, \tag{100}$$

where p_1, p_2, p_3 satisfy the conditions

$$p_1 + p_2 + p_3 = 0, \qquad p_1^2 + p_2^2 + p_3^2 = 1. \tag{101}$$

A parametric representation for the p's is given by

$$p_1 = -\frac{u}{1+u+u^2}, \quad p_2 = \frac{1+u}{1+u+u^2}, \quad p_3 = \frac{u(1+u)}{1+u+u^2}. \tag{102}$$

This model has singularity at $t = 0$, although there is contraction along two axes and elongation along the third.

8.6.2. *The Bianchi Type I Model*

This is the simplest of all homogeneous (but anisotropic) cosmologies. Its line element is given by

$$\mathrm{d}s^2 = \mathrm{d}t^2 - X_1^2(t)\,(\mathrm{d}x^1)^2 - X_2^2(t)\,(\mathrm{d}x^2)^2 - X_3^2(t)\,(\mathrm{d}x^3)^2. \tag{103}$$

For a dust-filled universe it is easy to give a solution of Einstein's equations for the above line element. We have

$$X_\mu(t) = Q(t)\left(\frac{t^{2/3}}{Q(t)}\right)^{n_\mu}, \qquad n_\mu = 2\sin\left(\alpha + \frac{2\mu}{3}\right), \tag{104}$$

where α is a constant and

$$Q^3(t) \equiv X_1 X_2 X_3 = At(t + \Sigma). \tag{105}$$

Here A and Σ are also constants and are related to the matter density ϱ by

$$\varrho = \frac{1}{6\pi G t(t + \Sigma)}. \tag{106}$$

This model has singularity at $t = 0$ for $\Sigma > 0$. Notice that for suitable choices of α we can eliminate particle horizons in one of the three directions – but not in all three. In the late 1960s C. W. Misner used this idea to investigate the 'mixmaster model' of the universe in which the horizon-free direction is rotated rapidly and at random. Would such a device allow mixing of different parts of the universe? D. M. Chitre showed that the answer to the question is in the negative. The mixmaster does not mix and the particle horizon problem remains.

To sum up: going over to anisotropic models does not solve the problems of singularity and particle horizons nor does it throw any light on the flatness problem. We therefore need fresh inputs into classical relativistic cosmology, and this need provides the motivation for the ideas described in Parts III and IV.

Notes and References

1. A classic text book on cosmology which covers many topics in this chapter is:

Weinberg: S. 1972, *Gravitation and Cosmology*, Wiley, New York.

A more recent text which covers all topics in greater detail·is:

Narlikar, J. V.: 1983, *Introduction to Cosmology*, Jones & Bartlett, Boston.

2. Physical cosmology at an elementary level is discussed by:

Peebles, P. J. E.: 1970, *Physical Cosmology*, Princeton, New Jersey.

3. For a discussion of nucleosynthesis in the primordial context see the review article:

Schramm, D. N., and Wagoner, R. W.: 1977, *Ann. Rev. Nucl. Sci.* **27**, 37.

4. For the problems connected with the photon/baryon ratio see the discussion by:

Steigmam, G.: 1979, *Ann. Rev. Nucl. Sci.* **29**, 313.

5. The problem of flatness is highlighted by A. Guth in his article:

Guth, A. H.: 1981, *Phys. Rev.* D**23**, 347.

6. Anisotropic cosmologies have been discussed by:

Ryan, M. P., and Shapley, L.: 1975, *Homogeneous Relativistic Cosmologies*, Princeton, New Jersey.

Part III
Quantization in Curved Spacetime

Chapter 9

Quantum Theory in Curved Spacetime

9.1. Quantum Theory in a Curved Background: Why?

The previous chapters of the book have developed the various features of quantum theory and general relativity separately. The next logical step would have been to construct a quantum theory of gravity interacting with matter fields. Unfortunately, this grand synthesis still remains a distant dream.

There are many reasons why a quantum theory of gravity is intrinsically much more complicated than other quantum field theories. These problems will be discussed at length in the next part of the book. Here we shall merely note that we do not have a complete formalism for quantum gravity at our disposal at present. Thus the logical procedure of treating gravity on the same footing as any other quantum field just cannot be implemented.

Granted our ignorance of quantum gravity, how can we make any worthwhile investigation of the interface between quantum theory and gravity?

History provides one suggestion. Before the full quantum theory of the electromagnetic field was developed, people used the picture of a classical electromagnetic field interacting with atomic and molecular systems to study various spectroscopic results. This can be considered to be an approximation in which a system obeying quantum rules (say, an atom) interacts with another system obeying classical physics to 'sufficient accuracy'. From the theoretical point of view, it was shown by N. Bohr and L. Rosenfeld in 1933 that such an approach can *only* be an approximation, albeit a good one. By the very nature of the approximation, we neglect the changes produced in the classical system by the quantum system.

In an analogous fashion, we may consider studying quantum systems interacting with a classical gravitational field. For example, we would expect the external gravitational field to create particles in much the same way as an external electric field (as discussed in Chapter 4). We are thus led to the study of quantum fields in curved spacetime, which is the subject of this chapter.

Several cautionary remarks are in order.

To begin with, there is no compelling reason to believe in the validity of such an approximation. It is possible that the basic nature of quantum gravity makes such an approximation meaningless. For example, incorporating special relativistic invariance into quantum theory necessitated such a drastic change in the formalism which no amount of approximation in powers of some suitably chosen parameter, like (v/c), would have revealed. There are indications that incorporating general coordinate invariance (which is in some sense the 'kinematic part' of general relativity) will lead to more drastic changes in the formalisms than suggested by approximations like that in Example 9.1.

Normally, quantum phenomena are associated with length scales of the order of 10^{-13} cm. If the curvature of the spacetime varies significantly at such a short length scale, then we would expect gravity to significantly *alter the nature* of quantum phenomena.

Example 9.1. To get a feel for the number involved, consider the following simple example: a black hole creating particles. A pair of virtual particles of energy ω can exist with a lifetime ω^{-1}. In this span of time they can be separated, at most by the distance ω^{-1} (we are using natural units: $c = \hbar = 1$). Suppose this process takes place at a distance r from a gravitating body of mass M. The *tidal force* separating the particles when they are apart by a distance x is

$$F \sim \frac{GM\omega}{r^3}\, x, \qquad r \gg 2MG, \quad x \ll r.$$

The virtual particles can absorb gravitational energy and become a real pair only if the following condition is satisfied: Work done by the gravitational tidal force within the lifetime of the particles must be at least of the order of energy of the pair. Thus,

$$\int_0^{\omega^{-1}} F\, dx \gtrsim \omega; \qquad \frac{GM}{r^3}\, \omega^{-1} \gtrsim \omega.$$

Further, the particles can escape to infinity only if $r > GM$. Thus we expect significant creation for modes which satisfy

$$\omega \lesssim (GM)^{-1}.$$

We shall later see that such processes can actually occur. □

Quantum gravitational effects, on the other hand, will be important only at length scales of the order of Planck length $\sim 10^{-33}$ cm. It appears as though we are left with a terrain of 20 orders of magnitude (10^{-13} to 10^{-33}) on which our approximations can be valid.

Such arguments, to say the least, are naïve. It is not entirely correct to say that, in field theory, quantum mechanical effects are important in a particular range of scales. We saw in Chapter 4 that divergences in field theory are closely related to the small distance behaviour of the Green's functions. If gravitational effects alter this small-scale behaviour, then the formalism of field thoery will become very different from what we are used to. Since quantum field theory 'sums over' virtual states of arbitrarily high energies, it is not entirely clear whether a definite energy or length scale can be associated with field theory phenomena.

Added to this consideration is the fact that gravity couples to matter *and itself* with equal strength. Thus a photon and a graviton (of the same energy) couples to an external gravitational field with equal strength. Creation of photons by a changing gravitational field will have a counterpart of creation of gravitons. Clearly the 'classical background' gravitational field itself is experiencing first-order perturbations. This is an additional complication not present in a linear field theory.

The above problems are, in some sense, technical in nature. The most serious difficulty in discussing quantum field theory in curved spacetime is the following: *There does not exist a quantum field theory formalism in an arbitrary curved spacetime.* This problem is deep rooted and arises from the fact that standard formalisms of field theory require a preferential slicing in spacetime. In other words, quantum field theory, as we know today, is Lorentz invariant; but it is *not* generally covariant. This conceptual problem introduces a new level of observer dependence in quantum theory, the implications of which are not yet well understood. It should be noted that such a difficulty exists even in flat space when curvilinear coordinates are used. We shall say more about this in the next section.

The above remarks were intended to convince the reader that the so-called topic 'quantum fields in curved spacetime' is far from a well-defined area of study. Many fundamental questions still remain to be answered. For the rest of the chapter, however, we shall forget these 'questions of principle' and proceed bravely ahead. We shall begin by discussing field theory in an accelerated frame in order to point out certain major new features. Following this we shall consider three kinds of background spacetimes which have been well analysed in the literature: viz. Robertson–Walker spacetimes, de Sitter spacetime, and black-hole spacetimes.

9.2. General Covariance and the Particle Concept

Intuitively we always consider a 'particle' to be a sufficiently small 'billiards ball'. From such a naïve, classical point of view we expect the concept of a particle to be coordinate independent. However, the quantum field theoretic concept of a particle is far different from that of a 'structureless billiards ball'. Thus, we have to examine from first principles the general covariance or otherwise of the particle concept.

Let us consider a massless scalar field described by the generally covariant action in the Minkowski spacetime

$$S = \tfrac{1}{2} \int \varphi^i \varphi_i \sqrt{-g}\; \mathrm{d}^4x. \tag{1}$$

Clearly, the action leads to a generally covariant propagator kernel

$$K[\varphi_2, \Sigma_2; \varphi_1, \Sigma_1] = \int \exp[iS]\, \mathcal{D}\varphi, \tag{2}$$

describing the propagation of the field φ from one hypersurface to another.

Thus the *dynamics* of the field *is* generally covariant and independent of the coordinates used. However, this is not all. In order to obtain numbers, the kernel must be integrated over initial and final states. Coordinate dependence can arise through the possibility that these states are coordinate dependent. We must also investigate this 'kinematic' aspect of the theory.

The standard Fock states (vacuum, one-particle state, etc.) are constructed by introducing the creation and annihilation operators (see Chapter 4). More formally this may be done as follows: consider a complete set of orthonormal solutions to the Klein–Gordon equation

$$\Box f_\alpha(x) = 0, \tag{3}$$

where α labels the solution; the orthonormality implies, for example (note that α is *not* a spacetime index)

$$(f_\alpha, f_\beta) \equiv -i \int f_\alpha \overleftrightarrow{\partial}_i f_\beta^* \, d\Sigma^i \equiv -i \int (f_\alpha \partial_i f_\beta^* - f_\beta^* \partial_i f_\alpha) \, d\Sigma^i = \delta_{\alpha\beta}. \tag{4}$$

Being a complete set, $\{f_\alpha(x)\}$ can be used to expand any real field operator $\varphi(x)$ as

$$\varphi(x) = \sum_\alpha \left(a_\alpha f_\alpha(x) + a_\alpha^\dagger f_\alpha^*(x)\right). \tag{5}$$

Here a_α (and $a_\alpha^\dagger$) are operators (independent of x^i), thereby maintaining the operator nature of $\varphi(x)$. The commutation rules for $\varphi(x)$ then imply that

$$[a_\alpha, a_\beta] = 0; \quad [a_\alpha^\dagger, a_\beta^\dagger] = 0; \quad [a_\alpha, a_\beta^\dagger] = \delta_{\alpha\beta}. \tag{6}$$

Usually the plane wave solutions of (3) are used, which are

$$f_{\mathbf{k}}(\mathbf{x}, t) = \frac{1}{\sqrt{2\omega_{\mathbf{k}}}\,\sqrt{(2\pi)^3}} \exp i(\mathbf{k}\cdot\mathbf{x} - \omega_{\mathbf{k}} t).$$

The index label α is nothing but the wave vector $\mathbf{k}$. One can easily verify that (4) is satisfied by this choice. These commutation rules suggest that we use a basis labelled by a set integers of $\{n_\alpha\}$. In particular, we define a 'vacuum' state to be

$$a_\alpha |0\rangle = 0 \quad \text{for all } \alpha. \tag{7}$$

A general state $|\{n_\alpha\}\rangle$ is constructed by the usual procedure

$$|{}^1n_{\alpha_1}, {}^2n_{\alpha_2} \ldots\rangle = ({}^1n!\,{}^2n! \ldots)^{-1/2} (a_{\alpha_1}^\dagger)^{1n}(a_{\alpha_2}^\dagger)^{2n} \ldots |0\rangle \tag{8}$$

It is now clear that these states are very dependent on the choice of operators a_α, *or, in other words, the choice of the mode functions* $f_\alpha(x)$. For example, two different choice for the mode functions $f_\alpha(x)$ will lead to different vacua even in flat space!

This result can be seen as follows: assume that we have chosen some orthonormal basis set $\{f_\alpha\}$ and a corresponding set of operators $\{a_\alpha\}$. Let us denote by $|0\rangle^f$ the 'vacuum' state corresponding to this particular choice $\{f_\alpha\}$. Now consider a new set of mode functions

$$F_\alpha(x) \equiv \sum_\beta (M_{\alpha\beta} f_\beta + N_{\alpha\beta} f_\beta^*). \tag{9}$$

The orthonormality conditions on F imply the following constraints on M and N,

$$\sum_\alpha \, M_{\beta\alpha} M^*_{\gamma\alpha} - N_{\beta\alpha} N^*_{\gamma\alpha}) = \delta_{\beta\gamma}, \tag{10}$$

$$\sum_a (M_{\beta\alpha} N_{\gamma\alpha} - N_{\beta\alpha} M_{\gamma\alpha}) = 0. \tag{11}$$

Such transformations are called 'Bogolubov transformations'. M and N are called the 'Bogolubov coefficients' for the transforms. Now we can expand $\varphi(x)$ in terms of $F_\alpha(x)$ with a new set of creation and annihilation operators $(A_\alpha, A_\alpha^\dagger)$:

$$\varphi(x) = \sum_\alpha \; (a_\alpha f_\alpha + a_\alpha^\dagger f_\alpha),$$

$$= \sum_\varphi \; (A_\alpha F_\alpha + A_\alpha^\dagger F_\alpha). \tag{12}$$

Using (9) in (12), we find,

$$a_\alpha = \sum_\beta (M_{\alpha\beta} A_\beta + N^*_{\beta\alpha} A_\beta^\dagger) \tag{13}$$

and

$$a_\alpha^\dagger = \sum_\beta (M^*_{\beta\alpha} A_\beta^\dagger + N_{\beta\alpha} A_\beta). \tag{14}$$

The 'vacuum state' corresponding to the new set $\{F_\alpha\}$ is defined by

$$A_\alpha \, |0\rangle^F = 0. \tag{15}$$

These two vacua $|0\rangle^f$ and $|0\rangle^F$ are entirely different. The vacuum $|0\rangle^f$ appears to be a many-particle state if we use the set $\{F_\alpha\}$. More specifically, the number operator corresponding to the choice F does not vanish in $|0\rangle^f$:

$${}^f\langle 0| \, A_\alpha^\dagger A_\alpha \, |0\rangle^f = \sum_\beta \; |N_{\alpha\beta}|^2. \tag{16}$$

Thus the Fock states depend very much on the choice of the mode functions. Before we talk about particles, we must specify a criterion for the choice of a particular set of mode functions.

It is at this juncture that the 'particle concept' ceases to be generally covariant. We usually impose the criterion that 'the mode functions must be positive frequency functions with respect to the global Minkowski time coordinate'. In other words, we take, $f_\alpha(\mathbf{x}, t)$ to have a time dependence of the form

$$f_\alpha(\mathbf{x}, t) = \exp(-i\omega_\alpha t) \, g_\alpha(\mathbf{x}), \quad \omega_\alpha > 0. \tag{17}$$

The vacuum state defined with reference to these mode functions will be the usual field theory vacuum state. 'Particles' constructed on this basis are taken to represent physical particles.

In order to choose mode functions of the type given by (17), the Klein–Gordon equation must be separable into time and space parts. In other words, the spacetime must be stationary in the particular coordinate system. The standard inertial coordinates, in which the metric takes the form

$$ds^2 = dt^2 - dx^2 - dy^2 - dz^2 \tag{18}$$

is definitely a natural choice. However, this is by no means a unique choice! For example, consider the coordinates, (τ, ξ), defined via

$$x = \xi \cosh g\tau;\ t = \xi \sinh g\tau;\ y = y;\ z = z \tag{19}$$

in terms of which the metric reads as

$$ds^2 = g^2\xi^2\, d\tau^2 - d\xi^2 - dy^2 - dz^2. \tag{20}$$

Quite clearly, the metric is stationary with respect to the new time coordinate τ. The coordinate system in (20) represents the proper coordinate system for a uniformly accelerated observer in Minkowski space whom we earlier encountered in Chapter 6. The time coordinate τ represents the proper time of the accelerated observer.

Example 9.2. Consider the observer following a uniformly accelerated trajectory (i.e. $a^i a_i = g^2$ =constant; where $a^i = d^2x^i/d\tau^2$) along the x axis. Denoting the 4-velocity by u^i; we have the relations

$$\frac{dt}{d\tau} = u^0; \qquad \frac{dx}{d\tau} = u^1; \qquad \frac{du^0}{d\tau} = a^0; \qquad \frac{du^1}{d\tau} = a^1.$$

Writing out the conditions

$$u^i u_i = 1; \qquad u^i a_i = 0; \qquad a^i a_i = g^2$$

we get

$$a^0 = \frac{du^0}{d\tau} = gu^1, \qquad a^1 = \frac{du^1}{d\tau} = gu^0.$$

Therefore, the trajectory is given by

$$t = g^{-1} \sinh g\tau, \qquad x = g^{-1} \cosh g\tau.$$

The coordinate frame of such an observer is described by a tetrad (i.e. a set of four 4-vectors), which is

$$(e_0)^i = (\cosh g\tau, \sinh g\tau, 0, 0), \qquad (e_2)^i = (0, 0, 1, 0)$$

$$(e_1)^i = (\sinh g\tau, \cosh g\tau, 0, 0), \qquad (e_3)^i = (0, 0, 0, 1).$$

More specifically, the accelerated observer uses (i) τ as the time coordinate and (ii) $\xi^\mu (\mu = 1, 2, 3)$ for space coordinates such that

$$x^i = \xi^\mu [e_\mu(\tau)]^i + x_0^i(\tau),$$

where $x_0^i(\tau)$ is the trajectory of the observer. Solving these equations, we find the coordinate transformations

$$t = \xi^1 \sinh g\tau;\ x = \xi^1 \cosh g\tau;\ y = \xi^2;\ z = \xi^3.$$

There are same as (19) with $\xi^1 = \xi$. □

Let us see how the field theory looks in the new coordinate system (τ, ξ, y, z). First, notice that the transformation given above covers only one-fourth of the Minkowski space – the region marked I in Figure 9.1.

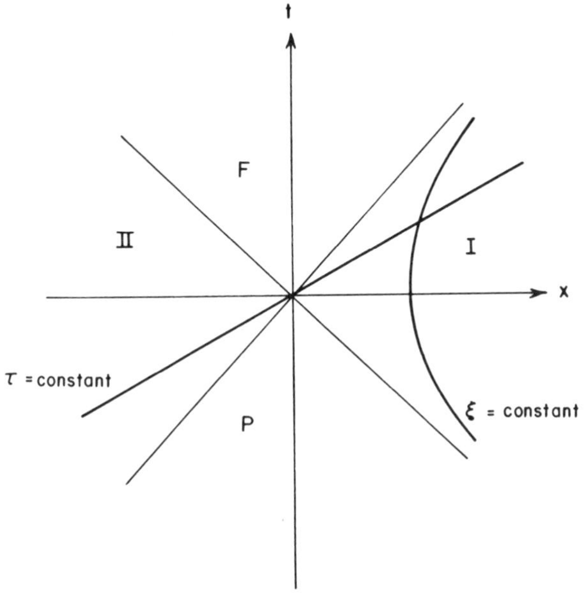

Fig. 9.1. Minkowski spacetime and the coordinate transformations to the accelerated frame.

We supplement it with the transformations:

$$\left.\begin{aligned} x &= \xi \sinh g\tau \\ t &= \xi \cosh g\tau \end{aligned}\right\} \text{ in } F \qquad \left.\begin{aligned} x &= -\xi \sinh g\tau \\ t &= -\xi \cosh g\tau \end{aligned}\right\} \text{ in } P. \qquad \left.\begin{aligned} x &= -\xi \cosh g\tau \\ t &= -\xi \sinh g\tau \end{aligned}\right\} \text{ in II.} \tag{21}$$

so that

$$ds^2 = \pm\, \xi^2 g^2\, d\tau^2 \mp d\xi^2 - dy^2 - dz^2 \tag{22}$$

with the upper sign holding in I and II and the lower sign holding in F and P. Since the metric is stationary in the (τ, ξ) coordinate system the Klein–Gordon equation can be separated into positive and negative frequency mode functions with respect to the τ coordinate. These mode functions are (in region I)

$$U_{\nu\mathbf{k}} = N_{\nu\mathbf{k}} \exp(-i\nu\tau + i\mathbf{k}\cdot\mathbf{y})\, K_{i\nu}(kg\xi). \tag{23}$$

Here $\mathbf{k} = (k_y, k_z)$ stands for the two-dimensional vector and $k = |\mathbf{k}|$. The function $K_{i\nu}(x)$ is the Bessel function of imaginary order and argument. The normalization constant $N_{\nu\mathbf{k}}$ is given by

$$N_{\nu\mathbf{k}} = \frac{(\sinh \nu\pi)^{1/2}}{2\pi^2}, \tag{24}$$

We can now expand our scalar field in terms of the $u_{\nu\mathbf{k}}$,

$$\varphi(x) = \int_0^\infty \int [a(\nu, \mathbf{k})\, u_{\nu\mathbf{k}}(x) + \text{h.c.}]\, \mathrm{d}\nu\, \mathrm{d}^2\mathbf{k} \tag{25}$$

or in terms of the usual Minkowski modes, $f_{\mathbf{p}}(t, \mathbf{x})$:

$$\varphi(x) = \int \left[A(\mathbf{P}) \frac{1}{\sqrt{2\omega_{\mathbf{k}}}\,\sqrt{(2\pi)^3}}\, \mathrm{e}^{\mathrm{i}(\mathbf{P}\cdot\mathbf{x} - \omega_{\mathbf{p}}t)} + \text{h.c.} \right] \mathrm{d}^3\mathbf{P}. \tag{26}$$

where by 'h.c.' we mean the Hermitian conjugate of the previous term. Vacuum states and 'particles' can be defined using either basis, and in general the concepts will not be the same. For example, it can be shown that the expectation value of the number of 'particles' defined (based on $a(\nu, \mathbf{k})$) in the Minkowski vacuum state is given by

$$^f\langle 0|\, N_{\nu\mathbf{k}}\, |0\rangle^f \equiv {}^f\langle 0|\, a^\dagger\, (\nu, \mathbf{k})\, a(\nu, \mathbf{k})\, |0\rangle^f$$

$$= \frac{1}{(\exp\,(2\pi\nu/g) - 1)} \tag{27}$$

[see example below]. In other words, the vacuum state of the Minkowski observer will appear to be a thermal state in the uniformly accelerated frame with a 'temperature' given by $T = g/2\pi$.

Example 9.3. The derivation of (27) proceeds along the following lines. First, note that we need the mode functions in regions I and II in order to have proper Cauchy data. (Any spacelike hypersurface τ = constant extends into both regions.) Since the positive direction of τ is different in I and II, it is $u^*_{\nu\mathbf{k}}$ which is a positive frequency mode in II. Let us define, $u^{\pm}_{\nu\mathbf{k}}$ by

$$u^{+}_{\nu\mathbf{k}}(x) \equiv \begin{cases} u_{\nu\mathbf{k}}(x) & \text{in I} \\ 0 & \text{in II} \end{cases} \qquad u^{-}_{\nu\mathbf{k}}(x) = \begin{cases} 0 & \text{in I} \\ u_{\nu\mathbf{k}} & \text{in II.} \end{cases}$$

Let us call the surfaces $H^+ = (\tau \to \infty, \ln\xi \to -\infty, \tau + \ln\xi$ finite) and $H^- = (\tau \to -\infty, \ln\xi \to \infty, \tau - \ln\xi$ finite) the 'horizons'. Note that the combination, $u^{+}_{\nu\mathbf{k}} + \mathrm{e}^{-\nu\pi}\, u^{-}_{\nu\mathbf{k}}$ goes as $(x + t - i\delta)^{-i\nu}$ near H^+, and hence, is analytic on H^+. Moreover, for complex values of $\tau + \ln\xi = t + x$ this combination is analytic in the lower half $(t + x)$ plane. In other words, $(u^{+}_{\nu\mathbf{k}} + \mathrm{e}^{-\nu\pi}\, u^{-}_{\nu\mathbf{k}})$ is expressible completely in terms of the positive frequency Minkowski modes. Similarly, $(u^{-}_{\nu\mathbf{k}} + \mathrm{e}^{-\nu\pi}\, u^{+}_{\nu\mathbf{k}})$ is a negative frequency Minkowski mode. Therefore, a Minkowski decomposition of the field is given by

$$\varphi(x) = \int_0^\infty \mathrm{d}\nu \int \frac{\mathrm{d}^2\mathbf{k}}{(2\pi)^2} \{b^+_{\nu\mathbf{k}} v^+_{\nu\mathbf{k}} + b^-_{\nu\mathbf{k}} v^{-*}_{\nu\mathbf{k}} + \text{Hermitian conjugate}\},$$

where

$$v^{(\varepsilon)}_{\nu k} = (1 - \mathrm{e}^{-2\pi\nu})^{-1/2}\,[u^{(\varepsilon)}_{\nu k} + \mathrm{e}^{-\pi\nu}\, u^{(-\varepsilon)}_{\nu k}];\ \varepsilon = \pm.$$

Comparing with the decomposition in terms of the accelerated frame modes

$$\varphi(x) = \int_0^\infty \mathrm{d}\nu \int \frac{\mathrm{d}^2\mathbf{k}}{(2\pi)^2} \{a^+_{\nu\mathbf{k}} u^+_{\nu\mathbf{k}} + a^-_{\nu\mathbf{k}} u^{-*}_{\nu\mathbf{k}} + \text{h.c.}\}$$

we get

$$b^{(\varepsilon)}_{\nu\mathbf{k}} = (1 - \mathrm{e}^{-2\pi\nu/g})^{-1/2}\,[a^{(\varepsilon)}_{\nu\mathbf{k}} - \mathrm{e}^{-\pi\nu/g}\, a^{(-\varepsilon)\dagger}_{\nu\mathbf{k}}].$$

The inertial and accelerated vacua are defined, respectively, by

$$b^{(\varepsilon)}_{\nu\mathbf{k}}\, |0\rangle^f = 0; \qquad a^{(\varepsilon)}_{\nu\mathbf{k}}\, |0\rangle^A = 0.$$

Using the above expressions we see that

$${}^f\langle 0|\, a^\dagger_{\nu\mathbf{k}} a_{\nu\mathbf{k}}\, |0\rangle^f = \frac{\mathrm{e}^{-2\pi\nu/g}}{(1 - \mathrm{e}^{-2\pi\nu/g})} = \frac{1}{(\mathrm{e}^{2\pi\nu/g} - 1)}. \qquad \square$$

Example 9.4. The above result can be given the following heuristic interpretation. Consider a classical scalar plane wave along the x axis represented by

$$\varphi(x, t) = \frac{1}{\sqrt{2\,\omega}} \exp i\omega(x - t).$$

An inertial observer in the (t, x) frame will consider φ to be a monochromatic wave of frequency ω. An accelerated observer, with the trajectory $x = g^{-1}\cosh g\tau$ and $t = g^{-1}\sinh g\tau$ will find φ to vary with τ in a complicated manner:

$$\varphi(\tau) = \frac{1}{\sqrt{2\,\omega}} \exp\left(\frac{i\omega}{g}\,\mathrm{e}^{-g\tau}\right).$$

This may be interpreted as a superposition of monochromatic waves in the form

$$\varphi(\tau) = \int \frac{1}{2\pi\sqrt{2\omega}} \exp(i\nu\tau)\cdot f(\nu)\,\mathrm{d}\nu,$$

where

$$f(\nu) = \int_{-\infty}^{+\infty} \left(\frac{\nu}{\omega}\right)^{1/2} \exp\left(i\,\frac{\omega}{g}\,\mathrm{e}^{-\nu\tau} - i\nu\tau\right)\mathrm{d}\tau$$

$$= \left(\frac{\nu}{\omega}\right)^{1/2} \frac{1}{g}\left(\frac{g}{\omega}\right)^{i\nu} \int_0^\infty x^{(i\nu/g\, -\, 1)}\mathrm{e}^{ix}\,\mathrm{d}x; \quad x = \mathrm{e}^{-\nu\tau}.$$

This integral can be evaluated by rotating the contour to the positive imaginary axis. Using further the relation: $\Gamma(z)\,\Gamma(1 - z) = \pi \operatorname{cosec} \pi z$, we get

$$|f(\nu)|^2 \propto \frac{1}{(\exp(2\pi\nu/g) - 1)}.$$

Thus the Planck spectrum represents the intensity of negative frequency components ($e^{i\nu\tau}$) in the positive frequency wave ($e^{-i\omega t}$). □

The above analysis would have convinced the reader that one cannot have an unambiguous definition of the particle even in flat spacetime. The general formalism of quantum theory *does not* provide a natural definition of mode functions. For every choice of mode function there emerges a particular kind of 'particle'. An extra physical input is required to choose any one of these mode functions.

In flat spacetime one may arbitrarily decide to use the positive frequency part that corresponds to a geodesic observer, thereby recovering the 'usual particle'. In flat spacetime, all geodesic observers agree on the particle concept. However, this is not found to be true in an arbitrary curved spacetime. Thus, even the geodesic definition has only a limited validity.

The Minkowski time coordinate assures the existence of a global timelike Killing vector field in flat spacetime (see Appendix D for a discussion of Killing vectors). Whenever a global timelike Killing vector field is available (in a given curved spacetime) it may be used to provide a natural definition for particles. Once again, not all spacetimes permit this luxury.

If our spacetime asymptotically approaches static limits, once again particle definition can be based on these asymptotic domains. An arbitrary curved spacetime which does not possess any of the above-mentioned features defies a field theory formalism.

While the above fact may be somewhat discouraging, it should not be surprising. Even in nongravitational field theories the particle content of the field can be defined only for free fields. Since quantum theory in a curved spacetime is equivalent to a quantum theory interacting with a gravitational field, it is only natural that the particle concept is not well defined.

We often resort to the construction of simple 'toy particle detectors' to demonstrate (or otherwise) the particle content of the theory. While particle detectors can certainly provide yet another definition of a particle, in general such a procedure will not be equivalent to using the field theory formalism. Essentially, particle detectors respond to the fluctuation pattern of the vacuum rather than to particles themselves.

Example 9.5. The simplest model for a particle detector can be constructed along the following lines: consider a system with a 'monopole moment' $m(\tau)$ coupled to the scalar field φ by the Lagrangian

$$L_I = m(\tau)\varphi(x(\tau)).$$

(Here $x(\tau)$ denotes the trajectory of the detector.) The amplitude for the detector to make a transition from a state $|E_0\rangle$ at $\tau = -\infty$ to some state $|E\rangle$ at time T while the field makes the transition from the vacuum state $|0\rangle$ to some one-particle state $|\psi\rangle$ is given by

$$\mathcal{A} = i\,\langle E|\; m(0)\; |E_0\rangle \int_{-\infty}^{T} e^{-i(E\,-\,E_0)\tau}\,\langle\psi|\;\varphi(x(\tau))\;|0\rangle\,\mathrm{d}\tau.$$

Therefore, the rate of transition to the energy level E (after summing over all $|\psi\rangle$) is

$$R = \frac{d}{dT} |\mathcal{A}|^2 \propto \int_{-\infty}^{+\infty} e^{-i(E - E_0)\tau} \langle 0| \varphi(x(T + \tau))\varphi(x(T)) |0\rangle \, d\tau.$$

Now for a free scalar field,

$$\langle 0| \varphi(x)\varphi(x') |0\rangle = \frac{1}{4\pi^2} \frac{1}{[(t - t' - i\varepsilon)^2 - (\mathbf{x} - \mathbf{x}')^2]} .$$

Clearly, if the trajectory is inertial, the integral in R gives

$$R \propto \int_{-\infty}^{+\infty} e^{-i(E - E_0)\tau} \frac{d\tau}{(\tau - i\varepsilon)^2} \propto \theta(E_0 - E)$$

which vanishes for $E > E_0$. However, for the uniformly accelerated trajectory,

$$\langle 0| \varphi(x(\tau + T))\varphi(x(T)) |0\rangle = - \frac{g^2}{16\pi^2 \sinh^2 g(\frac{1}{2}\tau - i\varepsilon)} .$$

Evaluating the Fourier transform of $\operatorname{cosech}^2 g(\frac{1}{2}\tau - i\varepsilon)$ by the method of residues, we get

$$R \propto \left(\exp \frac{2\pi(E - E_0)}{g} - 1 \right)^{-1} .$$

Thus, in the case of uniformly accelerated detectors this result agrees with (27) which was obtained from the field theory formalism.

However, there are many other cases in which results based on the detector model do not agree with the field theoretic concept of particles. One simple case corresponds to the rotating observer with the trajectory

$$x^i(\tau) = (A\tau, R \cos \alpha\tau, R \sin \alpha\tau, 0)$$

and metric

$$ds^2 = dt^2 - dr^2 - r^2(d\theta + \Omega \, dt)^2 - dz^2, \quad \Omega = \frac{\alpha}{A} .$$

The mode functions corresponding to this observer are given by

$$f_{\omega m k_z} = \frac{1}{2\pi[2(\omega + m\Omega)]^{1/2}} e^{-i\omega t} e^{im\theta} e^{ik_z z} J_m(\alpha r).$$

It is seen directly that $f_{\omega m k_z}$ is a superposition of positive frequency Minkowski modes. Therefore, a uniformly rotating detector should not see any particles in the Minkowski vacuum. However, the correlation function $\langle 0| \varphi(x(\tau + T))\varphi(x(T)) |0\rangle$ in the rotating coordinates is given by

$$G(\tau) = \frac{1}{2\pi^2} \frac{1}{A^2(\tau - i\varepsilon)^2 - 4R^2 \sin^2 \alpha/2(\tau - i\varepsilon)} .$$

The Fourier transform of $G(\tau)$ is nonvanishing (though it cannot be expressed in terms of elementary functions), thereby showing that a rotating detector will be *excited* in the inertial vacuum.

We shall now proceed to discuss field theory in a few specific background spacetimes, keeping the above reservations in mind.

9.3. Field Theory in Robertson–Walker Spacetime

We saw in Chapter 8 that the present-day structure of the universe is very well described by the maximally symmetric metric

$$ds^2 = dt^2 - Q^2(t)\left(\frac{dr^2}{1-kr^2} + r^2(d\theta^2 + \sin^2\theta\, d\varphi^2)\right). \tag{28}$$

The early epochs of the universe are very well described by the above metric with $Q(t) \propto t^{1/2}$ and $k = 0$. Clearly, the metric values in time very rapidly near $t = 0$, since

$$\left(\frac{\dot{Q}}{Q}\right) \sim \frac{1}{2t} \quad (\text{near } t = 0). \tag{29}$$

Thus, we expect significant effects on quantum fields in the early epochs of the universe. In particular, the rapidly varying metric will create various species of particles during this epoch. We would like to estimate the effect of this process in the normal cosmological scenarios.

To make quantitative predictions with a realistic evolution for $Q(t)$, however, is extremely difficult because of the following reasons:

(i) When the background metric is not even asymptotically static, the particle concept loses its meaning.

(ii) All realistic models lead to a singularity at t = 0, thereby spoiling any possibility of initial conditions being prescribed.

Therefore we shall proceed in a less ambitious manner, making some idealizations and approximations.

Let us begin by considering a massive scalar field described by the action

$$S = \int [g^{ik}\, \partial_i\varphi\partial_k\varphi - (m^2 + \xi R)\varphi^2]\sqrt{-g}\, d^4x, \tag{30}$$

in a spacetime described by

$$ds^2 = dt^2 - Q^2(t)\,[dx^2 + dy^2 + dz^2]. \tag{31}$$

(Specializing to the $k = 0$ case, however, is done only to keep the discussion simple. Generalization to $k \neq 0$ is straightforward.) In the action, we have included a curvature coupling of the form $(-\xi R\varphi^2)$. The case $\xi = 0$ corresponds to 'minimal coupling' and $\xi = 1/6$ corresponds to what is called 'conformal coupling'. Other values of ξ do not seem to possess any special significance.

The really important assumption relates to the form of the function $Q(t)$. We shall assume that $Q(t)$ is constant asymptotically: i.e.

$$Q(t) = \begin{cases} Q_1 & \text{for } t < t_1, \\ Q_2 & \text{for } t > t_2. \end{cases} \tag{32}$$

In the above expressions t_1 and t_2 are taken to be large compared to any other time scale in the theory and we may eventually take the limits $t_1 \to -\infty$, $t_2\, +\infty$. Obviously, realistic metrics in the universe do not satisfy these requirements.

Let us try to see, in a semiclassical description, how particles are created by the time-dependent metric. Consider the field equation for φ in this metric,

$$\Box\varphi + (m^2 + \xi R)\varphi = 0. \tag{33}$$

If the scale factor Q were constant, then this equation would possess the normal modes

$$f_{\mathbf{k}} \sim \exp(i\mathbf{k}\cdot\mathbf{x})\exp(\pm i\omega_k t), \tag{34}$$

where

$$\omega_{\mathbf{k}} = \left(\frac{|\mathbf{k}|^2}{Q^2} + m^2\right)^{1/2}. \tag{35}$$

An arbitrary solution, of course, can be constructed by a superposition of these modes. Let us assume that at early times $t < t_1$ (when $Q = Q_1$), $\varphi(\mathbf{x}, t)$ was a plane wave travelling in the direction of $\mathbf{k}$:

$$\varphi(\mathbf{x}, t) = \frac{1}{\sqrt{2\omega_1 Q_1^3}}\exp(i\mathbf{k}\cdot\mathbf{x})\exp(-i\omega_1 t), \tag{36}$$

where

$$\omega_1 \equiv \left(\frac{|\mathbf{k}|^2}{Q_1^2} + m^2\right)^{1/2}. \tag{37}$$

During the expansion $\varphi(\mathbf{x}, t)$ will have a complicated form, described by the equation

$$\frac{1}{Q^3(t)}\frac{\partial}{\partial t}\left(Q^3(t)\frac{\partial\varphi}{\partial t}\right) - \frac{1}{Q^2}\delta^{\mu\nu}\frac{\partial}{\partial x^\mu}\frac{\partial}{\partial x^\nu}\varphi + (m^2 + \xi R)\varphi = 0. \tag{38}$$

This equation can be separated into plane wave modes

$$\varphi(\mathbf{x}, t) = \frac{1}{\sqrt{Q^3(t)}}\exp(i\mathbf{k}\cdot\mathbf{x})h_{\mathbf{k}}(t), \tag{39}$$

where $h_{\mathbf{k}}(t)$ satisfies the equation

$$\left[\frac{d^2}{dt^2} + \frac{k^2}{Q^2} + m^2 - \frac{3}{4}\left(\frac{\dot{Q}}{Q}\right)^2 - \frac{3}{2}\frac{\ddot{Q}}{Q} + \xi R\right]h_{\mathbf{k}}(t) = 0. \tag{40}$$

Comparing (39) with (36) we find that $h_{\mathbf{k}}$ must satisfy the boundary condition

$$h_{\mathbf{k}}(t) = (2\omega_1)^{-1/2}\exp(-i\omega_1 t) \quad \text{for } t \leq t_1. \tag{41}$$

It also follows from the Klein–Gordon equation that the quantity

$(h_\mathbf{k}\dot{h}_\mathbf{k}^* - h_\mathbf{k}^*\dot{h}_\mathbf{k})$ is a constant of motion. The boundary condition fixes the value of this constant to be

$$h_\mathbf{k}\dot{h}_\mathbf{k}^* - h_\mathbf{k}^*\dot{h}_\mathbf{k} = i. \tag{42}$$

Now let us see what happens at very late times. From (40) we see that $h_\mathbf{k}(t)$ will have a form $\propto \exp(\pm\, i\omega_2 t)$ with $\omega_2 = [(|\mathbf{k}|^2/Q_2^2) + m^2]^{1/2}$. Of course, we should take the general solution, which is the superposition of both the positive and negative energy solutions:

$$h_\mathbf{k}(t) = (2\omega_2)^{-1/2}\,[\alpha_\mathbf{k} \exp(-i\omega_2 t) + \beta_\mathbf{k} \exp(+i\omega_2 t)]\ ; \quad t > t_2, \tag{43}$$

where $\alpha_\mathbf{k}$ and $\beta_\mathbf{k}$ are complex numbers. The constancy of the left-hand side of (42) implies that,

$$|\alpha_\mathbf{k}|^2 - |\beta_\mathbf{k}|^2 = 1. \tag{44}$$

Though the detailed form of $\alpha_\mathbf{k}$ and $\beta_\mathbf{k}$ depends on the form of $Q(t)$ [and can be determined only by solving Equation (40)], it is clear that, in general, $\beta_\mathbf{k}$ will not be zero.

This situation has a very simple interpretation even at the semiclassical level. We started with a plane wave travelling along $+\mathbf{k}$ direction. Now, at late times $t > t_2$, we have a superposition

$$\varphi = (2\omega_2 Q_2^3)^{-1/2}\,\{\alpha_\mathbf{k} \exp i(\mathbf{k}\cdot\mathbf{x} - \omega_2 t) + \beta_\mathbf{k} \exp i(\mathbf{k}\cdot\mathbf{x} + \omega_2 t)\} \tag{45}$$

consisting of two waves, one along the $(+\mathbf{k})$ direction and the other along the $(-\mathbf{k})$ direction. The normalization factor $(2\omega_2 Q_2^3)^{-1/2}$ merely represents the kinematic change in wavelength due to expansion. In addition to this trivial fact, we notice the following: the intensity of the component travelling along the $+\mathbf{k}$ direction has increased from 1 (originally) to $|\alpha_\mathbf{k}|^2 = 1 + |\beta_\mathbf{k}|^2$. In addition, a wave of intensity $|\beta_\mathbf{k}|^2$ has been created in the direction of $(-\mathbf{k})$. If one makes the semiclassical identification that the wave intensity is proportional to the mean number of particles, then for every 1-particle present initially, $|\beta_\mathbf{k}|^2$ *pairs* of particles have been created in the process. Thus, one expects

$$\left\{\begin{matrix}\text{Mean number of particles}\\ \text{created}\end{matrix}\right\} = \langle N_\mathbf{k}\rangle = |\beta_\mathbf{k}|^2. \tag{46}$$

We shall now see how these expectations are borne out in a full quantum field theoretic picture.

In this approach, using the Heisenberg picture we shall treat $\varphi(\mathbf{x}, t)$ to be an operator satisfying the evolution equation (for simplicity, we have put $\xi = 1/6$)

$$(\Box + m^2 + \tfrac{1}{6}R)\,\varphi(\mathbf{x}, t) = 0, \tag{47}$$

and the commutation relations

$$[\boldsymbol{\varphi}(\mathbf{x}, t), \boldsymbol{\varphi}(\mathbf{y}, \mathrm{t})] = [\boldsymbol{\pi}(\mathbf{x}, t), \boldsymbol{\pi}(\mathbf{y}, t)] = 0. \tag{48}$$

$$[\boldsymbol{\varphi}(\mathbf{x}, t), \boldsymbol{\pi}(\mathbf{y}, \mathrm{t})] = i\delta(\mathbf{x} - \mathbf{y}); \tag{49}$$

where, as usual, the momentum is defined to be

$$\boldsymbol{\pi}(\mathbf{x}, t) = \frac{\partial \mathscr{L}}{\partial \dot{\varphi}} = \sqrt{-g}\ \dot{\boldsymbol{\varphi}}(\mathbf{x}, t). \tag{50}$$

We can always expand $\varphi(\mathbf{x}, t)$ in terms of a complete set of solutions of the wave equation (47),

$$\varphi(\mathbf{x}, t) = \sum_{\mathbf{k}} \left(\frac{1}{LQ^3} \right)^{1/2} [\mathbf{A}_{\mathbf{k}} \exp(i\mathbf{k} \cdot \mathbf{x}) \cdot h_{\mathbf{k}}(t) + \mathrm{h.c.}]. \tag{51}$$

Here we have assumed that the quantization is in a large box with periodic boundary conditions $\boldsymbol{\varphi}(\mathbf{x} + L\mathbf{n}) = \varphi(\mathbf{x})$. (This somewhat simplifies the mathematics.) The commutation rules in (49) imply that

$$[\mathbf{A}_{\mathbf{k}}, \mathbf{A}_{\mathbf{k}'}] = [\mathbf{A}^{\dagger}_{\mathbf{k}}, \mathbf{A}^{\dagger}_{\mathbf{k}'}] = 0 \tag{52}$$

$$[\mathbf{A}_{\mathbf{k}}, \mathbf{A}^{\dagger}_{\mathbf{k}'}] = \delta_{\mathbf{k}\mathbf{k}'}. \tag{53}$$

Now at early times ($t < t_1$), $\mathbf{A}_{\mathbf{k}}$ is the annihilation operator for particles with momentum $(\mathbf{k}/Q_1)$ and energy $\omega_1 = ((\mathbf{k}/Q_1)^2 + m^2)^{1/2}$. The vacuum state at early times (usually called the 'in' vacuum) is defined by

$$\mathbf{A}_{\mathbf{k}} |0, \mathrm{in}\rangle = 0. \tag{54}$$

Let us assume that the quantum field is in this state $|0, \mathrm{in}\rangle$. Being a state vector in the Heisenberg picture, this state does *not* evolve with time.

The field operator $\boldsymbol{\varphi}(\mathbf{x}, t)$, of course, changes with time. At later times, we know that $h_{\mathbf{k}}(t)$ is given by (43). Substituting (43) into (51) and regrouping the terms, we get, for $t > t_2$,

$$\boldsymbol{\varphi}(\mathbf{x}, t) = \left[\frac{1}{LQ_2^3} \right]^{1/2} \sum_{\mathbf{k}} [\mathbf{a}_{\mathbf{k}} \exp i(\mathbf{k} \cdot \mathbf{x} - \omega_2 t) + \mathrm{h.c.}], \tag{55}$$

where

$$\mathbf{a}_{\mathbf{k}} = \alpha_{\mathbf{k}} \mathbf{A}_{\mathbf{k}} + \beta^*_{\mathbf{k}} \mathbf{A}^{\dagger}_{\mathbf{k}}. \tag{56}$$

From (52), (53), (56) and (44) it can easily be verified that

$$[\mathbf{a}_{\mathbf{k}}, \mathbf{a}_{\mathbf{k}'}] = [\mathbf{a}^{\dagger}_{\mathbf{k}}, \mathbf{a}^{\dagger}_{\mathbf{k}'}] = 0, \quad [\mathbf{a}_{\mathbf{k}}, \mathbf{a}^{\dagger}_{\mathbf{k}'}] = \delta_{\mathbf{k}\mathbf{k}'}. \tag{57}$$

In other words, $(\mathbf{a_k}, \mathbf{a_k^\dagger})$ are the annihilation and creation operators at late times. The state of the field, being independent of time, is still $|0, \text{in}\rangle$. But this state is no longer a vacuum state with respect to the set $(\mathbf{a_k}, \mathbf{a_k^\dagger})$! The number of particles present at late times is easily computed. We get,

$$\langle \mathbf{N}_k \rangle \equiv \langle 0, \text{in}|\, \mathbf{a_k^\dagger}\, \mathbf{a_k}\, |0, \text{in}\rangle = |\beta_{\mathbf{k}}|^2. \tag{58}$$

This agrees completely with the semiclassical description given before. (Compare (46) with (58).)

Let us work out the details for a simple case, first discussed by C. Bernard and A. Duncan in 1977. We shall take our metric to be

$$ds^2 = Q^2(t)\,[dt^2 - dx^2 - dy^2 - dz^2] \tag{59}$$

with

$$Q(t) = A + B \tanh \varrho t. \tag{60}$$

Clearly, $Q(t)$ becomes constant for $|t| \gg \varrho$, so that our Q_1 and Q_2 are given by

$$Q_1 = A - B; \qquad Q_2 = A + B. \tag{61}$$

The advantage of keeping the metric in this form is that the equation for $h_{\mathbf{k}}(t)$, which now reads as (we have taken the case of minimal coupling with $\xi = 0$)

$$\frac{d^2}{dt^2}\, h_{\mathbf{k}}(t) + (\mathbf{k}^2 + Q(t)m^2)h_{\mathbf{k}} = 0, \tag{62}$$

can be solved explicitly. The solution, which behaves as a positive frequency Minkowski mode at early times ($t \to -\infty$) is given by

$$\begin{aligned} U_{\mathbf{k}}^{\text{in}}(t, \mathbf{x}) &= (4\pi\omega_1)^{-1/2} \exp\left\{i\mathbf{k}\cdot\mathbf{x} - i\omega_+ t - \frac{i\omega_-}{\varrho}\ell\text{n}[2\cosh(\varrho t)]\right\} \\ &\times {}_2F_1\left(1 + \left(\frac{i\omega_-}{\varrho}\right);\ \frac{i\omega_-}{\varrho};\ 1 - \frac{i\omega_1}{\varrho};\ \tfrac{1}{2}(1 + \tanh \varrho t)\right) \\ &\xrightarrow[t\to-\infty]{} (4\pi\omega_1)^{-1/2} \exp i(\mathbf{k}\cdot\mathbf{x} - \omega_1 t). \end{aligned} \tag{63}$$

Here, we have used the notation

$$\begin{aligned} \omega_1^2 &= \mathbf{k}^2 + m^2(A - B), \\ \omega_2^2 &= \mathbf{k}^2 + m^2(A + B), \\ \omega_\pm &= \tfrac{1}{2}(\omega_1 \pm \omega_2); \end{aligned} \tag{64}$$

and ${}_nF_m$ are the confluent hypergeometric functions. On the other hand, the modes that behave as positive frequency solutions as $t \to +\infty$ are

$$U_{\mathbf{k}}^{\text{out}} = (4\pi\omega_2)^{-1/2} \exp\left\{i\mathbf{k}\cdot\mathbf{x} - i\omega_+ t - \frac{i\omega_-}{\varrho}\, ln[2\cosh\varrho t)]\right\} \times$$

$$\times\ {}_2F_1\left(1 + \frac{i\omega_-}{\varrho};\ \frac{i\omega_-}{\varrho};\ 1 + \frac{i\omega_2}{\varrho};\ \tfrac{1}{2}(1 - \tanh\varrho t)\right)$$

$$\xrightarrow[t\to+\infty]{} (4\pi\omega_2)^{-1/2} \exp i(\mathbf{k}\cdot\mathbf{x} - \omega_2 t) \qquad (65)$$

Since $U_{\mathbf{k}}^{\text{in}}$ and $U_{\mathbf{k}}^{\text{out}}$ are not the same, we shall have nonzero $\beta_{\mathbf{k}}$ coefficients. Writing

$$U_{\mathbf{k}}^{\text{in}}(t, \mathbf{x}) = \alpha_{\mathbf{k}} U_{\mathbf{k}}^{\text{out}}(t, \mathbf{x}) + \beta_{\mathbf{k}} U_{\mathbf{k}}^{\text{out}}(t, \mathbf{x})^* \qquad (66)$$

and using the linear transformation formulae for hypergeometric functions, we obtain,

$$\alpha_{\mathbf{k}} = \left(\frac{\omega_2}{\omega_1}\right)^{1/2} \frac{\Gamma(1 - i\omega_1/\varrho)\,\Gamma(-i\omega_2/\varrho)}{\Gamma(-i\omega_+/\varrho)\,\Gamma(1 - i\omega_+/\varrho)}, \qquad (67)$$

$$\beta_{\mathbf{k}} = \left(\frac{\omega_2}{\omega_1}\right)^{1/2} \frac{\Gamma(1 - i\omega_1/\varrho)\,\Gamma(i\omega_2/\varrho)}{\Gamma(i\omega_-/\varrho)\,\Gamma(1 - i\omega_-/\varrho)}. \qquad (68)$$

In particular,

$$|\beta_{\mathbf{k}}|^2 = \frac{\sinh^2(\pi\omega_-/\varrho)}{\sinh(\pi\omega_1/\varrho)\,\sinh(\pi\omega_2/\varrho)}, \qquad (69)$$

which represents the number of particles created in the expansion.

Example 9.6. Let us define an 'out' vacuum and an 'out' one-pair state by the relations

$$a_{\mathbf{k}}\,|0, \text{out}\rangle = 0, \qquad |1_{\mathbf{k}}, \text{out}\rangle \equiv a_{\mathbf{k}}^\dagger a_{-\mathbf{k}}^\dagger\,|0, \text{out}\rangle.$$

Consider the probability of finding a pair with momentum ($\mathbf{k}$, $-\mathbf{k}$). We have

$$|\langle 1_{\mathbf{k}}, \text{out}\mid 0, \text{in}\rangle|^2 = \left|\left\{\frac{\beta_{\mathbf{k}}^*}{\alpha_{\mathbf{k}}^*}\langle 0, \text{out}\mid 0, \text{in}\rangle\right\}\right|^2 = \left|\frac{\beta_{\mathbf{k}}}{\alpha_{\mathbf{k}}}\right|^2 |\langle 0, \text{out}\mid 0, \text{in}\rangle|^2.$$

Similarly, the amplitude for a state with n_{k_1} pairs of momentum $\mathbf{k}_1$, n_{k_2} pairs of momentum $\mathbf{k}_2$, etc., is

$$\langle\{n_{\mathbf{k}}\}, \text{out}\mid 0, \text{in}\rangle = \prod_{\mathbf{k}}\left[\frac{\beta_{\mathbf{k}}^*}{\alpha_{\mathbf{k}}^*}\right]^{n_{\mathbf{k}}} \langle 0, \text{out}\mid 0, \text{in}\rangle.$$

Notice that

$$1 = \langle 0, \text{in}\mid 0, \text{in}\rangle = \sum_{\{n_{\mathbf{k}}\}} \langle 0, \text{in}\mid\{n_{\mathbf{k}}\}, \text{out}\rangle\langle\{n_{\mathbf{k}}\}, \text{out}\mid 0, \text{in}\rangle$$

$$= |\langle 0, \text{out}\mid 0, \text{in}\rangle|^2 \prod_{\mathbf{k}} \sum_{n=0}^{\infty} \left|\frac{\beta_{\mathbf{k}}}{\alpha_{\mathbf{k}}}\right|^{2n}$$

$$= |\langle 0, \text{out} \mid 0, \text{in}\rangle|^2 \prod_{\mathbf{k}} |\alpha_{\mathbf{k}}|^2 \quad (\text{since } |\alpha_{\mathbf{k}}|^2 - |\beta_{\mathbf{k}}|^2 = 1).$$

Therefore,

$$|\langle 0, \text{out} \mid 0, \text{in}\rangle|^2 = \prod_{\mathbf{k}} |\alpha_{\mathbf{k}}|^{-2}.$$

Using this result, we can write the probability for observing $n_{\mathbf{k}}$ pair of particles in the mode ($\mathbf{k}$, $-\mathbf{k}$) as

$$P_{n_{\mathbf{k}}} = \left| \frac{\beta_{\mathbf{k}}}{\alpha_{\mathbf{k}}} \right|^{2n_{\mathbf{k}}} |\alpha_{\mathbf{k}}|^{-2}.$$

Example 9.7. A time-dependent background metric need not always create particles. This would be the case if there exists some additional symmetry, like the conformal invariance, in the theory. We know that (see Section 6.11) the equation

$$(\Box + m^2 + \xi R)\varphi = 0$$

is conformally invariant for $m = 0$, $\xi = \frac{1}{6}$. Equation (40) now becomes

$$\left[\frac{d^2}{dt^2} + \frac{|\mathbf{k}|^2}{Q^2} + \frac{1}{4}\left(\frac{\dot{Q}}{Q}\right)^2 - \frac{1}{2}\frac{\ddot{Q}}{Q} \right] h_{\mathbf{k}}(t) = 0,$$

where we have used the fact that $R = 6(\dot{Q}^2 Q^{-2} + \ddot{Q} Q^{-1})$. This is solved by the function

$$h_{\mathbf{k}}(t) = [2\omega(t)]^{-1/2} \exp(-i\int^t \omega(t')\, dt'),$$

with $\omega(t) = |\mathbf{k}|/Q(t)$. Clearly, a positive frequency solution at early times evolves into a positive frequency solution at late times, making $\beta = 0$. Thus, there is no particle creation.

Even in the case of Robertson–Walker models with $k \neq 0$, there will not be any particle creation if the field φ is conformally invariant. This can easily be seen from the fact that

$$ds^2 = dt^2 - Q^2(t)\gamma_{\mu\nu}\, dx^\mu\, dx^\nu = W^2(\eta)\,[d\eta^2 - \gamma_{\mu\nu}\, dx^\mu\, dx^\nu]$$

(where $\eta = \int dt/Q(t)$). If the system is conformally invariant, then the physics must be the same in the two metrics, ds^2 and $Q^{-2}\, ds^2$. In the second case, because of the static nature of the metric, there can be no particle creation. If follows that no particle creation can take place in the Robertson–Walker universe if φ obeys a conformally invariant equation. □

The above analysis has been carried out in the idealized spacetimes with asymptotically well-defined 'in' and 'out' states. It seems 'reasonable' to assume that the particles were actually created between t_1 and t_2. In that case, it would be gratifying to define a 'particle number function $N_{\mathbf{k}}(t)$' which starts at zero for $t < t_1$ and reaches $|\beta_{\mathbf{k}}|^2$ at $t = t_2$. Unfortunately, deeper analysis shows that such a simple-minded idea will not work, except in very special cases.

The problems are twofold. We have already shown in the previous section that the particle number is not a generally covariant concept. Thus, in a general time-dependent metric, a unique particle number function $N_{\mathbf{k}}(t)$ does not exist. Various functions of this type can be defined for various physical situations. None of them will have any deep significance.

The second reason why such a function loses credibility stems from the uncertainty principle. If the average creation rate is C, then the uncertainty in $N(t)$ (due to creation) in a small time interval Δt is

$$\Delta N_{\text{creation}} \sim C\,\Delta t. \tag{70}$$

On the other hand, if the particles have mass m, the uncertainty principle ($\Delta E\,\Delta t \sim 1$) demands

$$\Delta N_{QM} \sim (m\,\Delta t)^{-1}. \tag{71}$$

Thus,

$$\Delta N_{\text{total}} \sim (m\,\Delta t)^{-1} + C\,\Delta t. \tag{72}$$

Clearly, this has a minimum value $2(C/m)^{1/2}$ when $\Delta t \sim (mC)^{-1/2}$. In other words, $N(t)$ will have an inherent uncertainty and cannot be represented by a simple classical function. Only when (i) the creation rate is low (small C) and (ii) the mass of the particle is high, can a useful $N_{\mathbf{k}}(t)$ be defined. In such circumstances we can resort to adiabatic approximations or to an instantaneous diagonalization of the Hamiltonian. The interested reader can consult the literature cited at the end of the chapter for a detailed discussion of this topic.

9.4. Field Theory in de Sitter Spacetime

De Sitter spacetime has been the subject of extensive analysis in the recent years. Besides possessing a large class of symmetries, this spacetime metric has been invoked in a so-called 'inflationary scenario' which we shall consider in the next chapter. Nevertheless, it is not quite clear whether the physics of quantum field theory in de Sitter spacetime is well understood. We shall discuss only the simplest features, which themselves defy an intuitive understanding.

Before we proceed to discuss de Sitter spacetime, it is probably worthwhile to point out the distinction between 'particle creation' and 'particle detection'. It seems clear enough that there should be no particle creation in the flat Minkowski spacetime. If we use the flat Lorentzian coordinates, and take the vacuum state to be $|0, \text{in}\rangle$ and $|0, \text{out}\rangle$ as $t \to \pm\infty$, then clearly,

$$|0, \text{in}\rangle = |0, \text{out}\rangle. \tag{73}$$

However, this does not mean that particle detectors will not click in this spacetime. The behaviour of particle detectors depend on the state of motion of the detector. In fact, we have seen that an accelerated detector will see a nonzero particle spectrum in the Minkowski vacuum $|0, \text{in}\rangle$. Mathematically, the spectrum seen by a detector depends on the Fourier transform of the correlation function (see Example 9.5)

$$P(E) = \int_{-\infty}^{+\infty}\int_{-\infty}^{+\infty} \mathrm{e}^{-iE(\tau - \tau')}\,\langle 0|\,\varphi(x(\tau))\varphi(x(\tau'))\,|0\rangle\,\mathrm{d}\tau\,\mathrm{d}\tau'$$

$$\equiv \int_{-\infty}^{+\infty} \int_{-\infty}^{+\infty} e^{-iE(\tau - \tau')} G(\tau, \tau') \, d\tau \, d\tau', \tag{74}$$

which can be nonvanishing for many $x^i(\tau)$ trajectories. Thus, even without particle creation, there can be particle detection.

This fact leads to an interesting phenomenon in the Robertson–Walker spacetimes, when we consider conformally invariant fields. Robertson–Walker spacetimes can be expressed in conformally flat form with the metric (see Chapter 12, Example 12.13):

$$g_{ik} = \Omega^2(x)\eta_{ik}. \tag{75}$$

Therefore, the conformally invariant field equation

$$(\Box + \tfrac{1}{6}R)\, \varphi(\mathbf{x}, T) = 0 \tag{76}$$

can be at once solved in the coordinate system $(T, \mathbf{x})$ in which the line element has the form

$$ds^2 = \Omega^2(\mathbf{x}, T)\, [dT^2 - |d\mathbf{x}|^2], \tag{77}$$

giving rise to the decomposition

$$\varphi(\mathbf{x}, T) = \sum_{\mathbf{k}} \frac{1}{\Omega(\mathbf{x}, T)} \, [a_{\mathbf{k}} \, e^{i(\mathbf{k} \cdot \mathbf{x} - \omega_k T)} + \text{h.c.}] \, . \tag{78}$$

Using these mode functions we can define a vacuum (usually called the 'conformal vacuum') by

$$\mathbf{a}_{\mathbf{k}} \, |0, \text{conf vac}\rangle \equiv \mathbf{a}_{\mathbf{k}} \, |0, c\rangle = 0. \tag{79}$$

Clearly, this vacuum state will remain a faithful vacuum state all through the evolution. [We are only restating the result of Example 9.7; in a conformally flat spacetime no conformally invariant particles are created.]

However, the story is different with respect to geodesic observers in this spacetime. The geodesic (or comoving) observer uses the familiar, isotropic coordinate system in which the metric takes the form

$$ds^2 = dt^2 - Q^2(t)\, \gamma_{ij} \, dx^i \, dx^j. \tag{80}$$

What these observers will see depends on the Fourier transform of the Green's function G with respect to the time coordinate t. Since t and T are related by a complicated coordinate transformation, a comoving geodesic observer will detect particles in the $|0, c\rangle$ state.

For example, consider the $k = 0$ Robertson–Walker spacetime written in two forms

$$\begin{aligned} ds^2 &= \Omega^2(T)\, [dT^2 - dX^2 - dY^2 - dZ^2] \\ &= dt^2 - Q^2(t)\, [dX^2 + dY^2 + dZ^2], \end{aligned} \tag{81}$$

where

$$t \equiv t(T) \equiv \int \Omega(T)\,\mathrm{d}T; \qquad \mathrm{Q}(t) = \Omega(T(t)). \tag{82}$$

The Green's function, in terms of $(T, \mathbf{X})$ coordinate, is easily obtained by transforming from flat space, since the scalar field $\varphi(X)$ transforms to $(\varphi(X)/\Omega(X))$, we have

$$G(X, X') = \Omega^{-1}(X)\, G_{\text{flat}}(X, X')\Omega^{-1}(X'). \tag{83}$$

We are interested in $G(X, X')$ when X and X' lie in the geodesic $\mathbf{X}$ = const, so that $X = (T, \mathbf{X})$ and $X' = (T', \mathbf{X})$. Then,

$$G(T, T') = \Omega^{-1}(T)\,\Omega^{-1}(T')\,\frac{1}{4\pi^2(T - T' - i\varepsilon)^2}\,. \tag{84}$$

The geodesic detector will therefore see a particle spectrum

$$P(E) = \frac{1}{4\pi^2}\int_{-\infty}^{+\infty}\int_{-\infty}^{+\infty} \frac{\exp\{-iE\int_{T'}^{T}\Omega(\alpha)\,\mathrm{d}\alpha\}}{(T - T' - i\varepsilon)^2}\,\mathrm{d}T\,\mathrm{d}T'. \tag{85}$$

Because of the Ω factor this will not generally vanish. Thus a geodesic detector will 'see' particles in the conformal vacuum.

Normally, the conformal vacuum $|0, c\rangle$ does not possess any other special properties. However, the situation is different in the de Sitter spacetime to which we shall now turn our attention.

The de Sitter manifold is best defined to be the four-dimensional hyperboloid

$$Z_0^2 - Z_1^2 - Z_2^2 - Z_3^2 - Z_4^2 = -\alpha^2 \tag{86}$$

in five-dimensional 'Minkowski space' with the metric

$$\mathrm{d}s^2 = \mathrm{d}Z_0^2 - \mathrm{d}Z_1^2 - \mathrm{d}Z_2^2 - \mathrm{d}Z_3^2 - \mathrm{d}Z_4^2. \tag{87}$$

Clearly the manifold possesses all the symmetries of 'Lorentz transformations' in five dimensions. This is represented by the SO(1, 4) group with ten parameters. This group is usually called the 'de Sitter group'. In order to do anything in the manifold, we must introduce coordinate patches. Various coordinate systems are possible, leading to different quantization scenarios.

The most useful way of introducing coordinates into the de Sitter spacetime is done by choosing

$$Z_0 = \alpha \sinh\left(\frac{t}{\alpha}\right) + \frac{1}{2\alpha}\,\mathrm{e}^{t/\alpha}|\mathbf{x}|^2,$$

$$Z_4 = \alpha \cosh\left(\frac{t}{\alpha}\right) - \frac{1}{2\alpha}\,\mathrm{e}^{t/\alpha}|\mathbf{x}|^2,$$

$$Z_\mu = \mathrm{e}^{t/\alpha}\,x_\mu \quad (\mu = 1, 2, 3), \quad -\infty < t, \quad x_\mu < \infty. \tag{88}$$

In terms of t and $\mathbf{x}$, the metric becomes

$$ds^2 = dt^2 - \exp\left(\frac{2t}{\alpha}\right)[dx_1^2 + dx_2^2 + dx_3^2]. \tag{89}$$

This is a spacetime of constant curvature, with the curvature scalar given by

$$R = \frac{12}{\alpha^2}, \tag{90}$$

and is a solution to Einstein's equations for the vacuum with a cosmological constant. We can also write

$$ds^2 = \left(\frac{\alpha^2}{T^2}\right)[dT^2 - |d\mathbf{x}|^2], \tag{91}$$

where

$$T = -\alpha \exp\left(-\frac{t}{\alpha}\right), \quad -\infty < T < 0. \tag{92}$$

The coordinate system in (88) covers only half the de Sitter manifold ($Z_0 + Z_4 > 0$). The other half can be covered by changing the sign of Z_0 and Z_4 in the right-hand side of (88). This would allow us to consider the full range of values for $T : (-\infty < T < \infty)$. Consider the scalar wave equation for the wave modes $U_{\mathbf{k}}$,

$$(\Box + m^2 + \xi R)U_{\mathbf{k}} = 0 \tag{93}$$

in the coordinate system of Equation (91). Putting

$$U_{\mathbf{k}} = (2\pi)^{-3/2}\Omega^{-1}(T)\,\chi_{\mathbf{k}}(T) \tag{94}$$

we get

$$\frac{d^2\chi_{\mathbf{k}}}{dT^2} + \{\mathbf{k}^2 + \Omega^2(T)[m^2 - (\xi - \tfrac{1}{6})R]\}\,\chi_{\mathbf{k}} = 0. \tag{95}$$

This equation has the exact solution

$$\chi_{\mathbf{k}}(T) = \tfrac{1}{2}(\pi T)^{1/2} H_\nu^{(2)}(kT), \tag{96}$$

where

$$\nu^2 = \frac{9}{4} - 12(m^2R^{-1} + \xi) \tag{97}$$

and $H_\nu^{(2)}(x)$ is the Hankel function of the second kind. From the properties of the Hankel function, we can see that

$$\chi_{\mathbf{k}}(T) \to \exp(-i|\mathbf{k}|T) \tag{98}$$

at both $T \to -\infty$ and $T \to +\infty$. In other words, the definition of creation and annihilation operators coincide asymptotically. There is no particle creation in the de Sitter spacetime when the quantization is performed in the conformal coordinate system.

Notice that this, itself, is a strong result. It is quite trivial that conformally invariant fields are not excited in conformally flat spacetime. But we have *not* put $m = 0$ or $\xi = \frac{1}{6}$ in (93)! That the de Sitter spacetime possesses a vacuum state which is faithful to general quantum fields arises from the high degree of symmetry possessed by the spacetime. The vacuum state defined in this coordinate system is invariant under the de Sitter group of motions.

A more interesting result is obtained if we consider the response of geodesic detectors in this conformal vacuum. The geodesic detectors use the coordinate system in (89). The Green's function between two events $X = (T, \mathbf{X})$ and $X' = (T', \mathbf{X})$ in the geodesic $\mathbf{X}$ = constant is given [see Equation (83)] by

$$G(X, X') = \frac{-TT'}{4\pi^2\alpha^2[(T - T' - i\varepsilon)^2]} . \tag{99}$$

The power spectrum seen by these detectors can be calculated with the help of (85). We obtain

$$\begin{aligned} P(E) &= \frac{1}{16\pi^2\alpha^2}\int_{-\infty}^{+\infty}\int_{-\infty}^{+\infty}\frac{e^{-iE(t-t')}}{\sin^2[(i(t-t')/2\alpha)+\varepsilon]}\,d\left(\frac{t+t'}{2}\right)d(t-t') \\ &= \frac{1}{16\pi^2\alpha^2}\int_{-\infty}^{+\infty}\sum_{n=-\infty}^{\infty}\int_{-\infty}^{+\infty}dx\,\frac{e^{-iEx}}{[(ix/2\pi\alpha)+\varepsilon-n]^2}\,d\left(\frac{t+t'}{2}\right) \\ &= \int\left(\frac{E}{2\pi}\right)[\exp(2\pi\alpha E)-1]^{-1}\,dT. \end{aligned} \tag{100}$$

In other words, the detection rate is Planckian

$$\frac{P(E)}{(\text{Unit time})} = \left(\frac{E}{2\pi}\right)\frac{1}{(e^{2\pi\alpha E} - 1)} , \tag{101}$$

with a temperature

$$T = \frac{1}{(2\pi\alpha)} . \tag{102}$$

It must be noted that geodesic observers using the coordinates in (89) exist only in part of the de Sitter manifold and are separated by a particle horizon from the other regions. The situation is thus very similar to that of accelerated observers with Minkowski spacetime. This has led to the popular myth that the presence of event or particle horizons is necessary for 'particle detection'. This, of course, is not true. Particle horizon does not play any fundamental role in our derivation above. In fact, as we showed in (85), similar phenomenon can exist in any Robertson–Walker model. What makes the result for the de Sitter spacetime important is the presence of an invariance group assuring

(i) the lack of any particle creation and (ii) a special status to the conformal vacuum. For the sake of completeness we mention two other coordinates frequently used in de Sitter spacetime. Taking

$$\begin{aligned} Z_0 &= \alpha \sinh (t/\alpha), \\ Z_1 &= \alpha \cosh (t/\alpha) \cos \chi, \\ Z_2 &= \alpha \cosh (t/\alpha) \sin \chi \cos \theta, \\ Z_3 &= \alpha \cosh (t/\alpha) \sin \chi \sin \theta \cos \varphi, \\ Z_4 &= \alpha \cosh (t/\alpha) \sin \chi \sin \theta \sin \varphi. \end{aligned} \tag{103}$$

the metric can be written as

$$ds^2 = dt^2 - \alpha^2 \cosh^2 \left(\frac{t}{\alpha}\right) [d\chi^2 + \sin^2\chi(d\theta^2 + \sin^2 \theta\, d\varphi^2)], \tag{104}$$

where the coordinate ranges are

$$-\infty < t < \infty, \quad 0 \leq r \leq \pi, \quad 0 \leq \theta \leq \pi, \quad 0 \leq \varphi < 2\pi. \tag{105}$$

Quantum theory in these coordinates is not easy to develop. (The interested reader is referred to original literature at the end of this chapter.)

Lastly, the de Sitter spacetime can be represented in static form by introducing the coordinates

$$\begin{aligned} Z_0 &= (\alpha^2 - r^2)^{1/2} \sinh(t/\alpha), \\ Z_1 &= (\alpha^2 - r^2)^{1/2} \cosh(t/\alpha), \\ Z_2 &= r \sin \theta \cos \varphi, \\ Z_3 &= r \sin \theta \cos \varphi, \\ Z_4 &= r \cos \theta, \quad 0 \leq r < \infty. \end{aligned} \tag{106}$$

(Note that this also covers only half the manifold $Z_0 + Z_1 > 0$.) In these coordinates the metric will read as

$$ds^2 = \left(1 - \frac{r^2}{\alpha^2}\right) dt^2 - \frac{dr^2}{(1 - r^2/\alpha^2)} - r^2(d\theta^2 + \sin^2 \theta\, d\varphi^2). \tag{107}$$

The metric has a coordinate singularity at $r = \alpha$. An observer at $r = 0$ will have an event horizon at $r = \alpha$. The vacuum state defined in this coordinate system is very peculiar compared to the conformal vacuum even though the comoving observer will see a thermal spectrum in this vacuum as well. It turns out that this vacuum is not even translation invariant – let alone de Sitter invariant! Qualitatively we can see this effect arising from the fact that, in the above coordinate system, the event horizon is not invariant under the translation $r \rightarrow r + a$. Once again, we refer the reader to original literature for details.

9.5. Euclideanization and the Thermal Green's Functions

In analysing the field theory in flat spacetime, we found that analytic continuation to imaginary values of the time coordinate serves as a useful tool. People are often tempted to try out this method in the curved spacetime. It should, however, be kept in mind that the analytic continuation is not guaranteed to succeed in an arbitrary spacetime.

The reason for this is twofold:

(i) In an arbitrary curved spacetime, g_{ik}'s will be functions of time. There is no assurance that $g_{ik}(-i\tau, \mathbf{x})$ will be a *real* function of time, let alone a positive definite function. In fact, there exist Lorentzian metrics which will not have positive definite Euclidean extensions in any coordinate system, whatsoever.

(ii) The simple looking prescription $t \to -i\tau$ very often changes the topology of the spacetime drastically. It is not clear whether the results will then have any intrinsic validity.

In spite of the above facts, analytic continuation techniques, used with caution and common sense, can illustrate some of the previously derived results in a simple fashion. We shall briefly indicate this procedure.

Recall from Chapter 4 that quantum theory at a finite temperature is developed using the density matrix

$$\varrho = \mathrm{e}^{-\beta H}. \tag{108}$$

For example, the expectation value of an operator $\hat{\mathbf{A}}$ involves computing

$$\langle \mathbf{A} \rangle \equiv \sum_E \frac{\mathrm{e}^{-\beta E} \langle E |\, \hat{\mathbf{A}}\, | E \rangle}{\Sigma E\ \mathrm{e}^{-\beta E}} = \frac{\mathrm{Tr}\ \varrho \mathbf{A}}{\mathrm{Tr}\ \varrho} \ . \tag{109}$$

Of central importance are the finite temperature Green's functions, defined to be (for simplicity we consider the case with $G(x, x') = G(x', x)$)

$$G_\beta(x, x') \equiv \langle \varphi(x) \varphi(x') \rangle \,. \tag{110}$$

These Green's functions have an important property. They are periodic in the imaginary time coordinate (see Chapter 4, Equation (190))

$$G_\beta(t, \mathbf{x}; t', \mathbf{x}') = G_\beta(t + i\beta, \mathbf{x}; t', \mathbf{x}'). \tag{111}$$

It can often be shown, without much computation, that a particular Green's function is periodic in imaginary time. This then establishes the thermal nature of the particle spectrum.

As a simple example, consider the flat spacetime in the uniformly accelerated frame

$$-\mathrm{d}s^2 = -g^2 \xi^2\, \mathrm{d}t^2 + \mathrm{d}\xi^2 + \mathrm{d}y^2 + \mathrm{d}z^2. \tag{112}$$

Transforming to a complex time coordinate $t = -i\tau$, the metric becomes

$$-ds^2 = ds^2_{\text{Euclidean}} = (g^2\xi^2\, d\tau^2 + d\xi^2) + dy^2 + dz^2. \tag{113}$$

The quantity in the bracket has the same form as the two-dimensional plane in polar coordinates ($dr^2 + r^2\, d\theta^2$). Regularity at $\xi = 0$ demands that τ must be periodic with the period

$$P_\tau = \left(\frac{2\pi}{g}\right). \tag{114}$$

Since the periodicity of the Green's function in τ directly gives $\beta = 1/T$ we obtain the temperature

$$T = \frac{g}{2\pi}. \tag{115}$$

This agrees with the result obtained above (see (27)).

Example 9.8. We have seen that the Green's function in the uniformly accelerated frame is given (see Example 9.5) by

$$G(\tau, 0) = -\frac{g^2}{16\pi^2}\frac{1}{\sinh^2 g(\frac{1}{2}\tau - i\varepsilon)}.$$

Making the analytic continuation to imaginary time τ_E ($\tau = -i\tau_E$) we get,

$$G(-i\tau_E, 0) \equiv G^E(\tau_E, 0) = +\frac{g^2}{16\pi^2}\frac{1}{\sin^2 g(\tau_E/2)}.$$

We have taken $\varepsilon = 0$, since there is no ambiguity in the Euclidean domain. Clearly $G^E(\tau_E, 0)$ is periodic in τ_E with the period $(2\pi/g)$. □

Example 9.9. The transformation from the inertial frame to the uniformly accelerated frame is given by

$$\begin{aligned} ds^2 &= dT^2 - dX^2 - dY^2 - dZ^2 \\ &= g^2\xi^2\, dt^2 - d\xi^2 - dy^2 - dz^2 \end{aligned}$$

with

$$T = \xi \sinh gt, \qquad X = \xi \cosh gt.$$

Under analytic continuation, $T = -iT_E$, $t = -it_E$, these equations go over to

$$T_E = \xi \sin gt_E; \quad X = \xi \cos gt_E.$$

Notice that we have tacitly identified the points (t_E, ξ) and $(t_E + (2\pi/g)n, \xi)$ for $n = 0, \pm 1, \ldots$. In other words, the topology of the manifold (which was R^4 in the (T_E, X, Y, Z) coordinates) has changed to $S^1 \times R^3$ in the (t_E, ξ, y, z) coordinates. □

As a second example, consider the de Sitter line element written in the form

$$-ds^2 = -dt^2 + \cosh^2\left(\frac{t}{\alpha}\right)[d\chi^2 + \sin^2\chi(d\theta^2 + \sin^2\theta\, d\varphi^2)]. \tag{116}$$

Making an analytic continuation (to $t \to -i\tau$), we obtain the Euclidean metric

$$ds_E^2 = -ds^2 = -d\tau^2 + \cos^2\left(\frac{\tau}{\alpha}\right)[d\chi^2 + \sin^2\chi(d\theta^2 + \sin^2\theta\, d\varphi^2)]. \quad (117)$$

Clearly, the τ coordinate is periodic with a period of

$$P_\tau = 2\pi\alpha; \quad (118)$$

leading to a temperature

$$T = \frac{1}{2\pi\alpha}, \quad (119)$$

which agrees with our previous result (102). The present example also illustrates the need for a 'clever' choice of coordinate system before analytic continuation is performed. If we had used the metric in the form

$$-ds^2 = -dt^2 + e^{2t/\alpha}[dx^2 + dy^2 + dz^2] \quad (120)$$

an analytic continuation would have led to a complex metric. In other words, analytic continuation is not a covariant procedure.

9.6. Field Theory in the Black-Hole Spacetime

From a naïve point of view, we may have expected a static spacetime, like the Schwarzschild spacetime, not to participate in any particle creation. Such a reasoning will be incorrect. It is true that the Schwarzschild geometry, expressed as

$$ds^2 = \left(1 - \frac{2M}{r}\right)dt^2 - \frac{dr}{(1 - (2M/r))} - r^2(d\theta^2 + \sin^2\theta\, d\varphi^2) \quad (121)$$

has metric coefficients which are independent of t. However, t is a timelike coordinate only in the region $r > 2M$. For $r < 2M$, it is the r coordinate which is timelike. The metric definitely depends on r and, hence, is not 'static' in the $r < 2M$ region.

Because of the dubious nature of the t coordinate it is better to introduce the Kruskal coordinates (cf. Section 7.5.1)

$$\frac{u}{4M} = \left(\frac{r}{2M} - 1\right)^{1/2} e^{r/4M}\cosh\left(\frac{t}{4M}\right),$$

$$\frac{v}{4M} = \left(\frac{r}{2M} - 1\right)^{1/2} e^{r/4M}\sinh\left(\frac{t}{4M}\right), \quad (122)$$

in terms of which the metric becomes

$$ds^2 = \frac{2M}{r}\, e^{-r/2M}(dv^2 - du^2) - r^2(d\theta^2 + \sin^2\theta\, d\varphi^2). \quad (123)$$

Here r is supposed to be an implicit function of u, v given by

$$\left(\frac{r}{2M} - 1\right) e^{-r/2M} = u^2 - v^2. \tag{124}$$

As explained in Section 7.5, the coordinate v is a global time coordinate. Since the metric does depend on v, the spacetime is not static globally.

From a realistic point of view, a black hole is formed from stellar collapse and thus the Schwarzschild metric is only applicable in the outside region. In the region occupied by the collapsing matter, the metric is definitely time dependent. Thus, there is no *a priori* reason to expect the Schwarzschild metric not to create particles.

It turns out that the metric actually does create particles. Suppose that a spherical star begins to collapse at $t = 0$. Assume that a distant particle detector (i.e. a detector in the asymptotically flat region $r \to \infty$) does not see any particles at $t \leq 0$. We can then show, following Hawking, that at late times ($t \to +\infty$), the asymptotic particle detectors will see a thermal spectrum of particles with temperature

$$T = \frac{1}{8\pi M}. \tag{125}$$

We shall try to indicate how this result comes about.

The most elegant way of deriving this result uses the analytic continuation technique which was discussed in the previous section. Clearly, an analytic continuation of $t \to -i\tau$ in (121) does not lead to a positive definite metric because of the $(1 - (2M/r)$ factor. However, a positive definite metric can be obtained by analytically continuing the v coordinate in (123) to $-iv$. The metric becomes

$$-\mathrm{d}s^2 = \mathrm{d}s_E^2 = \left(\frac{2M}{r}\right) e^{r/2M} (\mathrm{d}u^2 + \mathrm{d}v^2) + r^2(\mathrm{d}\theta^2 + \sin^2\theta\, \mathrm{d}\varphi^2). \tag{126}$$

A comparison with (122) shows that the analytic continuation has the effect of using $\tau = it$, and the transformations

$$u = 4M\left[\frac{r}{2M} - 1\right]^{1/2} \cos\left(\frac{\tau}{4M}\right), \tag{127}$$

$$v = 4M\left[\frac{r}{2M} - 1\right]^{1/2} \sin\left(\frac{\tau}{4M}\right). \tag{128}$$

Clearly the imaginary time coordinate τ has a periodicity of $8\pi M$. Identifying this with $\beta = 1/T$, we recover the result of (125).

What is the physical reason for this particle creation? Consider a spherical wave front that falls on a *static* star. (Let us assume that the star is completely transparent; in other words, the stellar material does not interact with the field. This assumption is, however, not necessary for what follows.) While falling towards the star, the waves will be blueshifted. After passing through the star the wave still re-emerge and travel to $r \to \infty$ as an outgoing wave. During this phase the waves will be redshifted. Clearly, as long as the star is static the redshift and the blueshift will cancel out. The situation is quite different, however, if the star is collapsing during this process. As the wave traverses through the star, the star will be contracting. This has the effect of producing larger redshift when the wave comes out than the blueshift that occurred on the inward trip. In other words, the wave modes will come out with a different frequency than the ones that went in. This mismatch between the incoming and outgoing wave modes appears as created particles. Let us now make the above observation more quantitative. The wave equation for the scalar field,

$$\Box\varphi = 0 \tag{129}$$

can be separated in the Schwarzschild coordinates (t, r, θ, φ) in the form of mode functions,

$$g_{\omega lm} \propto \frac{R_{\omega l}(r)}{r} Y_{lm}(\theta, \varphi)\, \mathrm{e}^{-i\omega t}, \tag{130}$$

where $Y_{lm}(\theta, \varphi)$ are the spherical harmonics and $R_{\omega l}(r)$ satisfies the radial equation

$$\frac{\mathrm{d}^2 R_{\omega l}}{\mathrm{d}r^{*2}} + \{\omega^2 - [l(l+1)r^{-2} + 2Mr^{-3}]\,(1 - 2Mr^{-1})\}\, R_{\omega l} = 0 \tag{131}$$

with

$$r^* = r + 2M \ln\left[\frac{r}{2M} - 1\right] . \tag{132}$$

Notice that the effective potential in (131) vanishes as $r^* \to -\infty$ and $r^* \to +\infty$ (i.e. for $r \to 2M, \infty$) the solutions in those regions will be of the form $\mathrm{e}^{\pm i\omega r^*}$. Thus the solutions $g_{\omega lm}$ will be those obtained from a superposition of 'ingoing' and 'outgoing' modes,

$$\left(\frac{1}{r}\, \mathrm{e}^{-i\omega V}\, Y_{lm}\right) \text{ and } \left(\frac{1}{r}\, \mathrm{e}^{-i\omega U}\, Y_{lm}\right),$$

where

$$U = t - r^*; \qquad V = t + r^*. \tag{133}$$

Let us denote a complete set of wave modes at early times ($t \to -\infty$, before the collapse has begun) by $f_{\omega lm}(r, \theta, \varphi, t)$. This is chosen to be a positive frequency mode at early times. A wave packet can be constructed out of the incoming waves at early times. Writing the field in terms of the creation and annihilation operators

$$\varphi = \sum_{lm} \int (a_{\omega lm} f_{\omega lm} + \text{h.c.})\, d\omega, \tag{134}$$

we can define the 'in' vacuum to be the state satisfying the condition

$$a_{\omega lm} \, |0, \text{in}\rangle = 0. \tag{135}$$

The behaviour of the wave modes are schematically shown in Figure 9.2. Consider the three rays numbered 1, 2, 3. Ray 1 converges to the spherical collapsing ball but does not succeed in escaping to $r = +\infty$ again. It is trapped by the event horizon and hits the singularity. Ray 2 is the critical null ray which for ever hovers on the event horizon. (In fact, the event horizon is defined in terms of such null rays.) Ray 3 passes through the collapsing material and manages to 'escape out' to $r = +\infty$, though it spends a large fraction of its life around $r \gtrsim 2M$. This is an example of a wave mode which started out to be an ingoing ray and ended as an outgoing ray at late times.

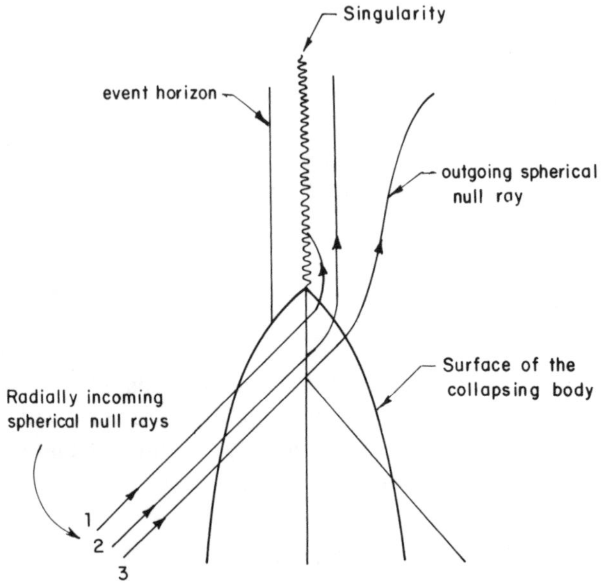

Fig. 9.2. Behaviour of light rays in the spacetime metric of a collapsing star.

It should be borne in mind that the above picture of rays is valid only in the high-frequency approximation. In addition to rays that reach $r = +\infty$ after passing through the collapsing matter, there will always be those scattered by the background spacetime. This scattering is an irrelevant complication which we shall ignore.

The above picture shows that, at late times, the field is described by the superposition of two types of modes. First, there are the outgoing modes ($1/r$

$e^{-i\omega U}\, Y_{lm}$). Let us call them $p_{\omega lm}$. These are positive frequency modes at late times, at large r_*. But, in addition, there will always be (even at late times) modes which are incoming at the event horizon. We shall call these modes $q_{\omega lm}$. (It should be clear that the exact form of $q_{\omega lm}$ should not affect physics at $r \to \infty$ at late times.) Thus, the field at late times is expanded as

$$\varphi = \sum_{lm} \int d\omega \, \{ b_{\omega lm} p_{\omega lm} + \text{h.c.} + c_{\omega lm} q_{\omega lm} + \text{h.c.} \} \,. \tag{136}$$

The quantity we are after is the expectation value

$$N_{\omega lm} = \langle \text{in}, 0 | \, b^{+}_{\omega lm} b_{\omega lm} \, | 0, \text{in} \rangle \,. \tag{137}$$

Introducing the Bogolubov expansion

$$p_{\omega lm} = \int d\omega' (\alpha_{\omega lm \; \omega' lm} f_{\omega' lm} + \beta_{\omega lm \; \omega' lm} f^{*}_{\omega' lm}) \,, \tag{138}$$

we immediately set

$$N_{\omega lm} = \int d\omega' \, | \, \beta_{\omega\omega' lm} |^2 \tag{139}$$

with

$$\beta_{\omega\omega' lm} \equiv \beta_{\omega lm \; \omega' lm} = (p_{\omega lm}, f^{*}_{\omega' lm}) \tag{140}$$

To evaluate this scalar product, consider a wave mode $p_{\omega lm}$ which is of the form $r^{-1} \exp(-i\omega U) Y_{lm}$. At early times it must have originated at infinity as a linear combination of positive and negative frequency modes which were incoming, i.e. $\sim r^{-1} \exp(\pm\, i\omega' V) Y_{lm}$. Moreover, since the mode has reached $r = \infty$ at late times with a finite frequency ω (in spite of a large redshift) the contributions must come from very high frequency ω' modes at early times. This justifies using the ray approximation. The outgoing ray at late times, of the form $e^{-i\omega U}$ can now be traced back along the null path. At early times, the ray will be moving along constant V lines. But since the phase of the wave remains constant it will still have the numerical value $e^{-i\omega U(V)}$ where the function $U(V)$ has to be determined. An involved argument originally due to Hawking shows that

$$U = -4M \ln[V_0 - V] + \text{const}, \tag{141}$$

where V_0 is the value of the ray surface that forms the event horizon.

Example 9.10.[1] We outline here the argument leading to (141). Consider a null vector n^i outside the collapsing body, at the horizon. Choose ε such that $(-\varepsilon n^i)$ connects the horizon to a ray with an asymptotically large U value (see Figure 9.3).

We determine $\varepsilon(U)$ by the following argument. Transport $(-\varepsilon n^i)$ parallelly along a null generator of the future horizon to the point where past and future horizons meet. On the past horizon, choose an affine parameter

[1] This example uses concepts like the Penrose diagram which are beyond the scope of this book.

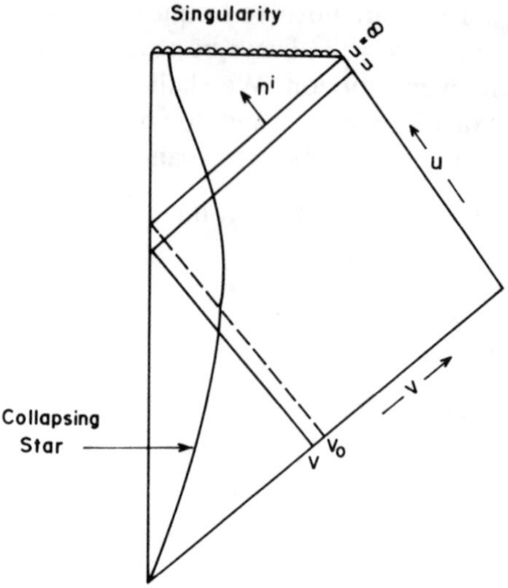

Fig. 9.3. Behaviour of two neighbouring null geodesics in the spacetime metric of a collapsing star.

$$\lambda = -c \exp(-\varkappa U) \quad ; \qquad \varkappa = (4M)^{-1}.$$

Integrating $n^i \, d\lambda = dx^i$ along the past horizon from $\lambda(0)$ to $\lambda(U)$, we get

$$x^i(\lambda) - x^i(0) = \lambda n^i.$$

But since $(-\varepsilon n^i)$ connects horizon to the ray at U, we also have $x^i(\lambda) - x^i(0) = -\varepsilon n^i$. Thus,

$$\varepsilon = -\lambda(U) = C \, e^{-\varkappa U}.$$

Now, in parallel, transport $(-\varepsilon n^i)$ along (i) the null generator of the past horizon and then along (ii) the path of radially incoming null ray at V_0 out to early times and large distances from matter. The vector will connects the rays with values V and V_0. Thus in (U, V) coordinates,

$$V - V_0 = -\varepsilon n^v.$$

But at infinity space is nearly flat, so that n^v is a constant (say, D). Thus,

$$V - V_0 = -\varepsilon D = -cD \, e^{-\varkappa U}, \quad \varkappa = (4M)^{-1}.$$

Taking logarithm, we get $U = -4M \ln[(V_0 - V)/cD]$, which is the same as (141). □

Thus, at early times, we have

$$p_{\omega lm} = \begin{cases} N\omega^{-1/2} r^{-1} \exp\left[4iM\omega \ln\left(\dfrac{V_0 - V}{c}\right)\right] Y_{lm}, & \text{for } V < V_0, \\ 0, & \text{for } V > V_0. \end{cases} \tag{142}$$

($p_{\omega lm}$ vanishes for $V > V_0$ because these modes do not reach $r = \infty$.) The function $f_{\omega' lm}$, at early times, has the form

$$f_{\omega' lm} = N(\omega')^{-1/2}\, r^{-1} \exp(-i\omega' V)\, Y_{lm}. \tag{143}$$

Substituting Equations (142) and (143) into (140) we get

$$\beta_{\omega\omega' lm} = \frac{1}{2\pi}\int_{-\infty}^{V_0} \mathrm{d}V \left(\frac{\omega'}{\omega}\right)^{1/2} \mathrm{e}^{-i\omega' V} \exp\left[4iM\omega \ln\left(\frac{V_0 - V}{c}\right)\right],$$

and, using (140) with f replacing f^*,

$$\alpha_{\omega\omega' lm} = \frac{1}{2\pi}\int_{-\infty}^{V_0} \mathrm{d}V \left(\frac{\omega'}{\omega}\right)^{1/2} \mathrm{e}^{+i\omega' V} \exp\left[4iM\omega \ln\left(\frac{V_0 - V}{c}\right)\right].$$

These integrals are readily evaluated in terms the Γ-functions. We get

$$|\alpha_{\omega'\omega}|^2 = \exp(8\pi M\omega)\, |\beta_{\omega'\omega}|^2.$$

Because of our normalization

$$1 = \int \mathrm{d}\omega' (|\alpha_{\omega\omega'}|^2 - |\beta_{\omega\omega'}|^2).$$

Therefore,

$$\langle N_\omega\rangle = \int \mathrm{d}\omega' \, |\beta_{\omega\omega'}|^2 = [\exp(8\pi M\omega) - 1]^{-1},$$

which was the result advertised earlier.

Example 9.11. Most of the redshift comes from the region near $r = 2M$. This fact suggests the following alternative interpretation of the result. Consider an observer at constant r, $r = r_0$ with $r_0 \gtrsim 2M$. The trajectory of such an observer will be

$$x^i = (t, r_0, 0, 0, 0).$$

Therefore,

$$u^i = \frac{\mathrm{d}x^i}{\mathrm{d}\tau} = \frac{1}{\sqrt{g_{00}}}\frac{\mathrm{d}x^i}{\mathrm{d}t} = \frac{1}{\sqrt{g_{00}}}(1, 0, 0, 0)$$

and the proper acceleration is

$$a^i = \Gamma^i{}_{kl}u^k u^l = \frac{1}{g_{00}}\Gamma^i{}_{00} = -\frac{1}{2g_{00}} g^{rr} \frac{\partial}{\partial r}\left(1 - \frac{2M}{\mathrm{r}}\right)\delta^i_r$$

$$= -\frac{M}{r^2}\delta^i_r \Rightarrow |a^l a_l| \equiv g^2 = \frac{M^2}{r^4}\,\frac{1}{(1 - (2M/r))}.$$

If we assume that an observer at r_0 detects particles with temperature $T = g/2\pi$, then

$$T(r_0) = \frac{M}{2\pi r_0^2}\left(1 - \frac{2M}{r_0}\right)^{-1/2}.$$

At infinity, these particles will be described by the redshifted temperature,

$$T(\infty) = T(r_0)\left(1 - \frac{2M}{r_0}\right)^{+1/2} = \frac{M}{2\pi r_0^2} \approx (8\pi M)^{-1}.$$

In arriving at the last equation we have put $r_0 \cong 2M$. The physical meaning of the above analysis is, however, not clearly understood. □

Example 9.12. The temperature of the black-hole can also be motivated by a thermodynamic argument. Suppose a black hole of mass M is formed out of N particles of mass m each. Attributing unit information content to each particle, the minimum information loss in black-hole formation is

$$\text{Information loss} = \text{Entropy} = N = \left(\frac{M}{m}\right).$$

Since the particles are lost inside the black hole, we expect (with $\hbar, c, G \neq 1$)

$$\left(\frac{\hbar}{mc}\right) \leq \frac{2GM}{c^2}$$

or with $\hbar, c, G = 1$,

$$m \geq (2M)^{-1}.$$

Thus the black-hole entropy has the limiting value

$$\text{Entropy} \equiv s = 2M^2$$

corresponding to a temperature

$$T = \left(\frac{\partial s}{\partial M}\right)^{-1} = (4M)^{-1}.$$

The result differs by a factor of $(2\pi)^{-1}$ from the correct expression. □

The 'temperature' of the black hole is related in a straightforward way with the 'surface gravity' discussed in Chapter 8. The Newtonian value of gravitational acceleration at the surface of the black hole is

$$\varkappa = \frac{GM}{(2GM/c^2)^2} = \frac{c^4}{4GM} = \frac{1}{4M},$$

where in the last step we have set $c = G = 1$. The thermodynamic argument in Example 9.12 leads to $\varkappa$ as the temperature.

Notes and References

1. There exists a monograph on physics in a curved spacetime by Birrell and Davies:

Birrell, J. D., and Davies, P.C.W.: 1982 *Quantum Fields in Curved Spacetime*, Cambridge.

This book discusses various topics of the chapter in much more detail.

2. There are large number of review articles in this subject. An extensive list of references can be found in the book referred to above. Two particularly lucid reviews are:

Parker, L.: 1976, 'The Production of Elementary Particles by Strong Gravitational Fields', *Asymptotic Properties of Space-Time*, Plenum, New York.

Sciama, D. W., Candelas, P., and Deutsch, D.: 1983, 'Thermal Properties of Accelerated States in Quantum Field Theory', *Adv. Phys.* **30** 327.

3. Some original references of relevance to Section 2 are:

Letaw, J. R.: 1981, *Phys. Rev.* **D23** 1709; Padmanabhan, T.: 1982, *Astrophys. Space Sci.* **83**, 247.

Candelas, P., and Sciama, D. W.: 1983, *Phys. Rev.* **D27**, 1715.

4. The model discussed in Section 3 (see Equations (59–69)) is taken from:

Bernard, C., and Duncan, A.: 1977, *Ann. Phys.* **107**, 201.

Chapter 10

The Very Early Universe

10.1. Symmetry Breaking in the Early Universe

We have seen in Chapter 5 that the concept of spontaneous symmetry-breaking plays an important role in the description of gauge theories. It turns out that this process also plays a crucial role in the dynamics of the early universe.

The spontaneous breakdown of the symmetry which is present in the Lagrangian is usually achieved by introducing a Higgs scalar field with the action (in flat spacetime) given by

$$S_{\text{Higgs}} = \int [\tfrac{1}{2}\varphi_i\varphi^i - V(\varphi)]\, d^4x. \tag{1}$$

The potential $V(\varphi)$ is chosen such that it possesses a minimum at some nonzero value φ_0 of φ. That is,

$$V'(\varphi_0) = 0; \qquad V''(\varphi_0) > 0; \qquad \varphi_0 \neq 0. \tag{2}$$

The 'ground state' will then have a nonzero vacuum expectation value for $\langle\varphi\rangle = \varphi_0$, which will break the symmetry.

In realistic (e.g. SU(5)) gauge theories, the scalar field will not be a single entity. One normally works with a set of Higgs scalar fields φ^A, $A = 1, 2, \ldots, N$ where N is the number of generators in the group. It is then preferable to work with the matrix $\varphi = \varphi^A\tau_A$, where τ_A are the generators of the group in some particular matrix representation. The potential $V(\varphi)$, of course, is a scalar number and is usually taken to have the form (see Equation (5.79))

$$V = -\tfrac{1}{2}\mu^2 \operatorname{Tr}\varphi^2 + \tfrac{1}{4}a(\operatorname{Tr}\varphi^2)^2 + \tfrac{1}{2}b\operatorname{Tr}\varphi^4 + \tfrac{1}{3}c\operatorname{Tr}\varphi^3, \tag{3}$$

where μ^2, a, b, c are constants appearing in the theory. The vacuum expectation value now has to be specified for each of the components φ^A. The detailed minimum structure of $V(\varphi)$ then depends on the values of the particular constants (μ^2, a, b, c) in $V(\varphi)$.

For example, in the SU(5) models, people use two different symmetry-breaking scenarios. By a suitable choice of coefficients in (3) we can arrange for the vacuum expectation values of φ to take either of the following two forms (Φ and σ are scalar constants):

$$\langle\varphi\rangle = \Phi \operatorname{diag}(1, 1, 1, -\tfrac{3}{2}, -\tfrac{3}{2}), \tag{4}$$

or

$$\langle\varphi\rangle = \sigma \operatorname{diag}(1, 1, 1, 1, -4). \tag{5}$$

When the ground state takes the form given in (4), the original SU(5) invariance is lost and is replaced by an SU(3) × SU(2) × U(1) invariance. When the ground state takes the form in (5), we are left with an SU(4) × U(1) invariance. The coefficients in $V(\varphi)$ can be so arranged that one of these two minima is preferred. Usually the symmetry-breaking as given by (4) is preferred.

In a realistic model, we have to cope up with some new complications. The potential $V(\varphi)$ is the classically specified 'tree level' potential. We saw in Chapter 4 that quantum fluctuations will change the form of $V(\varphi)$. We should calculate the effective potential and use that $V_{\text{eff}}(\varphi)$ in determining the minima. These quantum corrections – in the SU(3) × SU(2) × U(1) phase, say – change the potential to a form like

$$V_{\text{eff}} = \alpha\varphi^2 - \beta\varphi^4 + \gamma\varphi^4 \ln\left(\frac{\varphi^2}{\sigma^2}\right). \tag{6}$$

(We have discussed in Chapter 4 the derivation of $V_{\text{eff}}(\varphi)$.) The values of α, β, γ, of course, depend critically on the theory under consideration. We shall see later in Sections 10.4 and 10.5 that V_{eff} and V can lead to very different physics.

A more direct complication arises from the fact that in the early phases of the universe we have to contend with very high temperatures. The potential energy of the Higgs field acquires a contribution from thermal energy. This correction must be taken into account in the $V_{\text{eff}}(\varphi)$. In general (see Chapter 4), this correction adds to $V_{\text{eff}}(\varphi)$ the temperature-dependent term

$$V_{\text{thermal}}(\varphi) = \frac{18T^4}{\pi^2}\int_0^\infty \mathrm{d}x\, x^2 \ln\left[1-\exp\left(-\left(x^2 + a\frac{\varphi^2}{T^2}\right)^{1/2}\right)\right]. \tag{7}$$

The constant a depends on the parameters of the theory under consideration, and T is the temperature measured in energy units.

The reader may wonder why we have not been very specific as regards, for example, the parameters of the theory in the discussion so far. The reason is twofold:

(i) The subject we are discussing at present is far from being closed, and the details are likely to change in the course of future investigations.
(ii) The results we wish to highlight turn out to be remarkably independent of the details and will only depend on general considerations.

The potential $V(\varphi) = V_{\text{eff}}(\varphi) + V_{\text{thermal}}(\varphi)$ is shown in Figure 10.1 A glance at the figure shows the following features:

(i) At $T = 0$, the potential is completely decided by $V_{\text{eff}}(\varphi)$. This potential has a true minimum at some $\varphi = \varphi_0$ and a local maximum at $\varphi = 0$.

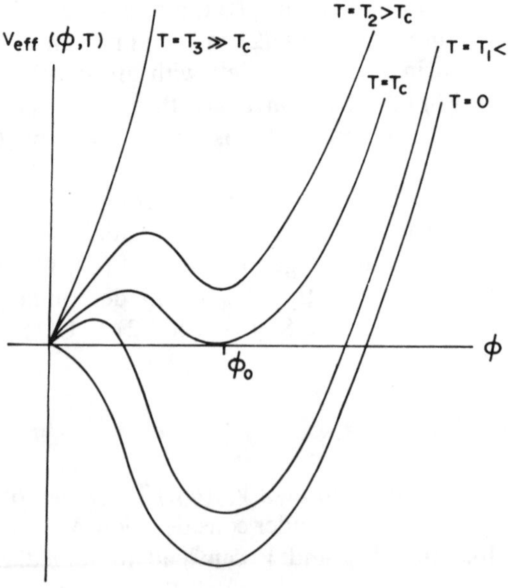

Fig. 10.1. Schematic diagram showing the minima of $V_{\text{eff}}(\varphi, T)$ at various temperatures. Since $V_{\text{eff}}(\varphi, T) = V_{\text{eff}}(-\varphi, T)$ the curves are shown only for $\varphi > 0$.

(ii) As the temperature increases, thermal energy contributes to the potential and the true minimum rises (see the curve for $T = T_1$).
(iii) Clearly, there will be some temperature $T = T_c$ at which both the minima have the same energy. At temperatures below T_c, the ground state will have $\langle\varphi\rangle \neq 0$. At temperatures above T_c, $\langle\varphi\rangle = 0$.
(iv) At temperatures $T_2 > T_c$, the global minimum is at $\varphi = 0$; the secondary local minimum is not energetically favoured.
(v) At very high temperatures $T_3 \gg T_c$, the only minimum that exists is for $\varphi = 0$.

Such a behaviour of the potential energy is well known in solid state physics and indicates the presence of a phase transition in the model. Such a phase transition also occurs in the hot universe. Before we go into the details, let us qualitatively see what kind of behaviour we expect.

We know that the 'energy gap' V_0 at $T = 0$ is what leads to the masses of the particles in the broken symmetric phase. In the GUTs phase transition we are dealing with masses of the order of 10^{14}–10^{15} GeV. Thus, it is reasonable to assume that $T_c \sim 10^{14}$ GeV. At temperatures above 10^{14} GeV, the effective potential dictates $\langle\varphi\rangle = 0$ for the ground state. We are then in the SU(5) invariant, symmetric phase. As the temperature goes down, the secondary minimum appears and tends to become energetically more and more favourable. When $T < 10^{14}$ GeV, it is energetically favourable for φ to develop a nonzero vacuum expectation value ($\langle\varphi\rangle \neq 0$). This would correspond to the 'broken symmetric phase'. Thus, we expect a phase transition to occur at around $T \sim 10^{14}$ GeV.

The situation may be compared to steam condensing into water. Suppose steam is kept in a container at a temperature $T > 373$ K. The water molecules at this temperature are in the symmetric phase. If the temperature is lowered below 373 K, the minimum energy configuration changes; it is preferable to be in the liquid state, and *usually* a phase transition occurs condensing the steam to water. However, in very special circumstances it is possible to 'supercool' the steam; that is, it is possible to keep water molecules in the gaseous unstable state even at $T < 373$ K. Such a supercooled state, however, is unstable to a process known as 'bubble creation'. Random fluctuations produce 'bubbles' of water in the supercooled steam. Since energy is released when steam condenses to water in a local region, the bubble expands rapidly and precipitates a condensation of water phase.

We are therefore faced with two possible kinds of behaviour in the early universe. One way assumes that the minimum of the scalar field smoothly goes from $\varphi = 0$ to $\varphi = \varphi_0$ as the universe cools. In this case, the phase transition is accomplished everywhere in the universe as the temparature falls through T_c. Alternatively, we may assume that the universe remains at the $\varphi = 0$ minimum even when T falls below T_c. In this case, we may say that the universe is supercooling in the metastable phase. As with the steam–water example, bubbles of the second (broken symmetric) phase will form in the universe and will expand. The phase transition will be completed by the collision and coalescence of the bubbles. Of course, 'bubbles' in this situation are regions of spacetime with $\langle\varphi\rangle = \varphi_0$.

Between the above two possibilities, it appears to be attractive to choose the latter. The phase transitions in the early universe, as we shall show, lead to the appearance of a huge number of magnetic monopoles. The second scenario seems to be more successful in weeding out these monopoles. In addition, the second scenario seems to produce a mechanism which solves the flatness and horizon problems of cosmology. In the next two sections we discuss the magnetic monopole problem and the so-called *inflationary* scenarios.

10.2. Cosmological Monopoles

In Chapter 5 we showed that gauge theories may possess monopole solutions when certain conditions are satisfied. The SU(5) models in the broken symmetric phase satisfy these conditions. In fact, any kind of gauge model which breaks into a U(1) part at low energies will exhibit monopole solutions. Thus, we should consider seriously the number of magnetic monopoles which could have been produced in the early universe.

Let us recall that the SU(5) monopoles will have masses of the order of

$$m_M \approx \alpha^{-1} M_X \approx 10^{16} \text{ GeV}, \tag{8}$$

where α is the coupling constant and M_X is the X-boson mass (again, we must remember that these are rough estimates). In principle we can produce monopole–antimonopole pairs in collisions like particle + antiparticle → monopole + antimonopole. However, monopole configurations exist in the theory only after the symmetry breaking has taken place, i.e. for $T \lesssim 10^{14}$ GeV $= T_c$. But we see from (8) that $M \sim 100T_c$. Thus monopoles will never be present in equilibrium numbers. Using the formulae of Chapter 8 it is easy to see that the present monopole to photon ratio is like

$$\frac{n_M}{n_\gamma} \simeq 10^3 \left(\frac{m_M}{T_c}\right)^3 \exp\left(-\frac{2m_M}{T_c}\right). \tag{9}$$

If $m/T_c \approx 10^2$ then $n_M/n_\gamma \approx 10^{-70}$ which is a negligible number! However, note that the number is exponentially sensitive to the ratio (m_M/T_c). Neither the monopole mass nor the critical temperature is known with any great precision. An uncertainty of a factor 3 to 10 in this ratio – which can easily arise – will change the (n_M/n_γ) ratio drastically.

There is, however, another process that produces monopoles in the early universe, and it has made the thermal production comparatively unimportant. To understand this mechanism, we should go back to the nature of the monopole solution. Essentially, monopoles are classical, solitonic solutions to the field equations, with a particular Higgs scalar configuration. Notice that the potential $V(\varphi)$ depends only on the 'magnitude' of φ^A (in combinations like Tr φ^2, etc.) and does not, in any way, determine the 'direction' in which φ^A is pointing in the group space. The monopole solution arises from choosing a particular kind of configuration $\varphi^A(\mathbf{x})$ in space.

In flat space, of course, we can correlate the $\varphi^A(\mathbf{x})$ all over the space and produce a monopole configuration. However, in a Big Bang universe we cannot do this! In the radiation-dominated universe there is a maximum distance (horizon) to which signals could have propagated since $t = 0$. The proper distance to the horizon $\bar{r}_H$ is obtained by multiplying r_H in (8.82) by $Q(t)$. That is,

$$\bar{r}_H(t) \equiv Q(t) \int_0^t \frac{dt'}{Q(t')} = 2t = \left(\frac{8\pi^3 N T^4}{45 m_p^2} \right)^{-1/2}, \tag{10}$$

where N is the number of degrees of freedom which is of the order of 10^2, and $m_p \sim 10^{19}$ GeV.[1] The existence of the horizon implies that the field $\varphi^A(\mathbf{x})$ cannot be correlated over distances larger than $\bar{r}_H(t)$. In other words, monopole configurations are produced within each horizon volume. Assuming every correlation volume $\frac{4}{3}\pi\bar{r}_H^3$ contributes at least one monopole at the time of symmetry breaking, we get the bound on the number density of the monopoles to be

$$n_M \gtrsim \left(\frac{4}{3}\pi \bar{r}_H^3 \right)^{-1} \sim N^{-3/2} \left(\frac{T_c^2}{m_p} \right)^3 . \tag{11}$$

We have tacitly assumed that, as the phase transition occurs, the Higgs field within each correlation volume obediently goes into monopole configuration. More rigorously, the right-hand side of (11) must be multiplied by a factor p which denotes the probability for the field configuration to be that of a monopole. We need not, however, worry because rigorous computation shows that $p \approx \frac{1}{8}$ in the SU(5) model. Though the magnetic monopole configuration is somewhat special, it must be remembered that it is a solution to classical equations and thus minimizes the action. This is a crude way of understanding why $p \sim 10^{-1}$ rather than $p \sim 10^{-10}$. Thus, in an order of magnitude estimate like (11), p may be ignored. Assuming monopole–antimonopole annihilations to be negligible at these epochs, we can calculate the present monopole–photon ratio to be

$$\frac{n_M}{n_\gamma} \geq 10\, N^{1/2} \left(\frac{T_c}{m_p} \right)^3 . \tag{12}$$

Since the monopoles are believed to have a mass of ~10^{16} GeV, this calculation leads us to an energy density in the present universe of the order of

$$\varrho_M \gtrsim 5 \times 10^{-18}\ \text{g cm}^{-3}, \tag{13}$$

which is 10^{11} times larger than the critical density (cf. Chapter 8). This is clearly an unacceptable result, and hence has come to be called the 'monopole problem'.

No definite solution to this problem is known. The inflationary universe scenarios, to be discussed in the next section, provide one possible solution except that the inflationary scenarios themselves suffer from difficulties of a different nature.

[1] In arriving at the last inequality we have also used the time–temperature relationship (cf. (8.63)) generalized for N degrees of freedom.

Example 10.1. Consider a region of linear coordinate size L in the early universe. To use flat spacetime field theory and special relativistic statistical mechanics, we must have

$$L \ll \bar{r}_H(t).$$

Further, to use thermodynamical description for particles, we must have sufficiently large number of particles inside a region L^3. That is, we need

$$N \equiv L^3 n(t) \gg 1.$$

Let $L/\bar{r}_H(t) \approx L/ct = \varepsilon$. Then, since

$$n(t) \cong \frac{g}{\pi^2}\left(\frac{kT}{c\hbar}\right)^3 \quad \text{and} \quad \left(\frac{t}{t_P}\right) \cong \left(\frac{45}{16\pi^3 g}\right)^{1/2}\left(\frac{T_P}{T}\right)^2$$

we get

$$N \cong \frac{1}{300}\left(\frac{1}{g}\right)^{1/2}\left(\frac{\varepsilon T_P}{T}\right)^3; \qquad T_P = 10^{19}\ \text{GeV}.$$

If we take $\varepsilon \sim 10^{-2}$ and $N \sim 10^6$ then flat spacetime statistical mechanics ceases to be valid for $T \gtrsim 10^{14}$ GeV! At 10^{16} GeV, for $g \sim 100$ and $\varepsilon \sim 10^{-2}$ we get the ridiculously low value of $N \sim \frac{1}{3}$. No clear solution is known to this problem, though particle physicists working on the early universe continue to talk in the flat spacetime language. The result given above must be kept in mind. □

10.3. Cosmological Inflationary Scenarios

In this section we shall describe an idea which has of late caught the fancy of many workers in the field of cosmology. This popularity is mainly due to the intrinsic simplicity of the idea; it prevails in spite of the fact that – at the time of writing – all inflationary scenarios suffer from one difficulty or another.

The main motivation for the idea stems from a desire to explain the 'flatness' and 'horizon' problems discussed in Chapter 8. In standard cosmological models the expansion factor is a polynomial in t:

$$Q(t) \propto t^n. \tag{14}$$

In the radiation-dominated phase $n = \frac{1}{2}$ and in the matter-dominated era, $n = \frac{2}{3}$ for the $k = 0$ model. Let us assume for the moment that the universe followed a different kind of evolution:

$$Q(t) = \begin{cases} Q_0\left(\dfrac{t}{t_1}\right)^{1/2} & 0 \leq t \leq t_1, \\ Q_0 \exp H(t - t_1) & t_1 \leq t \leq t_2, \\ Q_0\left[\exp H(t_2 - t_1) + \left(\dfrac{t - t_2}{\alpha}\right)^{1/2}\right], & t \geq t_2. \end{cases} \tag{15}$$

In other words, the expansion factor follows a power law for $t < t_1$ and $t > t_2$. In between $(t_1 < t < t_2)$ the universe expands exponentially.

A little thought shows that if the exponential expansion can be sustained for some 60 e-folding time scales, then both horizon and flatness problems can be solved. The proper horizon distance within the time interval $(0, t)$ during the exponential phase has the value

$$r_H(t) = e^{Ht} \int_0^t e^{-Ht'} dt'$$

$$\sim \frac{1}{H} e^{Ht} \quad (\text{for } t \gg H^{-1}). \tag{16}$$

Thus the horizon size will be inflated by a huge amount in this process if $t \sim 60H^{-1}$ (say). The whole observed universe can come out of a single causally connected domain at the epoch when the temperature was around 10^{17} GeV. This is the inflationary scenario.

The flatness of the present universe can also be explained in the above process. In the Einstein equations,

$$\frac{\dot{Q}^2}{Q^2} + \frac{k}{Q^2} = \frac{8\pi G}{3} \varrho. \tag{17}$$

Consider the relative behaviour of the two terms on the left-hand side. In a power law behaviour $Q \propto t^n$, $(\dot{Q}/Q)^2$ goes as t^{-2} while k/Q^2 goes at t^{-2n}. Essentially both vary only as powers of t and hence it is difficult to produce large disparities between the two terms. However, if $Q \propto \exp(Ht)$, then the first term $(\dot{Q}/Q)^2$ remains constant at H^2 while the second term falls exponentially as $k\, e^{-2Ht}$. In other words, an exponential phase of expansion will completely kill the curvature term in comparison with the $(\dot{Q}/Q)^2$ term. Thus, a universe after inflation will be physically indistinguishable from a $k = 0$ model.

Hence, introducing an exponentially growing phase in the early universe solves both the flatness and the horizon problems. It should be noted that this inflation must take place sufficiently early in the universe, especially before the baryon number is generated at $\sim 10^{14}$ GeV as any conserved charge will be exponentially diluted out during the inflation. Thus, if any form of inflation takes place after the baryon nonconserving interaction has become inoperational, then the baryon density will be diluted tremendously.

We shall now describe some attempts to produce the exponential expansion in a natural way.

10.4. The Guth Inflation

In his original attempt in 1981, A. Guth suggested that a supercooled phase transition might do the trick. The idea is as follows. Consider the universe which is populated by various species of particles at temperatures of the order of 10^{17} GeV (say). The Higgs scalar fields are governed by potentials as shown in Figure 10.2.

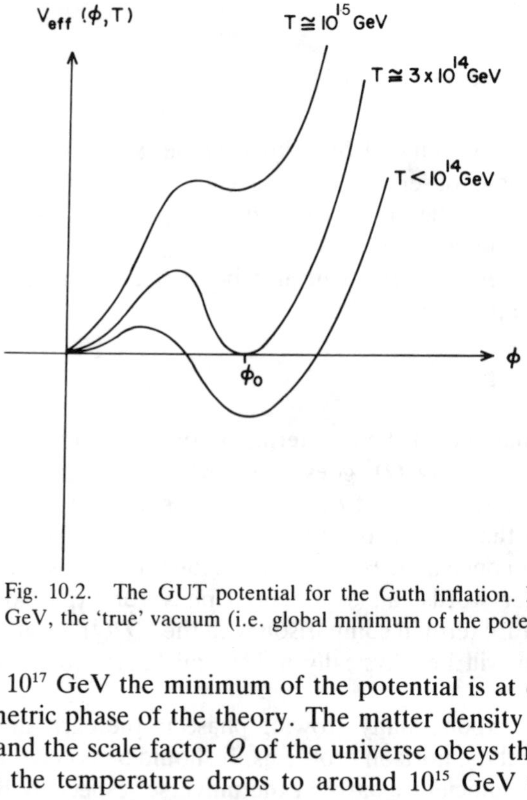

Fig. 10.2. The GUT potential for the Guth inflation. For temperature below ~ 10^{14} GeV, the 'true' vacuum (i.e. global minimum of the potential) is at $\varphi = \varphi_0$.

At 10^{17} GeV the minimum of the potential is at $\varphi = 0$ and we are in the symmetric phase of the theory. The matter density behaves like pure radiation and the scale factor Q of the universe obeys the $t^{1/2}$ law.

As the temperature drops to around 10^{15} GeV the minimum at $\varphi = \varphi_0$ becomes energetically favourable. However, as Guth argued, let us consider the effect of the universe 'supercooling' in the metastable vacuum at $\varphi = 0$. (This is in complete analogy with the supercooling of steam.) The evolution equation now reads

$$\frac{\dot{Q}^2}{Q^2} + \frac{k}{Q^2} = \frac{8\pi G}{3} (\varrho(\varphi) + \varrho_{\text{other}}). \tag{18}$$

Here the scalar field energy density has a value $V(\varphi = 0) = \varepsilon_0$ (see Figure 10.2) while ϱ_{other} corresponds to the rest of the matter and varies as $1/Q^4$. It is easy to see that once the universe stays in this metastable state for a sufficiently long time, the behaviour of $Q(t)$ is given by

$$Q(t) \cong Q_0 \exp(Ht); \qquad H^2 = \frac{8\pi G}{3}\,\varepsilon_0; \qquad t \gtrsim H^{-1}. \tag{19}$$

(Clearly, for $t \gtrsim H^{-1}$, k/Q^2 and $\varrho_{\text{other}} \propto Q^{-4}$ make negligible contribution.) We have achieved the exponential expansion in a very natural way.

The supercooled phase, of course, cannot last indefinitely. Notice that the two vacua at $\varphi = 0$ and $\varphi = \varphi_0$ are separated in Figure 10.2 by a potential barrier. There is a nonzero probability for the scalar field to tunnel from the $\varphi = 0$ value to $\varphi = \varphi_0$. This tunnelling is a random process and can happen at any event. These 'bubbles' of the asymmetric vacuum $\varphi = \varphi_0$, when formed, release the energy density ε_0 to the system. This release of energy allows the bubbles to grow rapidly with the bubble walls approaching the speed of light.

Let us assume that λ_0 denotes the probability per unit time per unit proper volume for the bubbles to be formed, i.e. the bubble nucleation rate. A bubble originating at a time t_1 will have a coordinate volume $V(t, t_1)$ at time t, where

$$V(t, t_1) = \frac{4\pi}{3}\left[\int_{t_1}^{t} \frac{\mathrm{d}t}{Q(t)}\right]^3 . \tag{20}$$

Since the bubble nucleation is an uncorrelated event, it is clear that the probability $p(t)$ for any given point *not* to be engulfed by a bubble is given (see Example 10.2 below) by

$$p(t) = \exp\left\{ -\int_0^t \mathrm{d}t_1\, \lambda_0 R^3(t_1)\, V(t, t_1)\right\} . \tag{21}$$

Using the fact that $Q \propto \mathrm{e}^{Ht}$, we get

$$p(t) = \exp\left(-\frac{t}{\tau}\right), \tag{22}$$

where

$$\tau = \frac{3H^3}{4\pi\lambda_0} . \tag{23}$$

Example 10.2. Consider some arbitrary coordinate volume V and time interval $(t, t, + \mathrm{d}t)$. The physical volume is $[\exp(3Ht)]\,V$ so that the number of bubbles that will nucleate is given by $\lambda_0 V\,\mathrm{e}^{3Ht}\,\mathrm{d}t$. The coordinate volume occupied by the bubbles can be taken to be $(4\pi/3)(1/H\,\mathrm{e}^{-Ht})^3$ for each bubble.[2] Thus a fraction $(4\pi/3)\,\lambda_0 H^{-3}\,\mathrm{d}t)$ of the volume is covered with new bubbles in an interval $\mathrm{d}t$. Since bubble creation is random process we immediately obtain that the probability for a point to be in a region outside bubbles at time t is:

$$p(t) = \lim_{N\to\infty}\left(1 - \frac{4\pi}{3}\,\lambda_0 H^{-3}\,\frac{t}{N}\right)^N = \exp\left(- \frac{4\pi}{3}\lambda_0 H^{-3}\right);$$

which is the same as Equation (22). The same argument applies to (21). □

[2] It follows from (20) that $V(t', t) = (4\pi/3)\, H^{-3}(\mathrm{e}^{-Ht} - \mathrm{e}^{-Ht'})^3$. Thus the size of the bubble asymptotically (i.e. $t' \to \infty$) approaches this value.

In other words, when $t \gtrsim \tau$, $p(t) \approx 0$. Thus the probability for any given point to be in the original phase (i.e. outside any bubble) dies down fast with time. We can therefore assume that the phase transition is complete in a time scale of the order of τ. In this span of time the universe would have expanded by a factor

$$Z \equiv \exp H\tau = \exp\left(\frac{3H^4}{4\pi\lambda_0}\right). \tag{24}$$

The value of H is determined by ε_0 which can be expressed in terms of the GUTs mass scale. Thus the problem reduces to estimating λ_0.

We have already encountered such problems in Chapter 4 while discussing the double-well potential. The transition amplitude between two minima of a potential can always be expressed in the form

$$\lambda_0 = A \exp(-B).$$

Here $A \propto \varepsilon_0^4$ and B is the Euclidean action evaluated for the instanton solution which tunnels through the potential. Unfortunately, in the present case, B has to be determined by numerical integration. A crude dimensional estimate, which turns out to be exact, is given by

$$B \cong \frac{\pi^2}{6}\, \alpha^2 \varepsilon_0^4 R_c^4. \tag{25}$$

Here R_c is a characteristics radius of the bubble which satisfies the relation $H^{-1} \gg R_c \gg m_H^{-1}$. Substituting this into Z shows that even values as great as $Z \sim 10^{10^{10}}$ are easily obtainable in this scenario. We give in Figure 10.3 the

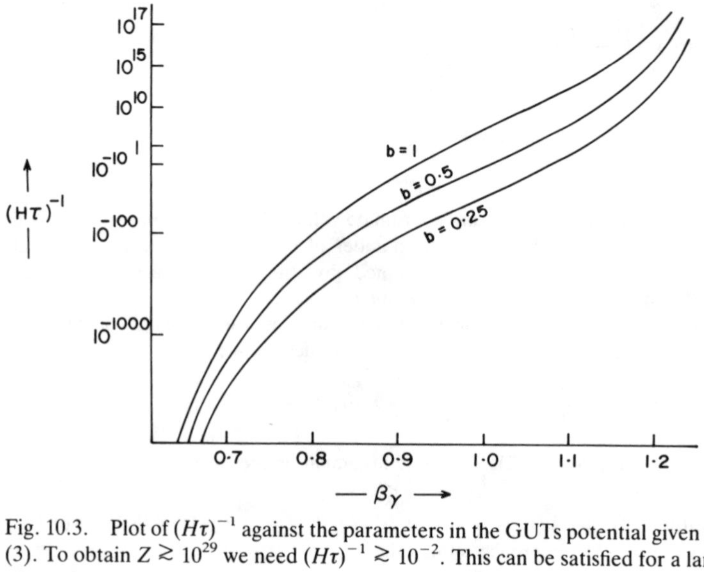

Fig. 10.3. Plot of $(H\tau)^{-1}$ against the parameters in the GUTs potential given in Equation (3). To obtain $Z \gtrsim 10^{29}$ we need $(H\tau)^{-1} \gtrsim 10^{-2}$. This can be satisfied for a large range of parameters.

plot of the parameter $(H\tau)^{-1}$ for various reasonable values of parameters in GUTs potentials. The x axis is the ratio (β/γ), where (see Equation (3))

$$\beta = \frac{\mu^2 b}{c^2}; \qquad \gamma = \frac{a}{b} + \frac{7}{15}.$$

Clearly, there is no problem in obtaining the necessary inflation.

Example 10.3. The entropy density of microwave photons in given by

$$s = \frac{4\pi^2}{45} T^2.$$

Since $Q \propto T^{-1}$, the total entropy sQ^3 is conserved in the early universe. For our present universe, Q is definitely larger than 10^{28} cm, giving

$$S \equiv sQ^3 \gtrsim 10^{85}.$$

The flatness problem (see Section 8.5) can be alternatively posed as follows: Why is a conserved quantity in our universe, which is a pure number, so large? Notice that an inflation by a factor of $\sim 10^{28}$ will change Q^3 by a factor $\sim 10^{84}$ and will produce a large S at the present epoch. □

Now we come to the main difficulty in the otherwise ideal formalism. During the exponential expansion, the temperature (of the material content) of the universe is falling as

$$T(t) \propto \frac{1}{Q(t)} \propto \exp(-Ht).$$

Clearly, after the phase transition is over, we must somehow reheat the universe to temperatures of the order of at least 10^{14} GeV so that baryons can be synthesized. This has to be achieved through the thermalization that takes place when the bubbles collide. (The 'latent heat' released in the phase transition goes to accelerate the bubble wall; only through bubble collisions can be universe heat up.) Very unfortunately, the bubble collisions do not efficiently thermalize the universe. The reason is not very difficult to see. If a bubble nucleates at $(t_0, \mathbf{x}_0)$ and expands even with the velocity of light, its surface at t is given by

$$|\mathbf{x} - \mathbf{x}_0| \equiv r(t, t_0) = \int_{t_0}^{t} \mathrm{d}t' \frac{1}{Q(t')}$$

$$= \frac{1}{H} [\mathrm{e}^{-Ht_0} - \mathrm{e}^{Ht}]. \tag{26}$$

Even as $t \to \infty$, the bubble wall never goes beyond

$$D(t_0) = \frac{1}{H} \exp(-Ht_0). \tag{27}$$

Thus a pair of bubbles which are separated in physical distance (coordinate distance $\times$ $Q(t)$) by more than $2H^{-1}$ at the time of creation will never collide! In other words, the endpoint of the inflationary universe will be populated by a series of bubbles clustered randomly in space. This is definitely not the kind of model universe we have in mind.

What about the monopole problem in this inflationary model? Notice that the monopoles come into being only in the broken symmetric phase of the theory – i.e. after the phase transition is completed. The original argument which led to too many monopoles is rendered invalid because the horizon size grows to a large value before the phase transition gets over. To obtain a reliable estimate of the present monopole density we, of course, need a detailed model for the phase transition.

10.5. Inflation with the Coleman–Weinberg Potential

The mechanism suggested above produces sufficient inflation for a wide choice of parameters in the GUTs potential. From this point of view, the above mechanism is very natural. When it was clear that this procedure does not work, investigators looked for alternative scenarios. All these alternative scenarios involve more stringent assumptions, and, hence, cannot really be considered as 'natural' as the original Guth model. We shall briefly discuss one such model in this section.

In the original Guth model we used a general form for $V(\varphi)$. Instead we shall now assume that symmetry breaking occurs through radiative corrections and choose the effective potential to be the Coleman–Weinberg type. (This form of potential was discussed in Chapter 5.) The form of the potential is given (see Figure 10.4) by

$$V(\varphi) = \frac{25}{16}\alpha^2\left[\varphi^4 \ln\left(\frac{\varphi^2}{\sigma^2}\right) + \tfrac{1}{2}(\sigma^4 - \varphi^4)\right]. \tag{28}$$

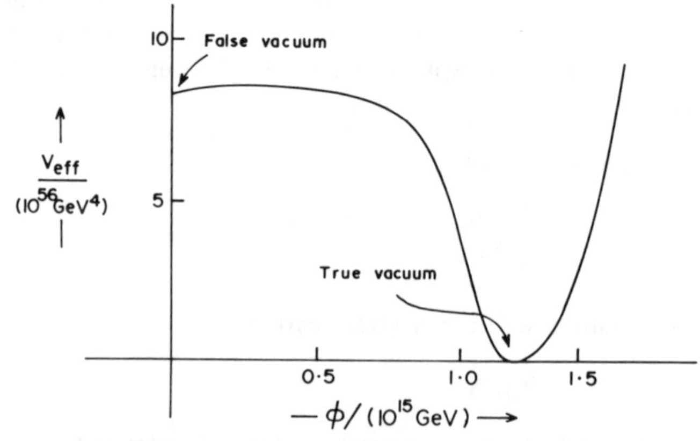

Fig. 10.4. The Coleman–Weinberg potential used in the modified inflationary scheme.

Here α^2 is the strong coupling constant with the value $\sim 1/45$ and σ is the broken symmetric minimum value of $V(\varphi)$ which is of the order of 1.2×10^{15} GeV. As compared to the GUTs potential used in the last section, this potential is extremely smooth near $\varphi = 0$, with the first three derivatives of V vanishing.

At finite temperatures, the potential is modified by the finite temperature corrections. For small T the $\varphi = 0$ point is made into a local minimum with a potential barrier of height $\sim T^4$ and width T.

Let us again begin with the universe which is in the radiation-dominated phase at $T > 10^{15}$ GeV. The present mechanism differs from the original idea in the following way. Let us assume that thermal or quantum fluctuations allow the field to tunnel through the 'bump', produced by finite temperature effects, at some stage. Since the potential barrier is small and thermal fluctuations are present at this high-temperature phase, this seems to be a reasonable assumption. The point is that, since the potential is very flat near $\varphi = 0$, it takes a long time for the field φ to 'roll down' along the potential slope to reach the $\varphi \sim \sigma$ region. This rolling is governed by the classical equation

$$\ddot{\varphi} + 3 \frac{\dot{Q}}{Q} \dot{\varphi} = - \frac{\partial V}{\partial \varphi} . \tag{29}$$

Unfortunately, this equation has to be integrated numerically. However, if we take $\varphi \sim H$, then the potential can be approximated to be

$$\frac{\partial V}{\partial \varphi} \approx - \lambda \varphi^3, \quad \lambda \approx \tfrac{1}{2} . \tag{30}$$

Then, neglecting the $\ddot{\varphi}$ term, the 'slow rolling' can be described by the solution

$$\varphi^2(t) \approx - \frac{3H}{2\lambda t} + \text{const.} \tag{31}$$

An exact numerical integration shows the behaviour depicted in Figure 10.5. Clearly φ remains constant for nearly 40 or 50 e-folding times. All throughout this phase, the potential $V(\varphi)$ remains at almost constant value at $V(0)$, driving the inflation.

As the field 'reaches' the value $\sim\sigma$ it rapidly 'plunges' down to zero value and then oscillates about the minimum $\varphi = \sigma$. In the classical picture for φ which we are using, φ must be considered to be some kind of coherent state for the Higgs field. The decay of φ to other particles acts as a damping mechanism for these oscillations. Thus the plunge of φ to zero value and the damping of the oscillations reheat the universe to 10^{14} GeV.

In addition to horizon and flatness problems, the scenario also solves the monopole problem. This is because the Higgs field is completely correlated from the beginning. (Another way of looking at it is to think of our whole universe as having originated from a single bubble.)

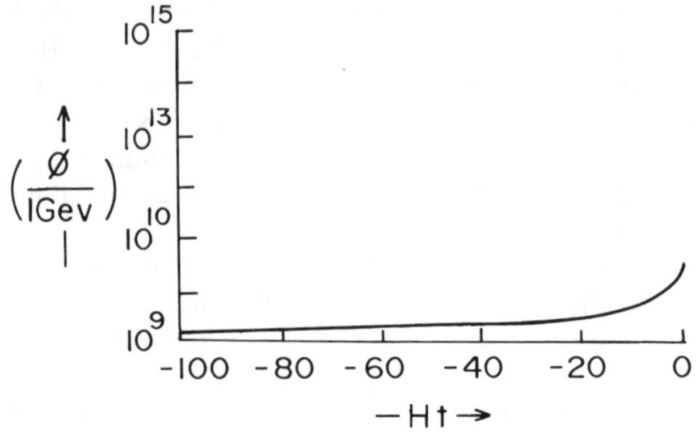

Fig. 10.5. The evolution of the scalar field in the inflationary scenario with Coleman–Weinberg potential. This curve is obtained by the numerical integration of Equation (29).

It should be made clear that the above picture is rather too naïve. Many open questions still remain in this scheme. First, the slow rolling down can be achieved only if the φ and $\dot{\varphi}$ fluctuations are small. While small fluctuations are more probable than large ones, this does *not* rule out large fluctuations. Second, the oscillations and damping of a φ field near $\varphi = \sigma$ have to be understood better in a completely quantum mechanical language. Lastly, one requires additional evidence for favouring the Coleman–Weinberg potential over the more general GUTs potential.

These are not, however, the reasons why this mechanism also has fallen into disfavour. The main problem with this scenario also turns out to be the inhomogeneities it produces. We shall discuss this difficulty briefly.

Suppose that the initial unverse had perturbations in the scalar field φ. This would imply fluctuations in the energy density. As the universe evolves, we can follow the perturbations using linearized Einstein equations. When the field φ falls to zero and thermalizes the matter, its fluctuations carry over to fluctuations in matter density. Once again linearized Einstein's equations can be used to propagate these fluctuations to the present epoch, to be compared with observations.

An important quantity in studying the fluctuations is the ratio of the physical wavelength of a perturbation (i.e. the characteristic distance over which the physical variable changes), Ql, to the horizon size H^{-1}:

$$\eta(t) \equiv Q(t)H(t)l. \tag{32}$$

(Here l is the coordinate scale of fluctuations.) In the de Sitter phase $Q \propto \exp Ht$ and H is a constant so that $\eta(t)$ grows exponentially. The epoch at which $\eta = 1$ shall be known as t_1 (we say that fluctuation enters the horizon at $t = t_1$). During the radiation-dominated phase ($Q \propto t^{1/2}$, $H \propto t^{-1}$), $\eta(t)$ decreases as $t^{-1/2}$. Thus η will again reach a value of unity at some later epoch t_2.

When $\eta < 1$ fluctuations are inside the horizon and a fluctuation of the order $\delta\varphi$ will produce

$$\frac{\delta\varrho}{\varrho} = H\,\dot{\varphi}\,\frac{\delta\varphi}{V_0} \equiv \varepsilon. \tag{33}$$

To a good approximation $\dot{\varphi}$ may be taken to be the 'rolling down' velocity. Before we go any further we need an estimate for $\delta\varphi$. We shall assume that $\delta\varphi \sim H/2\pi$. (We discuss this assumption later.) As long as $\eta < 1$, $(\delta\varrho/\varrho)$ is determined by local physics and is given by (33):

$$\varepsilon = \frac{H^2}{2\pi V_0}\,\dot{\varphi}, \quad t \leq t_1.$$

After $t > t_1$ the fluctuations are bigger than the horizon and must be evolved using Einstein's equations. It can be shown that in the de Sitter spacetime, $\varepsilon(t)$ dies down as

$$\varepsilon(t) = \alpha \exp(-2Ht).$$

This form of evolution is valid just up to an epoch t_f at which the field plunges into $\varphi = 0$. Since Q $\propto \exp(Ht)$ we can write

$$\varepsilon(t_f) = \left(\frac{H^2\,\dot{\varphi}}{2\pi V_0}\right)\left[\frac{H(t_1)Q(t_1)}{H(t_f)Q(t_f)}\right]^2 .$$

At the time of fall into $\varphi = 0$, the universe changes from the de Sitter form to a radiation-dominated form. It is quite a complicated problem to analyse how the de Sitter fluctuations change over to radiation fluctuations. Crudely, we may take the ϱ(radiation phase) $\sim \dot{\varphi}^2$ and ϱ(de Sitter phase) $\sim V_0$. Then we would expect

$$\left(\frac{\delta\varrho}{\varrho}\right)_{\text{radiation}} \sim \left(\frac{\delta\varrho}{\varrho}\right)_{\text{de Sitter}} \times \left(\frac{\varrho_{\text{de Sitter}}}{\varrho_{\text{radiation}}}\right) = \left(\frac{\delta\varrho}{\varrho}\right)_{\text{de Sitter}} \times \left(\frac{V_0}{\dot{\varphi}^2}\right).$$

An exact calculation using the junction conditions at the transition gives four times this value; that is,

$$\begin{aligned}\varepsilon(t < t_f) &= \varepsilon(t > t_f)\,\frac{4V_0}{\dot{\varphi}^2} .\\ &= \left(\frac{H^2\dot{\varphi}}{2\pi V_0}\right)\left\{\frac{H(t_1)Q(t_1)}{H(t_f)Q(t_f)}\right\}^2\left(\frac{4V_0}{\dot{\varphi}^2}\right) .\end{aligned}$$

After t_f, ε is a perturbation in the radiation-dominated universe and grows as t. In this epoch $Q(t)H(t) \propto t^{-1/2}$, and so we can write, until $t \leq t_2$,

$$\varepsilon(t_2) = \left(\frac{H^2\dot{\varphi}}{2\pi V_0}\right)\left[\frac{H(t_1)Q(t_1)}{H(t_f)Q(t_f)}\right]^2\left(\frac{4V_0}{\dot{\varphi}^2}\right)\left[\frac{H(t_f)Q(t_f)}{H(t_2)Q(t_2)}\right]^2 .$$

Notice that, by definition, $H_1Q_1 = 1$ and $H_2Q_2 = 1$. Thus we are left with the simple result

$$\varepsilon(t_2) = \frac{2H^2}{\pi\dot{\varphi}} = \frac{16G}{3}\left(\frac{V_0}{\dot{\varphi}}\right).$$

Physically, $\varepsilon(t_2)$ is the value of the perturbation $(\delta\varrho/\varrho)$ at an epoch when the perturbation re-enters the horizon. Cosmological observations on the inhomogeneities of the microwave background suggest that

$$\varepsilon(t_2) < 10^{-2} \quad \text{(observation)}$$

and most likely $\varepsilon \lesssim 10^{-4}$. From Figure 10.5 we can calculate that, in the flat rolling region, φ varies from $0.17H$ to $1.7H$ as (Ht) goes from (-100) to (-1). Thus,

$$\langle\dot{\varphi}\rangle \sim \frac{H^2}{100}.$$

This gives,

$$\varepsilon(t_2) \approx \frac{2}{\pi}\cdot\frac{H^2}{\langle\dot{\varphi}\rangle} = \frac{200}{\pi} \sim 60 \quad \text{(theory)}.$$

Clearly, theory and observation definitely disagree at least by three orders of magnitude and probably by as much as five orders of magnitude! Our inflationary mechanism produces a universe too inhomogeneous to be acceptable.

The analysis assumed that the initial perturbation of $\delta\varphi$ was of the order of $(H/2\pi)$. The reason for the particular value is the popular belief that the de Sitter spacetime must have quantum fluctuations with a characteristic temperature of this order. We saw in the previous chapter that this result was derived in very special circumstances using the global properties of spacetime. It is far from clear as to whether such a result has any validity in the present context. However, if the Hawking thermal fluctuation is rejected, there is no other natural choice for $\delta\varphi$ and the model loses its predictive power.

10.6. Fine-Tunings in the Early Universe

The discussions in the previous two sections show that we have failed to achieve a consistent inflationary scenario capable of solving the flatness, horizon, and monopole problems. In view of this failure, it would be proper to critically examine the whole attempt.

Let us begin with the horizon problem. The mathematical reason for the existence of the horizon is the finiteness of the integral

$$r_H(t) = \int_0^t \frac{\mathrm{d}t'}{Q(t')} \,.$$

However, we are already assuming that the behaviour of the integrand $Q^{-1}(t')$ near $t' \sim 0$ is completely determined by classical theory! Nothing can be farther from the truth. It is well known that all classical models lead to a singularity at $t = 0$. If quantum gravity generates models that avoid the singularities, it will necessarily modify the form of $Q(t')$ near $t' = 0$. The question of horizon must be re-examined with the modified form of $Q(t)$, taking quantum gravitational corrections into account.

We wish to stress the fact that cosmological singularity is by far a more 'real' problem than the question of horizon or flatness. Horizon and flatness problems can be completely circumvented by imposing suitable initial conditions. These are problems only in the sense that we feel these initial conditions to be contrived. The issue of singularity, however, is independent of our personal prejudices.

Next, consider the flatness problem. The fine-tuning of ϱ and ϱ_c remains a mystery only when we believe that there is no special reason for the universe to be a $k = 0$ model. As Guth himself remarks in his original paper, flatness ceases to be a problem if we take the universe to be a strictly $k = 0$ model. Now it is clear that the quantum gravitational process that 'created' the universe would also have decided on one of the three choices $k = 0, \pm 1$. Classical evolution cannot, thereafter, change this value. If the quantum gravitational processes prefer $k = 0$ over the other values we have once again obtained a natural solution to the problem.

It is quite true that we would have preferred a solution to the above difficulties without resorting to quantum cosmology. However, our experience with the inflationary universe shows that this may not be feasible. Horizon and flatness problems are purely gravitational in nature; probably it is impossible to produce a nongravitational solution to these problems.

We now turn our attention to the most enigmatic fine tuning that exists in the present universe: the vanishing of the cosmological constant. Observational evidence indicates that the cosmological constant Λ of our universe satisfies the bound

$$\Lambda L_P^2 < 10^{-120};$$

that is,

$$\Lambda < 10^{-54} \text{ cm}^{-2}.$$

(Probably Λ is the only physical quantity that is known to be zero to such a fantastic accuracy!) Considering the fact that Λ and L_P^2 are the two constants that enter gravitational physics, the smallness of ΛL_P^2 is quite astonishing.

Before the days of spontaneous symmetry breaking, the vanishing of Λ was considered to be peculiar, but not disturbing. With the advent of gauge theories a new dimension has been added to the problem. In the early hot phase of the universe the Higgs field has zero expectation value. When the universe cools, the ground state becomes a scalar condensate with huge vacuum energy. If the present phase of the universe should have zero cosmological constant, it is necessary to cancel the vacuum energy with a correspondingly large cosmological constant of opposite sign. In other words, all Higgs potentials must be chosen such that

$$V(\varphi_0) = 0 + O(10^{-108}\ \text{cm}^{-4}),$$

where φ_0 is the absolute minimum of the potential. (The zero point value of V is a completely undetermined quantity in flat space theory.) Probably no other parameter of $V(\varphi)$ will ever be measured to this level of accuracy. Quite clearly either some deep physical principle is operating here, or we are on an entirely wrong track.

Here we conclude our discussions in which gravity is treated classically. In the remaining sections of the book we shall attempt to study these problems in a quantum gravitational context.

Notes and References

1. A few original papers discussing phase transitions in the early universe are:

   Cook, G. P., and Mahanthappa, K. T.: 1982, *Phys. Rev.* **D9**, 3320.
   Guth, A. H., and Tye, S. H.: 1980, *Phys. Rev. Lett.* **45**, 1131.
   Guth, A. H., and Weinberg, E. J.: 1981, *Phys. Rev.* **D23**, 876.

2. For a discussion of the monopole problem, see:

   Kibble, T. W. B.: 1976, *J. Phys.* **A9**, 1387.
   Preskill, J. P.: 1979, *Phys. Rev. Lett.* **43**, 1365.

   The probability estimate relevant to Equation (11.11) is made in:

   Einhorn, M. B., Stein, D. L., and Toussaint, D.: 1980, *Phys. Rev.* **D21**, 3295.

3. Example 10.1 is based on the paper:

   Padmanabhan, T., and Vasanthi, M. M.: 1983, *Phys. Lett.* **89A**, 327.

4. The original inflationary model was proposed by Guth in:

   Guth, A. H.: 1981, *Phys. Rev.* **D23**, 347.

   The difficulties in this model were discussed in detail in:

   Guth, A. H., and Weinberg, E. J.: 1983, *Nuc. Phys.* **B212**, 321.

5. The tunnelling probability and bubble creation rate are calculated in:

   Callan, C. G., and Coleman, S.: 1977, *Phys. Rev.* **D16** 1762.

   See also:

   Coleman, S.: 1977, *Phys. Rev.* **D15**, 2929.

6. The 'new' inflationary scenario was proposed in:

   Linde, A. D.: 1982, *Phys. Lett.* **108B**, 389.

Albrecht, A., and Steinhardt, P. J.: *Phys. Rev. Lett.* **48**, 1220.

7. The fluctuations in the inflationary universe are calculated in:

Guth, A. H., and Pi, S. Y.: 1982, *Phys. Rev. Lett.* **49**, 1110.
Hawking, S. W.: 1982, *Phys. Lett.* **115B**, 295.
Starobinskii, A.: 1982, *Phys. Lett.* **117B**, 175.

8. A more detailed discussion of many of the topics in this chapter can be found in:

Gibbons, G. W., Hawking, S. W., and Siklos, S. T. C. (eds): 1983, *The Very Early Universe*, Cambridge.

Part IV
Quantum Cosmology

Chapter 11

Approaches to Quantum Cosmology

11.1. Introduction

We have now reached the final part of our trek which began with quantum theory and classical relativistic cosmology. In Part III we dealt with attempts to describe quantum field theory in the curved spacetime of general relativity. The discussions of the early universe in Chapter 9 and 10 showed how the dynamics of the universe can be affected by such events as particle creation and phase transition. These discussions took us very close to the singular epoch $t = 0$; but not close enough to be labelled 'quantum cosmology'. As indicated earlier, quantum cosmology is expected to be relevant prior to or around the Planck epoch

$$t_P = \sqrt{\frac{G\hbar}{c^5}}. \tag{1}$$

None of the techniques so far described can help us in understanding this phase in the history of the universe because, here, we are concerned wtih the more difficult notion of quantization of spacetime geometry. To explain this difference further, consider the classical Einstein equations

$$R_{ik} - \tfrac{1}{2} g_{ik} R = -\varkappa T_{ik}. \tag{2}$$

These may be written symbolically as

$$\text{Geometry} \sim \text{Matter}, \tag{3}$$

where matter, in its broader sense, also includes radiation.

By quantizing matter in curved spacetime we keep the left-hand side intact and classical. The right-hand side is treated semiclassically; that is, whatever the quantum effects, they lead to a new energy momentum tensor which serves as a back reaction of quantization on spacetime geometry. Although considerable progress has been made in understanding how this is done, the complete answer still eludes us.

Still less progress in understanding can be claimed in our efforts to quantize gravity; that is, in understanding what happens to the left-hand side of (2) in the quantum domain. In this chapter the reader will be presented with glimpses of the major attempts in this area. We do not promise a discussion which does full justice to these efforts, nor do we aim at a critical comparison of techniques which, at this stage, appear so different from each other. The interested reader should follow up any particular approach by studying the literature cited at the end of the chapter.

Since quantum cosmology (like its classical counterpart) belongs to quantum gravity we are concerned here essentially with approaches towards a theory of quantum gravity.

11.2. The Linearized Theory

In Chapter 7 we saw that the classical phenomenon of gravitational radiation could be discussed in the linearized weak field approximation. There we found that the radiation was described by a tensor field of rank 2. Can we gain some insight into the quantum gravity problem by quantizing this field?

The flat space quantization was essentially solved by R. P. Feynman in the early 1960s. Feynman's method is briefly as follows.

Write, with $c = 1$,

$$g_{ik} = \eta_{ik} + \varkappa^{1/2} h_{ik}; \quad \varkappa = 8\pi G, \tag{4}$$

where, in the weak field approximation,

$$|\varkappa^{1/2} h_{ik}| \ll 1. \tag{5}$$

Further, define for any tensor A_{ik} another tensor $\bar{A}_{ik}$ by

$$\bar{A}_{ik} = \tfrac{1}{2}\{A_{ik} + A_{ki} - \eta_{ik} A^{l}_{l}\}. \tag{6}$$

For a symmetric tensor $\bar{\bar{A}}_{ik} = A_{ik}$. Since tensors frequently occur in the specific combination (6) in this linearized theory, the above notation helps to compact long expressions.

Consider φ as a scalar field of mass m in interaction with gravity, for which the Hilbert action is

$$S = \frac{1}{2\varkappa}\int R\sqrt{-g}\;\mathrm{d}^4x + \tfrac{1}{2}\int(\varphi^i\varphi_i - m^2\varphi^2)\sqrt{-g}\;\mathrm{d}^4x, \tag{7}$$

where, as before, $\varphi_i = \varphi_{,i}$.

In the weak field approximation (7) becomes

$$S \approx \int(\bar{h}_{ik,l}\bar{h}^{ik,l} - 2\bar{h}^{ik}{}_{,k}\,\bar{h}^{l}_{i,l})\,\mathrm{d}^4x + \tfrac{1}{2}\int(\varphi_i\varphi^i - m^2\varphi^2)\mathrm{d}^4x + $$
$$+ \varkappa\int(\bar{h}_{ik}\varphi^i\varphi^k - \tfrac{1}{2}\,m^2\bar{h}^k{}_k\,\varphi^2)\,\mathrm{d}^4x. \tag{8}$$

Higher order terms have been neglected in this approximation. As discussed in Chapter 7 (Equation (7.48)) we still have the freedom to make a gauge transformation which can be utilized to ensure that

$$\bar{h}^{ik}{}_{,k} = 0. \tag{9}$$

The classical equations given by $\delta S = 0$ then become two interlinked wave equations

$$\Box \bar{h}_{ik} = \varkappa \bar{W}_{ik}(h, \varphi) \tag{10}$$

$$(\Box + m^2)\varphi = \varkappa\chi(h, \varphi), \tag{11}$$

where $\bar{W}$ and χ are known combinations of h, φ.

Example 11.1. Set $\bar{h}^{ik}{}_{,k} = 0$ according to (9). Then the Lagrangian density for (8) is simplified to

$$\mathscr{L} = \bar{h}_{ik,l}\bar{h}^{ik,l} + \tfrac{1}{2}(\varphi_i\varphi^i - m^2\varphi^2) + \varkappa(\bar{h}_{ik}\varphi^i\varphi^k - \tfrac{1}{2}m^2\bar{h}^k{}_k\varphi^2).$$

The field equations then become:
For $\bar{h}_{ik}$:

$$2\Box\bar{h}_{ik} - \varkappa(\varphi_i\varphi_k - \tfrac{1}{2}\, m^2\eta_{ik}\varphi^2) = 0,$$

that is,

$$\bar{W}[h, \varphi] = \tfrac{1}{2}\, \{\varphi_i\varphi_k - \tfrac{1}{2}\, \eta_{ik}m^2\varphi^2\}.$$

For φ:

$$\Box\varphi + m^2\varphi + \varkappa\, [(\bar{h}^{ik}\varphi_i)_{,k} + m^2\bar{h}^k{}_k\varphi] = 0,$$

that is,

$$\chi(h, \varphi) = -\, \{\bar{h}^{ik}\varphi_{,ik}) + m^2\bar{h}^k{}_k\varphi\, \}.$$

□

In Feynman's weak field theory (10) and (11) serve as the basis for setting up quantum gravitodynamics. Thus the propagator for the free field φ is given in momentum space by

$$K^{(\varphi)} = \frac{1}{p^2 - m^2 + i\varepsilon}\ , \tag{12a}$$

while that for the free tensor-gravitational field is given by

$$K^{(h)}{}_{ik|lm} = \eta_{il}\,\eta_{km}\quad \frac{1}{k^2 + i\varepsilon}\ . \tag{12b}$$

The quantum versions of Equations (10) and (11) are then obtained by perturbation expansions. Notice, however, that the constant $\varkappa$ goes as m^{-2}. It can be shown that perturbation expansion in $\varkappa$ will generate a series which will diverge very badly. The theory is therefore nonrenormalizable. Further, the number of Feynman diagrams at each order of perturbation expansion is far more than for the corresponding terms in quantum electrodynamics.

Example 11.2. Some typical diagrams at the tree level are shown in Figure 11.1. These do not have counterparts in quantum electrodynamics.

□

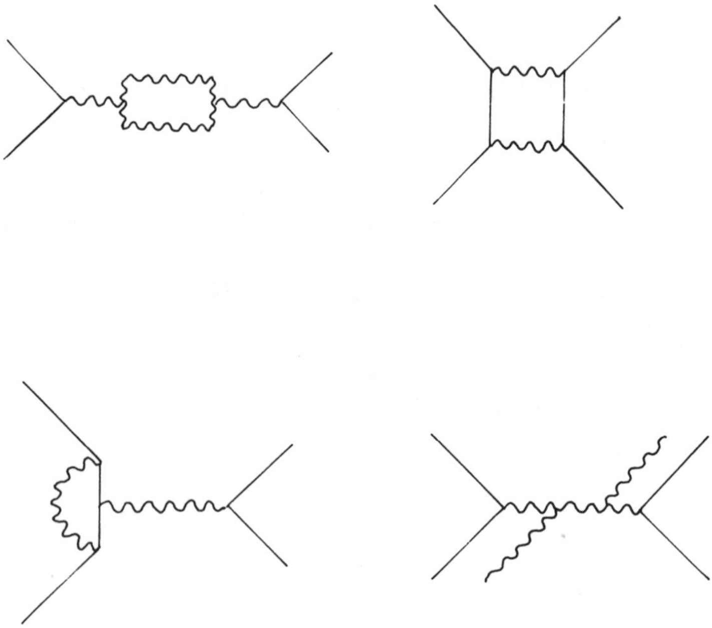

Fig. 11.1. Some typical Feynman diagrams encountered in the linearized version of quantum gravity.

Apart from the problem of renormalization, the first-order theory can be studied diagram by diagram. However, such studies are of academic interest since (unlike the experimentally verifiable calculation of the Lamb–Shift in electrodynamics) there are no experimental results in weak gravity theory for which such computations would be relevant.

For quantum cosmology, especially in its early universe applications, this weak field theory is again of no use. At $t \lesssim t_P$ the curvature effects are so large that the approximations of weak field gravity are not valid. For example, at $t = t_P$, the typical linear size of the region over which locally inertial coordinates are usable is much smaller than $ct_P \sim 10^{-33}$ cm.

We therefore leave this topic and turn to a more general approach to quantum gravity, which is not restricted to weak fields.

11.3. Canonical Quantization

In Section 7.4 we discussed the formulation of classical geometrodynamics. Based on that description, R. Arnowitt, S. Deser and C. W. Misner developed a comprehensive approach to quantum gravity and their formalism is often referred to as the ADM formalism. The method aims at identifying the 'true' degrees of freedom of the gravitational field – freedoms which may otherwise be hidden under various symmetries. Before proceeding to the actual formulation we first make some general comments.

Consider a theory described by an action S and suppose that $\psi_A(x)$ are the dynamical field variations ($A = 1, 2, 3, \ldots, N$, N being the number of such variables) contained in S. In the path integral quantization described in Part I, we would write the propagator formally as

$$K = \int \ldots \int \exp \frac{iS}{\hbar} \prod_A \mathcal{D}\psi_A \,. \tag{13}$$

Suppose there exists a symmetry transformation

$$T[\{\psi_A\}] = \psi'_A \tag{14}$$

which leaves S invariant. Then in the functional integral (13), both ψ_A and ψ'_A will generate the same value of S. In the normal course of such integrations we therefore expect to get a divergent value for K. To make the answer finite, we somehow need to pick one representative (or a finite number of representatives) from all the ψ's related by (14).

Alternatively, in the presence of symmetries we must reduce the variables down to the true degrees of freedom before quantization is attempted. Although we encountered this problem in a manifest form in the path integral (13), it might be present in a hidden form in other approaches to quantization.

Suppose that $\varphi_B (B = 1, 2, \ldots, M < N)$ are the true degrees of freedom in the above problem and that ψ_A can be expressed in terms of φ_B. In that case we may re-express S in terms of φ_B and avoid the above difficulty. Suppose, however, that it is not possible to identify φ_B and the best that is possible is to identify a set of constraint relations for ψ's. There will be $N-M$ such constraint relations which may not be explicitly solvable to give us the M independent φ's.

We now return to general relativity where such a situation does exist. Because of the invariance of the theory under the four coordinate transformations, the ten Einstein equations:

$$R_{ik} - \tfrac{1}{2}\, g_{ik}\, R = -\varkappa\, T_{ik} \tag{15}$$

are not independent but are connected by the four Bianchi identities

$$(R^{ik} - \tfrac{1}{2}\, g^{ik}R)_{;k} \equiv 0 \equiv T^{ik}{}_{;k}. \tag{16}$$

Thus, instead of ten variables only six are independent; but they cannot be explicitly identified.

We now recall Equation (7.59) of Section 7.4 giving the action for the gravitational field in the form

$$S = \int \{\pi^{\mu\nu}\dot{g}_{\mu\nu} - NX^0 - N_\alpha X^\alpha\}\, \mathrm{d}^3\mathbf{x}\, \mathrm{d}t, \tag{17}$$

where N and N_α are the Lagrange multipliers for the constraints

$$X^0 \equiv \sqrt{-{}^3g}\, \{{}^3R + (\mathrm{Tr}\, K)^2 - \mathrm{Tr}\, K^2\} = 0, \tag{18}$$

$$X^\alpha \equiv -2\, \{\sqrt{-{}^3g}\, (K^{\mu\alpha} - g^{\mu\alpha}K)\}_{;\mu} = 0. \tag{19}$$

The constraints are on the metric tensor g_{ik}, reducing their effective (independent) number to six. We remind the reader what the ten g_{ik} are, from the line element (7.60):

$$g_{00} = N^2 - N_\alpha N^\alpha, \qquad g_{0\alpha} = N_\alpha, \quad g_{\mu\nu}. \tag{20}$$

For the action (17) the simplest quantization scheme is described through the commutators given below:

$$[g_{\alpha\beta}(x), g_{\alpha\beta}(x')] = 0, \qquad [\pi^{\alpha\beta}(x), \pi^{\alpha\beta}(x')] = 0$$

$$[g_{\alpha\beta}(x), \pi^{\mu\nu}(x')] = \tfrac{1}{2} i(\delta^\mu_\alpha\, \delta^\nu_\beta - \delta^\nu_\alpha\, \delta^\mu_\beta)\delta(x, x'). \tag{21}$$

The operators $\pi^{\alpha\beta}$ are therefore given by the following metric representation in terms of functional derivatives

$$\pi^{\alpha\beta} = -i\, \frac{\delta}{\delta g_{\alpha\beta}}\,. \tag{22}$$

We recall from Equation (7.58) the classical definition of metric momentum:

$$\pi^{\alpha\beta} = \frac{\delta S}{\delta \dot{g}^{\alpha\beta}} = \sqrt{-{}^3g}\, \{g^{\alpha\beta}(\mathrm{Tr}\, K) - K^{\alpha\beta}\}. \tag{23}$$

In the 'Schrödinger representation' we think of a wave functional $\Psi(g_{\alpha\beta})$ representing a quantum state for spacetime geometry. Ψ allocates different probabilities to different metric functions $g_{\alpha\beta}$ and the 'Schrödinger equation' for quantum gravity is supposed to tell us how Ψ varies with respect to the variation of $g_{\alpha\beta}$ in the function space. Our next job is to write down formally this equation.

11.3.1. *Superspace*

Omitting for the time being the constraint equations, we concentrate on the action given by Equation (7.57) which has the integrand

$$I \equiv \sqrt{-{}^3g}\,[{}^3R - (\mathrm{Tr}\,K)^2 + \mathrm{Tr}\,K^2]$$

$$= \sqrt{-{}^3g}\;{}^3R + \sqrt{-{}^3g}\,[K_{\alpha\beta}K^{\beta\alpha} - K^{\alpha}_{\alpha}K_{\beta}{}^{\beta}]$$

$$= V + \sqrt{-{}^3g}\,[g^{\mu\beta}g^{\nu\alpha} - g^{\mu\nu}g^{\alpha\beta}]\,K_{\alpha\beta}K_{\mu\nu},$$

where, as we shall shortly see,

$$V = \sqrt{-{}^3g}\;{}^3R \tag{24}$$

serves the role of the driving potential. Notice, however, that for the special line element of Equation (7.52), Equation (7.54b) gave

$$K_{\mu\nu} = \tfrac{1}{2}\dot{g}_{\mu\nu}. \tag{25}$$

Hence, if we define the fourth-rank tensor G by

$$G^{\mu\nu\alpha\beta} = \tfrac{1}{2}\sqrt{-{}^3g}\,(g^{\mu\alpha}g^{\nu\beta} + g^{\mu\beta}g^{\nu\alpha} - 2g^{\mu\nu}g^{\alpha\beta}), \tag{26}$$

we can write the above integrand as

$$I = V + \tfrac{1}{4}\,G^{\mu\nu\alpha\beta}\dot{g}_{\mu\nu}\dot{g}_{\alpha\beta}. \tag{27}$$

Definition (26) is due to B. S. De Witt.

At this stage we introduce the concept of 'superspace'. This is an abstract space whose typical points are given by the metric $\{g_{\mu\nu}\}$, modulo coordinate transformations. In other words, each point of the superspace describes a 3-geometry of ordinary space. Viewed in this way, the second term of I in (27) suggests that we introduce a 'metric' for the superspace by the functional relation

$$\mathrm{d}L^2 = \int G_{\mu\nu\alpha\beta}\,\mathrm{d}g^{\mu\nu}\,\mathrm{d}g^{\alpha\beta}\,\mathrm{d}^3\mathbf{x}$$

$$= \iint G_{\mu\nu\alpha\beta}(\mathbf{x}, \mathbf{x}')\,\mathrm{d}g^{\mu\nu}\,\mathrm{d}g^{\alpha\beta}\,\mathrm{d}^3\mathbf{x}\,\mathrm{d}^3\mathbf{x}'. \tag{28}$$

Here we have defined the 'superspace metric tensor' by

$$G_{\mu\nu\alpha\beta}(\mathbf{x}, \mathbf{x}') = G_{\mu\nu\alpha\beta}(\mathbf{x})\,\delta_3(\mathbf{x}, \mathbf{x}') \tag{29}$$

Notice the similarity between (28) and the ordinary spacetime metric

$$\mathrm{d}s^2 = g_{ik}\,\mathrm{d}x^i\,\mathrm{d}x^k. \tag{30}$$

As in the compact summation convention of (30) we can write (28) as

$$dL^2 = G_{\mu\nu\alpha\beta}(\mathbf{x}, \mathbf{x}')\, dg^{\mu\nu}(\mathbf{x})\, dg^{\alpha\beta}(\mathbf{x}'), \tag{31}$$

where it is assumed that not only do we sum over repeated indices, but we also integrate over repeated spatial variables $\mathbf{x}$ and $\mathbf{x}'$.

With this notation the action becomes

$$S = \iint I\, d^3\mathbf{x}\, dt = \iint V\, d^3\mathbf{x}\, dt + \int \frac{1}{4}\left(\frac{dL}{dt}\right)^2 dt, \tag{32}$$

where the time integrals are over the time range $[t_1, t_2]$, as in Chapter 7.

Example 11.3. Suppose we make a transformation from the chosen time coordinate t to another time coordinate t' related to t by

$$dt = f^2\, dt'$$

where f is a function of t'. Then (32) changes to

$$S = \iint Vf^2\, d^3\mathbf{x}\, dt' + \frac{1}{4}\int \left(\frac{dL}{dt'}\right)^2 f^{-2}\, dt'.$$

Thus a reparametrization of time coordinate is equivalent to making a change from (V, dL^2) to $(Vf^2, dL^2 f^{-2})$.

□

To make the analogy between spacetime and superspace more exact we relabel $g^{\alpha\beta}$ by a single suffix A to write them as g^A, say. Thus $A = 1, 2, \ldots, 6$ correspond to $(\alpha, \beta) = (1, 1,), (1, 2), (1, 3), (2, 2), (2, 3), (3, 3)$. The action may then be written as

$$S = \iint V\, d^3\mathbf{x}\, dt + \frac{1}{4}\iint G_{AB}\dot{g}^A \dot{g}^B\, dt. \tag{33}$$

The variation of g^A gives the forced geodetic equation in superspace:

$$\frac{d^2 g^A}{dt^2} + \Gamma^A_{BC}\frac{dg^B}{dt}\frac{dg^B}{dt} = V^{:A}. \tag{34}$$

These equations are equivalent to the space–space parts of the Einstein equations. On the right-hand side, $V^{:A}$ denotes the functional derivative of V with respect to g^A. The Γ^A_{BC} are similarly defined by

$$\Gamma^A_{BC} = \tfrac{1}{2} G^{AD}\,[G_{DB.C} + G_{DC.B} - G_{BC.D}]. \tag{35}$$

If we can choose f in Example 11.3 such that $f^2 V$ = constant, then with respect to the new time coordinate t' (34) becomes just the geodetic equation:

$$\frac{d^2 g^A}{dt'^2} + \Gamma^A_{BC}\frac{dg^B}{dt'}\frac{dg^B}{dt'} = 0. \tag{36}$$

It is also possible to express these equations through a *super-Hamiltonian* H, defined by (where p_A is the momentum conjugate to g_A),

$$H = G^{AB}p_Ap_B - \sqrt{-{}^3g}\ {}^3R. \tag{37}$$

The geodetic equation (36) is then equivalent to the two sets of Hamilton's equations

$$\frac{\mathrm{d}g^A}{\mathrm{d}t'} = \frac{\partial H}{\partial p_A}, \quad \frac{\mathrm{d}p^A}{\mathrm{d}t'} = -\frac{\partial H}{\partial g_A}. \tag{38}$$

What about the constraint equations? A little calculation will show that the constraint equation (18) is simply

$$H = 0. \tag{39}$$

The time–space constraints (19) are more difficult to interpret and we shall not consider that problem here. For fuller discussions see the references listed at the end.

From these classical equations we now turn to their quantum gravitational analogue. The quadratic nature of the Hamiltonian (37), the constraint condition (39), together with the operator definition (22) suggest the following form for the Schrödinger equation satisfied by $\Psi[g_A]$:

$$-G_{AB}\frac{\delta^2\Psi}{\delta g_A\ \delta g_B} + V\Psi = 0. \tag{40}$$

Here the operator δ indicates functional differentiation as in (22).

It can be shown that the constraint equations (19) are expressible as:

$$\left(\frac{\delta\Psi}{\delta g_{\beta\alpha}}\right)_{;\beta} = 0. \tag{41}$$

Equations (40) and (41) contain the necessary dynamical information for the quantum gravity situation. However, the formal structure so far erected cannot be put to any practical demonstration in its most general form, largely because we are dealing with *functional* rather than ordinary derivatives. We can, however, *reduce* the infinite number of degrees of freedom of $g_{\alpha\beta}$ to a finite number by going over to symmetric cosmological models like the Robertson–Walker models or the Bianchi type (I–IX) models described in Chapter 8. There the metric tensor becomes describable by a smaller number of functions of t only. In the superspace language we say that the infinite dimensional superspace has shrunk to a 'minisuperspace' of finite dimensions. We shall illustrate this procedure next.

Example 11.4. It may come as a surprise that the constraint equations (40) and (41) are able to describe the dynamics of the spacetime geometry. A simple example from quantum mechanics will illustrate how this works in practice. Consider the action S_p for a spinless relativistic particle:

$$S_p = \tfrac{1}{2}m\int_0^\lambda \dot{x}^i\dot{x}_i\,\mathrm{d}\lambda + \tfrac{1}{2}m\lambda,$$

where $\dot{x}^i = dx^i/d\lambda$. On quantization this action leads to the Schrödinger equation

$$2mi \frac{\partial \psi}{\partial \lambda} = -(\Box + m^2)\psi, \qquad H = \tfrac{1}{2}m\left[\frac{P_i P^i}{m^2} - 1\right]$$

Since m is the rest mass of the particle, H must satisfy the constraint equation $H = 0$. This gives us the Klein–Gordon equation for the spacetime dependent ψ:

$$(\Box + m^2)\psi = 0.$$

□

11.3.2. *Minisuperspace*

To illustrate the concept of minisuperspace consider the diagonal Bianchi model described in Section 8.6. It uses only three time-dependent functions $\lambda(t)$, $\beta_1(t)$, and $\beta_2(t)$ to describe the spacetime metric (cf. Equations (8.97)–(8.99)). Misner has discussed this model in some detail. After a few simple calculations we are led to the line element for the minisuperspace in the form

$$dL^2 = -24 d\lambda^2 + 6(d\beta_1^2 + d\beta_2^2). \tag{42}$$

Notice that in contrast to (31) the minisuperspace has only three dimensions corresponding to the time-dependent functions λ, β_1, and β_2. [In Misner's formulation the superspace metric does not contain the $\sqrt{-{}^3g}$ factor that is present in (26), the definition used by De Witt. Hence, our expression (42) differs from that originally derived by Misner.] If we normalize the spatial integration of (32) to unity, we get

$$S = \int_{t_1}^{t_2} [R\, e^{3\lambda} - \{6\dot{\lambda}^2 - \tfrac{3}{2}(\dot{\beta}_1^2 + \dot{\beta}_2^2)\}]\, dt. \tag{43}$$

The Schrödinger equation corresponding to this action is given by spacializing (40) to the minisuperspace form:

$$\frac{\partial^2 \psi}{\partial \beta_1^2} + \frac{\partial^2 \psi}{\partial \beta_2^2} - \frac{1}{4}\frac{\partial^2 \psi}{\partial \lambda^2} + \frac{1}{24} V\psi = 0. \tag{44}$$

This has a Klein–Gordon type form, and at first sight we may ask: "What is the interpretation of the negative probability densities that such an equation would lead to for some wave functions?"

The answer is that whereas, in real spacetime, negative probabilities imply time going backwards and, hence, they have to be reinterpreted in terms of antiparticles, here the timelike role is played by λ which is related not to real time but to the overall expansion factor of the universe. Thus in (8.93), e^λ represents the average linear scale factor and the increasing of λ describes expansion of the universe. Likewise, a decrease of λ describes contracting phases of the universe. In a general cosmology both expanding and contracting solutions are possible. This is reflected in the quantum treatment here by having some ψ with positive probability densities and some with negative probability densities.

From the above example it is clear that the problem of the Robertson–Walker spacetimes will lead to a one-dimensional superspace. To introduce extra degrees of freedom it is necessary to consider the matter variables appearing on the right-hand side of Einstein's equations. Example 11.5 below describes De Witt's quantized version of the Robertson–Walker universe containing material particles.

Example 11.5. Consider the $k = +1$ Friedmann model with the line element given by

$$ds^2 = \alpha^2\, dt^2 - S^2(t)\lambda^0_{\mu\nu}\, dx^\mu\, dx^\nu,$$

where $\lambda^0_{\mu\nu}$ is the metric on the 3-sphere of unit radius and α, S depend on t. The ${}^3R$ on the unit sphere is $6/S^2$, and the effective Lagrangian for the gravitational field is given by

$$L_g \equiv \int \alpha \sqrt{-\lambda}\ \{K_{ij}K^{ij} - K^2 + {}^{(3)}R\}\ d^3x$$

$$= 12\pi^2\,[\alpha S - \alpha^{-1}S\dot{S}^2],$$

where $\dot{S} = dS/dt$. Let L_m denote the effective particle Lagrangian for dust. If the canonical coordinates of a typical particle are $\{q\}$ and the proper velocities are $\alpha^{-1}\dot{q}$, then its Lagrangian has the form $\alpha l(q, \alpha^{-1}\dot{q})$. If there are N such particles in the universe, we write

$$L_m = n\alpha l(q, \alpha^{-1}\dot{q}).$$

Since there is no explicit dependence in $L_g + L_m$ on $\dot{\alpha}$ we get

$$\bar{\pi} \equiv \frac{\partial(L_g + L_m)}{\partial\dot{\alpha}} = 0.$$

Thus α is an ignorable coordinate. We expect, therefore, that the wave function of the Friedmann universe Ψ will depend on S only. The Hamiltonian is

$$\mathcal{H}_g + \mathcal{H}_m = \bar{\pi}\dot{\alpha} + \Pi\dot{S} + P\dot{q} - L_g - L_m$$

$$= \bar{\pi}\dot{\alpha} + \alpha(\mathcal{H}_g + \mathcal{H}_m),$$

where

$$\Pi = \frac{\partial L_g}{\partial\dot{S}} = -24\pi^2\,\alpha^{-1}\,S\dot{S}$$

$$P = \frac{\partial L_m}{\partial\dot{q}} = N\alpha\,\frac{\partial l}{\partial\dot{q}}$$

$$\mathcal{H}_g = -\frac{\Pi^2}{48\pi^2 S} - 12\pi^2 S$$

$$\mathcal{H}_m = Nm, \qquad m \equiv \frac{1}{\alpha}\,p\dot{q} - l.$$

Note that the kinetic energy term in $\mathcal{H}_g$ has the opposite sign of conventional Hamiltonians. We shall encounter this situation again in Chapter 13. The reason is that the only degree of freedom is $S(t)$ which behaves in a timelike fashion. Since $\bar{\pi} = 0$, we get the wave equation as

$$(\mathcal{H}_g + \mathcal{H}_m)\,\Psi(S) = 0.$$

There is an ambiguity of factor ordering in the first term of $\mathcal{H}_g$. De Witt chooses the ordering which gives

$$\frac{1}{48\pi^2}\, S^{-1/4}\, \frac{\partial}{\partial S}\, S^{-1/2}\, \frac{\partial}{\partial S}\, S^{-1/4}\Psi - (12\pi^2 S - Nm)\Psi = 0.$$

There is no obvious correspondence principle to relate this wave equation to the classical Friedmann solution.

The transformation

$$X = S^{3/2};\ \Phi \equiv \left(\frac{\partial S}{\partial X}\right)^{1/2} \Psi = \sqrt{\frac{2}{3}}\,.\, S^{-1/4}\Psi$$

converts this equation to

$$-\frac{3}{64\pi^2}\,\frac{\partial^2\Phi}{\partial X^2} + 12\pi^2 X^{2/3}\,\Phi = Nm\Phi.$$

Treating m as a c-number (so that the dust particles are in their eigenstates) and using the boundary conditions

$$\Phi = 0 \text{ at } X = 0 \Leftrightarrow \Psi = 0 \text{ at } S = 0,$$

we look upon the above equation as the Schrödinger equation for a particle of mass $32\pi^2/3$ moving with energy Nm in a one-dimensional potential given by

$$V = 0 \qquad \text{for } X < 0,$$

$$V = 12\pi^2 X^{2/3} \quad \text{for } X > 0.$$

If we insist on normalizable solutions of the wave equations, then in the WKB approximation we obtain the 'energy' spectrum

$$Nm = \left[\, 48\pi^2\left(n + \frac{3}{4}\right)\right]^{3/2}, \quad n = 0, 1, 2, \ldots$$

while S is bounded above by

$$S_{\max} = \frac{Nm}{12\pi^2}\,.$$

For the solution to apply to the actual universe, n has to be as high as $\sim 10^{120}$. Even so, the wave function being static describes neither the expanding nor the contracting phase. This standing wave solution may be looked upon as a superposition of waves travelling in opposite directions – in the direction of expansion (S increasing) and contraction (S decreasing).

For a discussion of the more general case of particle motions (instead of particles being in stationary states as assumed here) see the literature cited at the end of the chapter. □

Do quantum universes get round the problem of spacetime singularity? If the wave functional Ψ vanished at the classical singularity we could argue that the probability of singular solutions is zero. However, the converse is not true. For example, in the ground state of the H atom the wave function does not vanish at the origin. Yet, the probability of finding the electron at the origin is zero. It is therefore necessary to take into account the volume measure factors in the superspace before deciding whether singular solutions are improbable or not. As yet a general answer yes/no to the singularity question is not available within this canonical framework.

11.4. Manifestly Covariant Quantization

The main conceptual difficulty of quantizing relativity is to separate the gravitational *field* effects of the spacetime metric from the geometrical effects. In the flat spacetime approach of Feynman described in Section 11.2 the latter are ignored altogether and the nonlinear Einstein equations are linearized in the flat spacetime. As we mentioned earlier, this approach suffers from lack of applicability in the strong field limit.

To get round this difficulty, De Witt developed in a series of papers an elaborate formalism which is manifestly covariant. Rather than use the flat spacetime as a background, De Witt used a general but adjustable *c*-number background geometry in which the various gravitational propagators are calculated. Consider first the linearized version of this theory.

Let g_{ik} denote the *c*-number metric tensor of the background spacetime. To this add an operator part $\boldsymbol{\varphi}_{\mathbf{ik}}$ denoting quantum fluctuations. Let $\boldsymbol{\psi}$ denote, in operator form, any matter/radiation field assumed to be present in the spacetime region under consideration. The conventional action functional S depends on $\boldsymbol{\psi}$ and $g_{ik} + \boldsymbol{\varphi}_{\mathbf{ik}}$. Introduce another functional $\bar{S}$ through the definition

$$\bar{S}[g, \boldsymbol{\varphi}, \boldsymbol{\psi}] = S[g + \boldsymbol{\varphi}, \boldsymbol{\psi}] - S[g, \mathbf{0}] - \int \frac{\partial S[g, \mathbf{0}]}{\partial g_{ik}} \boldsymbol{\varphi}_{ik}\, \mathrm{d}^4x. \tag{45}$$

If no external sources are present we put $\boldsymbol{\psi} = \mathbf{0}$.

If the as yet arbitrary g_{ik} are chosen to satisfy the classical Einstein equations in empty spacetime, then the operator field equations become

$$\frac{\delta\bar{S}[g, \boldsymbol{\varphi}, \boldsymbol{\psi}]}{\delta\boldsymbol{\psi}} = \mathbf{0} \tag{46}$$

and

$$F^{iklm}\boldsymbol{\varphi}_{lm} = -\tfrac{1}{2}\mathbf{T}^{ik}, \tag{47}$$

where

$$F^{iklm}\boldsymbol{\varphi}_{lm}(x) = \int \frac{\delta^2 S[g, \mathbf{0}]}{\delta g_{ik}(x)\, \delta g_{lm}(x')} \boldsymbol{\varphi}_{lm}(x')\, \mathrm{d}^4x' \tag{48}$$

and

$$\mathbf{T}^{ik} = \frac{2\delta\bar{S}[g, \boldsymbol{\varphi}, \boldsymbol{\psi}]}{\delta g_{ik}} . \tag{49}$$

Because $\boldsymbol{\varphi}$ acts as its own source, the equations are more involved than they appear at first glance. The operator conservation law follows, however, as in the classical theory:

$$\mathbf{T}^{ik}{}_{;k} = \mathbf{0}. \tag{50}$$

(The covariant derivatives are to be taken with respect to the background metric.)

In the linearized approximation the $\boldsymbol{\varphi}_{ik}$ are regarded on the same footing as $\boldsymbol{\varphi}$ and $\bar{S}[g, \boldsymbol{\varphi}, \boldsymbol{\psi}]$ and the $\mathbf{T}^{ik}$ expanded as functional power series in $\boldsymbol{\varphi}_{ik}$ and $\boldsymbol{\psi}$. These power series then form the basis of a perturbation theory. In the 'one loop' approximation we first consider the operator solutions of the equations

$$F^{iklm}\boldsymbol{\varphi}_{lm} = \mathbf{0} \tag{51}$$

and

$$\left[\frac{\delta\bar{S}[g, \mathbf{0}, \boldsymbol{\psi}]}{\delta\boldsymbol{\psi}} \right]_{\text{Lin}} = \mathbf{0}, \tag{52}$$

where 'Lin' indicates that only linear terms in $\boldsymbol{\varphi}$ are kept. Thus $\boldsymbol{\varphi}_{ik}$ and $\boldsymbol{\psi}$ do not interact with one another but freely propagate in the spacetime with metric g.

For such solutions we compute $\mathbf{T}^{ik}$ up to quadratic order in $\boldsymbol{\varphi}$ and $\boldsymbol{\psi}$. The resulting value of $\langle\mathbf{T}^{ik}\rangle$ is known as the expectation value of the energy tensor in one-loop approximation. This expectation value generates the back reaction on the metric through the relation

$$F^{iklm} \langle\boldsymbol{\varphi}_{lm}\rangle = \langle\mathbf{T}^{ik}\rangle. \tag{53}$$

Considerably deeper discussion is needed, than will be given here, of the appropriate average implied in $\langle\ \rangle$. We saw in Chapter 9 that when there is particle creation *à-la*-Hawking, the vacuum for incoming states $|\text{in, vac}\rangle$ is different from the vacuum for outgoing states $|\text{out, vac}\rangle$. Further, the existence of these vacua is normally based on the spacetime being stationary[3] in the asymptotic past and asymptotic future. If such 'in' and 'out' vacua exist then the meaning to be attached to the expectation value of any operator $\mathbf{A}$ is its Schwinger average:

$$\langle\mathbf{A}\rangle = \frac{\langle\text{out, vac}|\ T(\mathbf{A})\ |\ \text{in, vac}\rangle}{\langle\text{out, vac}\ |\ \text{in, vac}\rangle} . \tag{54}$$

[3] Stationarity may be formally identified with the existence of a timelike Killing vector. See Appendix D.

If the operator **A** is local then it need not be time ordered as implied in $T(\mathbf{A})$. Further, if there is no asymptotic particle creation, the 'in' and 'out' vacua are indentical and $\langle \mathbf{A} \rangle$ becomes the usual vacuum expectation value.

The computation of $\langle \mathbf{T}^{ik} \rangle$ calls for all the tricks learnt in the renormalization programme of quantum electrodynamics. The details show that the integrals diverge and have to be 'made convergent' by the introduction of arbitrary cut-off. Moreover, unlike electrodynamics, the theory is nonrenormalizable with successive perturbation terms producing increasingly divergent terms. In any case a perturbative approach can hardly do justice to the full nonlinearity of relativity.

De Witt has argued that in the nonperturbative approach needed to get round these problems, the Feynman functional integral should play a key role. The following description is intended to give the reader only a flavour of De Witt's technique rather than the full formal arsenal called upon to tackle this intractable problem.

To compact the notation, introduce a single index μ to denote[4] the discrete indices $i, k, \ldots$, and the continuous position coordinates x^k. In particular, the metric tensor $g_{ik}(x^l)$ will be compacted simply as $\boldsymbol{\varphi}_\mu$.

Consider an infinitesimal displacement of spacetime points so that a point P with coordinates x^i goes to the place Q with coordinates: $(x^i + \xi^i)$, where $\xi^i(x)$ are infinitesimal quantities. (We have discussed such displacements in Appendix D.) Denote by $\delta \mathbf{g}_{ik}$ the difference between the metric tensor at P in its new position and the metric tensor that existed at Q before displacement. This is worked out in Appendix D, and we find

$$\begin{aligned} \delta \mathbf{g}_{ik} &= [\mathbf{g}_{lk}\, \xi^l{}_{,i} + \mathbf{g}_{il}\, \xi^l{}_{,k} - \mathbf{g}_{ik,l}\, \xi^l\,] \\ &= -\int [\mathbf{g}_{ik,l}(x)\, \delta(x, x') + \mathbf{g}_{lk}(x)\, \delta(x, x')_{,i} + \\ &\qquad + \mathbf{g}_{il}(x)\, \delta(x, x')_{,k}\,]\, \xi^l\, \mathrm{d}^4x'. \end{aligned} \tag{55}$$

This is a gauge transformation and the operative group is the gauge group. In the compact notation this becomes

$$\delta \boldsymbol{\varphi}_\mu = Q_{\mu l}[\boldsymbol{\varphi}]\, \xi^l, \tag{56}$$

where μ is the group index for i, x. The group structure tells us that the Q's satisfy the relation

$$Q^\mu{}_{i,\lambda}\, Q^\lambda{}_k - Q^\mu{}_{k,\lambda}\, Q^\lambda{}_i = C^l{}_{ik}\, Q^\mu{}_l. \tag{57}$$

[4] De Witt uses Latin indices for this purpose since he has Greek indices for vectors, tensors, etc. Our notation is the opposite of his.

Here a raised Greek index also implies raised Latin indices. Further, the commas in $Q^{\mu}{}_{i,\lambda}$, etc., denote functional derivatives with respect to $\boldsymbol{\varphi}$. The structure constants C satisfy the cyclic Jacobi identity

$$C^p{}_{qr}C^r{}_{st} + C^p{}_{sr}C^r{}_{tq} + C^p{}_{tr}C^r{}_{qs} = 0. \tag{58}$$

Suppose, the field is completely specified (e.g. in terms of a complete set of commuting variables) in states labelled $|\text{in}\rangle$ and $|\text{out}\rangle$. Then by the Feynman prescription the amplitude from 'in' to 'out' is given by the functional integral

$$\langle \text{out}|\text{in}\rangle = N \int \exp\{iS[\boldsymbol{\varphi}]\}\, M(\boldsymbol{\varphi})\, \mathcal{D}\boldsymbol{\varphi}. \tag{59}$$

Here N is the normalizing constant and $M(\boldsymbol{\varphi})$ is a functional measure over $\{\boldsymbol{\varphi}\}$. $S(\boldsymbol{\varphi})$ is the classical action functional.

To determine $M(\boldsymbol{\varphi})$, De Witt considers something like the superspace of the previous section. This involves looking for a metric tensor $\lambda_{\mu\nu'}$ for the superspace of fields $\boldsymbol{\varphi}^{\mu}$, for which the actions of the gauge group do not produce any effective change. In other words, the $\lambda_{\mu\nu'}$ should not change under infinitesimal displacements in the superspace. This requirement is called *isometry* and is discussed, in the context of ordinary spacetime, in Appendix D. The simplest $\lambda_{\mu\nu'}$ are given by

$$\lambda_{\mu\nu'} = \tfrac{1}{2}(-g)^{-1/2}\{g_{il}g_{km} + g_{im}g_{kl} - \lambda g_{ik}g_{lm}\}\ \delta(x, x'), \tag{60}$$

where $\mu = (i, k, x)$, $\nu' = (l, m, x')$ and $\lambda \neq \frac{1}{2}$. For this $\lambda_{\mu\nu'}$, the measure M, which is proportional to the determinant of $\lambda_{\mu\nu'}$, is a constant and can be taken to be unity.

Example 11.6. It is easy to see that $\|\mu_{\mu\nu'}\|$ is a square matrix of order 10×10 in its discrete indices:

$$\mu, \nu' \equiv 1(0,0), 2(0,1), 3(0,2), 4(0,3), 5(1,1), 6(1,2),$$
$$7(1,3), 8(2,2), 9(2,3), \text{ and } 10(3,3).$$

Let us compute the determinant of the matrix in { } of (60). Writing it as $G_{\mu\nu}$ with $x' = x$ (because of the multiplying delta function)

$$G_{\mu\nu} = \{g_{il}g_{km} + g_{im}g_{kl} - \lambda g_{ik}g_{lm}\}\ .$$

first note how $G_{\mu\nu}$ transforms under coordinate transformation $x^i \to x'^i$ in spacetime. We find that

$$G'_{\mu\nu} = T_{\mu\sigma}T_{\nu\varrho}G_{\sigma\varrho},$$

where $T_{\nu\varrho}$ is the matrix

$$T_{\nu\varrho} \equiv T_{(ik)(lm)} \equiv \frac{\partial x^i}{\partial x'^l}\ \frac{\partial x^k}{\partial x'^m}\ .$$

Thus

$$\det\|G'_{\mu\nu}\| = \det\|G_{\nu\varrho}\| \cdot \{\det\|T_{\nu\varrho}\|\}^2 .$$

But determinant of $\|T_{\nu\varrho}\|$ is a sum of 10! product expressions with each expression having factors $(\partial x^i/\partial x'^l) \times (\partial x^k/\partial x'^m)$, ten in number. It will be seen that permutations of ν, ϱ across the ten values 1–10 can be expressed as products of permutations of i, k, etc., over 0, 1, 2, 3. There are five such products so that

$$\det\|T_{\nu\varrho}\| = J^5, J = \frac{\partial(x^0, x^1, x^2, x^3)}{\partial(x'^0, x'^1, x'^2, x'^3)} .$$

Therefore,

$$\det\|G'_{\mu\nu}\| = J^{10} \det\|G_{\nu\varrho}\| .$$

However, from (6.25) we have

$$\sqrt{-g'} = \sqrt{-g} \times J.$$

Therefore,

$$\det\|(-g')^{-1/2}G'_{\mu\nu}\| = \det\|(-g)^{-1/2}G_{\mu\nu}\| .$$

Hence, without loss of generality we consider the case where g_{ik} are diagonal. Then it is easy to see that the only nonzero components of $\|G_{\mu\nu'}\|$ are:

$$G_{11} = g_{00}^2(2 - \lambda), \quad G_{55} = g_{11}^2(2 - \lambda), \quad G_{88} = g_{22}^2(2 - \lambda), \quad G_{1010} = g_{33}^2(2 - \lambda)$$

$$G_{15} = G_{51} = -\lambda g_{00}g_{11}, \quad G_{18} = G_{81} = -\lambda g_{00}g_{22}, \quad G_{110} = G_{101} = -\lambda g_{00}g_{33}$$

$$G_{58} = G_{85} = -\lambda g_{11}g_{22}, \quad G_{510} = G_{105} = -\lambda g_{11}g_{33}, \quad G_{810} = G_{108} = -\lambda g_{22}g_{33}$$

$$G_{22} = 2g_{00}g_{11}, \quad G_{33} = 2g_{00}g_{22}, \quad G_{44} = 2g_{00}g_{33},$$

$$G_{66} = 2g_{11}g_{22}, \quad G_{77} = 2g_{11}g_{33}, \quad G_{99} = 2g_{22}g_{33}.$$

Therefore,

$$G \equiv \det\|G_{\mu\nu}\| = 2^6(g_{00}g_{11}g_{22}g_{33})^5\,\Delta,$$

where

$$\Delta = \begin{vmatrix} 2-\lambda & -\lambda & -\lambda & -\lambda \\ -\lambda & 2-\lambda & -\lambda & -\lambda \\ -\lambda & -\lambda & 2-\lambda & -\lambda \\ -\lambda & -\lambda & -\lambda & 2-\lambda \end{vmatrix} = 16(1 - 2\lambda).$$

Hence we get

$$G = 2^{10}g^5(1 - 2\lambda),$$

$$f(x) \equiv \det\|\tfrac{1}{2}(-g)^{-1/2}G_{\mu\nu}\| = -(1 - 2\lambda).$$

Because of the delta function $\delta(x, x')$, the continuum part of the metric $||\lambda_{\mu\nu'}||$ is diagonal and its determinant is simply the product of $f(x)$ over all x. Further, since f is independent of x, the determinant is a constant and nonzero provided that $\lambda \neq \frac{1}{2}$.

Here we end our superficial discussion of De Witt's manifestly covariant approach. The formalism is far from being in a complete state; nor can it yet be applied to any specific problem of quantum gravity and cosmology. Its value lies mainly in its attempts to clarify the conceptual difficulties of quantum gravity in a way that is manifestly independent of any special choice of coordinates.

11.5. Path Integrals in Euclidean Spacetime

G. Gibbons and S. W. Hawking have suggested a novel approach towards quantization of gravity that also involves the use of Feynmans path integrals. The novelty lies in the circumstance that the 'spacetimes' over which path integrals are evaluated are made Euclidean, i.e. with a metric of signature (−, −, −, −). Euclideanization as a technique in field theory has already been discussed in Part I. Here we consider it afresh in the gravitational context.

The usual path integral (in real spacetime) has the form

$$K = \int \exp(iS)\, \mathcal{D}\Gamma, \tag{61}$$

where the action S evaluated for any path Γ is real. Thus the integrand in K is an oscillating exponential and it is not clear that the integral would converge and have any physical significance.

Suppose we make a transformation

$$t \to -i\tau, \tag{62}$$

where τ is real. The time is thus purely imaginary and so is S. Therefore, if we write $S = i\hat{S}$, the path integral (61) becomes

$$K = \int \exp(-\hat{S})\, \mathrm{D}\Gamma. \tag{63}$$

If $\hat{S}$ is positive, the integrand is a damped exponential and hence (most likely) convergent).

Clearly, as discussed towards the end of Chapter 10, in a general spacetime the time coordinate is arbitrary and the Euclidean space obtained by (62) is also to that extent arbitrary. Thus the process is not covariant, nor is it automatically possible in all spacetimes. If, however, the original spacetime has a timelike Killing vector (see Appendix D) the corresponding time has a special status and could be used in the transformation (62) with some justification. Still, this justification disappears when we consider spacetime fluctuations which do not have the isometry of the original spacetime.

With these reservations in mind we give a brief outline of the Gibbons–Hawking approach. Its advantage of convergent path integrals (provided they can be evaluated) has already been mentioned. The classical problem of $\delta S = 0$ yields, in real spacetime, hyperbolic differential equations. Such equations may not have solutions for boundary conditions which are not well posed on initial and final spacelike hypersurfaces. By contrast, the Euclidean version gives elliptic differental equations which do have well-defined solutions in most cases.

There is, however, another difficulty. The Euclidean action for scalar or Yang-Mills fields is positive definite while that of classical (relativistic) gravity, not so. Thus $\hat{S}$ is not positive in all cases and convergence of (63) is not guaranteed. (The Euclidean action for fermion fields is also not positive definite, but there the anticommutation relations guarantee convergence.)

In gravity, a conformal transformation

$$\tilde{g}_{ik} = \Omega^2 g_{ik} \tag{64}$$

changes the classical Hilbert action

$$\hat{S} = \frac{-1}{16\pi} \int R \sqrt{-g}\, \mathrm{d}^4x \tag{65}$$

to the form

$$\tilde{\hat{S}} = -\frac{1}{16\pi} \int (\Omega^2 R + 6\Omega_i \Omega^i) \sqrt{-g}\, \mathrm{d}^4x, \tag{66}$$

where $\Omega_i = \partial/\partial x^i$. By choosing a rapidly varying Ω, $\tilde{S}$ can be made arbitrarily large and negative. The integral over $\exp(-\hat{S})$ cannot therefore converge.

To get round this problem the path integral is split into two components. In one component all metrics conformal to a given metric are included, while in the other we have all different metrics, no two of which are conformal to each other. (This is possible because conformal transformation is algebraically an equivalence relation.) The path integration is done separately over the two components, of which the former (which may blow up because of the above effect) may be artificially controlled.

The stationary phase approximation or the equivalent of the WKB method is used to evaluate the path integral approximately. In this approximation, metrics in the neighbourhood of the classical metric (i.e. the solution of Einstein's equations) are included in the path integral. In the one-loop approximation only departures up to second order from the classical metric are included.

Example 11.7. Suppose the Euclidean action is written as $\hat{S}[g, \varphi]$, where g is the symbolic representation of spacetime geometry and φ is the representative matter field. Denote by $\hat{S}[g_0, \varphi_0]$ the solution of Einstein's equations and assume, for simplicity, that $\varphi_0 = 0$. Because of the stationary action condition at g_0, φ_0, if we expand $\hat{S}[g, \varphi]$ as a Taylor series around g_0, φ_0 the first-order term in perturbations $\bar{g}$, $\bar{\varphi}$ will be zero. Thus, we have

$$\hat{S}[g, \varphi] = \hat{S}_0[g_0, 0] + \hat{S}_2[\bar{g}, \bar{\varphi}] + \cdots$$

In the stationary phase approximation we neglect the higher order terms. Further, since $\varphi_0 = 0$, we may decouple $\bar{g}$ from $\bar{\varphi}$ and write

$$\hat{S}_2[\bar{g}, \bar{\varphi}] = \hat{S}_2[\bar{g}] + \hat{S}_2[\varphi].$$

Then the propagator is given by K, where

$$\ln K = -\hat{S}_0[g_0] + \ln \int \exp\{-\hat{S}_2[\varphi]\}\mathcal{D}\varphi + \ln \int \exp\{-\hat{S}_2[g]\}\,\mathcal{D}g.$$

Consider the second term only. Usually $\hat{S}_2(\varphi]$ is expressible as

$$S_2[\varphi] = \tfrac{1}{2}\int \varphi A \varphi \sqrt{g_0}\,\mathrm{d}^4x,$$

where A is a differential operator defined in the background spacetime.

Using techniques discussed in Chapter 4 (cf. Section 4.6), we can evaluate the Euclidean path integral as

$$\int \exp\{-\hat{S}_2[\varphi]\}\,\mathrm{D}\varphi = N(\det A)^{-1/2},$$

where N is a normalizing factor. Regularization techniques like the zeta-function method of Section 4.6 are needed to regularize the determinant of A.

11.5.1. *Spacetime Foam*

It was suggested by J. A. Wheeler that quantum fluctuations may also lead to changes of topology on the scale of the Planck length L_P. The idea is sometimes described under the title of 'spacetime foam' and is discussed briefly below.

Consider path integrals over metrics with a given 4-volume V, i.e.

$$\int \sqrt{g}\;\mathrm{d}^4x = V. \tag{67}$$

Using the stationary phase approximation we therefore introduce a Lagrangian multiplier λ for the above restriction, and write

$$Z(\Lambda) = \int \exp\{-\hat{S}[g] - \frac{\Lambda}{8\pi}\;V[g]\}\mathcal{D}g. \tag{68}$$

Stationarizing the exponent gives the classical theory with the Λ term for the Euclidean spacetime. The value of Λ, however, turns out to be large compared to Einstein's original cosmological constant since the only length scale in the theory is the Planck length.

The number of gravitational fields with 4-volumes between V and $V + \mathrm{d}V$ is given by $N(V)\,\mathrm{d}V$, where

$$N(V) = \frac{1}{16\pi^2 i}\int_{-i\infty}^{\infty} Z(\Lambda)\exp(\Lambda V)\,\mathrm{d}\Lambda. \tag{69}$$

As the choice of the letter Z implies, the function $Z(\Lambda)$ behaves like the thermodynamic *partition function*. Hawking calls $Z(\Lambda)$ the partition function for the volume canonical ensemble. The dominant contribution to $Z(\Lambda)$

comes from the solutions of classical Einstein equations with the given Λ as the Λ term.

The complexity of different topologies is measured by the Euler number χ which, for the above classical solutions, becomes

$$\chi = \frac{1}{32\pi^2} \int_{\mathcal{V}} (C_{abcd}C^{abcd} + \tfrac{8}{3}\Lambda^2)\sqrt{g}\,\mathrm{d}^4x, \tag{70}$$

where C_{abcd} is the Weyl tensor. Those familiar with the instanton jargon may refer to χ as the number of gravitational instantons in $\mathcal{V}$.

In the stationary phase approximation we can show that $\chi = hV$, where h is a constant, and the dominant contribution to $N(V)$ comes from the metrics with one gravitational instanton per volume h^{-1}. The collection of such metrics is called the *spacetime foam*.

In the following chapter we shall discuss the path integral quantization in real spacetime. The approach to be described there is different from the Gibbons–Hawking method outlined here. Since the Gibbons–Hawking method has so far been concerned with formal issues, it is not yet possible to compare it with the pragmatic approach of the following chapter.

11.6. Concluding Remarks

Our discussion of the various approaches to quantum gravity has been sketchy, to say the least. This was deliberately so because we do not feel that the reader encountering quantum gravity for the first time would gain much by being exposed to too many formal details. Nevertheless, those who wish to find further details of the approaches described in Sections 11.2–11.5 may consult the literature listed at the end of this chapter.

The references cited will also give details of other formalisms that we have not mentioned here, such as the twistor formalism of Penrose, the geometric approaches of Isham and others, etc. At the time of writing this book these formalisms are still being developed.

Notes and References

1. A classic discussion of linearized gravity and its quantization has been given by:

    Feynman, R. P.: 1963, *Acta Phys. Polon.* **24**, 697.

    See also his contribution in:

    Feynman, R. P.: 1972, *Magic without Magic* (ed. J. Klauder), Freeman.

2. The ADM formalism has been discussed in:

    Arnowitt, R., Deser, S., and Misner, C. W.: 1962, in *Gravitation – An Introduction to Current Research* (ed. L. Witten), Wiley, New York.

3. In three long papers published during the late 1960s, B. S. De Witt outlined the ideas on canonical quantization and superspace. The references are:

De Witt, B. S.: 1967, 'Quantum Theory of Gravity, I', 'II', and 'III', *Phys. Rev.* **160**, 1113; **162**, 1195; **162**, 1239.

4. For Misner's work on superspace see his article in the volume *Magic without Magic*:

Misner, C. W.: 1972, in *Magic without Magic* (ed. J. Klauder), Freeman.

5. Articles by G. Gibbons and S. W. Hawking in the Einstein centenary volume referred below outline their ideas on quantization:

Hawking, S. W., and Israel, W. (eds): 1979, *General Relativity – An Einstein Centenary Survey*, Cambridge.

6. More detailed discussion of many of the topics mentioned in this chapter can be found in:

Isham, C. J., Penrose, R., and Sciama, D. W. (eds); 1975, *Quantum Gravity – An Oxford Symposium*, Oxford.

Isham, C. J., Penrose, R., and Sciama, D. W. (eds): 1981, *Quantum Gravity 2 – Second Oxford Symposium*, Oxford.

Chapter 12

Quantum Conformal Fluctuations

12.1. Quantum Gravity via Path Integrals

In Chapter 7 we discussed some specific applications of Einstein's general relativity and also the general way of formulating the problem of classical geometrodynamics. In the present chapter we shall first examine how *quantum geometrodynamics* can be formulated with the help of Feynman's path integral formalism. The sketchy survey of other approaches to quantum gravity in the previous chapter will have convinced the reader that the problem is not simple, whatever means we may choose to adopt for its solution. We begin, therefore, by looking at the difficulties facing the path integral formalism.

Recalling the discussion of Section 7.4 we note that in the classical geometrodynamic problem we study the evolution of 3-geometries over a chronologically arranged series of spacelike hypersurfaces $\{\Sigma\}$. Thus, naïvely the classical problem describes how, starting from a given 3-geometry ${}^3G_i$ on an initial spacelike hypersurface Σ_i, the Einstein equations evolve the spacetime geometry through $\{\Sigma\}$ to tell us what the 3-geometry ${}^3G_f$ will be like on a final spacelike hypersurface Σ_f.

Figure 12.1 compares this situation with the simplest problem of Newtonian classical mechanics, the motion of a particle of mass m under given forces described by potential V. To know the trajectory of the particle we have to specify its position $\mathbf{r}_i$ and velocity $\dot{\mathbf{r}}_i$ at an initial instant t_i. The final position $\mathbf{r}_f$ at time $t_f > t_i$ is then determined by the stationary action principle

$$\delta \int [\tfrac{1}{2} m\dot{\mathbf{r}}^2 - V]\,\mathrm{d}t = 0. \tag{1}$$

Similarly, the Hilbert action principle

$$\delta \int_v \left[\frac{1}{16\pi} R + \mathscr{L}_m \right] \sqrt{-g}\;\mathrm{d}^4x = 0 \tag{2}$$

determines the march of ${}^3G$ from Σ_i to Σ_f. What do we have to specify on Σ_i to ensure a unique 'trajectory'? In Section 7.4 we mentioned that the work of

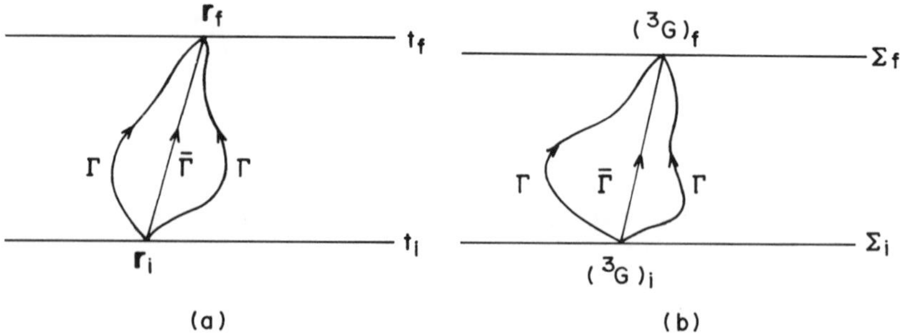

Fig. 12.1. Comparison of Newtonian mechanics with Einstein's gravity. In (a) the initial and final position ($\mathbf{r}_i$, $\mathbf{r}_f$) are specified. Classical physics selects the extremum trajectory $\bar{\Gamma}$. In quantum mechanics, we should sum over all Γ's. In (b) the corresponding situation for gravity is shown.

Isenberg and Wheeler tells us that the appropriate quantities are the conformal part of ^3G and the trace K of the extrinsic curvature tensor. Let us assume that this specification does result in a unique trajectory which we shall symbolically denote by $\bar{\Gamma}$. In Figure 12.1 $\bar{\Gamma}$'s are shown for the two situations being compared here.

To understand how quantum geometrodynamics would proceed let us look at the motion of the particle from the standpoint of quantum mechanics. We note the following difference. The uncertainty principle prevents us from specifying both $\mathbf{r}_i$ and $\dot{\mathbf{r}}_i$ at $t = t_i$. If we just specify $\mathbf{r}_i$ we do not, of course, expect to be able to tell where the particle will be at $t = t_f$. The meaningful question that we can ask in the quantum context is: "What is the probability amplitude that the particle will be at $\mathbf{r}_f$ at $t = t_f$ given that it was at $\mathbf{r}_i$ at $t = t_i$?" The answer, as we saw in Chapter 2, is expressible as a path integral:

$$K[\mathbf{r}_f, t_f; \mathbf{r}_i, t_i] = \int_\Gamma \exp\left\{ \frac{iS[\Gamma]}{\hbar} \right\} \mathscr{D}\Gamma, \tag{3}$$

where $S[\Gamma]$ is the action computed along any given trajectory Γ: $\mathbf{r} = \mathbf{r}(t)$. Paths other than the classical path $\bar{\Gamma}$: $\mathbf{r} = \bar{\mathbf{r}}(t)$, satisfying (1) are also included in the path integral (3). Only in the limit $S \gg \hbar$ does the classical path emerge as the favoured one.

Likewise, we may formally describe the quantum geometrodynamics through a propagator

$$K[^3G_f, \Sigma_f; {^3G_i}, \Sigma_i] = \int_\Gamma \exp\left\{ \frac{iS[\Gamma]}{\hbar} \right\} \mathscr{D}\Gamma \tag{4}$$

with S given by (2). In the limit $S \gg \hbar$ we expect the classical Einstein trajectory $\bar{\Gamma}$ to emerge from the path integral.

So far the formal analogy has helped us in visualizing what quantum gravity implies. In practice, however, the following conceptual and computational difficulties emerge.

12.1.1. *Uncountable Degrees of Freedom*

The path integrals discussed and evaluated in Chapter 2 contained, as in the above example from particle mechanics, a finite (or at most countable) number of degrees of freedom. Thus in (3) we are summing over $\mathbf{r}(t)$ which are only three functions of time. In (4), however, the time-dependent ${}^3G$ are also functions of space. Thus, the degrees of freedom form a continuum rather than a finite discrete set. A rigorous definition and evaluation of such path integrals are therefore going to be much more difficult.

12.1.2. *Constraint Conditions*

The discussions in Section 7.4 and in Chapter 11 show how tricky the role of constraint conditions can be in classical equations of relativity. How can we take due note of it in evaluating the path integral (4)? The problem here is thus more difficult than that of standard quantum field theory in flat spacetime.

12.1.3. *The Problem of Second Derivatives*

The path integrals discussed and evaluated in Chapter 2 relate to actions which contain time derivatives of first order only. This constraint is essential for such a variational principle as (1) to work, for there we require (1) to hold for all paths which have $\mathbf{r}_i$, $\mathbf{r}_f$ fixed. If the action S contained the second derivative $\ddot{\mathbf{r}}$, then the above constraint is not enough; we shall also need to specify $\dot{\mathbf{r}}_i$ and $\dot{\mathbf{r}}_f$. The transitivity condition (2.10) for the probability amplitude will then be inoperative. This was, in fact, the reason why, in the development of quantum mechanics, the wave equation for a particle was expected to contain only the first derivative with respect to time.

The action in (2) – the Hilbert part of it – contains time (and space) derivatives of up to second order. This was noted towards the end of Section 6.8, where we pointed out that to satisfy the condition (2), $\delta g_{ik,l}$ had to vanish on $\partial\mathcal{V}$ *in addition* to the required condition $\delta g_{ik} = 0$. While this additional requirement did not seem to matter in our derivation of Einstein's equations, it poses problems for the path integral approach.

In 1976 Gibbons and Hawking (whose path integral approach was described in Chapter 11) were the first to appreciate this difficulty and to suggest a resolution of it (see note 1 at the end of the chapter). Their prescription invovles adding an extra *surface* term to the action:

$$S_{\rm GH} = \frac{1}{8\pi} \int_{\partial\mathcal{V}} (\chi + \mathscr{L})\,(-h)^{1/2}\,\mathrm{d}^3x. \tag{5}$$

Here h is the determinant of the induced metric on $\partial\mathcal{V}$ and χ is the second fundamental form for $\partial\mathcal{V}$. $\mathscr{L}$ is an extra term whose form is understood only in asymptotically flat spacetimes but which could otherwise be absorbed in the measure of path integrals. The example below illustrates how the Gibbons–Hawking prescription operates in the variational principle. Basically, it produces terms containing $\delta g_{ik,l}$ which neatly cancel the $\delta g_{ik,l}$ terms that arise from the variation of the Hilbert term.

Example 12.1. The second fundamental form of a 3-surface Σ is defined as follows.
Let the unit normal to Σ be given by n_i; that is,

$$n_i n^i = e,$$

where $e = +1$ for timelike n_i and -1 for spacelike n_i. The metric tensor

$$h^i_k = g^i_k - en^i n_k$$

has the property that it projects any 4-vector A_i into Σ. That is, we may write

$$A_i = h_i^k A_k + e(A_k n^k)n_i,$$

where the first part denotes the component lying in Σ (since it is perpendicular to n_i) while the second part is the component lying along the normal n_i to Σ.

Although n_i is defined only on Σ, we can extend its definition further afield by preserving continuity and differentiability further afield. Our definitions need not follow the precise nature of this extension since we shall be concerned with components in Σ only.

Define the second fundamental form χ_{ik} of Σ by

$$\chi_{ik} = h_i^l h_k^m n_{l;m}.$$

The trace of χ_{ik} is given by

$$\chi = h^{ik}\chi_{ik}.$$

Since n_i is unit formal, we get by differentiation

$$n_{i;k}\, n^i = 0.$$

Hence,

$$\chi \equiv n_{i;k}(g^{ik} - en^i n^k) = n_{n;k}g^{ik}.$$

The Gibbons–Hawking term is defined by

$$S_{\mathrm{GH}} = \frac{1}{8\pi}\int_{\partial\mathcal{V}} n_{i;k}(g^{ik} - en^i n^k)\sqrt{-h}\,\mathrm{d}^3x,$$

plus an additional term containing $\mathcal{L}$ which we shall ignore in what follows.

Consider variations in $\mathcal{V}$ bounded by $\Sigma \equiv \partial\mathcal{V}$ of the form $g_{ik} \to g_{ik} + \delta g_{ik}$, where $\delta g_{ik} = 0$ on Σ. We shall not *assume* that $\delta g_{ik,l}$ also vanish on Σ. However, since $\delta g_{ik} = 0$ everywhere on Σ, $\delta g_{ik,l}\, h^{kl} = 0$ on Σ. Only the normal component of $\delta g_{ik,l}$ could therefore be nonzero on Σ.

It is easy to verify that under the above variations on Σ,

$$\delta n_i = 0, \quad \delta n_{i,k} = \tfrac{1}{2} e n_i n^p n^q\, \delta g_{pq,k}.$$

Hence, $h^{ik}\,\delta n_{i,k} = 0$. (All variations of g_{ik}, h_{ik}, n_i vanish on Σ by prescription of $\delta g_{ik} = 0$ on Σ.)

Hence,

$$\delta S_{\mathrm{GH}} = \frac{1}{8\pi}\int_{\Sigma}(-\,\delta\Gamma^l_{ik})n_l(g^{ik} - en^i n^k)\sqrt{-h}\,\mathrm{d}^3x.$$

Compare this term with the variation of Hilbert action (cf. Equation (6.112)):

$$\delta S_g = \frac{1}{16\pi}\int_{\Sigma}(g^{il}\,\delta\Gamma^k_{lk} - g^{lk}\,\delta\Gamma^i_{lk})\,n_i\sqrt{-g}\,\mathrm{d}\Sigma.$$

Further, it can be shown that $\sqrt{-h}\ d^3x = \sqrt{-g}\ d\Sigma$. To show that the two variations cancel consider

$$\delta S_H - \delta S_{GH} = \frac{1}{16\pi}\{n_i g^{il}\,\delta\Gamma^k_{lk} - 2en^i n^k n_l\,\delta\Gamma^l_{ik} + g^{lk}\,\delta\Gamma^i_{lk} n_i\}\sqrt{-g}\ d\Sigma.$$

After some manipulation this reduces to

$$\delta S_H - \delta S_{GH} = \frac{1}{16\pi}\int_\Sigma (g^{kl} - en^k n^l)\,\delta g_{ik,l}\, n^i \sqrt{-g}\ d\Sigma$$

$$= 0;$$

since all components of $\delta g_{ik,l}$ in Σ vanish by our earlier arguments. □

Note that such a term restores the desired property of 'paths'; namely, that they are specified only by the values of g_{ik} at their ends. The Gibbons–Hawking prescription is coordinate independent and hence is free from the defect of the alternative Hilbert action (6.113).

In what follows we shall not explicitly mention S_{GH}, nor shall we take note of $g_{ik,l}$ terms on $\partial\mathcal{V}$ while evaluating path integrals based on the Hilbert term only. The assumption is that the $g_{ik,l}$ terms will be cancelled by similar terms in S_{GH}.

12.2. Conformal Fluctuations

The above difficulties are considerably simplified if we confine ourselves to conformal fluctuations around the classical solution. The idea of a conformal fluctuation may be introduced directly in the following manner.

Let $\bar{\mathcal{M}}$ denote a spacetime manifold with metric $\{\bar{g}_{ik}\}$ which satisfies the Einstein equations. Denote by $t = t_i$ and $t = t_f$ the initial and final spacelike hypersurfaces Σ_i and Σ_f which sandwich the region $\mathcal{V}$. Consider the conformal transformation

$$g_{ik} = (1 + \varphi)^2\,\bar{g}_{ik}, \tag{6}$$

where φ is an arbitrary scalar function of spacetime coordinates. We shall denote the new manifold with $\{g_{ik}\}$ as the metric by $\mathcal{M}_\varphi$, and call φ the conformal fluctuation of $\mathcal{M}_\varphi$ from $\bar{\mathcal{M}}$.

In the language of path integrals described in the previous section, the geometrical description of $\bar{\mathcal{M}}$ in the sandwich region $\mathcal{V}$ is given by the path $\bar{\Gamma}$. A nonclassical path Γ_φ corresponds to a specified function φ. By varying φ over the entire range of C^2 functions we generate a range of nonclassical geometries which do not satisfy Einstein's equations in general. If we confine ourselves to only this range of geometries, we arrive at a simplified version of (4), viz.

$$K\,[\varphi_f, t_f; \varphi_i, t_i] = \int_{\Gamma_\varphi} \exp iS\,[\Gamma_\varphi]\mathcal{D}\varphi, \tag{7}$$

when, henceforth, we take $\hbar = 1$ unless otherwise stated.

Before proceeding further, let us pause and consider the implications of what we are going. The specification of φ_i and φ_f at $t = t_i$ and t_f means that in the specification of 3-geometries we are departing from the classical specifications ${}^3\bar{G}_i$ and ${}^3\bar{G}_f$ by only the conformal factors. We have already seen that in the Isenberg–Wheeler prescription only the conformal part of the 3-geometry needs to be specified. Thus, our specification at t_i does not differ from the classical prescription. It is, however, incomplete since the trace of the extrinsic curvature is not specified – and herein lies the freedom afforded by conformal fluctuations from the stranglehold of Einstein equations. As will be clear shortly (see Equation (9)) the above prescription tranfers the degrees of freedom from metric to the conformal fluctuation φ. Thus we are essentially quantizing φ against the background $\bar{\mathcal{M}}$.

This procedure is certainly not as general as that demanded by a full theory of quantum gravity. Rather than cover the full possible ranges of geometries, we are limiting ourselves to those which are conformal to the classical geometry. It this going to give us any insight into quantum gravity?

Our defence of this specialization rests on three arguments. First, our entire approach to quantum gravity is pragmatic, guided by the way quantum theory itself developed. Take the case of the hydrogen atom. Bohr's theory, primitive though it was, proved to have great hidden depth. The discreteness of the orbital angular momentum of the atomic electron was assumed *ad hoc*. Its significance was only later appreciated by the theories of Schrödinger and Heisenberg which were further formalized by Dirac. The next input was of relativistic quantum mechanics and the role of electron-spin. The more sophisticated applications of quantum electrodynamics followed later; and even today the purist could object to the entire approach on the grounds that the perturbation theory generates divergent integrals (see Chapter 4). Our limited approach here is dictated by 'simplicity of concept' with the hope that, like Bohr's theory, the results obtained by this approach can be given greater justification by a later more formal theory. In any case the present approach will point to specific effects of quantum gravity which reveal some of the ways in which classical gravity is inadequate in describing natural phenomena.

In this connection we may also mention that in the introduction to quantum mechanics the solutions of Schrödinger's equation for limited degrees of freedom (such as the one-dimensional square well potential, the simple harmonic oscillator, etc.) have provided useful insights into the working of quantum theory. Likewise, our limited approach of studying conformal fluctuations will be seen to provide glimpses of the effects of quantum gravity.

While the first argument sounds somewhat defensive, the second – which we shall give now – is not. Remembering that in Einstein's description the geometry of spacetime provides the background for the operation of other physical interactions (besides gravity), we have to ask the meaning that can be attached to such interactions during quantum fluctuations of spacetime geometry. In a general fluctuation the causal relation between two spacetime points is not invariant because the light cone structure is not preserved under the fluctuation. There is one exception, however, to this conclusion. *The global light cone structure of a spacetime is preserved in a conformal fluctuation*. Thus, the causal structure of the various propagators describing other interactions (like quantum electrodynamics) is not affected during a conformal fluctuation and a conceptually coherent picture of what is going on is preserved. In short, except where only the conformal degrees of freedom are quantized, the spacetime description of physical interactions is hard to interpret consistently.

Example 12.2. Consider the conformal transformation $\mathcal{M} \equiv \Omega^2\, \overline{\mathcal{M}}$:

$$g_{ik} = \Omega^2 \bar{g}_{ik}.$$

The Christoffel symbols transform in the following way:

$$\Gamma_{i/kl} = \Omega^2 \bar{\Gamma}_{i/kl} + \Omega \left\{ \Omega_k \bar{g}_{il} + \Omega_l \bar{g}_{ik} - \Omega_i \bar{g}_{kl} \right\} ,$$

$$\Gamma^i_{kl} = \bar{\Gamma}^i_{kl} + \frac{1}{\Omega} \left\{ \Omega_k \bar{g}^i_l + \Omega_l \bar{g}^i_k - \Omega^i \bar{g}_{kl} \right\} ;$$

where $\Omega \equiv \partial\Omega/\partial x^i$.

It is obvious that null intervals remain null under this transformation. What about null geodesics? The null geodesics in $\mathcal{M}$ are given, in terms of an affine parameter, by

$$\frac{d^2x^i}{d\lambda^2} + \Gamma^i_{kl} \frac{dx^k}{d\lambda} \frac{dx^l}{d\lambda} = 0.$$

This equation becomes, in terms of $\bar{g}_{ik}$,

$$\frac{d^2x^i}{d\lambda^2} + \bar{\Gamma}^i_{kl} \frac{dx^k}{d\lambda} \frac{dx^l}{d\lambda} + \frac{2}{\Omega} \frac{d\Omega}{d\lambda} \frac{dx^i}{d\lambda} = 0,$$

that is,

$$\frac{d^2x^i}{d\bar{\lambda}^2} + \bar{\Gamma}^i_{kl} \frac{dx^k}{d\bar{\lambda}} \frac{dx^l}{d\bar{\lambda}} = 0.$$

with

$$\bar{\lambda} = \int \frac{d\lambda}{\Omega^2}$$

as the affine parameter in $\overline{\mathcal{M}}$. Thus, even null geodesics are conformally invariant. □

The third justification is entirely practical: the conformal degrees of freedom are capable of being quantized exactly by the path integral approach. To see this let us consider the change in the Hilbert action under (6). We shall denote by an overbar the quantities defined in the geometry of $\bar{\mathcal{M}}$. The metric $\{\bar{g}_{ik}\}$ will be used to lower or raise tensor indices in any expression with barred quantities. Thus the scalar curvature transforms as

$$R = \frac{\bar{R}}{(1+\varphi)^2} + \frac{6\,\bar{\Box}\,\varphi}{(1+\varphi)^3}\,, \tag{8}$$

and the Hilbert action becomes

$$S_g = \frac{1}{16\pi}\int_{\mathcal{V}} [(1+\varphi)^2\bar{R} - 6\varphi_i\varphi^i]\sqrt{-\bar{g}}\;\mathrm{d}^4x. \tag{9}$$

Here $\varphi_i = \partial\varphi/\partial x^i$ and the second derivatives in $\bar{\Box}\,\varphi$ are transformed away with the help of the Gibbons–Hawking surface term (see Example 12.3 below).

Example 12.3. Let us consider what happens to the Gibbons–Hawking action

$$S_{\mathrm{GH}} = \frac{1}{8\pi}\int_{\partial\mathcal{V}} \chi\sqrt{-h}\;\mathrm{d}^3x$$

under a conformal transformation $g_{ik} = \Omega^2\bar{g}_{ik}$.

Recall from Example 12.1 that

$$h_{ik} = g_{ik} - e n_i n_k.$$

Since n_i is a unit normal vector to $\partial\mathcal{V}$ it transforms as

$$n_i = \Omega\,\bar{n}_i$$

and hence

$$h^i_k = \bar{h}^i_k,\quad h_{ik} = \Omega^2\bar{h}_{ik},\quad h = \Omega^6\bar{h},\text{ etc.}$$

Further, from Example 12.2,

$$\begin{aligned} n_{i;k} &= n_{i,k} - \Gamma^l_{ik}n_l \\ &= (\bar{n}_i\,\Omega)_{,k} - \Omega\bar{\Gamma}^l_{ik}\bar{n}_l - \bar{n}_l\,\{\Omega_k\bar{g}^l_i + \Omega_k\bar{g}^l_k - \Omega^l\bar{g}_{ik}\,\} \end{aligned}$$

Hence,

$$\begin{aligned} \chi &= \frac{1}{\Omega}\bar{\chi} - \frac{1}{\Omega^2}\,(\bar{g}^{ik} - e\bar{n}^i\bar{n}^k)\,(\bar{n}_k\Omega_i - \bar{g}_{ik}\,\Omega^l\bar{n}_l) \\ &= \frac{1}{\Omega}\bar{\chi} + 3\,\frac{\Omega^l n_l}{\Omega^2}\;. \end{aligned}$$

It is now easy to verify that the cancellation mentioned in the text indeed occurs. □

Notice that S_g contains φ, φ_i at most up to the quadratic power and hence by the result derived in Chapter 2 (cf. Section 2.4) the path integral

$$\int \exp iS_g[\varphi]\,\mathcal{D}\varphi \tag{10}$$

can be evaluated exactly. However, we still have to take into account the matter part of the action. We shall consider three special cases which are of interest to cosmology.

12.2.1. *Free Particles*

The action

$$S_m = \sum_a \int m_a \, \mathrm{d}s_a \tag{11}$$

discussed in Chapter 6 (cf. Examples 6.15 and 6.16) describes a system of noninteracting massive particles with the minimum coupling to gravity. In the early stages of the hot Big Bang universe around the Planck epoch, the principle of asymptotic freedom ensures that all particles will be essentially noninteracting and, hence, could be described by the above action.

Under the conformal transformation (6) we find that

$$S_m = \bar{S}_m + \sum_a \int m_a \varphi \, \mathrm{d}\bar{s}_a \, . \tag{12}$$

Thus the matter action contains φ only linearly.

12.2.2. *Radiation*

Since radiation describes zero rest mass photons whose interaction is conformally invariant, we get

$$S_m = \bar{S}_m \, . \tag{13}$$

Thus, in this case S_m is independent of φ.

12.2.3. *Scalar Field*

For a scalar field ψ of zero rest mass the action has the form

$$S_m = \tfrac{1}{2} f \int \psi_i \psi^i \sqrt{-g} \, \mathrm{d}^4 x, \tag{14}$$

where f is a coupling constant. In this case $\psi = \bar{\psi}$,

$$S_m = \bar{S}_m + \tfrac{1}{2} f \int \bar{\psi}_i \bar{\psi}^i (2\varphi + \varphi^2) \sqrt{-\bar{g}} \, \mathrm{d}^4 x, \tag{15}$$

and hence S_m depends on φ quadratically.

We further note that from (6.76) the general relationship between δS_m and $\underset{(\mathrm{m})}{T}{}^{ik}$ is given by

$$\delta S_m = -\tfrac{1}{2} \int \underset{(\mathrm{m})}{T}{}^{ik} \, \delta g_{ik} \sqrt{-g} \, \mathrm{d}^4 x. \tag{16}$$

For small *conformal* fluctuations $\delta g_{ik} = 2\varphi \bar{g}_{ik}$ and hence because of Einstein's equations

$$\delta S_m = S_m - \bar{S}_m = -\int \varphi \bar{T} \sqrt{-\bar{g}}\, \mathrm{d}^4x$$
$$= -\frac{1}{8\pi} \int \varphi \bar{R} \sqrt{-\bar{g}}\, \mathrm{d}^4x. \tag{17}$$

Combining (9) with (17) we see that for small φ

$$S = S_g + S_m = \bar{S}_g + \bar{S}_m + \frac{1}{16\pi} \int_{\mathcal{V}} [\varphi^2\bar{R} - 6\varphi_{;i}\varphi^{;i}] \sqrt{-\bar{g}}\, \mathrm{d}^4x. \tag{18}$$

This expression is, however, *exact* for all φ (not necessarily small) provided the matter action S_m varies at most linearly with φ under conformal fluctuations. In the three cases considered above this condition is satisfied for the first two.

In those two cases we may write (7) in the more explicit form

$$K[\varphi_f, t_f; \varphi_i, t_i] = A \int \exp\left[\frac{i}{16\pi} \int (\varphi^2\bar{R} - 6\varphi_{;i}\varphi^{;i}) \sqrt{-\bar{g}}\, \mathrm{d}^4x\right] \mathcal{D}\varphi, \tag{19}$$

where A is a constant depending on t_i and t_f only. The integrand tells us that the problem of quantum conformal fluctuations (*QCF in brief hereafter*) is thus reduced to the quantization of a scalar field satisfying the wave equation

$$\bar{\Box}\, \varphi + \frac{1}{6} \bar{R}\varphi = 0 \tag{20}$$

against the prescribed classical background $\bar{\mathcal{M}}$. In the rest of this chapter we shall consider the implications of this conclusion for quantum cosmology.

Note also in passing that the restriction to conformal degrees of freedom enables us to discuss the quantum effects in a fully covariant manner. In this respect the approach to be described below differs from the path integral approach of Gibbons and Hawking which, in its Euclideanization of time, departs from general covariance.

12.3. QCF of Friedmann Cosmologies

Consider first a simple application of the above result to the Friedmann models discussed in Chapter 8. We write the classical line element of $\bar{\mathcal{M}}$ as

$$\mathrm{d}\bar{s}^2 = \mathrm{d}t^2 - \bar{Q}^2(t) \left[\frac{\mathrm{d}r^2}{1 - kr^2} + r^2(\mathrm{d}\theta^2 + \sin^2\theta\, \mathrm{d}\psi^2)\right]. \tag{21}$$

(To avoid confusion with the conformal fluctuation φ we have changed the azimuthal angle in the above line element from φ to ψ.) Suppose we wish to consider those QCFs which preserve the homogeneity and isotropy of the universe with respect to the fundamental observers of constant r, θ, ψ. This requires the QCF function φ to depend on t only. We shall proceed here with this assumption which we shall subsequently generalize in later sections.

If φ is a function of t only we can make a time transformation $t \longrightarrow \tau$ such that

$$d\tau = (1 + \varphi)\, dt. \tag{22}$$

The line element for $\mathcal{M}_\varphi$ will then be manifestly in the Robertson–Walker form.

For the sandwich region $\mathcal{V}$ we take the spatial region bounded by a coordinate sphere $r \leq r_b$, say with a coordinate volume

$$V = \int_0^{r_b} \frac{4\pi r^2\, dr}{\sqrt{1 - kr^2}}, \tag{23}$$

between $[t_i, t_f]$. Then, as in Section 12.2.1. or 12.2.2.,

$$S = \bar{S} - \frac{3V}{8\pi} \int_{t_i}^{t_f} (\dot{\varphi} - \frac{1}{6} \bar{R}\, \varphi^2)\bar{Q}^3\, dt. \tag{24}$$

We shall anticipate that the really interesting epochs for quantum cosmology will be close to the classical singular epoch $t = 0$ Hence, we shall be interested in calculating how the QCFs behave as $t_f \rightarrow 0$. The situation is illustrated in Figure 12.2, where our sandwich region is made to approach the singular epoch.

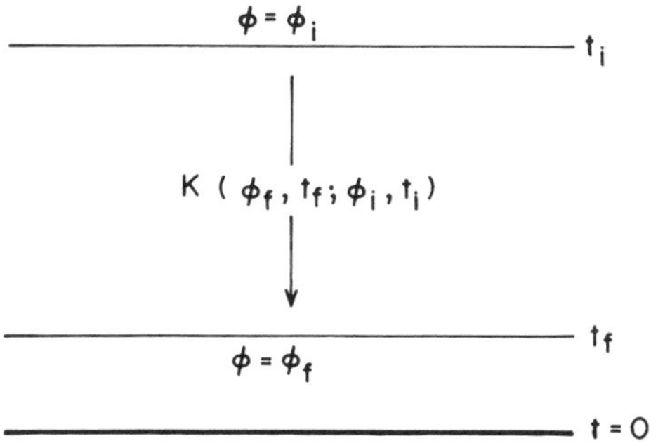

Fig. 12.2. The time evolution is reversed for the cosmological problem since we are now interested in the behaviour of K as $t_f \rightarrow 0$.

Because this epoch lies in the past we have to assume that $t_i > t_f > 0$; that is, we should discuss quantum evolution of the system *backwards in time* towards $t = 0$. This interpretational aspect will confront us towards the end of this section, but we might we might as well prepare for it now.

Accordingly, suppose that $\Psi(\varphi, t)$ denotes the wave function of the quantum universe. We shall suppose that at $t = t_i$ the wave function describes an almost classical situation, i.e. with φ strongly bunched round $\varphi = 1$. An ideal way to describe this wave function is by a wave packet.

$$\Psi_i(\varphi, t_i) = (2\pi\Delta_i^2)^{-1/4} \exp\left(-\frac{\varphi^2}{4\Delta_i^2}\right). \tag{25}$$

Then the quantum state of the universe at $t = t_f$ is given by

$$\Psi_f(\varphi_f, t_f) = \int K[\varphi_f, t_f; \varphi_i, t_i]\, \Psi_i(\varphi_i, t_i)\, \mathrm{d}\varphi_i. \tag{26}$$

Notice that because φ is a function of t only, the above integral is an ordinary rather than a functional integral. Because $t_f < t_i$ we interpret (26) in the reverse sense of the diffusion of a wave packet. We look upon (26) as telling us the state of the universe *in the past* from which it evolved to the observed state Ψ_i at $t = t_i$. The dispersion of the wave packet Δ_i at t_i can be related to the dispersion of Ψ_f at $t = t_f$. This is the problem that we shall examine next.

Using (26) we see that the propagator K is a simple path integral with a quadratic exponential. From Section 2.4 we can write the answer in the form

$$K[\varphi_f, t_f; \varphi_i, t_i] = F(t_f, t_i) \exp\left\{\frac{3iV}{8\pi}\chi\right\}, \tag{27}$$

where the integral

$$\chi = \int_{t_i}^{t_f} [\dot{\varphi}_c^2 - \tfrac{1}{6}\bar{R}\varphi_c^2]\bar{Q}^3\, \mathrm{d}t = 0, \tag{29}$$

is evaluated for that function φ which satisfies the 'classical' variational condition

$$\delta \int_{t_i}^{t_f} [\dot{\varphi}^2 - \tfrac{1}{6}\bar{R}\varphi^2]\bar{Q}^3\, \mathrm{d}t = 0, \tag{29}$$

together with the boundary conditions

$$\varphi_c(t_i) = \varphi_i, \qquad \varphi_c(t_f) = \varphi_f. \tag{30}$$

So we first solve the above variational problem.

From Chapter 8, by neglecting the curvature term, we get

$$\bar{R} = 6\left(\frac{\ddot{\bar{Q}}}{\bar{Q}} + \frac{\dot{\bar{Q}}^2}{\bar{Q}^2}\right). \tag{31}$$

The Euler–Lagrange equation for (29) to be satisfied by φ_c is

$$\frac{\mathrm{d}}{\mathrm{d}t}(\bar{Q}^3\dot{\varphi}_c) + (\bar{Q}^2\ddot{\bar{Q}} + \bar{Q}\dot{\bar{Q}}^2)\varphi_c = 0. \tag{32}$$

To solve this equation, define

$$\xi = \bar{Q}\,\varphi_c. \tag{33}$$

Then some simple calculus gives

$$\bar{Q}\,\ddot{\xi} + \dot{\bar{Q}}\;\dot{\xi} = 0. \tag{34}$$

This integrates to give

$$\xi = A + B\int \frac{\mathrm{d}t}{\bar{Q}(t)}\,. \tag{35}$$

With the boundary conditions (30) to be satisfied, we get

$$\bar{\varphi}_c(t) = \frac{[\varphi_i \bar{Q}_i T(t_f, t) + \varphi_f \bar{Q}_f T(t, t_i)]}{\bar{Q}(t)\; T(t_f, t_i)}\,, \tag{36}$$

where

$$T(x, y) = \int_y^x \frac{\mathrm{d}t}{\bar{Q}(t)}\,, \tag{37}$$

and $\bar{Q}_i = \bar{Q}(t_i)$, $\bar{Q}_f = \bar{Q}(t_f)$.

With $\varphi_c(t)$ given by (36), the computation of the integral χ in (28) is tedious but straightforward. The final value for χ is given by

$$\chi = -\frac{1}{T_{if}}\,\{(1 + T_{if}\,\bar{Q}_f\,H_f)\bar{Q}_f^2\,\varphi_f^2 + (1 - T_{if}\,\bar{Q}_i\,H_i)\bar{Q}_i^2\,\varphi_i^2 -$$

$$- 2\bar{Q}_i\,\bar{Q}_f \varphi_i \varphi_f\}\,, \tag{38}$$

where

$$H_i = \frac{\dot{\bar{Q}}\,(t_i)}{\bar{Q}_i}\,, \qquad H_f = \frac{\dot{\bar{Q}}\,(t_f)}{\bar{Q}_f}\,, \qquad T_{if} = T(t_i, t_f). \tag{39}$$

As shown in Example 12.4 below, the expression for $F(t_f, t_i)$ is given by

$$F(t_f, t_i) = \left(\frac{3Vi}{8\pi^2 T_{if}}\,\bar{Q}_i\,\bar{Q}_f\right)^{1/2}\,. \tag{40}$$

Example 12.4. We determine $F(t_f, t_i)$ from first principles in the following way. Consider the transitive relation for $t_i < t_2 < t_3$:

$$\int_{-\infty}^{\infty} K[\varphi_3, t_3; \varphi_2, t_2]\, K[\varphi_2, t_2; \varphi_1, t_1]\, \mathrm{d}\varphi_2 = K[\varphi_3, t_3; \varphi_1, t_1]\,.$$

We have from (38):

$$K[\varphi_2, t_2; \varphi_1, t_1] = F(t_2, t_1) \exp i(A\varphi_2^2 + 2B\varphi_1\varphi_2 + C\varphi_1^2)$$

$$K[\varphi_3, t_3; \varphi_2, t_2] = F(t_3, t_2) \exp i(D\varphi_3^2 + 2E\varphi_3\varphi_2 + H\varphi_2^2) \, .$$

where $A, \ldots, F$ are known constants. Substitute these into the left-hand side of the above relation and perform the φ_2 integral by completing the square:

$$i(A\varphi_2^2 + 2B\varphi_1\varphi_2 + H\varphi_2^2 + 2E\varphi_3\varphi_2) =$$

$$i(A + H)\left\{\varphi_2 + \frac{B\varphi_1 + E\varphi_3}{A + H}\right\}^2 - i\varphi_3 \frac{(B\varphi_1 + E\varphi_3)^2}{A + H} \, .$$

It is straightforward to verify that the exponential terms on both sides of the transitive relation agree. Hence, we need only compare their coefficients. So we get

$$F(t_3, t_2)\, F(t_2, t_1) \sqrt{\frac{-\pi}{i(A + H)}} = F(t_3, t_1).$$

From (38) we have

$$A = \frac{3V}{8\pi}\left\{-\frac{1}{T_{12}} - \dot{\bar{Q}}\right\}\bar{Q}_2^2 \, , \; H = \frac{3V}{8\pi}\left\{-\frac{1}{T_{23}} + \dot{\bar{Q}}\right\}\bar{Q}_2^2 \, ,$$

so that

$$\sqrt{\frac{-\pi}{i(A + H)}} = \frac{1}{\bar{Q}_2}\sqrt{\frac{8\pi^2}{3Vi}\left(\frac{1}{T_{12}} + \frac{1}{T_{23}}\right)^{-1}} = \frac{1}{\bar{Q}_2}\sqrt{\frac{8\pi^2}{3Vi}\,\frac{T_{12}\,T_{23}}{T_{13}}} \, .$$

Thus we have

$$F(t_3, t_2)\, F(t_2, t_1)\, \frac{1}{\bar{Q}_2}\sqrt{\frac{8\pi^2}{3Vi}\,\frac{T_{12}\,T_{23}}{T_{13}}} = F(t_3, t_1).$$

This relation is satisfied provided that

$$F(t_2, t_1) = \sqrt{\frac{3Vi}{8\pi^2}\,\frac{\bar{Q}_1\bar{Q}_2}{T_{12}}} \, .$$

□

We now use (26) to calculate Ψ_f. It is seen that Ψ_f has again the form of a wave packet centred on $\varphi = 0$ but with a dispersion Δ_f given by

$$\Delta_f = \frac{2\pi T_{if}}{3V\bar{Q}_i\bar{Q}_f}\left[1 + \frac{3V}{2\pi T_{if}\Delta_i^2}\;\bar{Q}_i^2(1 - T_{if}\bar{Q}_i H_i)^2\right]^{1/2} . \qquad (41)$$

For our purpose we need to know how Δ_f behaves as the classical singular epoch $t = 0$ is approached. We already know that $\bar{Q}_f \to 0$ as $t \to 0$. How does T_{if} behave?

From Example 12.5 below we see that T_{if} remains finite for all $\bar{Q}(t)$ which are solutions of the cosmological equations for positive pressure with $\varepsilon > 3p$. The finiteness of T_{if} is in fact a direct consequence of the property that all Friedmann models except the trivial case of the empty universe with $k = -1$ have particle horizons.

Example 12.5. From the cosmological equations (8.5)

$$\frac{\dot{\bar{Q}}^2 + k}{\bar{Q}^2} = \frac{8\pi}{3}\,\varepsilon, \qquad \frac{\mathrm{d}}{\mathrm{d}\bar{Q}}\,(\varepsilon\bar{Q}^3) + 3p\bar{Q}^2 = 0$$

we can prove the existence of particle horizon as follows. (Here ε = energy density and p = pressure.)

From the second equation we get, for $p > 0$, $\varepsilon\bar{Q}^3$ as a nonincreasing function of $\bar{Q}$. Thus, for $\bar{Q} < \bar{Q}_1 = \bar{Q}(t_1)$ we have $\bar{Q}^3\varepsilon > \bar{Q}_1^3\varepsilon_1$, where $\varepsilon_1 = \varepsilon(t_1)$. Therefore, from the first equation we get the inequality:

$$\dot{\bar{Q}}^2\bar{Q}^2 = -k\bar{Q}^2 + \frac{8\pi}{3}\,\varepsilon\bar{Q}^4 > -k\bar{Q}^2 + \frac{8\pi}{3}\,\varepsilon_1\,\bar{Q}_1^3\,\bar{Q}.$$

For t_1 sufficiently close to the Big Bang epoch, we may ignore the $-k\bar{Q}^2$ term on the extreme right. Hence, we get

$$\dot{\bar{Q}}\,\bar{Q} > \alpha\bar{Q}^{1/2} \quad \text{for } \bar{Q} < \bar{Q}_1,$$

for some constant α. Hence,

$$T(t_1, 0) = \int_0^{t_1} \frac{\mathrm{d}t}{Q(t)} = \int_0^{\bar{Q}_1} \frac{\mathrm{d}\bar{Q}}{\bar{Q}\dot{\bar{Q}}} < \alpha^{-1}\int_0^{\bar{Q}_1} \frac{\mathrm{d}\bar{Q}}{\sqrt{\bar{Q}}} < \infty.$$

That is, $T(t_1, 0)$ is finite and tends to zero as $t_1 \to 0$ and $\bar{Q}_1 \to 0$. □

With T_{if} finite in the limit of $t_f \to 0$, (41) tells us that in this limit Δ_f diverges as

$$\Delta_f \sim \bar{Q}_f^{-1}. \tag{42}$$

Figure 12.3 illustrates this backward diffusion of the wave packet. What it means is that because of the divergence of Δ_f, complete uncertainty prevails about the past state of the universe which led to the present near classical state. In particular, the classical solution $\varphi = 0$ has no special *locus standi* at the singular epoch in the sense that it is an average of a probability distribution whose dispersion is infinite.

The above conclusion implies that the classical solution cannot be trusted close enough to the singular epoch, say within a few Planck time intervals of $t = 0$. Nevertheless, can we also assert that the classical conclusion of space-time singularity still stands in the wider class of solutions offered by quantum cosmology? We answer this question in the following fashion.

For a typical nonclassical solution, the scale factor is given by

$$Q(t) = (1 + \varphi)\bar{Q}(t). \tag{43}$$

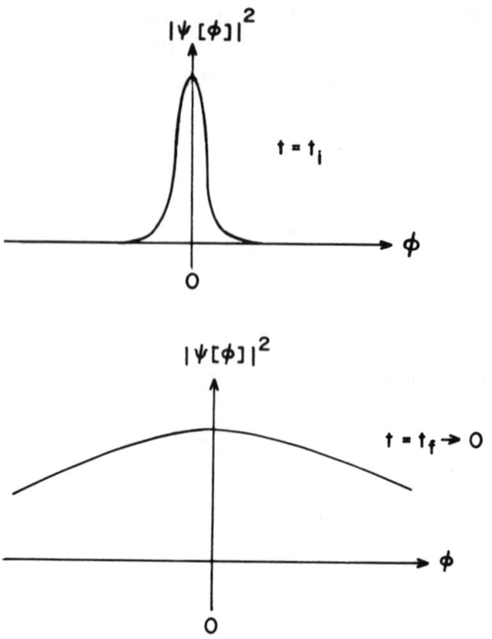

Fig. 12.3 The spread of the wave packet (for φ) as $t_f \to 0$.

A spacetime singularity at $t = 0$ implies that $Q(t) \to 0$ as $t \to 0$ (Example 12.6). Hence, we find that for all singular solutions

$$(1 + \varphi)\bar{Q}(t) \equiv Q(t) \to 0. \tag{44}$$

That is, for singular solutions

$$\frac{\varphi_f}{\Delta_f} \sim (Q_f - \bar{Q}_f) \to 0 \quad \text{as } t_f \to 0. \tag{45}$$

Hence, as shown in Figure 12.4, the singular solutions occupy the shaded central region of the normal probability distribution, a region which steadily shrinks to a line as the singular epoch is approached. Therefore we can assert that the probability that the universe came out of a singular state at $t = 0$ is vanishingly small.

Example 12.6. The Robertson–Walker line element is given by

$$ds^2 = dt^2 - Q^2 \left[\frac{dr^2}{1 - kr^2} + r^2(d\theta^2 + \sin^2\theta \, d\varphi^2) \right].$$

Here the spatial sections have constant curvature and as long as $Q \neq 0$ they are well-behaved homogeneous isotropic spaces. For $Q = 0$ only the case $Q \propto t$, $k = -1$ is nonsingular. In all other cases we have singularity. This can be seen by examining the scalar curvature R near $Q = 0$. We have

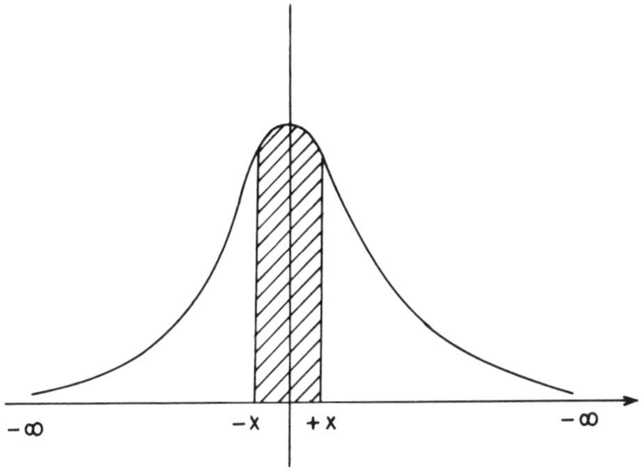

Fig. 12.4. The shaded region with $x = \varphi_f/\Delta_f$ represents the probability of singular models. As $t_f \to 0$, $x \to 0$. and hence the probability of singular origin tends to zero.

$$R = 6\frac{Q\ddot{Q} + \dot{Q}^2 + k}{Q^2}.$$

Suppose $Q \to 0$ as t^α. Then $\dot{Q} \sim \alpha t^{\alpha-1}$, $\ddot{Q} \sim \alpha(\alpha - 1)t^{\alpha-2}$

$$R \sim 6\,\frac{\alpha(2\alpha - 1)t^{2\alpha-2} + k}{t^{2\alpha}}.$$

The numerator vanishes in two cases: (i) $\alpha = 1$, $k = -1$ which is the nonsingular case mentioned above; (ii) $\alpha = \frac{1}{2}$, $k = 0$ in which case $R = 0$ but other curvature invariants diverge at $t = 0$. Thus except for (i) all solutions are nonsingular. □

To sum up: if we consider only the full range of homogeneous and isotropic then to obtain them from the classical Friedmann universe, we have to excite *only* the conformal degrees of freedom of the spacetime geometry. The QCFs in this case diverge at the classical singularity in such a way that the probability of solutions with singular origin is almost zero. Thus, as far as the quantum behaviour of these models is concerned, the conformal quantization is completely general.

12.4. Bianchi Type I Cosmologies

Let us consider a generalization of the isotropic universe to the Bianchi type I models described in Chapter 8. The line element of a model of this type is given by

$$ds^2 = dt^2 - X_1^2(t)\,dx_1^2 - X_2^2(t)\,dx_2^2 - X_3^2(t)\,dx_3^2. \tag{46}$$

For a dust-filled universe we have, from Section 8.6,

$$X_1X_2X_3 = \frac{9}{2} Mt(t + t_0), \quad M, t_0 > 0 \tag{47}$$

and

$$\bar{R} = \frac{4}{3t(t + t_0)} . \tag{48}$$

The singular epoch is at $t = 0$.

For a comoving coordinate 3-volume V sandwiched between $0 < t_f < t < t_i$ we have

$$S = \frac{VM}{16\pi} \int_{t_i}^{t_f} [6(\varphi^2 - 1) - 27t(t + t_0)\dot{\varphi}^2]\, dt. \tag{49}$$

As in Section 12.3 we have considered only those conformal fluctuations which preserve the Bianchi type of universe. Homogeneity again requires φ to depend on t only.

To evaluate the conformal propagator

$$K[\varphi_f, t_f; \varphi_i, t_i] = \int \exp\left\{ \frac{3iVM}{8\pi} \int_{t_i}^{t_f} \left[\varphi^2 - \frac{9}{2}t(t + t_0)\dot{\varphi}^2\right] dt \right\} \mathcal{D}\varphi, \tag{50}$$

we use the method of Section 2.4. and first solve for $\varphi = \varphi_c$ which stationarizes the integral in the above exponent and which takes the boundary values φ_i and φ_f at t_i and t_f. It is easy to verify that φ_c satisfies the differential equation

$$\frac{d}{dt} [t(t + t_0)\dot{\varphi}_c] + \frac{2}{9} \varphi_c = 0, \tag{51}$$

which has the general solution expressible in terms of the Legendre functions of fractional order

$$\varphi_c(t) = \alpha P_{-1/3}(\lambda) + \beta Q_{-1/3}(\lambda). \tag{52}$$

Here α and β are constants and $\lambda = 1 + 2t/t_0$.

We are really interested in the limiting situation of $t_f \to 0$. For small $t(\ll t_0)$, $\lambda \approx 1$ and we can use the approximations

$$P_n(\lambda) \sim 1 - \frac{n(n + 1)}{2} (1 - \lambda), \qquad Q_n(\lambda) \sim -\frac{1}{2} \ln\left(\frac{\lambda - 1}{2}\right) . \tag{53}$$

These approximations amount to the following approximate form for φ_c:

$$\varphi_c \sim A + B \ln t. \tag{54}$$

We shall assume that (54) holds throughout the sandwiched region $\mathcal{V}$. Then the boundary conditions on φ_c give

$$A = \frac{\varphi_i \ln t_f - \varphi_f \ln t_i}{\ln t_f - \ln t_i}, \qquad B = \frac{\varphi_f - \varphi_i}{\ln t_f - \ln t_i}. \tag{55}$$

We substitute (54) and (55) into (50) for $\varphi = \varphi_c$ and evaluate K by the method of Section 2.4. The final answer may be expressed in compact form:

$$K[\varphi_f, t_f; \varphi_i, t_i] = \sqrt{\frac{B}{i\pi\eta}} \exp \left\{ \frac{iB(\varphi_f - \varphi_i)^2}{\eta} \right\}, \tag{56}$$

where

$$B = \frac{27VMt_0}{16\pi} \tag{57}$$

and

$$\eta = \ln \left(\frac{t_i}{t_f} \right) \tag{58}$$

As in Section 12.3 we may investigate what happens to a wave packet of the form (25) by subjecting it to (26) with K given by (56) above. The calculation is straightforward but tedious. The result is that as in Section 12.3 the wave packet is centred on the classical value $\varphi = 0$ but with a larger dispersion Δ_f given by

$$\Delta_f \sim \frac{1}{4B\Delta_i} \ln \left| \frac{t_i}{t_f} \right|, \tag{59}$$

as $t_f \to 0$. Thus Δ_f diverges in the limit $t_f \to 0$.

12.5. Universes with Arbitrary Distribution of Massive Particles

The examples discussed in Sections 12.3 and 12.4 suggest that, as far as QCFs are concerned, the classical solutions are not 'reliable' in describing the state of the universe close to the singular epoch. Indeed, if Section 12.3 is any guide we should expect the singularity to be highly improbable. However, the result of Section 12.3 was derived for the highly special case of homogeneous isotropic cosmologies. It is not obvious that the result would carry over to more general spacetimes with no assumptions about homogeneity and isotropy. In this section we show that the result does indeed follow under assumptions more general than those of Section 12.3.

We shall assume that the universe can still be foliated by spacelike hypersurfaces $\{\Sigma\}$ of constant t, although the surfaces Σ are neither homogeneous nor isotropic. We shall also assume that the contents of the universe are noninteracting massive particles. The conformal fluctuation function φ is now an arbitrary C^2 function of spacetime coordinates and the propagator is given by (19).

Although the result given by Equation (2.35) in Section 2.4 would be applicable to φ a function of t only, the method used in its derivation can be generalized to the case where φ depends on x^μ and t. To see this result, define

$$\varphi = \varphi_c + \eta, \tag{60}$$

where

$$\eta(t_i) = \eta(t_f) = 0, \qquad \varphi_c(x^\mu, t_i) = \varphi_i(x^\mu),\ \varphi_c(x^\mu, t_f) = \varphi_f(x^\mu), \tag{61}$$

and φ_c satisfies the wave equation[1]

$$\bar{\Box}\, \varphi_c + \frac{1}{6}\bar{R}\, \varphi_c = 0. \tag{62}$$

Then we have

$$\begin{aligned}\int_v & (\varphi^2\bar{R} - 6\varphi_l\varphi^l)\sqrt{-\bar{g}}\ \mathrm{d}^4x \\ &= \int_v (\varphi_c^2\bar{R} - 6\varphi_{cl}\varphi_c^l)\sqrt{-\bar{g}}\ \mathrm{d}^4x + \\ &\quad + \int_v (2\eta\varphi_c\bar{R} - 12\eta_l\varphi_c^l)\sqrt{-\bar{g}}\ \mathrm{d}^4x + \\ &\quad + \int_v (\eta^2\bar{R} - 6\eta_l\eta^l)\sqrt{-\bar{g}}\ \mathrm{d}^4x\end{aligned} \tag{63}$$

The second integral on the right-hand side is zero for the following reason

$$\begin{aligned}\int_v & (2\eta\varphi_c\bar{R} - 12\eta_l\varphi_c^l)\sqrt{-\bar{g}}\ \mathrm{d}^4x \\ &= \int_v 2\eta(\varphi_c\bar{R} + 6\bar{\Box}\varphi_c)\sqrt{-\bar{g}}\ \mathrm{d}^4x - \int_{\Sigma_i}^{\Sigma_f} 12\eta\varphi_c^l\sqrt{-\bar{g}}\ \mathrm{d}\Sigma_l.\end{aligned}$$

The first integral on the right-hand side vanishes because of (62) and the second is cancelled by the Gibbons–Hawking term.

Thus we get, for (9),

$$\begin{aligned}K[\varphi_f, t_f; \varphi_i, t_i] = A \exp & \left\{\frac{i}{16\pi}\int_v (\varphi_c^2\bar{R} - 6\varphi_{cl}\varphi_c^l)\sqrt{-\bar{g}}\ \mathrm{d}^4x\right\} \times \\ & \times \int \exp\left\{\frac{i}{16\pi}\int_v (\eta^2\bar{R} - 6\eta_l\eta^l)\sqrt{-\bar{g}}\ \mathrm{d}^4x\right\}\mathcal{D}\eta.\end{aligned} \tag{64}$$

The path integral over η is, however, independent of endpoint values on t_i and t_f since $\eta = 0$ at these ends. Therefore the path integral will depend on t_i and t_f only. Also

[1] To avoid confusion we shall not use i and f as tensor indices in this section.

$$\int_{v} (\varphi_c^2 \bar{R} - 6\varphi_{cl}\varphi_c^l) \sqrt{-g}\ \mathrm{d}^4x = 6\int_{\Sigma_f}^{\Sigma_i} \varphi_c\varphi_c^l \sqrt{-g}\ \mathrm{d}\Sigma_l,$$

again, because φ_c satisfies the wave equation (62).

Putting all these results together we get the following formal solution for our problem:

$$K[\varphi_f, t_f; \varphi_i, t_i] = F(t_f, t_i) \exp\left(\frac{3i}{8\pi} \int_{\Sigma_f}^{\Sigma_i} \varphi_c\varphi_c^k \sqrt{-g}\ \mathrm{d}\Sigma_k\right). \tag{65}$$

The operational implications of (65) are straightforward. Given the boundary values φ_i and φ_f compute the solution of the wave equation (62) and for this solution evaluate the above difference of surface integrals.

12.5.1. *Green's Functions*

The solution of the wave equation (62) with specified boundary conditions can be expressed elegantly in the form of Green's functions. We define the retarded Green's function $G^R(X, B)$ for point B in spacetime as the solution of the wave equation

$$[\Box + \tfrac{1}{6}R]_X\, G(X, B) = [-g(X)]^{1/2}\, \delta_4(X, B), \tag{66}$$

which has support only on or inside the future light cone of B. Likewise, the advanced Green's function $G^A(X, B)$ is the solution of (66) with support on or inside the past light cone of B. As shown in Example 12.7 below,

$$G^R(X, B) = G^A(B, X). \tag{67}$$

Example 12.7. To prove (67) we proceed as follows. Define

$$G(X, B) = G^R(X, B) - G^A(X, B).$$

Then

$$\Box_X G(X, B) + \tfrac{1}{6} R(X)\, G(X, B) = 0.$$

Consider this surface integral on an arbitrary spacelike hypersurface Σ for any φ satisfying $(\Box + R/6)\varphi = 0$:

$$K = \int_\Sigma \sqrt{-g(X)}\,[\varphi(X)\, G_{,k}(X, Y) - \varphi(X)_{,k}\, G(X, Y)]\, \mathrm{d}\Sigma^k(X)$$

where the derivatives are with respect to X. By Green's theorem, we have for Y lying to the past of Σ

$$K = -\int_{v} \sqrt{-g(X)}\, g^{lk}\,[\varphi(X)\, G^R_{,k}(X, Y) - \varphi(X)_{,k}\, G^R(X, Y)]_{;l}\, \mathrm{d}^4X,$$

where v lies to the past of Σ. Since G^R satisfies the inhomogeneous equation (66) we get

$$K = \varphi(Y).$$

We get the same answer for Y lying to the future of Σ. Since the Cauchy data on Σ can be specified (in the form of φ and $\varphi_{,k}$) quite arbitrarily, the fact that $\varphi(Y)$ satisfies the wave equation

$$(\Box + \tfrac{1}{6} R)_Y \varphi = 0$$

enables us to deduce from the surface integral for K:

$$(\Box + \tfrac{1}{6} R)_Y G(X, Y) = 0.$$

However, from (66) we also see that $-G(Y, X)$ substituted for $G(X, Y)$ in the expression for K also gives $\varphi(Y)$. Hence, by the uniqueness of such solutions of the wave equations, we must have

$$G(X, Y) = -G(Y, X)$$

This leads to the result $G^R(X, Y) = G^A(Y, X)$ and $G^R(Y, X) = G^A(X, Y)$ by comparing the light cone supports from X and Y. □

We consider the application of the retarded Green's function to our sandwiched region $\mathcal{V}$.

By use of Green's theorem we get

$$\int_{\mathcal{V}} \{\varphi_c(X)\, \bar{\Box}_x \bar{G}^R(X, B) - \bar{G}^R(X, B)\bar{\Box}_x \varphi_c(X)\} \sqrt{-\bar{g}(X)}\, d^4x$$

$$= \int_{\partial\mathcal{V}} \{\varphi_c(X)\bar{G}^R(X, B)^{;k} - \bar{G}^R(X, B)\varphi_c(X)^{;k}\} \sqrt{-\bar{g}(X)}\, d\Sigma_k. \tag{68}$$

However, because of (62) and (66) the left-hand side equals $\varphi_c(B)$ provided B lies in $\mathcal{V}$ and is zero othewise. The right-hand side vanishes if X lies to the past of B. We shall anticipate that, as in our special examples of the previous two sections, the singularity of $\bar{\mathcal{M}}$ will lie in the past and hence we take $t_i > t_f$. Then for an interior point B of $\mathcal{V}$, the above consideration will rule out any contributions from Σ_f. Let us denote the spatial coordinates of a typical point on Σ_i by $\mathbf{r}_i$, the suffix i denoting the location of the point on Σ_i. The time coordinate is of course t_i. Further, let the space and time coordinates of B be $(\mathbf{b}, b)$. Then (68) gives

$$\varphi_c(\mathbf{b}, b) = \int \{\varphi_i(\mathbf{r}_i)\, \dot{\bar{G}}^R(\mathbf{r}_i, t_i; \mathbf{b}, b) - \bar{G}^R(\mathbf{r}_i, t_i; \mathbf{b}, b)\dot{\varphi}_i(\mathbf{r}_i)\} \sqrt{-\bar{g}(\mathbf{t}_i, t_i)}\, d^3\mathbf{r}_i, \tag{69}$$

where the overhead dot denotes derivatives with respect to t_i and $\dot{\varphi}_i(\mathbf{r}_i) = \dot{\varphi}_c(\mathbf{r}_i, t_i)$.

We take the limit of (69) as B approaches Σ_f. Then we get

$$\varphi_f(\mathbf{r}_f) = \int \Big\{\varphi_i(\mathbf{r}_i)\, \dot{\bar{G}}^R(\mathbf{r}_i, t_i; \mathbf{r}_f, t_f) - - \bar{G}^R(\mathbf{r}_i t_i; \mathbf{r}_f, t_f)\dot{\varphi}_i(\mathbf{r}_i)\Big\} \sqrt{-\bar{g}(\mathbf{r}_i, t_i)}\, d^3\mathbf{r}_i. \tag{70}$$

The above relation may be looked upon as an integral equation for $\dot{\varphi}_i(\mathbf{r}_i)$. A similar relation gives an integral equation for $\dot{\varphi}_f(\mathbf{r}_f)$ in terms of the advanced Green's function:

$$\varphi_i(\mathbf{r}_i) = -\int \Big\{\varphi_f(\mathbf{r}_f)\, \dot{\bar{G}}^A(\mathbf{r}_f, t_f; \mathbf{r}_i, t_i) - \\ - \bar{G}^A(\mathbf{r}_f, t_f; \mathbf{r}_i, t_i)\dot{\varphi}_f(\mathbf{r}_f)\Big\} \sqrt{-g(\mathbf{r}_f, t_f)}\, \mathrm{d}^3\mathbf{r}_f. \tag{71}$$

Here the dot denotes $\partial/\partial t_f$.

Let us recall (65) and note that to evaluate K we need φ_c, $\dot{\varphi}_c$ at Σ_i and Σ_f. In other words, we have to 'solve' the integral equations (70) and (71) to know $\dot{\varphi}_i(\mathbf{r}_i)$ and $\dot{\varphi}_f(\mathbf{r}_f)$. Let us consider (70) first.

Multiply (70) by a two-point function $\bar{H}^R(\mathbf{r}'_i, t_i; \mathbf{r}_f, t_f)$ and integrate over $\mathrm{d}^3\mathbf{r}_f$. Choose the function H to satisfy the relation

$$\int \bar{H}^R(\mathbf{r}'_i, t_i; \mathbf{r}_f, t_f)\, \bar{G}^R(\mathbf{r}_i, t_i; \mathbf{r}_f, t_f)\, \mathrm{d}^3\mathbf{r}_f = \delta(\mathbf{r}_i - \mathbf{r}'_i). \tag{72}$$

$\bar{H}$ is thus the 'inverse' of $\bar{G}$ in the operator sense in the Hilbert space of continuous variables $\mathbf{r}_i$, $\mathbf{r}_f$, etc. The above operation on Equation (70) is thus analogous to the solution of simultaneous linear equations in a finite number of variables. Example 12.8 below discusses the form of $\bar{H}$ in flat spacetime.

Example 12.8. In flat spacetime we have, with the help of Fourier transforms, the following relation:

$$G^R(\mathbf{r}_2, t_2; \mathbf{r}_1, t_1) \equiv \frac{\delta(t_2 - t_2 - |\,\mathbf{r}_2 - \mathbf{r}_1|)}{4\pi\,|\mathbf{r}_2 - \mathbf{r}_1|}$$

$$= \int \theta(t_2 - t_1)\, \frac{\sin[Q(t_2 - t_1)]}{4\pi Q}\, \exp[i\mathbf{Q}\cdot(\mathbf{r}_2 - \mathbf{r}_1)]\, \frac{\mathrm{d}^3\mathbf{Q}}{(2\pi)^3}\, .$$

The inverse function $H^R(\mathbf{r}_2, t_2; \mathbf{r}_1, t_1)$ has a Fourier transform that is simply the reciprocal of the Fourier transform of G^R. (This follows from definition (72).) Hence

$$H^R(\mathbf{r}_2, t_2; \mathbf{r}_1, t_1) = \int \left[\theta(t_2 - t_1)\, \frac{\sin[Q(t_2 - t_1)]}{4\pi Q}\right]^{-1} \times \\ \times\ \exp[iQ\cdot(\mathbf{r}_2 - \mathbf{r}_1)]\, \frac{\mathrm{d}^3\mathbf{Q}}{(2\pi)^3}\, .$$

This integral is divergent but it has to be looked upon as an operator. Thus if we take

$$\alpha(\mathbf{r}_2, t_2) = \int G^R(\mathbf{r}_2, t_2; \mathbf{r}_1, t_1)\, \beta(\mathbf{r}_1, t_1)\, \mathrm{d}^3\mathbf{r}_1,$$

then

$$\beta(\mathbf{r}_1, t_1) = \int H^R(\mathbf{r}_2, t_2; \mathbf{r}_1, t_1)\, \alpha(\mathbf{r}_2, t_2)\, \mathrm{d}^3\mathbf{r}_2,$$

For example, if $\beta(\mathbf{r}_1, t_1) = \exp(i\omega t_1 - i\mathbf{k}\cdot\mathbf{r}_1)$, then

$$\alpha(\mathbf{r}_2, t_2) = \frac{\sin[k(t_2 - t_1)]}{4\pi k} \exp(i\omega t_2 - i\mathbf{k}\cdot\mathbf{r}_2)$$

If we apply the operator H^R on it we get back $\beta(\mathbf{r}_1, t_1)$ after some manipulations with Fourier transforms. □

Then we get

$$\sqrt{-\bar{g}(\mathbf{r}_i', t_i)}\ \dot{\varphi}_i(\mathbf{r}_i') = \iint \bar{H}^R(\mathbf{r}_i', t_i; \mathbf{r}_f, t_f)\ \varphi_i(\mathbf{r}_i)\ \dot{\bar{G}}^R(\mathbf{r}_i, t_i; \mathbf{r}_f, t_f) \times$$
$$\times \sqrt{-\bar{g}(\mathbf{r}_i, t_i)}\ \mathrm{d}^3\mathbf{r}_i\ \mathrm{d}^3\mathbf{r}_f -$$
$$- \int \bar{H}^R(\mathbf{r}_i', t_i; \mathbf{r}_f, t_f)\ \varphi_f(\mathbf{r}_f)\ \mathrm{d}^3\mathbf{r}_f. \tag{73}$$

Similarly,

$$\sqrt{-\bar{g}(\mathbf{r}_f', t_f)}\ \dot{\varphi}_f(\mathbf{r}_f') = \iint \bar{H}^A(\mathbf{r}_f', t_f; \mathbf{r}_i, t_i)\ \varphi_f(\mathbf{r}_f)\ \dot{\bar{G}}^A(\mathbf{r}_f, t_f, \mathbf{r}_i, t_i) \times$$
$$\times \sqrt{-\bar{g}(\mathbf{r}_f, t_f)}\ \mathrm{d}^3\mathbf{r}_f\ \mathrm{d}^3\mathbf{r}_i -$$
$$- \int \bar{H}^A(\mathbf{r}_f', t_f; \mathbf{r}_i, t_i)\ \varphi_i(\mathbf{r}_i)\ \mathrm{d}^3\mathbf{r}_i. \tag{74}$$

Before proceeding further we economize on the various quantities used, by recalling the symmetry relation (67). We shall henceforth write

$$\left.\begin{aligned} \bar{G}^R(\mathbf{r}_i, t_i; \mathbf{r}_f, t_f) &\equiv \bar{G}^A(t_f, \mathbf{r}_f; t_i, \mathbf{r}_i) \equiv \bar{G}(\mathbf{r}_i, \mathbf{r}_f), \\ \bar{H}^R(\mathbf{r}_i, t_i; \mathbf{r}_f, t_f) &\equiv \bar{H}^A(t_f, \mathbf{r}_f; t_i, \mathbf{r}_i) \equiv \bar{G}(\mathbf{r}_i, \mathbf{r}_f)^{-1}, \end{aligned}\right\} \tag{75}$$

Also we will write

$$\left.\begin{aligned} \frac{\partial}{\partial t_i}\ \bar{G}^R(\mathbf{r}_i, t_i, \mathbf{r}_f, t_f) &\equiv \dot{\bar{G}}\ (\mathbf{r}_i, \mathbf{r}_f)_{(i)}, \\ \frac{\partial}{\partial t_f}\ \bar{G}^A(\mathbf{r}_f, t_f, \mathbf{r}_i, t_i) &\equiv \dot{\bar{G}}\ (\mathbf{r}_i, \mathbf{r}_f)_{(f)}. \end{aligned}\right\} \tag{76}$$

Now we consider the exponent in (65), viz.

$$i\mathcal{E} = \frac{3i}{8\pi} \int_{\Sigma_f}^{\Sigma_i} \varphi_c \varphi_c^k \sqrt{-\bar{g}}\ \mathrm{d}\Sigma_k \equiv \frac{3i}{8\pi} \left[\int \sqrt{-\bar{g}(\mathbf{r}_i', t_i)}\ \dot{\varphi}(\mathbf{r}_i')\varphi_i(\mathbf{r}_i')\ \mathrm{d}^3\mathbf{r}_i' - \right.$$
$$\left. - \int \sqrt{-\bar{g}(\mathbf{r}_f', t_f)}\ \dot{\varphi}_f(\mathbf{r}_f')\ \varphi_f(\mathbf{r}_f')\ \mathrm{d}^3\mathbf{r}_f' \right]. \tag{77}$$

Since all quantities on the right-hand side are now explicitly known because of (73) and (74), we can express $\mathcal{E}$ in terms of multiple integrals over spatial volumes. It is not difficult to verify that $\mathcal{E}$ has the form

$$\mathcal{E} = \iint A_{ii}(\mathbf{r}_i, \mathbf{r}_i') \, \varphi_i(\mathbf{r}_i) \, \varphi_i(\mathbf{r}_i') \, \mathrm{d}^3\mathbf{r}_i \, \mathrm{d}^3\mathbf{r}_i' \, +$$

$$+ \iint A_{ff}(\mathbf{r}_f, \mathbf{r}_f') \, \varphi_f(\mathbf{r}_f) \, \varphi_f(\mathbf{r}_f') \, \mathrm{d}^3\mathbf{r}_f' \, +$$

$$+ \, 2 \iint A_{if}(\mathbf{r}_i, \mathbf{r}_f) \, \varphi_i(\mathbf{r}_i) \, \varphi_f(\mathbf{r}_f) \, \mathrm{d}^3\mathbf{r}_i \, \mathrm{d}^3\mathbf{r}_f, \tag{78}$$

where

$$A_{ii}(\mathbf{r}_i, \mathbf{r}_i') = \frac{3}{8\pi} \int \dot{\bar{G}}\,(\mathbf{r}_i, \mathbf{r}_f)_{(f)} \bar{G}(\mathbf{r}_i', \mathbf{r}_f)^{-1} \sqrt{-\bar{g}(\mathbf{r}_i', t_i)} \, \mathrm{d}^3\mathbf{r}_f \tag{79}$$

$$A_{ff}(\mathbf{r}_f, \mathbf{r}_f') = \frac{3}{8\pi} \int \dot{\bar{G}}\,(\mathbf{r}_i, \mathbf{r}_f)_{(i)} \bar{G}(\mathbf{r}_i, \mathbf{r}_f')^{-1} \sqrt{-\bar{g}(\mathbf{r}_f', t_f)} \, \mathrm{d}^3\mathbf{r}_i \tag{80}$$

$$A_{if}(\mathbf{r}_i, \mathbf{r}_f) = \frac{3}{8\pi} \, \bar{G}(\mathbf{r}_i, \mathbf{r}_f)^{-1}. \tag{81}$$

And our final answer is the *master propagator*

$$K[\varphi_f, t_f; \varphi_i, t_i] = F(t_f, t_i) \exp i\mathcal{E}. \tag{82}$$

We next consider the application of this propagator: but to do that it is convenient to diagonalize $\mathcal{E}$.

12.5.2. *Diagonalization*

Our eventual application of the propagator will be considerably simplified if we diagonalize $\mathcal{E}$ in the following sense. We assume that linear transformations of the following kind exist:

$$\varphi_i(\mathbf{r}_i) = \int \alpha_i(\mathbf{r}_i, \mathbf{r}_i') \, \Phi_i(\mathbf{r}_i') \, \mathrm{d}^3\mathbf{r}_i', \tag{83}$$

$$\varphi_f(\mathbf{r}_f) = \int \alpha_f(\mathbf{r}_f, \mathbf{r}_f') \, \Phi_f(\mathbf{r}_f') \, \mathrm{d}^3\mathbf{r}_f', \tag{84}$$

where the α's satisfy the following conditions:

$$\iint A_{ii}(\mathbf{r}_i, \mathbf{r}_i') \, \alpha_i(\mathbf{r}_i, \mathbf{r}_i'') \, \alpha_i(\mathbf{r}_i', \mathbf{r}_i''') \, \mathrm{d}^3\mathbf{r}_i \, \mathrm{d}\mathbf{r}_i'$$
$$= B_{ii}(\mathbf{r}_i'') \, \delta_3(\mathbf{r}_i'' - \mathbf{r}_i'''), \tag{85}$$

$$\iint A_{ff}(\mathbf{r}_f, \mathbf{r}'_f)\, \alpha_f(\mathbf{r}_f, \mathbf{r}''_f)\, \alpha_f(\mathbf{r}'_f, \mathbf{r}'''_f)\, d^3\mathbf{r}_f\, d\mathbf{r}'_f$$
$$= B_{ff}(\mathbf{r}''_f)\, \delta_3(\mathbf{r}''_f - \mathbf{r}'''_f). \tag{86}$$

Then we get

$$\mathcal{E} = \int B_{ii}(\mathbf{r}_i)\, \Phi_i^2(\mathbf{r}_i)\, d^3\mathbf{r}_i + \int B_{ff}(\mathbf{r}_f)\, \Phi_f^2(\mathbf{r}_f)\, d^3\mathbf{r}_f +$$
$$+ \iint B_{if}(\mathbf{r}_i, \mathbf{r}_f)\, \Phi_i(\mathbf{r}_i)\, \Phi_f(\mathbf{r}_f)\, d^3\mathbf{r}_i\, d^3\mathbf{r}_f, \tag{87}$$

where

$$B_{if}(\mathbf{r}_i, \mathbf{r}_f) = \iint \alpha(\mathbf{r}'_i, \mathbf{r}_i)\, \alpha(\mathbf{r}'_f, \mathbf{r}_f)\, A_{if}(\mathbf{r}'_i, \mathbf{r}'_f)\, d^3\mathbf{r}'_i\, d^3\mathbf{r}'_f. \tag{88}$$

Expressed in this form, the propagator can be readily applied to study the diffusion of a wave packet.

12.5.3. *Diffusion of Wave Packet*

We are looking for a generalization of (26) to study the behaviour of QCF as we approach the singular epoch. Let us make the following assumption:

(a) The epoch $t = 0$ given by the hypersurface Σ_0 contains all the singularities of $\bar{\mathcal{M}}$.

We shall discuss the generality of this assumption later.

What happens to the Green's functions as one of the two points in their arguments approach Σ_0? To answer this question we make a further assumption:

(b) It is possible to find a conformal function Ω that diverges rapidly enough on Σ_0 such that the spacetime $\mathcal{M} \equiv \Omega^2\bar{\mathcal{M}}$ with metric $\Omega^2\bar{g}_{ik}$ is nonsingular at Σ_0

Work by S. Beem and A. K. Kembhavi and others suggests that this assumption is plausible and would cover physically reasonable spacetimes $\bar{\mathcal{M}}$. Some typical cases are discussed in Example 12.9 below. In all cases Ω diverges on Σ_0.

Example 12.9. The simplest examples of assumption (b) are the cases of flat Minkowski spacetime $\mathcal{M}$ obtained by a conformal transformation of Robertson–Walker models. Examples 12.12 and 12.13 show how this is done. The conformal function in the latter case is the reciprocal of Ω.

Another example is of Bianchi type I models. We illustrate by the case of the axially symmetric $\bar{\mathcal{M}}$.

$$ds^2 = dt^2 - \frac{a^2t^2}{(t + t_0)^{2/3}}\, dx^2 - b^2(t + t_0)^{4/3}\,(dy^2 + dz^2)$$

where a, b and t_0 are positive constants. The singular hypersurface is at $t = 0$.

To obtain $\mathcal{M}$ take

$$\Omega = \frac{t^2 + 1}{t^2} .$$

It can be verified that the new spacetime is nonsingular at $t = 0$, and indeed does not have a singularity anywhere else. The function Ω diverges at $t = 0$. □

The advanced and retarded Green's functions transform under the conformal transformation as (cf. Equation (6.129))

$$\bar{G}(X, B) = \Omega(X)\, \Omega(B)\, G(X, B), \tag{89}$$

where $G(X, B)$ is the Green's function in $\mathcal{M}$. Since $G(X, B)$ stays finite as X approaches Σ_0, $\bar{G}(X, B)$ will diverge as $\Omega(X)$. Consequently, by (81), A_{if} and B_{if} both tend to zero as t_f approaches t_0.

With this background we now investigate the formal generalization of (26) to

$$\Psi_f\, [\Phi_f] = \int K[\Phi_f,\, t_f;\, \Phi_i,\, t_i]\; \Psi_i[\Phi_i]\; \mathcal{D}\Phi_i. \tag{90}$$

What does (90) mean? For each specified function $\Phi_i(\mathbf{r}_i)$ we have a 3-geometry on Σ_i which is conformal to the classical 3-geometry of $\bar{\mathcal{M}}$. The functional Ψ_i denotes the quantum mechanical probability amplitude (in function space) for the 'state' Φ_i of the above 3-geometry. A state which is almost classical will have $\Phi_i \approx 0$ at all $\mathbf{r}_i$. If Ψ_i is peaked round the zero function, we have a quantum description of the universe that it nearly classical.

Notice, however, that the following important result applies whether or not Ψ_i is so peaked. For, as $t_f \to 0$, $B_{if} \to 0$ and the cross product of Φ_i and Φ_f in $\mathcal{E}$ vanishes. In the functional integral (90) this cross product was the only way of linking, through integration, the details of the initial state to those of the final state. All other terms lead to imaginary phase factors in the exponential so that no 'physical' information is carried by them. Thus, in this limit Ψ_f is completely decoupled from Ψ_i. The final state does not 'remember' the initial state.

Exactly the same situation pertained in the special cases discussed in Sections 12.3 and 12.4. As in those cases, we therefore arrive at the situation that quantum uncertainty diverges at the classical singularity. In practical terms our conclusion means that the present classical state of the universe need not necessarily have arisen from the classical singular state at $t = 0$. Rather, any state at $t = 0$ could have led to the present state at $t = t_i$ with some probability that it is nonzero.

Assumption (a) above can be generalized to where there are more than one hypersurfaces Σ_0 containing spacetime singularities. In that case we must proceed step by step in a chronological sense. Thus the above conclusion would apply to the first Σ_0 encountered by the final point $(t_f, \mathbf{x}_f)$. We then proceed to the next Σ_0 and so on.

Example 12.10. We shall consider the anisotropic and inhomogeneous model where the spacetime singularity is quite general and not restricted to the matter being in the form of noninteracting massive particles. In fact the dominant effects are geometrical and $T_{(m)}^{ik}$ may be set equal to zero. This model was proposed and discussed by V. A. Belinskii, I. M. Khalatnikov, and E. M. Lifshitz in the early 1970s. We shall refer to it as the BKL model. Its line element is

$$ds^2 = dt^2 - \gamma_{\mu\nu}\, dx^\mu dx^\nu,$$

$$\gamma_{\mu\nu} = \sum_{\alpha=1}^{3} t^{2p_\alpha} l_{\alpha\mu} l_{\alpha\nu},$$

where p_α and $l_{\alpha\mu}$ are functions of space coordinates. The p_α's satisfy the Kasner relations (8.101) encountered in Chapter 8:

$$p_1 + p_2 + p_3 = p_1^2 + p_2^2 + p_3^2 = 1.$$

We are interested in the form of the Green's function for the wave equation

$$\bar{\Box}\bar{G}(X, B) = [-\bar{g}(X)]^{-1/2}\, \delta_4(X, B).$$

Since we have neglected the matter terms, $\bar{R} = 0$ and we do not have the $\bar{R}/6$ term. From analysis similar to that in Section 12.4, we find that for t_i, t_f small

$$\bar{G}(\mathbf{r}_f, t_f; \mathbf{r}_i, t_i) = \frac{\delta_3(\mathbf{r}_i, \mathbf{r}_f)}{f(\mathbf{r}_f)} \ln \left| \frac{t_i}{t_f} \right| ,$$

where f is a function of space coordinates alone.

Since $\bar{G}$ is diagonal in the $(\mathbf{r}_i, \mathbf{r}_f)$ space, its inverse is also diagonal and tends to zero as $t_f \to 0$. Thus the QCF's diverge at the classical BKL singularity.

12.6. The Problems of Singularity and Horizons

In Section 12.3 we were able to estimate the probability amplitude for our universe to have had a singular origin, provided that it stayed homogeneous and isotropic throughout the quantum regime. This probability was found to be vanishingly small. We now consider the more difficult problem of estimating the probability of a singular origin of the universe in the general situation of Section 12.5.

We shall prove an important result under an assumption which seems intuitively plausible:

(c) Given a classical general relativistic spacetime $\bar{\mathcal{M}}$ singular on the hypersurface Σ_0, there exists a conformal function Ω_c such that:
- (i) $\mathcal{M}_c \equiv \Omega_c^2\, \bar{\mathcal{M}}$ is nonsingular on Σ_0;
- (ii) all conformally transformed manifolds $\mathcal{M} \equiv \Omega^2 \bar{\mathcal{M}}$ are singular on at least some points of Σ_0, provided that $\Omega/\Omega_c \to 0$, at least on those points of Σ_0; and
- (iii) all conformally transformed manifolds $\mathcal{M} \equiv \Omega^2 \bar{\mathcal{M}}$ are nonsingular on Σ_0, provided that $\Omega/\Omega_c \nrightarrow 0$ on Σ_0.

We shall denote by C_s the class of all manifolds $\mathcal{M}$ which are covered by (ii) above and by C_{NS} the class covered by (iii). By assumption (b) of Section 12.5, Ω_c itself must diverge on Σ_0.

Example 12.11. The rationale behind the above assumption may be understood with the help of the following example. Consider the invariant $\bar{R}$. Under the conformal transformation

$$g_{ik} = e^{2\zeta}\,\bar{g}_{ik}$$

$\bar{R}$ goes to R, where

$$R = e^{-2\zeta}\,R + 6e^{-2\zeta}\,(\Box\,\zeta + \zeta_i\zeta^i)$$

with $\zeta_i = \partial\zeta/\partial x^i$.

Suppose that at a curvature singularity, $\bar{R}$ goes to infinity. We then have to choose ζ such that $e^{2\zeta}$ also diverges at the singularity so fast that $R\,e^{-2\zeta}$ converges. What about $(\Box\zeta + \zeta_i\zeta^i)$? This term may also contribute to divergence but the exponential $e^{2\zeta}$ will diverge more rapidly to nullify its effect.

A similar consideration has to be given to all curvature invariants (cf. Chapter 8). The Ω_c in assumption (c) is chosen as the 'least' rapidly diverging function $2^{2\zeta}$ that ensures convergence of all curvature invariants. □

We now apply the master propagator given by (82) to a wave functional at t_i, representing the bunching of several geometrics near the classical one of $\bar{\mathcal{M}}$ given by $\varphi = 0$. Generalizing the notion of wave packet we write the wave functional describing the above state as

$$\Psi_i(\varphi_i) = \frac{1}{[2\pi\Delta_i]^{1/4}}\,\exp\left\{-\,\frac{\varphi_i^2}{4\Delta_i^2}\right\}, \tag{91}$$

where φ_i and Δ_i are functions of all three space coordinates $\mathbf{r}_i$. The fact that the present state of the universe is almost classical implies that the bunching near $\varphi_i = 0$ is sharp, i.e. $|\Delta_i| \ll 1$.

We now apply (90) to study the quantum state of the universe at $t = t_f$. Without loss of generality we shall assume the master propagator to be diagonal in the sense described by (87), with φ replacing Φ. The functional integral (90) can be easily computed in much the same way that Ψ_f was evaluated in Section 12.3. In analogy to (41) we get the result that $\Psi_f(\varphi_f)$ is again a wave packet centred on $\varphi_f = 0$ but with a dispersion Δ_f given by

$$\Delta_f^2 = 4\Delta_i^2\left[B_{ii}^2 + \frac{1}{16\Delta_i^4}\right]B_{if}^{-2}\,. \tag{92}$$

Thus Δ_f gives an indication of the spread of states from which the present near-classical state evolved. The exponential

$$\exp\left[-\,\frac{\varphi_f^2}{2\,\Delta_f^2}\right] \tag{93}$$

gives the relative probability of a group of manifolds $\mathcal{M}_\varphi$ with φ close to φ_f.

Imagine this probability distribution in function space with the point $\varphi_f = 0$ representing the classical singular solution. From Section 12.5 we know that, as $t_f \to 0$, $B_{if} \to 0$ and $\Delta_f \to \infty$. In fact, from assumption (c) and (89) we find that as $t_f \to 0$

$$\Delta_f \sim B_{if}^{-1} \sim \Omega_c. \tag{94}$$

Consider now any member $\mathscr{M}$ of C_s. It is given by a φ_f such that as $t \to 0$,

$$\frac{1 + \varphi_f}{\Omega_c} \to 0. \tag{95}$$

For such members (94) implies that as $t \to 0$,

$$\frac{\varphi_f}{\Delta_f} \to 0. \tag{96}$$

Hence, such spacetimes occupy a set of measure zero in the probability space computed according to (93). By contrast nonsingular solutions ($\mathscr{M} \varepsilon C_{NS}$) have finite probability which tends to unity as $t \to t_f$.

The reader will not have failed to notice the similarity between the general case discussed here and the homogeneous isotropic case of Section 12.3. The general case requires functional integrals whereas the special case uses ordinary integrals. We have neglected the rigorous justification of the technique used in this section and the previous section (Section 12.5): rather we have concentrated on the physical content of the theory. Since the functional integrals involved only quadratic terms in the exponential, we feel confident that a rigorous mathematical justification for these steps can eventually be given.

The physical content to emerge from these calculations is that the quantum mechanical probability that the present spacetime emerged from a singular state by conformal fluctuations is zero.

12.6.1 *Particle Horizons*

In Section 8.5 we identified three drawbacks of the canonical Big Bang picture. Of these, the most fundamental – that of spacetime singularity – appears to be the exception rather than the rule in the quantum regime. We end this section with a brief discussion of how the elimination of singularity might also lead to the elimination of particle horizons. Again, to get the feel of the general situation, we shall examine first the special case of the Friedmann universe.

We recall from Section 8.5 that in the Friedmann universe the existence of the particle horizon is due to the finiteness of the integral

$$T(t_1, 0) = \int_0^{t_1} \frac{dt}{\bar{Q}(t)}. \tag{97}$$

Further, the horizon problem arises because

$$\lim_{t_1 \to 0} T(t_1, 0) = 0. \tag{98}$$

In Example 12.5 we saw that all Friedmann models with $\varepsilon > 3p$ satisfy these conditions.

However, once we admit the QCF we have, for a nonsingular $\mathcal{M}_\varphi$.

$$[1 + \varphi(t)]\, Q(t) \to b > 0 \quad \text{as } t \to 0. \tag{99}$$

Does this alter the situation as far as the particle horizon is concerned?

The *prima facie* answer to this question is 'No', because the particle horizon is based on the concept of past light cone, which remains unaltered in conformal transformations. How then can we remove the difficulty of particle horizon by a conformal transformation alone?

To remove this apparent difficulty let us go back to the tranformation (22) which gives us the new cosmic time coordinate τ in the conformally transformed model satisfying (99). Let us assume that $t = t_1$ corresponds to τ_1 and for $\tau < \tau_1$.

$$\tau = \tau_1 - \int_t^{t_1} [1 + \varphi(u)]\, \mathrm{d}u. \tag{100}$$

Now consider what happens when $t \to 0$. Near $t_1 \cong 0$ we have

$$\tau_1 - \tau \approx \int_t^{t_1} \frac{b\, \mathrm{d}u}{\bar{Q}(u)} = bT(t_1, t). \tag{101}$$

Thus the finiteness of the right-hand side as $t \to 0$ tells us that τ is finite at $t = 0$.

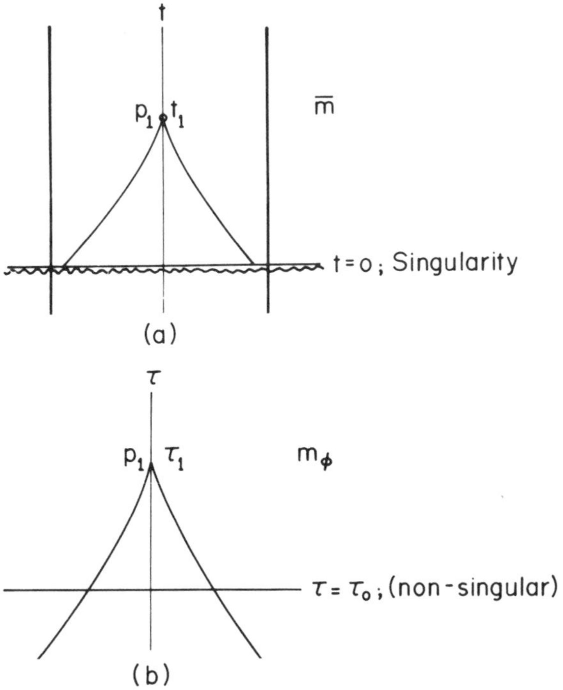

Fig. 12.5. In (a) the singularity at $t = 0$ terminates the past light cone of P_1. In (b) there is no singularity at $\tau = \tau_0$ and the light cone can be continued into $\tau < \tau_0$.

We thus have the situation illustrated in Figure 12.5. In (a) P_1 is the point in $\bar{\mathcal{M}}$ with time coordinate t_1. The past light cone from P_1 hits the singularity and terminates there. The thick lines parallel to the t axis from the boundary of this cone define the particle horizon of P_1 at $t = t_1$. In (b) the corresponding situation in $\mathcal{M}_\varphi$ is described. The cosmic time coordinate of P_1 is τ_1 and the τ coordinate of the surface $t = 0$ is τ_0. The $\tau = \tau_0$ surface is *not* singular, however, and we have no reason to terminate the light cone there. Thus we should extend the manifold $\mathcal{M}$ to $\tau < \tau_0$ as in Example 12.12 and also extend the past null cone of P_1 into this extended region. The particle horizon therefore need not arise in $\mathcal{M}$ as it does in $\bar{\mathcal{M}}$.

Example 12.12. Consider the Einstein–de Sitter model $\bar{\mathcal{M}}$:

$$d\bar{s}^2 = dt^2 - t^{4/3}[dr^2 + r^2(d\theta^2 + \sin^2\theta \, d\varphi^2].$$

Make a conformal transformation to a new spacetime $\mathcal{M}$:

$$ds = t^{-2/3} \, d\bar{s},$$

and define a time coordinate τ by

$$d\tau = t^{-2/3} \, dt, \quad \tau = 3t^{1/3}.$$

Then

$$ds^2 = d\tau^2 - dr^2 - r^2(d\theta^2 + \sin^2\theta \, d\varphi^2)$$

describes a Minkowski spacetime, but only the part $\tau > 0$. The hypersurface $\tau = 0$ is not singular as the same hypersurface $t = 0$ was in the original spacetime. Thus the spacetime $\mathcal{M}$ can be extended to $\tau < 0$.

The particle horizon in $\bar{\mathcal{M}}$ does not now have a counterpart in $\mathcal{M}$ since the past light cone from any point with $\tau > 0$ can be extended to arbitrarily large r by going into the extended part of $\mathcal{M}$. □

Our general spacetime $\bar{\mathcal{M}}$ also loses its singularity at Σ_0 in conformal transformation. The conformally transformed $\mathcal{M}_\varphi$ can likewise be extended beyond the singular surface and the particle horizon eliminated, provided that we can show that the new proper time interval from P_1 to Σ_0 is still finite. At the time of writing this book we do not know the additional conditions that this requirement imposes on $\mathcal{M}_\varphi$.

12.7. The Problem of Flatness

We next consider the problem of flatness highlighted in Section 8.5. To recall it briefly: if the overall global geometry of the universe was determined soon after the Big Bang, then the present observations of mean matter density suggest that the universe was very finely tuned close to the flat $k = 0$ Robertson–Walker model. The inflationary scenarios discussed in Chapter 10 offer arguments as to why this came about.

Here we take a different stand on the flatness problem. First we note that the standard Robertson–Walker line element

$$ds^2 = dt^2 - Q^2(t)\left[\frac{dr^2}{1 - kr^2} + r^2(d\theta^2 + \sin^2\theta\, d\varphi^2)\right] \tag{102}$$

is conformally flat. That is, there exists a conformal function Ω and coordinates $(\tau, \varrho, \theta, \varphi)$ such that (102) may be re-expressed in the form

$$ds^2 = \Omega^2\,[d\tau^2 - d\varrho^2 - \varrho^2(d\theta^2 + \sin^2\theta\, d\varphi^2)]. \tag{103}$$

Example 12.13. Infeld and Schild in 1946 gave the following series of transformation to take (102) to (103). For $k = 1$ define

$$T = \int^t \frac{dt}{Q(t)} \qquad r = \sin R$$

$$\xi = \tfrac{1}{2}(T + R) \qquad \eta = \tfrac{1}{2}(T - R)$$

$$X = \tan\xi \qquad Y = \tan\eta$$

$$\tau = \tfrac{1}{2}(X + Y) \qquad \varrho = \tfrac{1}{2}(X - Y).$$

The resulting line element (103) has

$$\Omega^2 = \frac{4Q^2}{(1 + X^2)(1 + Y^2)}.$$

The case $k = -1$ is similar with trigonometric functions replaced by hyperbolic ones. □

As shown in Examples 12.13 above, in the case $k = 0$

$$\tau = \tau(t),\ \varrho = r,\ \Omega = \Omega(t); \tag{104}$$

whereas in the cases $k = \pm 1$, τ, ϱ, and Ω all depend on *r and t.*

Consider now the hypothesis that the empty flat spacetime $\bar{\mathcal{M}}$, which is a trivial solution of Einstein's equations, is unstable to conformal fluctuations. This idea has been discussed earlier in different contexts by several authors (for detailed references, see Note 6 at the end of this chapter). We consider it here within the QCF framework discussed so far.

To begin with, let us write down the propagator for a transition between $\Omega = \Omega_i \equiv 1 + \varphi_i$ to $\Omega = \Omega_f \equiv 1 + \varphi_f$ from $t = t_i$ to $t = t_f$. Our previous work shows that since the background spacetime is the empty Minkowski spacetime $\bar{\mathcal{M}}$, we have

$$K[\varphi_f, t_f; \varphi_i, t_i] = \int \exp\left\{-\frac{3i}{8\pi}\int_{\mathcal{V}} \varphi_k \varphi^k\, d^4x\right\} \mathcal{D}\varphi, \tag{105}$$

where $\varphi_k = \partial\varphi/\partial x^k$ and the Minkowski metric is used for raising and lowering indices.

As we are discussing quantum fluctuations from the empty Minkowski universe we take our initial state Ψ_i to be a wave packet strongly peaked at $\varphi = 0$. Let the final state be a wave functional Ψ_f, strongly peaked about a function $\varphi_F(\mathbf{r}) \neq 0$. From path integral theory the transition amplitude $\Psi_i \rightarrow \psi_f$ is given by

$$\langle f|i\rangle = \iint \Psi_f{}^*[\varphi_f]\, K[\varphi_f, t_f; \varphi_i, t_i]\, \Psi_i[\varphi_i]\, \mathcal{D}\varphi_i\, \mathcal{D}\varphi_f. \tag{106}$$

Example 12.14 illustrates how (106) can be evaluated. We simply quote the result below.

Example 12.14. The result given in Equations (107) and (108) may be derived as follows. Suppose we write $\varphi(x^i)$ as a Fourier integral

$$\varphi(\mathbf{r}) = \int q_{\mathbf{k}}(t)\, e^{i\mathbf{k}\cdot\mathbf{r}} \frac{d^3\mathbf{k}}{(2\pi)^3}.$$

Then we get

$$\int_{\mathcal{V}} \varphi^{i}\varphi_i\, d^4x = \int \frac{d^3\mathbf{k}}{(2\pi)^3} \int_{t_i}^{t_f} \left\{ |\dot{q}_{\mathbf{k}}|^2 - |\mathbf{k}|^2\, |q_{\mathbf{k}}|^2 \right\} dt,$$

where $\mathcal{V}$ is the compact region of spacetime sandwiched between $t = t_i$ and $t = t_f$. This action is simply that for an infinite set of uncoupled harmonic oscillators. Let us suppose that the initial state $\Psi_i[\varphi_i]$ has φ_i strongly peaked at $\varphi_I(t, \mathbf{r})$, say and the final state $\Psi_f[\varphi_f]$ has φ_f strongly peaked at $\varphi_F(t, \mathbf{r})$. Let us write

$$\varphi_I(\mathbf{r}) = \int A^I_{\mathbf{k}}\, e^{i\mathbf{k}\cdot\mathbf{r}} \frac{d^3\mathbf{k}}{(2\pi)^3},$$

$$\varphi_F(\mathbf{r}) = \int A^F_{\mathbf{k}}\, e^{i\mathbf{k}\cdot\mathbf{r}} \frac{d^3\mathbf{k}}{(2\pi)^3}.$$

Saying that Ψ_i is peaked at φ_I is equivalent in Fourier space to having a wave functional $\Phi_i[q_{\mathbf{k}}]$ strongly peaked at $A^I_{\mathbf{k}}$. We shall assume that each of the harmonic oscillators is a minimum uncertainty wave packet (see Section 2.7). Thus the initial state in Fourier space is given by

$$\Phi_i[q_{\mathbf{k}}] = N_i \exp\left\{ -\frac{3}{8\pi} \int |q_k - A^I_{\mathbf{k}}|^2 \frac{d^3\mathbf{k}}{(2\pi)^3} \right\},$$

where N_i is a normalizing constant.

Now the transition amplitude in (106) is just the product of transition amplitudes for $A^I_{\mathbf{k}} \rightarrow A^F_{\mathbf{k}}$ (transition amplitude of the $\mathbf{k}$th harmonic oscillator) for all $\mathbf{k}$. Using this expression for transition amplitude from Chapter 2 we get

$$\langle f|i\rangle \propto \exp\left\{ -\frac{3}{8\pi} \int U_{\mathbf{k}}(A^F_{\mathbf{k}}; A^I_{\mathbf{k}}) \frac{d^3\mathbf{k}}{(2\pi)^3} \right\}.$$

where

$$U_{\mathbf{k}} = \tfrac{1}{2}|\mathbf{k}| \left\{ |A^I_{\mathbf{k}}|^2 + |A^F_{\mathbf{k}}|^2 - (A^I_{\mathbf{k}} \overset{*}{A}_{\mathbf{k}} - \overset{*}{A}{}^I_{\mathbf{k}} A^F_{\mathbf{k}})\, e^{-i|\mathbf{k}|(t_f - t_i)} \right\}$$

The real part of $U_{\mathbf{k}}$ which appears in $|\langle f|i\rangle|^2$ is thus positive definite.

This value of $\langle f|i\rangle$ can be expressed in terms of φ_I and φ_F by writing $A_{\mathbf{k}}^I$ and $A_{\mathbf{k}}^F$ as the inverses of the Fourier transforms. The final answer is

$$|\langle f|i\rangle|^2 = N\exp\left\{-\frac{3W}{8\pi}\right\},$$

where N is a normalizing constant and

$$W = \iint \left\{\frac{\nabla_1\varphi_I(\mathbf{r}_1)\cdot\nabla\varphi_I(\mathbf{r}_2)}{|\mathbf{r}_1-\mathbf{r}_2|^2} + \frac{\nabla_1\varphi_F(\mathbf{r}_1)\cdot\nabla_2\varphi_F(\mathbf{r}_2)}{|\mathbf{r}_1-\mathbf{r}_2|^2} - \right.$$

$$\left. - \frac{2\nabla\varphi_I(\mathbf{r}_1)\cdot\nabla\varphi_F(\mathbf{r}_2)}{|\mathbf{r}_1-\mathbf{r}_2|^2-(t_2-t_1)^2}\right\} \mathrm{d}^3\mathbf{r}_1\,\mathrm{d}^3\mathbf{r}_2.$$

If we assume that $\varphi_I = 0$ we get the result quoted in the text. □

The transition probability is given by

$$P_{if} = |\langle f|i\rangle|^2 = N\exp\left(-\frac{3W}{8\pi}\right), \tag{107}$$

where N is a normalization constant and

$$W = \iint \frac{\nabla_1\varphi_F(\mathbf{r}_1)\cdot\nabla_2\varphi_F(\mathbf{r}_2)}{|\mathbf{r}_1-\mathbf{r}_2|^2}\,\mathrm{d}^3\mathbf{r}_1\,\mathrm{d}^3\mathbf{r}_2. \tag{108}$$

The expression for W is positive definite so that P_{if} is maximum when $\nabla\varphi_F = 0$, i.e. when $\varphi_F = \varphi_F(t)$ only. But this is precisely the $k = 0$ case in the Robertson–Walker spacetime. In other words, if the universe arose out of QCFs from the empty Minkowski spacetime it will most likely go into the $k = 0$ type of the Robertson–Walker spacetime. Our finding that we live in a universe which is finely tuned at $k = 0$ is therefore not surprising; rather this to be expected from the above considerations of quantum cosmology.

12.8. Further Developments

This concludes our investigations of quantum conformal fluctuations. The discussion so far has been profitable on two counts. First, it has demonstrated that even with limited objectives quantum cosmology does lead to significant differences from classical cosmology. In particular, the probability measure of singular cosmological models in the full range of conformal transformations is zero. Thus, singularity is unlikely to be the menace that it had been in the classical context. Second, our approach has produced a working theory rather than an abstract formalism, and the examples discussed here show how explicit answers to path integrals can be obtained.

The puritanical cynic may ask: "How sure are we that the above conclusions will stand when all degrees of freedom of the spacetime geometry are quantized?" We have no answer to this question except to say: "Why don't you find out by quantizing the remaining degrees of freedom?"

Repartees apart, we can give an intuitive argument as to why the conformal degree of freedom is likely to be most crucial in deciding whether the quantum spacetime is singular or not, at any rate for curvature singularities. To follow the argument consider how the conformal degree of freedom separates out of any metric tensor g_{ik}. Write

$$g_{ik} = (-g)^{1/4}\, q_{ik}. \tag{109}$$

Under the conformal transformation

$$\hat{g}_{ik} = \Omega^2\, q_{ik}, \tag{110}$$

we have

$$(-\hat{g})^{1/4} = \Omega^2(-g)^{1/4}, \qquad \hat{q}_{ik} = q_{ik}. \tag{111}$$

Thus, the nonconformal degrees of freedom are contained in q_{ik}, while the conformal degrees are in $|g|$.

At the spacetime singularity we expect the proper volumes bound by comoving observers to shrink to zero, and this is reflected in $|g| \to 0$. (In the Friedmann models $Q \to 0$, $g \propto Q^6 \to 0$.) The crucial question, therefore, is whether the fluctuations of $|g|$ are large enough to wipe out the classical singularity.

A familiar analogy with the H atom can be given here once again. The classical theory tells us that the radial separation r between the electron and the proton tends to zero. It is the quantized version of r that tells us that quantum mechanics eliminates this singular fate of the classical H atom. The angular coordinates θ and φ are also quantized to give the various states of angular momentum. But their quantization is not as crucial to the above problem. The nonconformal degrees of freedom q_{ik} in (109) are analogous to the angular degrees of freedom of the H atom.

To proceed further, we shall try to answer the following question in the next chapter: Given the full range of QCFs, is it possible to formulate a semiclassical theory wherein some kind of average model can be obtained which takes note of *all* nonclassical models obtained by QCFs? To obtain such a model the classical Einstein equations have to be modified by the inclusion of an 'average' scalar field representing conformal fluctuations. Next, we shall see how it is done.

Notes and References

1. For the discussion of S_{GH} see:

   Gibbons, G. W., and Hawking, S. W.: 1977, *Phys. Rev.* **D15**, 2738.
   Hawking, S. W.: 1979, in *General Relativity – An Einstein Centenary Survey* (eds S. W. Hawking and W. Israel), Cambridge.

2. For the discussion of the boundary value problem in geometrodynamics, see:

   Isenberg J., and Wheeler, J. A.: 1979 in *Relativity, Quanta and Cosmology* (eds M. Pantaleo and F. De Finis), Johnson, New York, p. 267.

3. A discussion of Green's functions and their inverses will be found in:

   Morse, P. M., and Feschbach, H.: 1953, *Methods of Mathematical physics*, McGraw-Hill, New York.
   DeWitt, B. S., and Brehme, R. W.: 1960, *Ann. Phys.* **9**, 220.

4. The idea of transforming away the spacetime singularity through conformal transformations is discussed in different contexts by:

   Beem, J. K.: 1976, *Commun. Math. Phys.* **49**, 179.
   Kembhavi, A. K.: 1978, *Monthly Notices Roy Astron Soc.* **185**, 807.

3. The conformal flatness of Robertson–Walker models was first demonstrated by:

   Infeld, L., and Schild, A.: 1945, *Phys. Rev.* **68**, 250.

6. The instability of flat spacetime to conformal fluctuations has been discussed by:

   Atkatz, D., and Pagels, H.: 1982, *Phys. Rev.* **D25**, 2065.
   Brout, R., Englert, F., Frere, J. M., Gunzig, E., Naradone, P., and Truffin, C.: 1980, *Nucl. Phys.* **B170**, 228.
   Lindley, D.: 1981, *Nature* **291**, 392.
   Padmanabhan, T.: 1983, *Phys. Lett.* **93A**, 116.

Chapter 13

Towards a More Complete Theory

13.1. Towards a More Complete Theory

The analysis in the last chapter shows that conformal fluctuations make the classical solution completely insignificant near the spacetime singularity. In order to proceed further it is necessary to examine viable quantum mechanical descriptions which would replace the singular spacetime geometry. This is the aim of the present chapter.

It was seen in Chapter 3 that two equivalent descriptions are available for a quantum field. One may consider quantized gravitational field to be a system of 'gravitons'; the exchange of these virtual quanta will lead to the gravitational interaction between the particles. On the other hand, we may directly deal with the wave functional $\psi(\varphi(\mathbf{x}), t)$. The absolute value $|\psi|^2$ will give the probability to observe the field in the configuration $\varphi(\mathbf{x})$ at time t. In discussing the quantum cosmological aspects, it is more convenient and pertinent to use the latter description. We shall see that many useful analogies can be drawn between quantum cosmological models and simple atomic systems.

13.2. The Average Metric

In the last chapter we considered the conformal fluctuations around the Robertson–Walker metric, given (for $k = 0$) by

$$ds^2 = dt^2 - \bar{Q}^2 [dr^2 + r^2(d\theta^2 + \sin^2\theta \, d\varphi^2)]. \qquad (1)$$

This classical metric is singular at $t = 0$, because $\bar{Q} \to 0$ as $t \to 0$. It was seen that the spread of a Gaussian wave packet which describes the conformal factor behaves as

$$\Delta(t) \propto \bar{Q}(t)^{-1} \quad \text{as } t \to 0. \qquad (2)$$

Thus the quantum conformal fluctuations (QCFs) diverge near the classical singularity.

It is interesting to see what happens to the *spacetime metric* near the singularity. Since the state of the conformal factor Ω is described by a wave function

$$\psi(\Omega, t) = \left(\frac{1}{2\pi\Delta^2(t)}\right)^{1/4} \exp\left\{-\frac{(\Omega - 1)^2}{4\Delta^2(t)}\right\}, \tag{3}$$

we can make the following observation. All spacetime metrics of the type $g_{ik} = \Omega^2 \bar{g}_{ik}$ are allowed to exist with the probability distribution as given in Equation (3). Thus we may ask the question: "How does the expectation value of the metric evolve in this state?" We see that this 'average metric' is given by the expectation value

$$\begin{aligned}\langle g_{ik} \rangle &\equiv \int_{-\infty}^{+\infty} \psi^*(\Omega, t)\, \bar{g}_{ik}\Omega^2\, \psi(\Omega, t)\, d\Omega \\ &= [1 + \Delta^2(t)]\bar{g}_{ik}\,. \end{aligned} \tag{4}$$

This is equivalent to replacing the classical expansion factor $\bar{Q}(t)$ by an 'average expansion factor', where

$$\langle Q(t)^2 \rangle \equiv [1 + \Delta^2(t)]\, \bar{Q}^2(t). \tag{5}$$

The behaviour of $\langle Q^2 \rangle$ is very different from that of $\bar{Q}^2$ near the singularity. Near $t \cong 0$, since $\Delta \propto (\bar{Q})^{-1}$, we see that

$$\lim_{t \to 0} \langle Q^2 \rangle = \text{const} = b^2 \text{ (say)}. \tag{6}$$

In other words, the scale factor acquires a lower bound and avoids the singularity! The collapse does not proceed below the scale value of b.

Example 13.1. In the case of a radiation-filled universe we can take $\bar{Q}(t) = Bt^{1/2}$. The fluctuations at two times t_1 and t_2 ($t_1 < t_2$) are related by (with $L_P^2 = (G\hbar/c^3)$)

$$\Delta^2(t_1) = \Delta^2(t_2)\left[1 + \frac{L_P^2}{4a^2\Delta^4(t_2)B^6}\left(\frac{1}{t_2^{1/2}} - \frac{1}{t_1^{1/2}}\right)^2\right]$$

where $a = 3V/8\pi$. Near $t_1 \cong 0$, Δ behaves as

$$\Delta^2(t_1) \to \frac{L_P^2}{4a^2B^6\Delta^2(t_2)}\left(\frac{1}{t_1}\right)$$

and $\bar{Q}^2 = B^2 t_1$. Thus the average value $\langle Q^2 \rangle$ becomes

$$\langle Q^2 \rangle \cong \Delta^2 \bar{Q}^2 \to \frac{L_P^2}{4a^2B^4\Delta^2(t_2)} = \text{const.}$$

On the other hand, at late times, ($t_1 \gg t_2$) we recover the classical behaviour

$$\langle Q^2 \rangle \cong B^2 t\left(1 + \frac{L_P^2}{4a^2\Delta_2^4 B^6}\frac{1}{t_2}\right) \propto t. \qquad \square$$

This result is of crucial importance and may be interpreted in many ways. First, as discussed at the end of the last chapter, we can compare this result with the behaviour of simple atomic systems in classical and quantum mechanics. An electron and proton interacting via Coulomb forces constitute the simplest model of hydrogen atom. Classically, the trajectory $q(t)$ of the electron spirals down to the origin in a very short period of time. Quantum mechanics replaces this description in terms of a fixed trajectory $q(t)$ by a description in terms of the wave function $\psi(q, t)$. It is well known that these wave functions avoid the classical singularity. The conflict between the classical 'pull' (towards the origin) and the uncertainty principle leads to the appearance of a stable ground state. In the case of the collapsing universe the situation is identical. Classical singularity ($\bar{Q} = 0$) implies complete localization which leads to large uncertainty in the conformal factor. The balance between these two factors leads to a constant value of scale factor as $t \to 0$. From this point of view, we may expect the quantum universe to be described by a series of 'stationary geometrics' in complete analogy with the stationary states of a hydrogen atom, say. We shall develop this idea in detail in Section 13.4.

Example 13.2. Consider an electron in the hydrogen atom. If it is localized within a distance r from the proton, the uncertainty principle would imply a momentum of the order of

$$p \sim \frac{\hbar}{r},$$

which corresponds to a kinetic energy

$$\varepsilon = \frac{p^2}{2m} = \frac{\hbar^2}{2mr^2}.$$

The potential energy is of the order of $(-e^2/r)$ so that the total energy is

$$E = \frac{\hbar^2}{2mr^2} - \frac{e^2}{r}.$$

The ground state corresponds to the lowest possible value of E. When $\hbar \to 0$, E has a minimum of $(-\infty)$ at $r = 0$. In other words, the electron settles down at $r = 0$. But when $\hbar \neq 0$, the minimum value of E occurs at

$$\frac{\partial E}{\partial r} = 0, \qquad r = r_0 = \frac{\hbar^2}{me^2} \sim 10^{-8} \text{ cm}.$$

The minimum value of E is, in fact,

$$E_0 = -\tfrac{1}{2}\frac{me^4}{\hbar^2} \sim 10 \text{ eV}.$$

Notice that both the existence of the ground state *and* correct order of magnitude estimates for r_0, E_0 follow from the uncertainty principle. There exist analogous results in quantum cosmology.

There is an alternative way of looking at the above result. Consider the universe at any one time $t = t_1$ (say). The 3-space is described by the metric,

$$d\sigma^2 \equiv Q^2(t_1)\,(dx^2 + dy^2 + dz^2). \tag{7}$$

The proper distance between any two events $(\mathbf{r}, t_1)$ and $(\mathbf{r} + d\mathbf{r}, t_1)$ is given by,

$$dl^2 \equiv Q^2(t_1)\, d\mathbf{r} \cdot d\mathbf{r}. \tag{8}$$

In classical cosmology, as we take hypersurfaces closer and closer to singularity, $Q(t_1) \to 0$, and $dl^2 \to 0$ for the *same* pair of events. In quantum cosmology conformal fluctuations prevent this effect of infinite shrinking. Since for the 'averaged metric' $\langle Q^2 \rangle > 0$, two events which are originally separated by nonzero proper distance retain a nonzero proper separation distance at all times.

In the present context, we have assumed that the conformal factor depends only on time. When Ω is made to depend on space and time, the above result generalizes into a lower bound on proper length, which we shall consider in the next two sections.

13.3. Quantum Fluctuations and Proper Length

In classical gravity, the proper distance between two events x^i and $x^i + dx^i$ is given by the fundamental equation

$$ds^2 = \bar{g}_{ik}(x)\, dx^i\, dx^k. \tag{9}$$

The metric tensor $\bar{g}_{ik}$ is a fixed quantity in the classical theory and the proper distance between any two events is also fixed permanentaly. In particular, we are allowed to explore events arbitrarily close to a given event x^i; so much so that we can consider the limit of dx^i going to zero at a given value of $\bar{g}_{ik}(x)$. Clearly,

$$\lim_{dx^i \to 0} ds^2 = 0 \quad \text{(classical)}. \tag{10}$$

Quantum gravity changes this situation drastically. Since the metric is now a quantum variable, we cannot ascribe a definite value to the metric or to the line interval. Thus two events designated by x^i and $x^i + dx^i$ may be separated by any proper distance whatsoever; we can only talk about the probability for the distance to be equal to a particular value. It is this probability distribution in which we are interested.

Consider first the simplest case, viz. that of quantum conformal fluctuations around the flat space. *Flat space can alternatively be considered to be the gravitational vacuum, and we expect gravitational vacuum fluctuations to be present in the flat space*. The action governing the conformal fluctuations takes a very simple form when $\bar{g}_{ik} = \eta_{ik} \equiv \mathrm{diag}(1, -1, -1, -1)$:

$$\begin{aligned} S &= -\frac{1}{12L_P^2}\int [6\varphi^i\varphi_i - R(1+\varphi)^2]\sqrt{-g}\,\mathrm{d}^4x \\ &= -\frac{1}{2L_P^2}\int \varphi^i\varphi_i\,\mathrm{d}^4x; \qquad L_P^2 = \frac{4\pi}{3}G. \end{aligned} \tag{11}$$

This has the same form as the action for a massless scalar field except for an overall minus sign. Following the analysis of Chapter 4 (Example 4.5), we can write down the probability amplitude for the conformal fluctuations to have a value $\varphi(\mathbf{r})$ in the gravitational vacuum. This is given by

$$\begin{aligned} P[\varphi(\mathbf{r})] &= N\exp -\left\{\frac{1}{4\pi^2L_P^2}\int \frac{\nabla\varphi(\mathbf{r})\cdot\nabla\varphi(\mathbf{r}')}{|\mathbf{r}-\mathbf{r}'|^2}\,\mathrm{d}^3\mathbf{r}\,\mathrm{d}^3\mathbf{r}'\right\} \\ &= N\exp\left\{-\tfrac{1}{2}\int |\mathbf{k}|\,|\mathrm{q}_\mathbf{k}|^2\,\frac{\mathrm{d}^3\mathbf{k}}{(2\pi)^3}\right\}. \end{aligned} \tag{12}$$

where

$$\varphi(\mathbf{r}) = \int \mathrm{q}_\mathbf{k}\,\mathrm{e}^{i\mathbf{k}\cdot\mathbf{r}}\,\frac{\mathrm{d}^3\mathbf{k}}{(2\pi)^3}\,. \tag{13}$$

Now that the metric tensor is treated as a fluctuating quantum variable, it is necessary to measure the quantum field $\varphi(x)$ before the geometrical features of the spacetime can be determined. However, it is well known that, in quantum field theory, the expectation value of objects like $\varphi^2(\mathrm{x})$ are divergent. It is therefore necessary to approach the problem operationally. Let us consider a measurement of the field $\varphi(\mathbf{x}, t)$ by an 'apparatus' that has a spatial resolution limit of L (say). In other words, the apparatus does not distinguish $\mathbf{x}$ and $\mathbf{y}$ as different if $|\mathbf{x}-\mathbf{y}| < L$. (An ideal apparatus, of course, can be recovered in the limit of $L \to 0$.) Thus the apparatus will actually measure the 'smeared' (or 'coarse-grained') value of the field $\varphi(\mathbf{x})$ averaged over a region $\sim L^3$. If we denote by $f(\mathbf{r})$ the sensitivity profile of the apparatus, then the coarse-grained value is given by

$$\varphi_f(\mathbf{x}) \equiv \int \varphi(\mathbf{x}+\mathbf{r})\,f(\mathbf{r})\,\mathrm{d}^3\mathbf{r}. \tag{14}$$

Once again, the ideal experiment has the sensitivity profile of a delta function $f(\mathbf{r}) = \delta(\mathbf{r})$, leading to $\varphi_f(\mathbf{x}) = \varphi(\mathbf{x})$. In general, $f(\mathbf{r})$ will have a width of the order of L. That is, $f(\mathbf{r})$ will be of the order of unity for $|\mathbf{r}| \ll L$ and will drop to zero rapidly for $|\mathbf{r}| \gtrsim L$, Thus (14) defines a 'smeared' field $\varphi_f(\mathbf{x})$ as an average over a region of size L^3.

We wish to emphasize that the need to consider smeared values of fields, as in (14), has nothing to do with gravity *per se*, and arises purely from the general formalism for quantum fields. Also, the ideal experiment can always be recovered from this general case by taking $f(\mathbf{r}) = \delta(\mathbf{r})$, and thus nothing is lost in considering an arbitrary $f(\mathbf{r})$.

The apparatus when used to measure φ_f will come up with (in general) different values in different trials. (This is to be expected because the vacuum state is *not* an eigenstate of field operators.) The quantity of interest is the probability *amplitude* for the measurement to give φ_f the value η(say). This can easily be seen to be given by

$$A[\varphi_f = \eta] = \int \delta(\varphi_f - \eta)\, P[\varphi(\mathbf{x})]\, \mathcal{D}\varphi(\mathbf{x}), \tag{15}$$

where $P[\varphi(\mathbf{x})]$ is the vacuum functional in (12). This expression can be evaluated in the Fourier space in a straightforward manner (see Example 13.3 below). We get

$$A(\eta) = C \exp\left(-\frac{\eta^2}{4\Delta^2}\right) = \left(\frac{1}{2\pi\Delta^2}\right)^{1/4} \exp\left(-\frac{\eta^2}{4\Delta^2}\right), \tag{16}$$

where

$$\Delta^2 = L_{\mathrm{P}}^2 \int \frac{|f(\mathbf{k})|^2}{2|\mathbf{k}|} \frac{\mathrm{d}^3\mathbf{k}}{(2\pi)^3} \equiv \frac{L_{\mathrm{P}}^2}{4\pi^2 L^2}\,. \tag{17}$$

We have defined in (17) the 'resolution length' L for a distribution $f(r)$ by

$$L^{-2} \equiv \frac{1}{4\pi} \int \frac{|f(\mathbf{k})|^2}{|\mathbf{k}|}\, \mathrm{d}^3\mathbf{k}. \tag{18}$$

This definition is motivated by the fact that for a Gaussian $f(\mathbf{r})$ with a width σ:

$$f(\mathbf{r}) = \left(\frac{1}{2\pi\sigma^2}\right)^{3/2} \exp\left(-\frac{|\mathbf{r}|^2}{2\sigma^2}\right), \tag{19}$$

the resolution length L is equal to σ. Thus L essentially measures the width of $f(\mathbf{r})$ for any distribution.

Example 13.3. To derive (16) proceed as follows: Note that

$$\delta(\varphi_f - \eta) = \int_{-\infty}^{+\infty} \exp\left[i\lambda(\varphi_f - \eta)\right] \frac{\mathrm{d}\lambda}{2\pi} = \int_{-\infty}^{+\infty} \mathrm{e}^{-i\lambda\eta} \exp\left[i\lambda \int \varphi(\mathbf{r}+\mathbf{y}) f(\mathbf{y})\, \mathrm{d}^3\mathbf{y}\right] \frac{\mathrm{d}\lambda}{(2\pi)}$$

$$= \int_{-\infty}^{+\infty} \left\{\mathrm{e}^{-i\lambda\eta} \exp\left[i\lambda \int \frac{\mathrm{d}^3\mathbf{k}}{(2\pi)^3} \int \frac{\mathrm{d}^3\mathbf{p}}{(2\pi)^3}\, q_{\mathbf{k}}\, \mathrm{e}^{i\mathbf{k}\cdot(\mathbf{r}+\mathbf{y})} f_{\mathbf{p}}\, \mathrm{e}^{-i\mathbf{p}\cdot\mathbf{y}}\, \mathrm{d}^3\mathbf{y}\right]\right\} \frac{\mathrm{d}\lambda}{2\pi}$$

$$= \int_{-\infty}^{+\infty} \left\{\mathrm{e}^{-i\lambda\eta} \exp\left[i\lambda \int \frac{\mathrm{d}^3\mathbf{k}}{(2\pi)^3}\, q_{\mathbf{k}} f_{\mathbf{k}}\, \mathrm{e}^{i\mathbf{k}\cdot\mathbf{r}}\right]\right\} \frac{\mathrm{d}\lambda}{2\pi}\,.$$

In the above expressions $q_{\mathbf{k}}$ is the Fourier transform of $\varphi(\mathbf{x})$. Now substituting into (13.15) we get

$$A[\eta] = \int \delta(\varphi_f - \eta)\, P(\varphi)\, \mathcal{D}\varphi$$

$$= \prod_{\mathbf{p}} \int \mathrm{d}q_{\mathbf{p}} \int_{-\infty}^{+\infty} \left\{ \mathrm{e}^{-i\lambda\eta} \exp\left\{ - \int \frac{\mathrm{d}^3\mathbf{k}}{(2\pi)^3} \times \right.\right.$$

$$\left.\left. \times \left(\tfrac{1}{2}\,\omega_{\mathbf{k}}\, |q_{\mathbf{k}}|^2 + i\lambda q_{\mathbf{k}} f_{\mathbf{k}}\, \mathrm{e}^{i\mathbf{k}\cdot\mathbf{r}}\right)\right\}\right\} \frac{\mathrm{d}\lambda}{(2\pi)}$$

$$= \prod_{\mathbf{p}} \int_{-\infty}^{+\infty} \frac{\mathrm{d}\lambda}{2\pi} \mathrm{e}^{i\lambda\eta} \int_{-\infty}^{+\infty} \exp\left\{ - \sum_{\mathbf{k}} \left(\tfrac{1}{2}\,\omega_{\mathbf{k}} |q_{\mathbf{k}}|^2 + i\lambda q_{\mathbf{k}} f_{\mathbf{k}}\, \mathrm{e}^{i\mathbf{k}\cdot\mathbf{r}}\right)\right\} \mathrm{d}q_{\mathbf{p}}$$

Since $q_{\mathbf{p}}$ is a complex number, the symbol $\int \mathrm{d}q_{\mathbf{p}}$ actually represents integration over both real and imaginary parts of $q_{\mathbf{p}}$. Performing the Gaussian integrations for each $\mathbf{k}$, we get

$$A(\eta) = (\text{const}) \int_{-\infty}^{+\infty} \mathrm{e}^{-i\lambda\eta} \exp -\lambda^2 \left(\sum_{\mathbf{k}} \frac{|f(\mathbf{k})|^2}{2|\mathbf{k}|} \right) \frac{\mathrm{d}\lambda}{(2\pi)}$$

$$= (\text{const}) \exp\left\{ - \frac{\eta^2}{2\sum_{\mathbf{k}} |f_{\mathbf{k}}|^2/|\mathbf{k}|} \right\}.$$

Replacing $\sum_{\mathbf{k}}$ by $\int \mathrm{d}^3\mathbf{k}/(2\pi)^3$ and normalizing $A(\eta)$, we get the result (16) of text. □

As long as we consider length measurements averaged over many Planck lengths (i.e. for $L \gg L_P$), Δ is almost zero and the probability in (16) is sharply peaked at $\eta = 0$. In this case quantum fluctuations hardly affect the length measurements. However, as the resolution of the apparatus L goes to zero the fluctuations in η continue to increase. We can no longer talk about a unique proper distance between $(t, \mathbf{x})$ and $(t, \mathbf{y})$. When the conformal factor has a value η, the proper length between $(t, \mathbf{x})$ and $(t, \mathbf{y})$ is given by

$$R^2 = (1 + \eta)^2 R_0^2; \qquad R_0^2 = |\mathbf{x} - \mathbf{y}|^2. \tag{20}$$

Therefore, the probability for the events $(t, \mathbf{x})$ and $(t, \mathbf{y})$ to be separated by a proper length R is given by

$$P(R)\,\mathrm{d}R = P(\eta(R))\, \frac{\mathrm{d}\eta}{\mathrm{d}R}\, \mathrm{d}R. \tag{21}$$

Therefore,

$$P(R)\,\mathrm{d}R = \left(\frac{2}{\pi\sigma^2}\right)^{1/2} \exp\left\{ - \frac{(R - R_0)^2}{2\sigma^2} \right\} \mathrm{d}R; \quad R \geq 0 \tag{22}$$

with

$$\sigma^2 = \left(\frac{R_0^2}{L^2}\right)\left(\frac{L_P^2}{4\pi^2}\right); \quad R_0 = |\mathbf{x} - \mathbf{y}|. \tag{23}$$

Expression (22) demonstrates the effect of quantum fluctuations in a neat manner. Because of the vacuum fluctuations of the metric, we cannot attribute a unique proper distance between two events $(t, \mathbf{x})$ and $(t, \mathbf{y})$. The *probability* that this proper distance has a value R (when the measurement is performed with a resolution L) is given by (22). The proper distance is peaked at the classical value R_0. In order to have any confidence in this R_0 we must have,

$$\sigma^2 \ll R_0^2 . \tag{24}$$

This, in turn, implies that

$$L^2 \gg L_P^2 . \tag{25}$$

In other words, the concept of a unique distance between events ceases to have any meaning when $L \sim L_P$.

The previous analysis was confined to a spacelike hypersurface of $t =$ constant. We can trivially extend the analysis to four dimensions by taking a smeared value of $\varphi(x)$ over a spacetime region of 'volume' L^4. Since we shall consider the expectation values of proper distances in the next section, we shall not extend the above analysis now.

From the distribution (22) we obtain the mean square value of

$$\langle R^2 \rangle = R_0^2 + \sigma^2 = R_0^2 \left(1 + \frac{L_P^2}{4\pi^2 L^2} \right) ; \tag{26}$$

clearly to measure a distance R_0 we need $L \leq R_0$. As the events approach one another $R_0 \to 0$, $L \to 0$ keeping $R_0 \geq L$. In this limit, we get $\langle R^2 \rangle$ to be $(L_P/2\pi)^2$ if we assume $R_0 = L$ as R_0, $L \to 0$. In other words, we see that the fluctuations may lead to a lower bound to proper length. However, at this stage, L depends on the choice of $f(r)$ and the limiting value of (26) is not clear. We shall now consider this situation in more detail.

13.4. Lower Bound to Proper Length

Classically, there is nothing to prevent us considering two events which are arbitrarily close. However, the analysis in the previous section clearly shows that such considerations may have no physical relevance. In particular, Equations (16) and (17) can be written as an 'uncertainty principle'

$$\Delta\eta \, \Delta l \gtrsim L_P, \tag{27}$$

where Δl is the uncertainty in the measurement of proper length and $\Delta\eta$ is the uncertainty in the conformal factor. (In our case, $\Delta l \sim L$ and $\Delta\eta$ is $O(L_P/L)$. Therefore, the limit of zero proper separation ($\Delta l \to 0$) is operationally ill defined.

Let us consider the expectation value of the line element in the vacuum state:

$$\langle 0|\, \mathrm{d}s^2\, |0\rangle = \langle 0|\, g_{ik}\, |0\rangle\, \mathrm{d}x^i\, \mathrm{d}x^k = (1 + \langle \varphi^2(x)\rangle)\eta_{ik}\, \mathrm{d}x^i\, \mathrm{d}x^k.$$

Again $\langle \varphi^2 \rangle$ evaluated at a single event diverges. Also notice that $\mathrm{d}s^2$ involves for its definition two events x^i and $y^i \equiv x^i + \mathrm{d}x^i$. Since we are interested in the limit $x^i \to y^i$, it is more proper to consider (using the notation $l^2 = \eta_{ik}\, \mathrm{d}x^i\, \mathrm{d}x^k$) the limit

$$\begin{aligned} \operatorname*{Lim}_{x\to y}\ \langle \mathrm{d}s^2\rangle &\equiv \operatorname*{Lim}_{x\to y} \langle l^2(x, y)\rangle \equiv \operatorname*{Lim}_{x\to y} (1 + \langle \varphi(x)\varphi(y)\rangle)\eta_{ik}\, \mathrm{d}x^i\, \mathrm{d}x^k \\ &\equiv \operatorname*{Lim}_{x\to y} (1 + \langle \varphi(x)\varphi(y)\rangle) l^2. \end{aligned} \tag{28}$$

The 'average' value (vacuum expectation value) is defined via the usual path integral formula (see Chapter 4, Section 4.5)

$$\langle \varphi(x)\varphi(y)\rangle \equiv \frac{\int \varphi(x)\varphi(y) \exp iS(\varphi)\, \mathcal{D}\varphi}{\int \exp iS(\varphi)\, \mathcal{D}\varphi},$$

where the action S is given by (11). Defining, as in Chapter 4

$$Z[J] = \int \exp[iS + i\int J(x)\varphi(x)\, \mathrm{d}x]\, \mathcal{D}\varphi,$$

we get

$$\langle \varphi(x)\varphi(y)\rangle = -\frac{1}{2} \left.\frac{\delta^2 Z}{\delta J(x)\, \delta J(y)}\right|_{J=0} . \tag{29}$$

Straightforward calculation (noting the sign change in (11)) gives

$$\begin{aligned} \operatorname*{Lim}_{x^i\to y^i}\ \langle l^2(x, y)\rangle &= \operatorname*{Lim}_{x^i\to y^i} \left\{1 + \langle \varphi(x)\varphi(y)\rangle\right\} l^2 \\ &= \operatorname*{Lim}_{x^i\to y^i} \langle \varphi(x)\varphi)y)\rangle\, l^2 \\ &= \operatorname*{Lim}_{x\to y} \frac{L_P^2}{4\pi^2} \cdot \frac{1}{(x-y)^2} l^2 = \operatorname*{Lim}_{x\to y} \frac{L_P^2}{4\pi^2}\ \frac{1}{l^2} l^2 \\ &= \left(\frac{L_P^2}{4\pi^2}\right). \end{aligned} \tag{30}$$

In other words, the expectation value of the proper interval between any two events in the spacetime is bounded from below at $(L_P/2\pi)^2$. Quantum fluctuations produce a 'residual length' just as zero point vibrations of an oscillator lead to a residual ground state energy.

Example 13.4. Before we proceed with the discussion of the important result (30) we would like to clarify a technical point. The fact that the conformal degree of freedom comes up with the wrong sign in the kinetic energy term makes the *Euclidean* gravitational action unbounded from below. In our approach to quantum gravity, we do *not* consider the Euclidean section but work with the oscillating path integrals themselves. We believe this approach is more physical because of the ambiguities involved in defining the Euclidean section for an arbitrary curved Riemannian spacetime. In the Euclidean approach we must first transform from φ to another variable $\eta = i\varphi$, obtaining

$$A[\eta] = \frac{1}{2L_P^2} \int \eta^k \eta_k \, d^4x.$$

Now changing to the imaginary time coordinate $\tau = it$ we get the Euclidean path integral

$$K_E = \int \exp\left(-\int \frac{d^4x_E}{2L_P^2} \eta^k \eta_k\right) \mathcal{D}\eta.$$

This well defined path integral will lead to the Euclidean Green's function and expectation value

$$\langle \eta(x)\eta(y) \rangle_E = \frac{L_P^2}{4\pi^2} \frac{1}{(\tau_x - \tau_y)^2 + |\mathbf{x} - \mathbf{y}|^2} \equiv \frac{L_P^2}{4\pi^2} \frac{1}{(x - y)_E^2}.$$

Assuming (as is always done in the Euclidean approach to quantum gravity) that $\langle \varphi(x)\varphi(y) \rangle_E$ can be obtained as analytic continuation from η, we get

$$\langle \varphi(x_E)\varphi(y_E) \rangle = -\langle \eta(x_E)\eta(y_E) \rangle = -\frac{L_P^2}{4\pi^2} \frac{1}{(x_E - y_E)^2}.$$

Notice that we are still working in the Euclidean spacetime $(\tau, \mathbf{r})$. The proper distance in terms of this Euclidean coordinates has the limiting value

$$\begin{aligned}
\lim_{x \to y} l^2(x, y) &= \lim_{x_E \to y_E} \langle \varphi(x_E)\varphi(y_E) \rangle \left\{-(x_E - y_E)^2\right\} \\
&= \lim_{x_E \to y_E} \left\{-\frac{L_P^2}{4\pi^2} \frac{1}{(x_E - y_E)^2}\right\} \left\{-(x_E - y_E)^2\right\} \\
&= \left(\frac{L_P}{2\pi}\right)^2.
\end{aligned}$$

Thus this procedure, in this particular case, leads to the same conclusion as obtained in (30) by straightforward means. We have gone through the arguments in the Euclidean spacetime in order to show that mathematical ambiguities, which are always present with an oscillating path integral, do not introduce any spurious features into our discussions. □

Coming back to our basic result (30) it is clear that the conformal degree of freedom plays a special role in quantum gravity. Mathematically speaking, two factors have gone in crucially into the result (30): (i) conformal factor multiplies the spacetime line interval and (ii) the conformal factor appears in the action with quadratic dependence. Physically, we may consider the conformal factor as a 'conjugate variable' to proper length. The vacuum fluctuations of this conformal degree of freedom produce a 'zero point distance' in the spacetime.

Though we have been concentrating on the flat spacetime, result (30) happens to be valid for an arbitrary spacetime. In an arbitrary spacetime, the Green's function that determines $\langle \varphi(x)\varphi(y)\rangle$ satisfies the equation

$$(\Box + \tfrac{1}{6}R)G = 0. \tag{31}$$

However, the coincidence limit $(x \to y)$ of $G(x, y)$ again has the behaviour of s^{-2}, where s is the proper distance between x and y. This is because in the limit of $x^i \to y^i$ we can always construct a coordinate system that is locally flat and includes both x^i and y^i.

Example 13.5. A lower bound on proper length leads to interesting conclusions in field theory. In particular, the divergence problem can be avoided if we use the expectation value of the proper length. We shall explore some of these consequences in this example.

Let us recall from Chapter 4 that the propagator for a particle of mass m can be expressed (see Chapter 4, Example 4.9 and 4.3) as

$$\begin{aligned} G(x, y) &= -\int_0^\infty \frac{d\lambda}{2m} \int \exp\left\{-\frac{im}{2}\int_0^\lambda (\dot{x}_i\dot{x}^i + 1)\, ds\right\} \mathcal{D}x^i(s) \\ &= -\int_0^\infty \left(\frac{m}{2\pi i\lambda}\right)^2 \exp -\frac{im}{2}\left\{\frac{l^2(x, y)}{\lambda} + \lambda\right\} \cdot \frac{d\lambda}{2m}, \end{aligned}$$

where $l^2(x, y)$ is the proper distance between x^i and y^i. Averaging over the conformal fluctuations is equivalent to replacing $l^2(x, y)$ by

$$\begin{aligned} \langle l^2(x, y)\rangle &= [1 + \langle\varphi(x)\varphi(y)\rangle]\,(x - y)^2 \\ &= (x - y)^2 + \left(\frac{L_P}{2\pi}\right)^2 \quad (\text{near } x \cong y). \end{aligned}$$

Therefore, the averaged out value for the Green's function is

$$\langle G(x, y)\rangle = -\int_0^\infty \left(\frac{m}{2\pi i\lambda}\right)^2 \exp -\frac{im}{2}\left\{\frac{\langle l^2(x, y)\rangle}{\lambda} + \lambda\right\}\frac{d\lambda}{2m}.$$

Clearly,

$$\begin{aligned} \lim_{x\to y} \langle G(x, y)\rangle &= -\int_0^\infty \left(\frac{m}{2\pi i\lambda}\right)^2 \exp -\frac{im}{2}\left\{\frac{(L_P/2\pi)^2}{\lambda} + \lambda\right\}\frac{d\lambda}{2m} \\ &= \frac{m}{2\pi L_P} K_1\left(\frac{imL_P}{2\pi}\right); \end{aligned}$$

where K_1 is the modified Bessel function. Notice that quantum gravitational fluctuations have made the coincidence limit of $G(x, y)$ completely finite. The modified Green's function in the Fourier space may be taken to be

$$\begin{aligned} \langle G(k)\rangle &\equiv \int d^4x\, e^{-ik\cdot x}\langle G(x)\rangle \\ &= -\int d^4x\, e^{-ik\cdot x}\int_0^\infty \frac{d\lambda}{2m}\left(\frac{m}{2\pi i\lambda}\right)^2 \exp -\frac{im}{2}\left[\frac{x^2 + (L_P/2\pi)^2}{\lambda} + \lambda\right] \end{aligned}$$

$$= -\int_0^\infty \left(\frac{m}{2\pi i\lambda}\right)^2 \exp -\frac{im}{2}\left(\frac{L_P^2}{4\pi^2\lambda} + \lambda\right)\left(\frac{2\pi\lambda}{im}\right)^2 \exp\left(-\frac{k^2\lambda}{2im}\right)\frac{d\lambda}{2m}$$

$$= -\int_0^\infty \exp\left\{-\frac{imL_P^2}{8\pi^2\lambda} + \frac{i\lambda}{2m}(k^2 - m^2)\right\}\frac{d\lambda}{2m}$$

$$= \frac{L_P}{2\pi i}\frac{1}{(k^2 - m^2 + i\varepsilon)^{1/2}} K_1\left(i\frac{L_P}{2\pi}(k^2 - m^2 + i\varepsilon)^{1/2}\right).$$

The $(i\varepsilon)$ prescriptions are not necessary if we use the Euclidean 4-momentum k_E (so that $-k^2 = k_E^2$). The Euclidean Green's function is

$$\langle G(k_E)\rangle = \frac{L_P}{2\pi}\frac{1}{(k_E^2 + m^2)^{1/2}} K_1\left(\frac{L_P}{2\pi}(k_E^2 + m^2)^{1/2}\right).$$

The modified Bessel function $K_1(z)$ has the asymptotic behaviour

$$K_1(z) = \begin{cases} z^{-1}, & z \to 0, \\ \left(\dfrac{\pi}{2z}\right)^{1/2} e^{-z}, & z \to \infty; \end{cases}$$

so that

$$\langle G(k_E)\rangle = \begin{cases} (k_E^2 + m^2)^{-1} & (\text{as } L_P \to 0) \\ \exp\left\{-\dfrac{L_P}{2\pi}(k_E^2 + m^2)^{1/2}\right\} & (\text{as } k_E \to \infty). \end{cases}$$

Note that $\langle G(k_E)\rangle$ dies down exponentially for large k_E in contrast with the $G(k_E)$ which dies down only as k_E^{-2}.

This new feature leads to finite values for the quantum effective potential. We saw in Chapter 4 (see Equation (4.147)) that the 'one-loop' effective potential is given by

$$V_{\text{eff}} = \tfrac{1}{2}\int \ln(1 + G(k_E)V'')\frac{d^4k_E}{(2\pi)^4},$$

where V is the classical potential.[1] Since $G(k_E)$ goes as k_E^{-2} for large k_E the integral diverges. On the other hand, replacing $G(k_E)$ by $\langle G(k_E)\rangle$ the integrand is seen to behave as

$$\ln(1 + \langle G(k_E)\rangle V'') \cong \langle G(k_E)\rangle V''$$

$$\cong V'' \exp\left\{-\left(\frac{L_P}{2\pi}\right)k_E\right\}$$

[1] We have subtracted from V_{eff} in (4.147) a part independent of V to write it in the above form.

for large k_E. Clearly, the integral is convergent and can be approximated by

$$V_{\text{eff}} = \frac{4\pi^4}{L_P^2} V'' + \frac{\pi^2}{2} (V'')^2 \left[\ln \frac{V'' L_P^2}{4\pi^2} - \frac{1}{2} \right] + O(mL_P). \qquad \square$$

13.5. Quantum Stationary Geometries

The examples in the previous two sections show that quantum fluctuations lead to a 'ground state' in which the dispersion in length intervals is of the order of the Planck length. This fact motivates us to look for a description of quantum spacetime in terms of stationary states for quantized conformal factor. We shall call these stationary configurations 'quantum stationary geometrics' or QSGs, in short.

Evidently, the nature of QSGs will depend on the symmetries we impose on the metric. Of particular interest are the QSGs that are homogeneous and isotropic. Such geometrics may be described by the metric in either of the two coordinate systems

$$\overline{\mathrm{d}s}^2 = \mathrm{d}T^2 - \tilde{Q}^2(T)\left[\frac{\mathrm{d}r^2}{1 - kr^2} + r^2(\mathrm{d}\theta^2 + \sin^2\theta\, \mathrm{d}\varphi^2)\right]$$

$$= Q^2(t) \left[\mathrm{d}^2 t - \frac{\mathrm{d}r^2}{1 - kr^2} - r^2(\mathrm{d}\theta^2 + \sin^2\theta\, \mathrm{d}\varphi^2)\right];$$

$$t = \int \frac{\mathrm{d}T}{\tilde{Q}(T)}, \quad Q(t) = \tilde{Q}(\tau). \tag{32}$$

In considering all possible conformal fluctuations of this metirc, we shall have to take into account all functions of the form $\Omega(t)\, Q(t)$. When we consider the quantum structure of the spacetime, we may therefore set $Q = 1$ and consider all possible $\Omega(t)$. Clearly, when all possible values of $\Omega(t)$ are considered, it makes no difference in the combinations ΩQ, whether $Q = 1$ or not.

Henceforth, we consider the class of all *quantum geometries* in the form

$$\mathrm{d}s^2 = \langle \Omega^2 \rangle\, \overline{\mathrm{d}s}^2 \tag{33}$$

with

$$\overline{\mathrm{d}s}^2 = c^2\, \mathrm{d}t^2 - \frac{\mathrm{d}r^2}{1 - r^2/a^2} - r^2(\mathrm{d}\theta^2 + \sin^2\theta\, \mathrm{d}\varphi^2). \tag{34}$$

To illustrate the concept of QSG clearly we have reintroduced c in (34). We have also taken $k = +1$, and given r the dimension of length by bringing in a scale length a. In this notation Ω is a dimensionless variable. The gravitational part of the action, governing the conformal fluctuations is

$$S_g = \frac{Vc^4}{16\pi G}\int_{t_1}^{t_2} (\bar{R}\Omega^2 - 6\dot{\Omega}^2)\,\mathrm{d}t$$

$$= -\frac{1}{2}M\int_{t_1}^{t_2} (\dot{q}^2 - \omega^2 q^2)\,\mathrm{d}t, \tag{35}$$

where

$$q = a\Omega; \qquad M = \frac{3}{2}\pi\left(\frac{ac^2}{G}\right); \quad \omega = \frac{c}{a} \tag{36}$$

and V is the proper spatial volume of the closed universe

$$V = \int \sqrt{-g}\ \mathrm{d}^3x = 2\pi^2 a^3.$$

Notice that q, M, ω have the correct dimensions of length, mass, and frequency. The action S_g represents (except for an overall minus sign) a harmonic oscillator of mass M and frequency ω.

The total action for the system is obtained by adding to S_g the action for the matter. Since we are interested in the early universe phase, we shall assume that the matter is in the form of massless radiation. Therefore, the matter part of the action will be conformally invariant and independent of Ω. The complete dynamics of Ω is thus determined by S_g.

The classical dynamics of Ω is determined by the principle of stationary action $\delta S_g = 0$. This equation has the solution

$$q(t) = q_0 \sin \omega t \tag{37}$$

corresponding to the metric

$$\mathrm{d}s^2_{\text{classical}} = q_0^2 \sin^2 \omega t\left[c^2\,\mathrm{d}t^2 - \frac{\mathrm{d}r^2}{1 - r^2/a^2} - r^2(\mathrm{d}\theta^2 + \sin^2\theta\,\mathrm{d}\varphi^2)\right]$$

$$= A^2 \sin^2 \eta\,[\mathrm{d}\eta^2 - \mathrm{d}\chi^2 - \sin^2\chi\,(\mathrm{d}\theta^2 + \sin^2\theta\,\mathrm{d}\varphi^2)]\,. \tag{38}$$

(We have made a change of coordinates $r = a \sin \chi$ and $ct = a\eta$.)This is the correct form for the classical, radiation-filled $k = +1$ universe.

The quantum stationary geometries are obtained from the stationary states of the Schrödinger equation. For the system in (35) the Schrödinger equation

$$i\hbar\,\frac{\partial\psi}{\partial t} = \frac{\hbar^2}{2M}\,\frac{\partial^2\psi}{\partial q^2} - \tfrac{1}{2}M\omega^2 q^2\psi$$

can be separated by the choice

$$\psi(q, t) = \exp\left(\frac{i\varepsilon}{\hbar}\,t\right)\varphi(q); \tag{39}$$

so that, $\varphi(q)$ satisfies the equation

$$-\frac{\hbar^2}{2M}\frac{\mathrm{d}^2\varphi}{\mathrm{d}q^2}+\tfrac{1}{2}M\omega^2q^2\varphi=\varepsilon\varphi. \tag{40}$$

Equation (40) is the standard eigenvalue equation (with proper signs) for a harmonic oscillator. The overall minus sign in (35) has been taken care of in (39) by the $\exp(+(i/\hbar)\,\varepsilon t)$ factor. In standard quantum mechanics we would have taken $\exp(-(i/\hbar)\,Et)$ since E has the physical meaning of energy, and should be positive for the harmonic oscillator potential. In the present case, ε has no direct physical meaning and the sign in the exponent does not matter.

The normalizable solutions of (40) are labelled by the integers $n = 0, 1, 2, \ldots$ and are given by

$$\varphi_n(q)=(2^n n!)^{-1/2}\left(\frac{M}{\pi\hbar}\right)^{1/4}H_n\left(q\sqrt{\frac{M\omega}{\hbar}}\right)\exp\left\{-\frac{M\omega}{2\hbar}q^2\right\}. \tag{41}$$

In order to determine the metric in the quantum stationary geometries, we need the expectation value of Ω^2. From the theory of harmonic oscillators, we know that

$$\langle n|\,q^2\,|n\rangle=\frac{\hbar}{M\omega}(n+\tfrac{1}{2}). \tag{42}$$

Therefore,

$$\begin{aligned}\langle n|\,\mathrm{d}s^2\,|n\rangle &= a^{-2}\left(\frac{\hbar}{M\omega}\right)(n+\tfrac{1}{2})\Big[c^2\,\mathrm{d}t^2-\frac{\mathrm{d}r^2}{1-(r^2/a^2)}-\\ &\quad -\mathrm{r}^2(\mathrm{d}\theta^2+\sin^2\theta\,\mathrm{d}\varphi^2)\Big]\\ &=\left(\frac{2}{3\pi}\right)\left(\frac{L_\mathrm{P}^2}{a^2}\right)(n+\tfrac{1}{2})\Big[c^2\,\mathrm{d}t^2-\frac{\mathrm{d}r^2}{1-(r^2/a^2)}-\\ &\quad -r^2(\mathrm{d}\varphi^2+\sin^2\theta\,\mathrm{d}\varphi^2)\Big]; \end{aligned}\tag{43}$$

where we have used (36) and defined $L_\mathrm{P}^2=(G\hbar/c^3)$ (which differs by a factor $(4\pi/3)$ from the same symbol used previously.) By a coordinate transformation, $r = a\sin\chi$; $ct = a\eta$, the metric becomes

$$\langle \mathrm{d}s^2\rangle=\left(\frac{2}{3\pi}\right)L_\mathrm{P}^2(n+\tfrac{1}{2})\,[\mathrm{d}\eta^2-\mathrm{d}\chi^2-\sin^2\chi(\mathrm{d}\theta^2+\sin^2\theta\,\mathrm{d}\varphi^2)]. \tag{44}$$

Equation (44) for $n = 0, 1, 2, \ldots$ represents the various stationary states for the quantum universe. As discussed in Chapter 2, these stationary states contain all the information about the system. Any arbitrary state for the universe can be obtained by a superposition of these stationary states.

As an example, consider the classical limit in which $q(t)$ behaves as in (37). Such a near-classical behaviour can be described by the coherent state, studied in Chapter 2, Section 2.7. In a coherent state, the probability distribution is given (cf. (2.78) by :

$$\begin{aligned} |\psi(q, t)|^2 &= \left(\frac{M\omega}{\pi\hbar}\right)^{1/2} \exp - \left\{\frac{M\omega}{\hbar}(q - q_0 \sin \omega t)^2\right\} \\ &= \left(\frac{3}{2L_P^2}\right)^{1/2} \exp\left\{-\frac{3\pi}{2}\frac{1}{L_P^2}(q - q_0 \sin \omega t)^2\right\}. \end{aligned} \tag{45}$$

The line element in this coherent state is given by

$$\begin{aligned} \langle\psi|\, ds^2\, |\psi\rangle &= \frac{1}{a^2}\left(\frac{L_P^2}{3\pi} + q_0^2 \sin^2 \omega t\right) \times \\ &\quad \times \left[c^2\, dt^2 - \frac{dr^2}{1 - (r^2/a^2)} - r^2(d\theta^2 + \sin^2\theta\, d\varphi^2)\right] \\ &= (q_0^2 \sin^2 \eta + (3\pi)^{-1} L_P^2) \times \\ &\quad \times \left[d\eta^2 - d\chi^2 - \sin^2 (d\theta^2 + \sin^2\theta\, d\varphi^2)\right]. \end{aligned} \tag{46}$$

In arriving at (46) we have used

$$\int_{-\infty}^{+\infty} q^2|\psi|^2\, dq = q_0^2 \sin^2 \omega t + (3\pi)^{-1} L_P^2. \tag{47}$$

Equation (46) clearly shows the role of quantum fluctuations in cosmology. In the $\hbar \to 0$ (classical) limit the expansion factor goes as $(q_0 \sin \eta)$, which is to be expected (see Equation (38)). But when $\hbar \neq 0$, the expansion factor never vanishes:

$$(3\pi)^{-1} L_P^2 \leq q_0^2 \sin^2 \eta + (3\pi)^{-1} L_P^2 \leq (3\pi)^{-1} L_P^2 + q_0^2\,. \tag{48}$$

Quantum fluctuations stop the collapse at $(3\pi)^{-1} L_P^2$ and produce an oscillating universe, which is nonsingular. We have shown the classical and quantum behaviours in Figure 13.1

Example 13.6. Let us assume that our universe is described by a wave function in (45). Consider a spacelike hypersurface, t = constant. What is the probability that two events $(0, \theta, \varphi, t)$ and $(\chi, \theta, \varphi, t)$ are separated by a proper distance x? If the scale factor is q, then this distance is $q\chi$. So the probability in question is given by

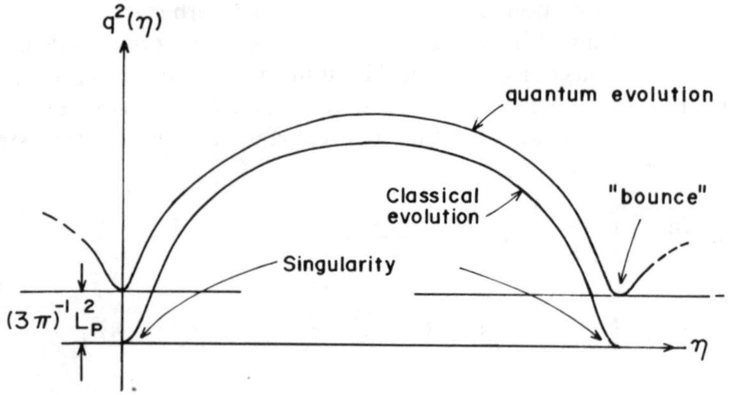

Fig. 13.1. The classical and quantum behaviour of the universe. Note that the quantum universe is nonsingular and reaches a minimum size $\sim L_P^3$. (The scales are exaggerated by a large factor.)

$$\tilde{P}(x)\,\mathrm{d}x = P(q)\left(\frac{\mathrm{d}q}{\mathrm{d}x}\right)\mathrm{d}x = \left(\frac{3}{2L_P^2}\right)^{1/2} \exp\left\{-\frac{3\pi}{2\chi^2 L_P^2}(x - q_0\chi\sin\omega t)^2\right\}\frac{\mathrm{d}x}{\chi}.$$

In other words, the proper length is peaked at the classical value of $(q_0 \sin \omega t) \cdot \chi$ with a dispersion $\sim L_P^2$. We see that because of this dispersion ($\sim L_P^2$) it is not really meaningful to talk about distances of the order of L_P in our model universe. □

We conclude this section with a brief discussion of QSGs in other spacetimes. As we said earlier (see p. 394) stationary states provide a perfectly consistent (though in many cases, messy) description of any quantum field theory. Therefore they exist even when other degrees of freedom are quantized. In particular, QSGs can be defined for any Bianchi spacetime (see Chapter 8). We recall that the metric for the diagonal Bianchi model is given by the equations

$$\begin{aligned} \mathrm{d}s^2 &= \mathrm{d}t^2 + g_{\mu\nu}\,\mathrm{d}x^\mu\,\mathrm{d}x^\nu \\ -g_{\mu\nu} &= \exp 2\lambda(t)\,[\exp - 2\beta(t)]_{\mu\nu} \\ \beta &= \mathrm{diag}(\beta_1 - \tfrac{1}{2}\beta_1 + \tfrac{1}{2}\sqrt{3}\,\beta_2, -\tfrac{1}{2}\beta_1 - \tfrac{1}{2}\sqrt{3}\,\beta_2). \end{aligned} \tag{49}$$

The action can be written, in terms of λ, β_1, β_2, as ($c = 1$, $G \neq 1$):

$$S = \frac{1}{16\pi G}\int\left\{-\,\mathrm{e}^{3\lambda}\left[6\dot{\lambda}^2 - \frac{3}{2}(\dot{\beta}_1^2 + \dot{\beta}_2^2)\right] + \mathrm{e}^{3\lambda} R^*\right\}\mathrm{d}t, \tag{50}$$

where R^* is given by (8.99). The Hamiltonian corresponding to (50) (for type B models; similar results also exist for type A models) is:

$$H = -\frac{\hbar^2}{4q^2}\left[\frac{\partial^2}{\partial u_1^2} + \frac{\partial^2}{\partial u_2^2}\right] + \frac{\hbar^2}{4}\frac{\partial^2}{\partial q^2} - c_1 q^{2/3\exp(-c_2 u_1)}. \tag{51}$$

where

$$q = \left(\frac{8}{3}\right)^{1/2} \exp\left(\frac{3}{2}\lambda\right) \tag{52}$$

$$u_1 = \left(\frac{8}{3}\right)^{1/2} \left[\frac{3}{2(1-3h)}\right]^{1/2} (\beta_1 + \sqrt{-3h}\,\beta_2) \tag{53}$$

$$u_2 = \left(\frac{8}{3}\right)^{1/2} \left[\frac{3}{2(1-3h)}\right]^{1/2} (\sqrt{-3h}\,\beta_1 - \beta_2); \tag{54}$$

and c_1, c_2, h (<0) are constants. The eigenvalue equation $H\varphi = E\varphi$ determines the stationary states. Near the classical singularity ($q = 0$) this eigenvalue equation becomes

$$q^2 \frac{\partial^2 \varphi}{\partial q^2} = \left(\frac{\partial^2}{\partial u_1^2} + \frac{\partial^2}{\partial u_2^2}\right)\varphi \tag{55}$$

which is solved by

$$\begin{aligned} \varphi(u_1, u_1, q) &= f_{kp}(q) \exp i(ku_1 + pu_2) \\ f_{kp}(q) &= q^{1/2}\,[Aq^{\sqrt{1-\alpha^2}/2} + Bq^{-\sqrt{1-\alpha^2}/2}]\,;\ \alpha^2 = 4(p^2 + k^2). \end{aligned} \tag{56}$$

Clearly, $f_{kp}(q) \to 0$ as $q \to 0$, irrespective of the value of α as long as $\alpha \neq 0$. In other words, the probability for the singular case ($q = 0$) is strictly zero in all quantum stationary states.

The above discussion shows that the QSGs can provide a viable alternative description to the singular classical cosmology. Also, the use of QSGs is in no way restricted to the conformal part of the metric. In what follows, however, we shall not discuss nonconformal degrees of freedom, but shall confine our attention to QCFs only.

13.6. QSG and the Back Reaction on the Metric

The analysis in the previous section shows that QSGs provide a simple description of the quantum universe. The formalism, as it stands, suffers from the one drawback: we have assumed *a priori* a background metric in (34). (In Chapter 12, the background metric was taken to be a *solution* to Einstein's equation, thereby justifying the choice. But (34) is not a solution of Einstein's equations.) A complete formalism should determine both the background metric and the quantum states of the conformal factor. We have to generalize our present formalism to achieve this goal.

Such a generalization, evidently, will have to include nonconformal degrees of gravity. Also, we do not expect to be able to treat all degrees of freedom exactly. We have already seen that when quantum fluctuations of nonconformal degrees of freedom are taken into account, the light cone structure of the spacetime becomes indeterminate, leading to serious conceptual problems. It is necessary to keep these points in mind while attempting a more ambitious theory of quantum gravity.

In this section we shall present a model for quantum gravity along the following lines: we shall quantize the conformal degree of freedom exactly and treat the background metric in a semiclassical limit. The back reaction of the QCF on the metric can be taken into account in the sense of expectation values.

To understand how such a model works consider the action

$$S = \frac{1}{16\pi G}\int R\sqrt{-g}\ \mathrm{d}^4x + \int \mathcal{L}_M\sqrt{-g}\ \mathrm{d}^4x, \tag{57}$$

where $\mathcal{L}_M$ is the matter Lagrangian. Let us separate out the conformal part Ω and the background $\bar{g}_{ik}$ in this equation by writing

$$g_{ik} = \Omega^2(x)\,\bar{g}_{ik}(x). \tag{58}$$

This will give

$$\begin{aligned} S &= \frac{1}{16\pi G}\int (\bar{R}\Omega^2 - 6\Omega^i\Omega_i)\sqrt{-\bar{g}}\ \mathrm{d}^4x + \int \mathcal{L}_M(\Omega, \bar{g}_{ik})\Omega^4\sqrt{-\bar{g}}\ \mathrm{d}^4x \\ &= \frac{1}{16\pi G}\int \Omega^2\bar{R}\sqrt{-\bar{g}}\ \mathrm{d}^4x + \frac{3}{4\pi G}\int \mathcal{L}_{NE}\sqrt{-\bar{g}}\ \mathrm{d}^4x + \\ &\quad + \int \mathcal{L}_M(\Omega, \bar{g}_{ik})\Omega^4\sqrt{-\bar{g}}\ \mathrm{d}^4x, \end{aligned} \tag{59}$$

where

$$\mathcal{L}_{NE} = -\tfrac{1}{2}\,\bar{g}^{ik}\,\partial_i\Omega\,\partial_k\Omega \tag{60}$$

has the form of the Lagrangian for a massless, *negative energy* scalar field. If the matter Lagrangian $\mathcal{L}$ is not conformally invariant ($\mathcal{L}_M\,\Omega^4$) will, in general, depend on Ω.

At the classical level, (59) has the same information content as (57); we have merely rewritten it. However, our aim is to treat Ω as a quantum variable while keeping $\bar{g}_{ik}$ as a *c*-number. To achieve this end we shall define an 'average action' $\bar{S}$ by replacing the quantum variables in (59) by their expectation values:

$$\begin{aligned} \bar{S} &= \frac{1}{16\pi G}\int \langle\Omega^2\rangle\,\bar{R}\sqrt{-\bar{g}}\ \mathrm{d}^4x + \frac{3}{4\pi G}\int \langle\mathcal{L}_{NE}\rangle\sqrt{-g}\ \mathrm{d}^4x \\ &\quad + \int \mathcal{L}_M(\langle\Omega\rangle, \bar{g}_{ik})\langle\Omega^4\rangle\sqrt{-\bar{g}}\ \mathrm{d}^4x. \end{aligned} \tag{61}$$

The field equations for $\bar{g}_{ik}$ are obtained by varying $\bar{g}_{ik}$ in the average action $\bar{S}$. We shall consider the variations term by term. For the first term, we have

$$\begin{aligned}\delta\bar{S}_1 &\equiv \frac{1}{16\pi G}\int \langle\Omega^2\rangle\delta(\bar{R}\sqrt{-g}\,)\mathrm{d}^4x\\ &= \frac{1}{16\pi G}\int \langle\Omega^2\rangle\,(\bar{R}_{ik} - \tfrac{1}{2}\bar{g}_{ik}\bar{R})\,\delta\bar{g}_{ik}\sqrt{-\bar{g}}\ \mathrm{d}^4x\,+\\ &+\frac{1}{16\pi G}\int \langle\Omega^2\rangle\bar{g}^{ik}\,\delta\bar{R}_{ik}\sqrt{-\bar{g}}\ \mathrm{d}^4x. \end{aligned}\tag{62}$$

While deriving Einstein's equations we neglect the last term because it can be converted into a surface integral. In the present case we cannot neglect this term because of $\langle\Omega^2\rangle$. We get

$$\begin{aligned}\delta I &\equiv \int \langle\Omega^2\rangle\,\bar{g}^{ik}\,\delta\bar{R}_{ik}\sqrt{-\bar{g}}\ \mathrm{d}^4x = \int \langle\Omega^2\rangle\partial_k(\sqrt{-\bar{g}}\ \bar{w}^k)\,\mathrm{d}^4x\\ &= -\int\sqrt{-\bar{g}}\ \bar{w}^k\partial_k\langle\Omega^2\rangle\,\mathrm{d}^4x;\end{aligned}\tag{63}$$

where

$$\bar{w}^k = \bar{g}^{ik}\,\delta\bar{\Gamma}^l{}_{il} - \bar{g}^{il}\,\delta\bar{\Gamma}^k{}_{il}.\tag{64}$$

Expressing $\bar{\Gamma}^i{}_{;k}$ in terms of $\bar{g}_{ik}$, we can write (with $\nabla_i A_k \equiv A_{k;i}$)

$$\delta I = -\int\left\{\delta\bar{g}^{ik}\left[\bar{g}_{ik}\Box - \nabla_i\nabla_k\right]\langle\Omega^2\rangle\right\}\sqrt{-g}\ \mathrm{d}^4x.\tag{65}$$

The variation of the $\mathscr{L}_{NE}$ term produces the energy momentum tensor for the negative energy field

$$t_{ik}^{(NE)} \equiv -\langle\partial_i\Omega\ \partial_k\Omega\rangle + \tfrac{1}{2}\bar{g}_{ik}\,\langle\partial_a\Omega\ \partial^a\Omega\rangle.\tag{66}$$

The variation of the matter term will, in general, be a complicated expression because of the presence of the Ω factors. However, when we confine our attention to the conformally invariant matter Lagrangians, this term will be independent of Ω. Therefore, the variation of $\bar{S}_m$ with respect to $\bar{g}_{ik}$ will only produce the energy momentum tensor for the matter field.

Equating the sum of all these variations to zero, we get the field equations

$$\langle\Omega^2\rangle(\bar{R}_{ik} - \tfrac{1}{2}\,\bar{g}_{ik}\bar{R}) + 6t_{ik}^{(NE)} - [\bar{g}_{ik}\Box - \nabla_i\nabla_k]\langle\Omega^2\rangle = -8\pi G T_{ik},\tag{67}$$

with $t_{ik}^{(NE)}$ given by (66). These equations determine the dynamics of the background metric. Notice that when $\langle\Omega^2\rangle = 1$ and $\langle\partial_i\Omega\ \partial_k\Omega\rangle = 0$, these equations go over to the familiar Einstein equations. As it stands, (67) incorporates the effect of QCF on the background metric in terms of the expectation values.

To have the complete story, of course, we should be able to quantize the conformal factor in the given background metric $\bar{g}_{ik}$. This can be done either by the path integral approach, using directly the action

$$S = \frac{1}{16\pi G} \int (\bar{R}\Omega^2 - 6\Omega^i\Omega_i) \sqrt{-\bar{g}}\, \mathrm{d}^4x, \tag{68}$$

or by solving the Schrödinger equation corresponding to (68). Notice that we have to ensure the self-consistency of the whole procedure. The quantum state for Ω, and hence $\langle\Omega^2\rangle$, $\langle\partial_i\Omega\ \partial_k\Omega\rangle$, etc., can be determined only after quantizing Ω in a given background geometry. On the other hand, this background geometry depends on expectation values like $\langle\Omega^2\rangle$. Thus (67), (68) together form a much more complicated set of equations than the classical Einstein equations. We shall now examine the cosmological solutions to these equations.

13.7. Solutions of Quantum Gravity Equations

In the case of homogeneous, isotropic spacetimes, Equations (67) and (68) simplify considerably, allowing a solution in closed form. *We shall first show that the QSGs discussed in* Section 13.5 *do indeed represent a solution of* (67) *and* (68). This demonstration, of course, is essential if we are to take QSGs as fundamental entities of spacetime.

We take our background metric to be ($c \neq 1$, $\hbar \neq 1$, and $G \neq 1$)

$$\overline{\mathrm{d}s}^2 = c^2\, \mathrm{d}t^2 - \frac{\mathrm{d}r^2}{1 - (r^2/a^2)} - r^2(\mathrm{d}\theta^2 + \sin^2\theta\, \mathrm{d}\varphi^2). \tag{69}$$

The full metric in the nth stationary state is

$$\mathrm{d}s_n^2 = \langle\Omega^2\rangle_n\, \overline{\mathrm{d}s}^2. \tag{70}$$

As in Section 13.5, the stationary states are determined by the Schrödinger equation (40). All the results represented by (41), (42), and (44) follow immediately. There is only one new feature: if the conformal factor is in the nth state, then (67) has to be satisfied with the expectation values evaluated in this state. We shall now see what this condition implies.

The expectation values are independent of space variables because of the homogeneity of the universe. Since they are independent of time in a stationary state, we have

$$(g_{ik}\Box - \nabla_i\nabla_k)\, \langle\Omega^2\rangle = 0, \tag{71}$$

So (67) becomes

$$\langle\Omega^2\rangle\, (\bar{R}^i{}_k - \tfrac{1}{2}\delta^i{}_k\bar{R}) + 6t^i{}_k + 8\pi G T^i{}_k = 0. \tag{72}$$

Now

$$t^i{}_k = \frac{1}{2}\langle\dot{\Omega}^2\rangle \, \mathrm{diag}(-1, 1, 1, 1). \tag{73}$$

Because of homogeneity, $T^i{}_k$ must have the form diag$(\varepsilon, -p, -p, -p)$ while conformal invariance forces $(\varepsilon - 3p)$ to be zero. Hence,

$$T^i{}_k = \varepsilon \, \mathrm{diag}(1, -\frac{1}{3}, -\frac{1}{3}, -\frac{1}{3}). \tag{74}$$

Substituting (73), (74) into (72) and using the known form of $\bar{R}^i{}_k$, we get the (0-0) component equation to be[2]

$$-\frac{3}{2}\frac{\langle\Omega^2\rangle}{a^2} - 3\langle\dot{\Omega}\rangle + 8\pi G\varepsilon = 0. \tag{75}$$

Because of homogeneity there are only two independent equations in (72) one of which is taken to be (75). The other may be obtained by taking the trace of (72):

$$\langle\Omega^2\rangle\left(-\frac{6}{a^2}\right) + 6\langle\dot{\Omega}^2\rangle = 0. \tag{76}$$

This equation is identically satisfied for a harmonic oscillator variable Ω in any stationary state. Equation (75), on the other hand, imposes a nontrivial restriction on the matter energy density ε:

$$\begin{aligned}\varepsilon_n &= \frac{9\hbar\omega c^2}{16\pi G M a^2}(n + \tfrac{1}{2}) \\ &= \frac{3}{8\pi^2}\left(\frac{\hbar c}{a^4}\right)(n + \tfrac{1}{2}).\end{aligned} \tag{77}$$

The QSGs exist as solutions to the quantum gravity equations as long as the matter density obeys the above quantization condition. Similar results exist with other choices for matter variables (dust, scalar field, etc.).

We shall now look for the most general homogeneous, isotropic background solution. The background metric (with $k = 0, \pm 1$) is

$$\overline{\mathrm{d}s}^2 = c^2\,\mathrm{d}t^2 - Q^2(t)\left\{\frac{\mathrm{d}r^2}{1 - k(r^2/a^2)} + r^2(\mathrm{d}\theta^2 + \sin^2\theta\,\mathrm{d}\varphi^2)\right\} \tag{78}$$

while the full metric is given by

$$\mathrm{d}s^2 = \langle\Omega^2\rangle\,\overline{\mathrm{d}s}^2. \tag{79}$$

[2] The dot denotes a derivative with respect to (ct).

Let us first consider the quantum dynamics of Ω governed by the action

$$S = -\frac{1}{2} M \int Q^3(t) \, [\dot{q}^2 - \omega^2(t) q^2] \, dt, \tag{80}$$

where

$$q = a\Omega \; ; \qquad M = \frac{2}{3}\pi\beta \left(\frac{ac^2}{G} \right) ; \qquad \nu = \frac{c}{a} \; ; \tag{81}$$

$$\omega^2(t) = \frac{\ddot{Q}}{Q} + \frac{1}{Q^2} (Q^2 + k\nu^2). \tag{82}$$

Here β denotes the ratio between volumes of the region of spacetime under consideration and the region $r \leq a$. Notice that S now represents a harmonic oscillator with variable mass and frequency. The quantity $\omega^2(t)$ is defined through (82) and need not be positive definite.

The Schrödinger equation corresponding to this action has the form

$$-i\hbar \frac{\partial \psi}{\partial t} = - \frac{\hbar^2}{2MQ^3(t)} \frac{\partial^2 \psi}{\partial q^2} + \tfrac{1}{2} M\omega^2(t) Q^3(t) q^2 \psi. \tag{83}$$

We are interested in the stationary state solutions of (83), with time dependence entirely separable as a phase

$$\psi(q, t) \sim [\exp if(t)] \varphi(q). \tag{84}$$

Such solutions exist when $Q(t)$ satisfies the condition

$$\omega^2(t) Q^6(t) = Q^6 \left\{ \frac{\ddot{Q}}{Q} + \frac{\dot{Q}^2 + k\nu^2}{Q^2} \right\} = \alpha^2, \tag{85}$$

where α is some real constant. It is easy to see that under condition (85), Equation (83) becomes

$$-i\hbar Q^3(t) \frac{\partial \psi}{\partial t} = - \frac{\hbar^2}{2M} \left(\frac{\partial^2 \psi}{\partial q^2} \right) + \tfrac{1}{2} M\alpha^2 q^2 \psi \tag{86}$$

and is solved by the choice

$$\psi(q, t) = \exp \left\{ \frac{iE}{\hbar} \int^t \frac{dt}{Q^3(t)} \right\} \cdot \varphi_n(q), \tag{87}$$

where $\varphi_n(q)$ is the nth harmonic oscillator eigenfunction. Thus, QSGs exist even for the time dependent case (in the sense of (84)) as long as (85) is satisfied. The expectation value of q^2 in the nth state is

$$\langle q^2 \rangle = \frac{\hbar}{M\alpha} (n + \tfrac{1}{2}). \tag{88}$$

The expectation value of $\dot{q}^2$ can be calculated by noting that the canonical momentum corresponding to q is, from (80),

$$p = \frac{\partial L}{\partial \dot{q}} = MQ^2(t)\dot{q}. \tag{89}$$

Since p is represented by the operator $(-i\hbar(\partial/\partial q))$ we have

$$\langle \dot{q}^2 \rangle = \frac{1}{M^2Q^6} \langle p^2 \rangle = \frac{1}{M^2Q^6} \int_{-\infty}^{+\infty} \psi^* \left(-\hbar^2 \frac{\partial^2}{\partial q^2} \right) \psi \, \mathrm{d}q$$
$$= \left(\frac{\hbar \alpha}{M} \right) \frac{1}{Q^2} \left(n + \tfrac{1}{2}\right). \tag{90}$$

Now we have to solve Equations (67) to determine $Q(t)$ and also to check for consistency. Since $\langle \Omega^2 \rangle$ is independent of time, we again have (71) and (72):

$$\langle \Omega^2 \rangle \left(R^i{}_k - \tfrac{1}{2}\, \delta^i{}_k R\right) + 6t^i{}_k = -8\pi G T^i{}_k. \tag{91}$$

Equations (73) and (74) also remain valid. Taking the trace of (91) we get

$$\langle \dot{\Omega}^2 \rangle = \omega^2(t) \langle \Omega^2 \rangle. \tag{92}$$

Using (88), and (90), and (85) we find that this equation is identically satisfied. Thus we only have to demand the validity of the (0-0) component equation of (91), viz.

$$3\, \frac{\dot{Q}^2 + kv^2}{Q^2} = \frac{8\pi G a^2}{c^2} \frac{\varepsilon(t)}{\langle q^2 \rangle} - 3\, \frac{\langle \dot{q}^2 \rangle}{\langle q^2 \rangle}. \tag{93}$$

This constraint on $\varepsilon(t)$ can be simplified further. Note that (85) can be integrated once, to give

$$\frac{\dot{Q}^2 + kv^2}{Q^2} = \frac{\varrho^2}{Q^4} - \frac{\alpha^2}{Q^6}, \tag{94}$$

where ϱ^2 is the new integration constant. Using (94) in (93) we get

$$\frac{8\pi G a^2}{\langle q^2 \rangle c^2}\, \varepsilon(t) = 3\, \frac{\dot{Q}^2 + kv^2}{Q^2} + 3\, \frac{\langle \dot{q}^2 \rangle}{\langle q^2 \rangle} = \frac{3\pi^2}{Q^4(t)}.$$

Therefore,

$$\varepsilon(t) = \left(\frac{9}{16\pi^2 \beta} \right) \left(\frac{h\varrho^2}{\alpha a^3} \right) \frac{1}{Q^4(t)} \left(n + \tfrac{1}{2}\right). \tag{95}$$

Formally speaking, this completes our solution. Equations (95), (94), and (87) completely specify our system.

The solutions involve two parameters ϱ and α (*three* parameters if a is also included). Note that ϱ, appearing in (95), sets the scale for energy density. The constant α denotes the deviation from the classical limit. For example, when $\alpha \to 0$, Equation (94) becomes

$$\frac{\dot{Q}^2 + kv^2}{Q^2} \cong \frac{\varrho^2}{Q^4}; \tag{96}$$

which is just the classical equation for the radiation-filled universe.

Let us look at the solutions in the various cases. When the conformal factor is in the nth stationary state, the metric has the form

$$\mathrm{d}s_n^2 = \langle\Omega^2\rangle_n \left\{c^2\,\mathrm{d}t^2 - Q^2(t)\left[\frac{\mathrm{d}r^2}{1 - k(r^2/a^2)} + r^2(\mathrm{d}\theta^2 + \sin^2\theta\,\mathrm{d}\varphi^2)\right]\right\}. \tag{97}$$

It is somewhat more convenient to rewrite this in the form

$$\mathrm{d}s_n^2 = \langle\Omega^2\rangle_n Q^2\left[c^2\,\mathrm{d}\tau^2 - \frac{\mathrm{d}r^2}{1 - k(r^2/a^2)} - r^2(\mathrm{d}\theta^2 + \sin^2\theta\,\mathrm{d}\varphi^2)\right]. \tag{98}$$

Equation (94) for Q becomes

$$\left(\frac{\mathrm{d}Q}{\mathrm{d}\tau}\right)^2 = \varrho^2 - \frac{\alpha^2}{Q^2} - kv^2Q^2. \tag{99}$$

Let us begin with a purely quantum gravitational solution with no matter; $\varrho = 0$, which implies $\varepsilon = 0$. From (99) it follows that $k = -1$ and

$$Q^2 = \left(\frac{\alpha}{v}\right)[1 + 2\sinh^2 v\tau]. \tag{100}$$

Writing (98) as

$$\mathrm{d}s_n^2 = \left(\frac{v}{\alpha}\right) L_n^2 Q^2 \{\mathrm{d}\eta^2 - \mathrm{d}\chi^2 - f^2(\chi)[\mathrm{d}\theta^2 + \sin^2\theta\,\mathrm{d}\varphi^2]\} \tag{101}$$

(where $f(\chi) = \sin\chi, \chi, \sinh\chi$ for $k = +1, 0, -1$, respectively), with

$$L_n^2 = \left(\frac{G\hbar}{c^3}\right)\left(\frac{3}{2\pi\beta}\right)(n + \tfrac{1}{2}), \tag{102}$$

we find

$$\mathrm{d}s_n^2 = L_n^2(1 + 2\sinh^2\eta)[\mathrm{d}\eta^2 - \mathrm{d}\chi^2 - \sinh^2\chi(\mathrm{d}\theta^2 + \sin^2\theta\,\mathrm{d}\varphi^2)]. \tag{103}$$

We see that this spacetime is nonsingular and that the expansion factor has a lower bound at L_n. For large η this metric mimics the form of a $k = -1$ radiation-filled model.

In models with nonzero energy density (i.e. $\varrho \neq 0$) solutions exists for all the three values of k. When $k = 0$, we have

$$\left(\frac{\mathrm{d}Q}{\mathrm{d}\tau}\right)^2 = \varrho^2 - \frac{\alpha^2}{Q^2} \tag{104}$$

with the solution

$$Q^2(\tau) = \frac{\alpha^2}{\varrho^2} + \varrho^2\tau^2 \tag{105}$$

leading to

$$ds_n^2 = \left(\frac{\alpha^2}{\varrho^2} + \varrho^2\tau^2\right)[d\tau^2 - dr^2 - r^2(d\theta^2 + \sin^2\theta\, d\varphi^2)]\ , \tag{106}$$

where we have rescaled (τ, r) absorbing constant factors. The model is again nonsingular and reaches minimum expansion at (α^2/ϱ^2).

For $k = \pm 1$ the solutions are more complicated and are given by

$$Q^2(\eta) = \left[\frac{\varrho^4}{4v^4} + \frac{\alpha^2}{v^2}\right]^{1/2} \cosh 2\eta - \frac{\varrho^2}{2v^2}\,, \quad (k = -1), \tag{107}$$

$$Q^2(\eta) = \frac{\varrho^2}{2v^2} - \left[\frac{\varrho^4}{4v^4} - \frac{\alpha^2}{v^2}\right]^{1/2} \cos 2\eta\ (k = +1). \tag{108}$$

These expressions go over to the corresponding classical expressions as $\alpha \to 0$. Note that the $k = +1$ model can exist only if there is 'sufficient matter', i.e. $\varrho^2 > 2\alpha v$. There is no such energy condition in classical models. These solutions are shown in Figure 13.2.

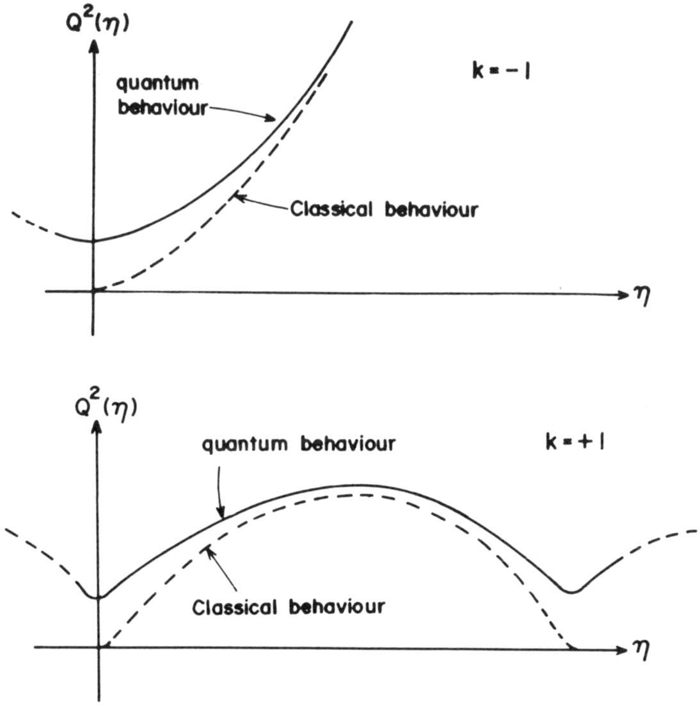

Fig. 13.2. The square of the scale factor of the quantum universe is plotted against the time parameter η.

In all the above models, the energy density of radiation goes as $Q^{-4}(t)$.

Example 13.7. The model universes described in this section are also free of the horizon problem. This is to be expected because any model that is nonsingular is very likely to be horizon-free. Consider, for example, the $k = 0$ model (106) which can be written as

$$ds^2 = (u^2 + v^2\tau^2)\,[d\tau^2 - dr^2 - r^2\,(d\theta^2 + \sin^2\theta\, d\varphi^2)].$$

Substituting

$$\tau = \frac{u}{v}\sinh\theta,$$

we can write

$$ds^2 = dt^2 - Q^2(t)[dr^2 + r^2(d\theta^2 + \sin^2\theta\, d\varphi^2)],$$

where

$$t = \int (u^2 + v^2\tau^2)^{1/2}\, d\tau = \frac{u^2}{4v}\,(2\theta + \sinh\theta),$$

$$Q = \frac{u}{v}\cosh\theta.$$

Because the model is nonsingular we are allowed the complete range $[-\infty < t < \infty;\ -\infty < \tau < +\infty,\ -\infty < \theta < \infty]$. Note that $t = 0$ at $\tau = 0$. Consider the integral

$$r_H(t_2, t_1) \equiv \int_{t_1}^{t_2} \frac{dt}{Q(t)}.$$

The model is horizon-free if

$$\lim_{t_1 \to \infty} r_H(t_2, t_1) = \infty \quad \text{(for any } t_2\text{)}.$$

We see that

$$r_H(t_2, t_1) = \int_{t_1}^{t_2} \frac{dt}{Q(t)} = \int_{\tau_1}^{\tau_2} \frac{d\tau}{Q(t)}\left(\frac{dt}{d\tau}\right) = \int_{\tau_1}^{\tau_2} d\tau = \tau_2 - \tau_1,$$

(where we have used $(dt/d\tau) = Q(t)$). Thus we need the limit,

$$L = \lim_{t \to -\infty} \tau(t)$$

which diverges because $\tau = (u/v)\sinh\theta$, $t = [u^2(2\theta + \sinh\theta)]/4v$. Thus the model is horizon-free. Similar results exist for other cases.

13.8. Cosmogenesis and Vacuum Instability

In the previous pages we have seen that quantization of the conformal degree of freedom allows us to discuss quantum cosmology in a meaningful way. Encouraged by this success we shall now try to attack the question: "How did the universe come into being?" Needless to say, the discussion in the present section will be somewhat speculative and naïve.

To begin with, let us note that the question "How did the universe come into being?" cannot even be posed – let alone answered – in classical relativity. This is because of the existence of singularity. Further, it is not possible in a classical theory to have the following features simultaneously:

(i) creation of matter;
(ii) conservation of $T^i{}_k$, i.e. validity of Einstein's equations;
(iii) positivity of energy density for all fields.

The Big Bang model, for example, violates (ii) at the singularity thereby 'creating' matter. Continuous creation models, for example, violate (iii) at all epochs by an unobservably small amount, and are nonsingular.

If quantum cosmology removes the classical singularity – and we saw that it does – we are in a better position to tackle the question of 'creation'. At the simplest level, any consistent, nonsingular model with matter has already answered the question of 'creation' in a somewhat disappointing way: there is no creation. The universe and the matter exist from the everlasting to the everlasting and evolve as dictated by the model (for example, say, by Equations (105) and (95)).

This answer is disappointing for two reasons:

(i) there exist a large number of such solutions and we have no criterion to choose one among them;
(ii) no explanation is forthcoming for the various (observed) fine tunings of the present universe that were mentioned in Chapters 8 and 10.

A somewhat different approach towards the creation of the universe – 'cosmogenesis' [see Note 4] – can be taken if we paraphrase the question "How did the universe come into being?" in a mathematically tractable form. Consider the equations of classical relativity

$$R^i{}_k - \tfrac{1}{2}\delta^i{}_k R = -8\pi G T^i{}_k. \tag{109}$$

These equations possess the following nonsingular solution:

$$\mathrm{d}s^2 = \mathrm{d}t^2 - \mathrm{d}r^2 - r^2(\mathrm{d}\theta^2 + \sin^2\theta\,\mathrm{d}\varphi^2), \quad T^i_k = 0, \tag{110}$$

and the following singular solution

$$\mathrm{d}s^2 = \mathrm{d}t^2 - Q^2(t)\left[\frac{\mathrm{d}r^2}{1 - kr^2} + r^2(\mathrm{d}\theta^2 + \sin^2\theta\,\mathrm{d}\varphi^2)\right] \tag{111}$$

$$T^i{}_k = \mathrm{diag}(\varrho, -p, -p, -p) \neq 0.$$

The question "How did the universe come into being?" is then equivalent to "Why does nature prefer (111) to (110)?" Normally, a choice of solution is made by the choice of initial conditions. Clearly no initial condition can be assumed at a singular event $t = 0$. However, there is an entirely different way by which one solution can be preferred over the other. If it turns out that the flat spacetime in (110) is unstable to quantum fluctuations, then we may consider the universe to have originated through this vacuum fluctuation.

Such a hope is reinforced by a closer look at our back reaction equation (67):

$$\langle\Omega^2\rangle(R_{ik} - \tfrac{1}{2}g_{ik}R) + 6t_{ik}^{NE} - (\bar{g}_{ik}\Box - \nabla_i\nabla_k)\langle\Omega^2\rangle = -8\pi G T_{ik}. \tag{112}$$

Because of vacuum fluctuations in Ω, flat, vacuum spacetime ($T_{ik} = 0$; $g_{ik} = \eta_{ik}$) is *not* a solution of this equation. We found, in fact, that the universe is nonflat even when there is no matter (see Equation (100)). Thus, vacuum fluctuations do seem to rule out the flat spacetime.

To produce a concrete, physical model demonstrating this instability is, however, a difficult task. The details of such models depend crucially on the nature of matter fields present in T_{ik}. We shall illustrate the concept of vacuum instability with two simple examples.

13.8.1. *Massive Scalar Field*

Consider a system described by the action

$$S = \frac{1}{16\pi G}\int R\sqrt{-g}\,d^4x + \frac{1}{2}\int (B^iB_i - \tfrac{1}{6}RB^2 - m^2B^2)\sqrt{-g}\,d^4x. \tag{113}$$

The first term is the gravitational action and the second term represents the action for a scalar field $B(\mathbf{r}, t)$ of mass m with an additional coupling $\frac{1}{6}RB^2$. In the limit of $m \to 0$ this action is conformally invariant. It is easier to proceed by writing the Robertson–Walker spacetimes in a conformally flat form

$$ds^2 = \Omega^2(\mathbf{r}, t)\,\eta_{ik}\,dx^i\,dx^k. \tag{114}$$

Thus we are left with two field degrees of freedom: $\Omega(\mathbf{r}, t)$ and $B(\mathbf{r}, t)$. Now we make a conformal transformation to

$$\varphi(\mathbf{r}, t) \equiv \Omega(\mathbf{r}, t)\,B(\mathbf{r}, t) \tag{115}$$

and write

$$\psi(\mathbf{r}, t) = \left(\frac{3}{4\pi G}\right)^{1/2}\Omega(\mathbf{r}, t), \tag{116}$$

so as to reduce (113) into the form

$$S = \tfrac{1}{2}\int (\varphi^i\varphi_i - \psi^i\psi_i - \alpha_G^2\varphi^2\psi^2)\,d^4x; \tag{117}$$

where $\alpha_G^2 = (4\pi Gm^2/3)$ is the analogue of 'fine structure constant' in gravity. The Euler–Lagrange equations $\delta S = 0$ for (117) are

$$\Box\varphi + \alpha_G^2 \psi^2 \varphi = 0, \tag{118}$$

$$\Box\psi - \alpha_G^2 \varphi^2 \psi = 0. \tag{119}$$

We shall assume that φ is a quantum field and that (118) is the operator equation for φ in the Heisenberg picture. On the other hand, ψ, which determines the metric via (116) and (114), will be treated as a c-number. We shall interpret (119) as

$$\Box\psi - \alpha_G^2 \langle 0| \varphi^2 |0\rangle \psi = 0. \tag{120}$$

The flat (vacuum) spacetime corresponds to the choice,

$$\psi = \text{const} = \left(\frac{m}{\alpha_G}\right), \tag{121}$$

$$\langle 0| \varphi^2 |0\rangle = 0. \tag{122}$$

Clearly, the equations are satisfied by this choice as long as we can ensure (122). We know that, formally, the left-hand side of (122) diverges because of zero point energy. However, we shall assume that this divergence is subtracted out in the quantity $\langle 0 | \varphi^2 |0\rangle$, ensuring (122).

Let us further suppose that the system was described by (121), (122) till $t = 0$. At $t = 0$, we perturb the spacetime slightly by putting

$$\psi(t) = \frac{m}{\alpha_G} + \varepsilon(t), \quad \text{for } t > 0. \tag{123}$$

When the field is quantized in this perturbed spacetime (123) via Equation (118), we shall get a nonzero value for $\langle 0 | \varphi^2 |0\rangle$ which will, in general, depend on $\varepsilon(t)$. Substituting back into Equation (120) we shall get an equation for the perturbation $\varepsilon(t)$. If this equation has exponentially growing solutions, then we can conclude that flat vacuum spacetime is unstable to perturbations of the kind in (123). We shall outline below how this is done.

Substituting (123) into (118) we get, to first order in $\varepsilon(t)$,

$$\Box\varphi + \alpha_G^2 \left[\frac{m^2}{\alpha_G^2} + \frac{2m}{\alpha_G}\varepsilon(t)\right]\varphi = 0. \tag{124}$$

Writing

$$\varphi(t, \mathbf{r}) = \int g_{\mathbf{k}}(t) \exp(i\mathbf{k}\cdot\mathbf{r}) \frac{d^3\mathbf{k}}{(2\pi)^3}, \tag{125}$$

we find that the functions $q_{\mathbf{k}}(t)$ satisfy the equations

$$\ddot{q}_{\mathbf{k}} + \omega_k^2 q_{\mathbf{k}} = -2m\alpha_G\varepsilon(t)q_{\mathbf{k}}(t):\ \omega_k^2 = |\mathbf{k}|^2 + m^2. \tag{126}$$

The zeroth-order solutions of these equations (when $\varepsilon = 0$) are

$$q_{\mathbf{k}}^{(0)}(t) = a_{\mathbf{k}}\, e^{i\omega_k t} + a_{\mathbf{k}}^{\dagger}\, e^{-i\omega_k t}, \tag{127}$$

where $a_{\mathbf{k}}^{\dagger}$ and $a_{\mathbf{k}}$ are the creation and annihilation operators. To obtain a solution $q_{\mathbf{k}}^{(1)}$ correct to first order in ε, we can use (127) on the right-hand side of (126) getting

$$\ddot{q}_{\mathbf{k}}^{(1)} + \omega_k^2 q_{\mathbf{k}}^{(1)} = -2m\alpha_G\varepsilon(t)\,[a_{\mathbf{k}}\,e^{i\omega_k t} + a_{\mathbf{k}}^{\dagger}\,e^{-i\omega_k t}]\;. \tag{128}$$

This is the familiar equation for a forced harmonic oscillator and can be solved in closed form, using the Green's functions discussed in Chapter 3:

$$\ddot{q}_{\mathbf{k}}^{(1)}(t) = q_{\mathbf{k}}^{(0)} + \int_{-\infty}^{+\infty} dt'\,G_k(t-t')\left\{-2m\alpha_G\varepsilon(t')\,[a_{\mathbf{k}}\,e^{i\omega_k t'} + a_{\mathbf{k}}^{\dagger}\,e^{-i\omega_k t'}]\right\}. \tag{129}$$

Substituting this into (125) we can calculate $\langle 0|\,\varphi^2\,|0\rangle$. Subtracting, the zero point term, we arrive at the final result:

$$\langle 0|\,\varphi^2\,|0\rangle = \frac{m\alpha_G}{(2\pi)^3}\int\frac{d^3\mathbf{k}}{2\psi_k}\int_0^t \dot{\varepsilon}(t')\cos[2\omega_{\mathbf{k}}(t-t')]\,dt'. \tag{130}$$

(We have integrated by parts, and used $\varepsilon(t) = 0$ for $t \leq 0$, to get an $\dot{\varepsilon}$ term inside the integral. It should be clear that only $\dot{\varepsilon}$ can contribute to the finite part of $\langle\varphi^2\rangle$ because a constant ε in (123) only represents a rescaling.) Substituting (130) into (120) and integrating once again by parts, we get

$$\ddot{\varepsilon}(t) = \left(\frac{m^2\alpha_G^2}{8\pi^2}\right)\dot{\varepsilon}(0)A(t) + \frac{m^2\alpha_G^2}{8\pi^2}\int_0^t \ddot{\varepsilon}(t')A(t-t')\,dt', \tag{131}$$

with

$$A(t) = \int_0^\infty \frac{k^2}{\omega_k^4}\sin(2\omega_k t)\,dk. \tag{132}$$

Equation (131) can be solved by the method of Laplace transforms. Taking the Laplace transform of both sides we get

$$\ddot{\varepsilon}(s) = \frac{m^2\alpha_G^2}{8\pi^2}\,\dot{\varepsilon}(0)L(s) + \frac{m^2\alpha_G^2}{8\pi^2}\,L(s)\,\ddot{\varepsilon}(s), \tag{133}$$

that is,

$$\ddot{\varepsilon}(s) = \frac{m^2\alpha_G^2}{8\pi^2}\,\dot{\varepsilon}(0)\left\{\frac{L(s)}{1-(m^2\alpha_G^2/8\pi^2)L(s)}\right\}, \tag{134}$$

where

$$L(s) = \int_0^\infty \frac{k^2\,dk}{\omega_k^4}\int_0^\infty e^{-st}\sin(2\omega t)\,dt = \int_0^\infty \frac{2k^2\,dk}{\omega_k^3(s^2+4\omega_k^2)}\;. \tag{135}$$

Though (135) can be evaluated in closed form, we do not need this result. It is clear from (135) that $L(s)$ is a monotonically decreasing function of $|s|$ and has a maximum at $s = 0$. The maximum value of L is

$$L(0) = \int_0^\infty \frac{k^2\, \mathrm{d}k}{2(k^2 + m^2)^{5/2}} = \frac{1}{6m^2}\,. \tag{136}$$

From the theory of Laplace transforms we know that $\dot{\varepsilon}(t)$ is determined by the poles of $\ddot{\varepsilon}(s)$ and *will* grow without bound if $\ddot{\varepsilon}(s)$ has poles for real values of s. Real poles can exist and produce instability if (using (136), (134))

$$\left(\frac{m^2\alpha_G^2}{8\pi^2}\right) L(0) = \left(\frac{\alpha_G^2}{48\pi^2}\right) > 1. \tag{137}$$

In other words, a flat vacuum is unstable to perturbations of the kind in (123), as long as $m^2 \geq 36\pi(\hbar c/G)$. This example clearly knows how critically the result depends on the parameters of the matter field.

13.8.2. *Quantum Fluctuations with Cosmological Constant*

As a second example of instability and cosmogenesis, consider a classical system described by the action

$$S = \frac{1}{12L_P^2} \int (R - 2\Lambda)\sqrt{-g}\, \mathrm{d}^4x;\; L_P^2 = \frac{4\pi}{3} G. \tag{138}$$

This action describes a system with cosmological constant Λ. At this stage we shall not worry about the origin and magnitude of Λ except to assume that it is positive. It can, for example, arise from vacuum expectation values of quantum fields, as in Chapter 10.

The classical solution, corresponding to the principle of stationary action, may be taken to be the de Sitter universe, represented by

$$\mathrm{d}s^2 = \mathrm{d}t^2 - \mathrm{e}^{2Ht}(\mathrm{d}r^2 + r^2(\mathrm{d}\theta^2 + \sin^2\theta\, \mathrm{d}\varphi^2));\; H^2 = \frac{1}{3}\Lambda \tag{139}$$

$$= \left(1 - \frac{\Lambda}{3}\, \bar{r}^2\right)\mathrm{d}\bar{t}^2 - \frac{\mathrm{d}\bar{r}^2}{(1 - (\Lambda \bar{r}^2/3))} - \bar{r}^2(\mathrm{d}\theta^2 + \sin^2\theta\, \mathrm{d}\varphi^2). \tag{140}$$

For nonzero Λ, this de Sitter universe plays the same role as the Minkowski spacetime for zero Λ. (When $\Lambda \neq 0$, Minkowski spacetime is not a solution of Einstein's equation.) Thus, we may take the de Sitter spacetime to be the 'ground state'. Equation (140) shows the static nature of this spacetime while cosmological observers use the comoving line element in (139). For this solution, the scalar curvature is given by

$$R = 4\Lambda. \tag{141}$$

Let us consider the quantum conformal fluctuations about this ground state.

We know that any local minimum will be a classical ground state, stable to small perturbations. Quantum fluctuations, however, can induce a tunnelling through potential barrier and render the local minimum unstable. This is exactly what happens to the ground state in (139) and (140). The QCFs make the de Sitter spacetime unstable and give it a finite lifetime τ. The universe tunnels out of the de Sitter phase after an inflation by factor ($\exp H\tau$) which we shall now compute.

The QCFs are governed by the action

$$S = -\frac{1}{2L_P^2} \int (\Omega^i\Omega_i - \frac{1}{6} R\Omega^2 + \frac{\Lambda}{3}\Omega^4)\sqrt{-g}\ \mathrm{d}^4x. \tag{142}$$

Defining

$$\varphi = L_P^{-1}\,\Omega, \qquad \lambda = \frac{1}{3}\Lambda L_P^2, \tag{143}$$

we get

$$S = -\frac{1}{2}\int \sqrt{-g}\ \{\varphi^i\varphi_i - V(\varphi)\}\ \mathrm{d}^4x, \tag{144}$$

with (using (141))

$$V(\varphi) = \frac{2\lambda}{L_P^2}\ \varphi^2(1 - \tfrac{1}{2}L_P^2\varphi^2). \tag{145}$$

This potential has the same form as that of the 'double-hump' potential discussed in Chapter 3, and is shown in Figure 13.3.

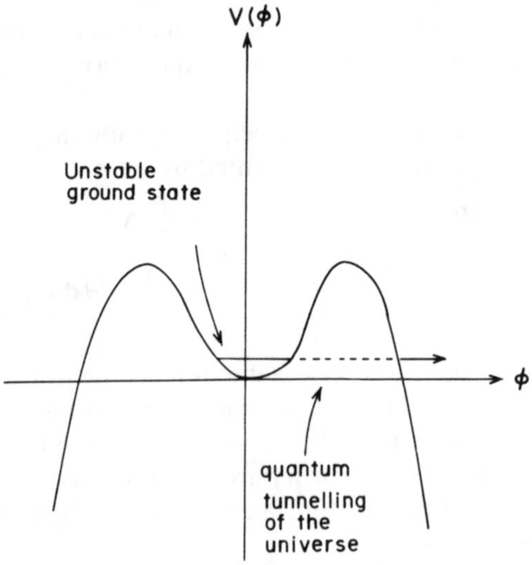

Fig. 13.3. The potential for the conformal factor in a universe with nonzero cosmological constant.

The classical ground state corresponds to the local minimum at $\varphi = 0$. Near $\varphi = 0$, the potential may be treated as a harmonic oscillator potential with $\omega^2 \sim V''(0)$ and $\langle \varphi^2 \rangle \sim \omega^2$. This ground state is separated from the regions of $|\varphi| \approx \infty$ by a potential barrier. Clearly, quantum tunnelling through this barrier renders the ground state unstable.

The tunnelling probability can be calculated by the instanton techniques discussed in Chapter 3 or by computing the WKB amplitude

$$T \propto \exp\left\{ -\int_{x_1}^{x_2} [V(x) - E]^{1/2}\, dx \right\} , \tag{146}$$

where, for $E \approx 0$ (bottom of the well), $x_1 = 0$ and $x_2 \cong 2/L_P^2$. We shall also assume that the tunnelling takes place homogeneously over the whole space of volume $(4\pi/3)\, H^{-3})$. Then the tunnelling probability per unit time is given by

$$P \cong \left(\frac{4\pi}{3} \right) \frac{1}{L_P \lambda^{3/2}} \exp\left(-\frac{32\sqrt{2}\,\pi}{3\lambda^{3/2}} \right) . \tag{147}$$

The reciprocal of P gives the lifetime τ of the metastable ground state:

$$\tau = \left(\frac{3}{4\pi} \right) L_P \lambda^{3/2} \exp\left(\frac{32\sqrt{2}\pi}{3\lambda^{3/2}} \right) . \tag{148}$$

Therefore, the inflation factor for the universe is

$$Z \equiv \exp(H\tau) \cong \exp\left[\frac{3\lambda^2}{4\pi} \exp\left(\frac{32\sqrt{2}\pi}{3} \frac{1}{\lambda^{3/2}} \right) \right] . \tag{149}$$

Being an exponential of an exponential, Z is huge for a wide range of λ. Let us assume that we require an inflation from a Planck size bubble ($\sim 10^{-33}$ cm) to the observed universe ($\sim 10^{28}$ cm). This requires a Z of $\gtrsim 10^{61} \cong \exp(140)$. This inequality is satisfied by *all* values of λ except in a small range of $8 \lesssim \lambda \lesssim 17$. (The minimum value of Z is $\cong \exp(106)$ and it occurs around $\lambda \cong 11$.) Two 'natural' choices for λ are ~ 1 (because of dimensional reasons!) and $\sim 10^{-8}$ (if λ arises from GUTs potentials.) These choices give Z values of $\exp(10^{16})$ and $\exp(10^{10})$, respectively.

The above models show that vacuum instability can lead to a tenable model of cosmogenesis. In particular, these models may provide natural inflationary scenarios based purely on gravity.

Example 13.8. We can give a plausibility argument favouring very small values of λ if we assume that the cosmological term is dynamical and is not a free parameter in the theory. Then the probability for having a configuration with a cosmological constant Λ may be taken to be proportional to the exponential of the Euclidean action

$$
\begin{aligned}
P &= \exp\left\{iS[\Lambda]\right\}\Big|_{t=-it_E} \\
&= \exp\left[\int \frac{\sqrt{g_E}\,(R-2\Lambda)}{12L_P^2}\,dt_E\,d^3x\right] \\
&= \exp\left[\frac{1}{6}\frac{\Lambda}{L_P^2}\int\sqrt{-g}\;dt_E\,d^3\mathbf{x}\right] \equiv \exp\left[\frac{1}{6}\frac{\Lambda}{L_P^2}V_E\right].
\end{aligned}
$$

Now, when $\Lambda > 0$ the Euclidean 4-volume V_E is finite and $\sim \Lambda^{-2}$. If $\Lambda < 0$, then $V_E = \infty$. Therefore.

$$
\begin{aligned}
P(\Lambda) &= \exp\left(\frac{1}{6L_P^2\Lambda}\right) \qquad (\text{for } \Lambda > 0), \\
&= 0 \qquad\qquad\qquad (\text{for } \Lambda < 0),
\end{aligned}
$$

If we accept this argument the probability for Λ values with $6L_P^2\,\Lambda > 1$, or equivalently, the probability for $\lambda > (1/18)$, is negligibly small. We note that in the range $(0 < \lambda < 1/18)$, $Z \geq \exp(10^{720}.)$ □

It is clear from the discussions in this chapter that quantization of the conformal degree of freedom does lead to a consistent, and, reasonably complete picture of a maximally symmetric spacetime. We believe that the conclusions of this chapter – in spite of the many simplifying assumptions – do contain the basic truths which will survive more elaborate model building.

Note and References

1. More detailed discussion on the physical significance of Planck length can be found in:

Padmanabhan, T.: 1984, 'Planck Length is the Lower Bound to All Physical Length Scales', Fifth award essay in Gravity Research Foundation Essay Contest.

Padmanabhan, T.: 1985, 'The Physical Significance of Planck Length', Ann. Phys. **165**, 38.

2. Some original references which are relevant to the concept of QSGs are:

Padmanabhan, T., and Narlikar, J. V.: 1981, 'Stationary States in a Quantum Gravity Model', *Phys. Lett.* **A84** 361.

Padmanabhan, T.: 1982, 'Quantum Stationary States in the Bianchi Universes', *Gen. Rel. Grav.* **14**, 549.

Padmanabhan, T.: 1984, 'Quantum Stationary Geometries and the Avoidance of Singularities', *Class. Quan. Grav.* **1**, 149.

3. The discussion in Sections 13.6 and 13.7 is based on the papers:

Padmanabhan, T.: 1983, 'An Approach to Quantum Gravity', *Phys. Rev.* **D28**, 745.

Padmanabhan, T.: 1983, 'Universe before Planck Time – A Quantum Gravity Model', *Phys. Rev.* **D28**, 756.

4. Cosmogenesis from quantum conformal fluctuations is discussed in:

Padmanabhan, T.: 1983, 'Instability of Flat Space and the Origin of Quantum Conformal Fluctuations', *Phys. Lett.* **93A**, 116.

Padmanabhan, T.: 1984, 'Inflation from Quantum Gravity', *Phys. Lett.* **A104**, 196.

Chapter 14

Epilogue

We have completed our programme of taking the reader from classical gravitational theory to quantum cosmology. The path has not been smooth and the fastidious critic may even question whether we have reached our desired destination. To such a critic (assuming that he is a baseball fan!) we say that if we have not completed the run, we have at least got to first base.

We began in Part I with a discussion of path integration as a technique and its applications to quantum mechanics and quantum field theory. Our purpose in doing so was twofold. First, these are areas which are well understood by theoretical physicists. Their experience with electromagnetic interaction has generated a measure of confidence in field theory as a satisfactory way of describing basic physics. Likewise, gauge field theories hold out the hope of ultimate unification of all basic interaction – a hope inspired by the success of the electro-weak interaction.

Our second purpose, as the reader will, by now, have appreciated, was to bring home to him the inadequacy of the present-day techniques in dealing with the problems of quantum gravity. True, as we have highlighted in Part III, the gauge theories and the grand unification programme have given us an interesting picture of the early universe while the quantization of fields in curve spacetime has revealed startling features about particle creation induced by gravity. However, these inputs fall short of telling us much about how to quantize gravity.

The reason for the inadequacy of the usual field theoretic techniques to quantize gravity lies in the unusual nature of the general theory of relativity. Part II described the classical theory and its applications to local source distributions as well as to cosmology. Relativity has a dual nature: it describes the geometry of spacetime and also the effects of gravity. In ordinary field theory the background spacetime is taken to be flat and no difficulties arise while studying the field dynamics since it does not affect the background geometry. This simplification is not possible for gravity. Even at the classical level the phenomenon of gravitational radiation is difficult to interpret except in the weak field approximation. These difficulties magnify when one tries to quantize gravity as given by general relativity. Against this background let us summarize what has been achieved in Part IV and then comment on what has still to be achieved.

First we start with a classical solution of Einstein's equations as given by the line element

$$\overline{ds}^2 = \bar{g}_{ik}\, dx^i\, dx^k. \tag{1}$$

We next consider nonclassical line elements which are conformal to (1) and are expressed in the form

$$ds^2 = (1 + \varphi)^2\, \bar{g}_{ik}\, dx^i\, dx^k \tag{2}$$

and then treat φ as a *quantum fluctuation* about the classical solution.

The quantum dynamics of φ is fixed by the Feynman path integral formula, symbolically written as

$$K[f, i] = \int \exp iS[\varphi]\, \mathcal{D}\varphi. \tag{3}$$

The advantage of (3) is that the path integral can be evaluated without ambiguity and we can use it to study the quantum evolution of the state of the universe from an initial wave functional $\Psi_i[\varphi]$ to a final wave functional $\Psi_f[\varphi]$:

$$\Psi_f[\varphi_f] = \int K[f, i]\, \Psi_i[\varphi_i]\, \mathcal{D}\varphi_i. \tag{4}$$

Again the above integral relation is not just a formal statement but has also a practical value. In particular, we can use it to investigate how probable was the origin of the universe in a spacetime singularity. We find that a singular origin was highly improbable. In quantum mechanical language, the transition amplitude from a singular state to the present state is vanishingly small. It is almost certain that the universe has had no singularity in the past (nor will it have one in the future). The singularity of spacetime which was considered inevitable in classical cosmology is thus seen more as an exception than a rule in the wider framework of quantum cosmology.

The attendant problem of vanishingly small particle horizon near the 'Big Bang' is also removed when the singularity itself is thus 'eliminated'. In the nonsingular conformal manifold the extension of past light cones to arbitrarily large proper distances becomes possible once the singular hypersurface is nonexistent.

The flatness problem of classical standard models is resolved if we regard the universe as arising from quantum fluctuations of the unstable empty Minkowski spacetime. It is seen that quantum fluctuations inevitably 'tune' the universe to the $k = 0$ (spatially flat) Robertson–Walker model. Inflationary scenarios do not have to be invoked for such fine tuning.

In this formalism it is also possible to set up semiclassical field equations for gravitation. These are the modified Einstein equations

$$\langle\Omega^2\rangle (R_{ik} - \tfrac{1}{2} g_{ik} R) + 6t_{ik}^{NF} - (\bar{g}_{ik}\Box - \nabla_i\nabla_k) \langle\Omega^2\rangle = -8\pi G T_{ik}, \quad (5)$$

in which the $\langle\ \rangle$ terms indicate the 'average' effect of quantum fluctuations. The metric $\{g_{ik}\}$ is *not* the classical metric but is the metric obtained by the dynamical back reaction of the quantum conformal fluctuations. The averaging of Ω^2 and its derivatives is supposed to be done in this spacetime treated as the background spacetime. The problem thus has a bootstrap nature but we can explicitly solve it in some simple cases. We can also obtain singularity free *stationary* states of the universe and demonstrate once again that the modified field equations (5) generate models free from the difficulties of horizon and flatness. Not surprisingly, the characteristic size of the ground state universe is approximately the Planck length L_P.

It appears, therefore, that this approach to quantum cosmology does yield valuable dividends in providing insights into how gravity and spacetime behave in extreme circumstances. Just as the classical conundrum of a stable hydrogen atom was resolved finally by quantum theory, so it seems that appeal to quantum cosmology resolves some of the outstanding problems of classical cosmology.

Having ended on this optimistic note it is essential to warn the reader that the problem of quantizing gravity is far from solved. In Chapters 13 and 14 we have reported the consequences of quantizing the conformal degrees of freedom only. These are essentially the (uncountably infinite) degrees of freedom contained in the scale of the metric tensor. Thus, given all possible, namely six (if we take into account the coordinate invariance) types of variations of g_{ik}, we are only concentrating on how the overall scale implicit in the volume element

$$d\tau = \sqrt{-g}\; d^4x \quad (6)$$

changes quantum mechanically. To what extent will the result of quantizing only this degree of freedom reflect the real state of affairs?

Since the Big Bang singularity arises from the vanishing of (6) in a global fashion, the conformal degree of freedom is clearly the most relevant one to be quantized in order to decide the issue of singularity in quantum cosmology. In the same way that simple examples of the harmonic oscillator and motion through a potential barrier in one dimension proved useful in understanding the true nature of quantum mechanics in earlier times, we feel that the results of Chapters 12 and 13 will prove helpful in highlighting the nature of quantum gravity.

All the same we have, at some stage, to face the more difficult problem of quantizing the nonconformal degrees of freedom. Since in such variations the global lightcone structure is invariably changed, all quantum propagators which embody causal relationships are themselves changed. It is this complicated issue that we have avoided by restricting ourselves to the conformal degrees of freedom only. None of the general formalisms sketchily described in Chapter 11 tells us how to deal with this problem.

Finally, we mention another issue that continues to trouble people concerned with the foundations of quantum mechanics, and which has its own peculiar difficulty for quantum cosmology. We did not discuss this problem earlier because we do not have any solution to offer, nor can we refer the reader to any other source for its enlightenment.

The issue relates to the role of the observer in quantum theory and was summed up by John Wheeler in the sentence: "No phenomenon is a phenomenon until it is an observed phenomenon." A quantum system prior to observation may be described by a wave function ψ which could be expressed as a superposition of stationary state wave functions $\{\psi_n\}$:

$$\psi = \sum_n a_n \psi_n. \tag{7}$$

When 'the observer looks at the system' he finds it in a particular stationary state ψ_n. Quantum mechanics tells us that the probability for this to happen is

$$P_n = |a_n|^2. \tag{8}$$

What is the role of the observer in making the wave function (7) collapse into a single component in this way? The probabilistic aspect of this collapse suggests a multifurcation of possibilities and it led H. Evrett in 1957 to propose the many-world picture. Thus, in one universe the observer finds the system in state ψ_1, in another in state ψ_2, and so on.

Much has been written on this and related ideas, the likely paradoxes arising from them and the different arguments for resolving them. We refer the reader to works of J. A. Wheeler, A. Shimony, H. Stapp, etc., for these discussions, to which we do not wish to add here except to highlight their implications for quantum cosmology.

What is the status of the observer in quantum cosmology? If the entire quantum universe is confined to linear dimensions of order L_P, what and where is the observer? What is the operational meaning of the stationary states of the universe and of transitions between them if we cannot bring the observer into the picture? To us, these problems seem unanswerable within the present framework of physics. By simply extrapolating the formalisms developed for quantum mechanics and quantum field theory to cosmology, the workers in this field (and in this respect we are no exception!) simply push these awkward questions under the rug.

We hope that with a better understanding of quantum theory and cosmology in the future someone way dare to lift the rug and sweep away these difficulties.

Part V
Appendices

Appendix A

Renormalization

In this appendix we introduce the reader to some simple features of the renormalization of $\lambda\varphi^4$ theory discussed in Chapter 4.

The concept of renormalization is best understood in the momentum space, using (what are called) Feynman diagrams and rules. Given a Lagrangian for the scalar field to be

$$L = \tfrac{1}{2}(\varphi_i\varphi^i - m^2\varphi^2) - \frac{1}{4!}\lambda\varphi^4, \tag{A.1}$$

there exists a systematic procedure for constructing the Feynman diagrams and rules for the theory. Since this procedure is discussed in detail in standard texts on field theory (see literature cited at the end of Chapter 4) we shall not attempt to derive them here and shall take these diagrams and rules for granted.

In the $\lambda\varphi^4$ theory, all diagrams are constructed from the basic ingredients given in Table A.1. The diagrams in the table are in the momentum space and the labels p, q, etc., are 4-momentum labels. Once all possible and distinct diagrams for a given process are drawn, the last column in the table allows us to translate the diagram into a meaningful algebraic expression.[1]

Let us see how these diagrams work in practice. Consider the simplest one

$$\xrightarrow[p]{\qquad} = \frac{i}{p^2 - m^2 + i\varepsilon}. \tag{A.2}$$

We notice (e.g. in Example 4.3) that the right-hand side is the Fourier transform of $-iG_F(x, x')$ and thus, using Equation (4.119), we can write

$$\xrightarrow[p]{\qquad} = \int e^{ip\cdot x}\langle 0|\, T\{\varphi(x)\varphi(0)\}\, |0\rangle\, d^4x. \tag{A.3}$$

The intergrand $\langle 0|\, T(\varphi(x)(\varphi(0)))|0\rangle$ was interpreted in Chapter 4 as the probability amplitude for a particle to propagate from origin to x. Thus one

[1] Actually, the dictionary in Table A.1 is incomplete because it neglects a numerical factor called the 'symmetry factor'. However, this extra complication does not affect the main points we intend to illustrate in this appendix.

TABLE A.1

Name	Feynman diagram	Equivalent algebraic expression/comment
Internal line	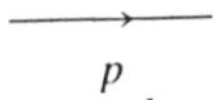	$i/(p^2 - m^2 + i\varepsilon)$; propagator in the Fourier space.
Internal loop	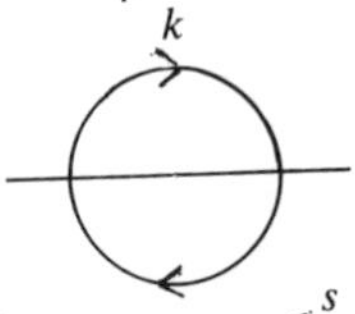	$\int d^4k/(2\pi)^4$; this prescription implies that one must integrate over the momentum variable associated with the closed loop.
Vertex	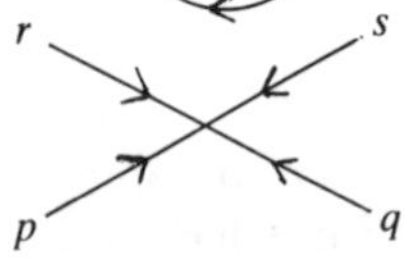	$-i\lambda$; the momentum labels p, q, r, s at a vertex are so arranged that the momentum is conserved; i.e. $p + q + r + s = 0$.

may consider the Fourier transform in (A.2) to be the probability amplitude in the momentum space, for the free particle.

This amplitude will be modified in the presence of the $\lambda\varphi^4$ interaction. Let us now draw the next order Feynman diagram to study this behaviour. From the ingredients of Table A.1 we can draw the following diagram:

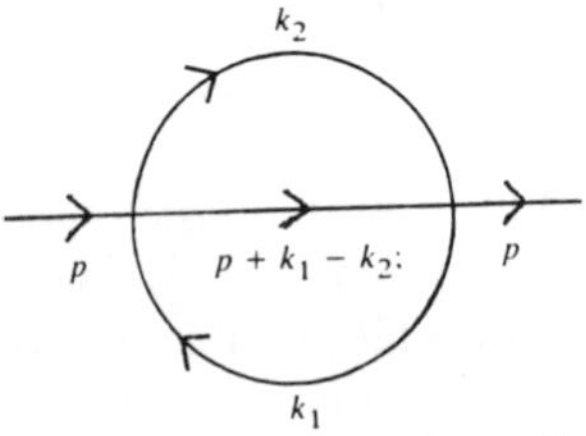

(A.4)

Note the construction of the diagram carefully. We have two vertices, at each of which four lines meet. At each vertex we have labelled the lines in such a way as to ensure momentum conservation. Further, this diagram has the same external line structure as (A.2) (both have the form 'p in' and 'p out').

Using the dictionary in Table A.1 we can translate (A.4) into the following expression:

$$\left\{\frac{i}{(p^2 - m^2 + i\varepsilon)}\right\}^2 \int \frac{dk_1}{(2\pi)^4} \int \frac{dk_2}{(2\pi)^4} (-i\lambda)^2 \times$$

$$\times \frac{1}{(k_1^2 - m^2)} \frac{1}{(k_2^2 - m^2)} \frac{1}{\{(p + k_1 - k_2)^2 - m^2\}} \qquad \text{(A.5)}$$

The factor outside the integral arises from the external legs carrying momentum p. The two vertices contribute $(-i\lambda)^2$. The three factors in the integrand are due to internal lines carrying the momenta k_1, k_2, $p + k_1 - k_2$. As required in Table A.1 we have integrated over the internal momenta k_1 and k_2. For future convenience we shall write this expression as

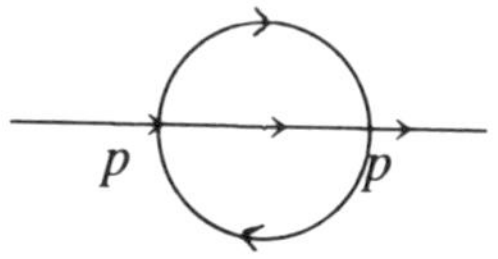

$$= \left\{ \frac{1}{(p^2 - m^2 + i\varepsilon)} \right\}^2 \Pi(p), \tag{A.6}$$

where $\Pi(p)$ stands for the integral expression over k_1, k_2. Thus to second order in λ, the probability amplitude described by (A.3) has changed to

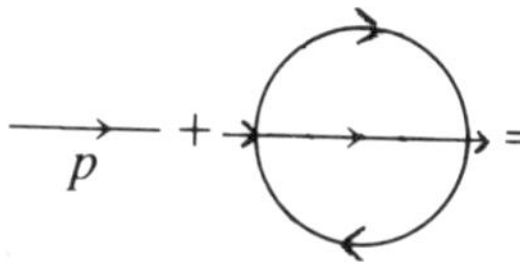

$$= \frac{i}{(p^2 - m^2 + i\varepsilon)} \left\{ 1 + \frac{i\Pi(p)}{(p^2 - m^2 + i\varepsilon)} \right\} . \tag{A.7}$$

We can now see how the concept of mass renormalization arises in the field theory. The mass of the quanta are given by the poles of the propagator $G_F(p)$ in momentum space. Thus in the absence of any interaction ($\lambda = 0$) our field φ has mass m. The propagator has now changed from (A.2) to (A.7) and it is not clear as to what happens to the pole at $p^2 = m^2$. To investigate this more closely we proceed further and calculate the infinite sum of diagrams

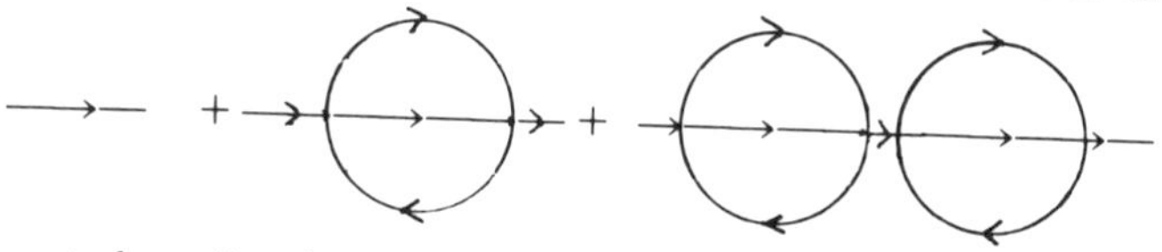

$$+ \cdots \tag{A.8}$$

by repeated application of our techniques. We find that sum to be

$$\frac{i}{(p^2 - m^2)} \left\{ 1 + \frac{i\Pi}{(p^2 - m^2)} + \left(\frac{i\Pi}{(p^2 - m^2)} \right)^2 + \cdots \right\}$$

$$= \frac{i}{(p^2 - m^2)} \frac{1}{1 - [i\Pi/(p^2 - m^2)]} = \frac{i}{p^2 - m^2 - i\Pi(p)} . \tag{A.9}$$

Thus, the new pole is at a value of p^2 for which

$$p^2 - m^2 - i\Pi(p) = 0. \tag{A.10}$$

We shall denote the value of p^2 for which (A.10) is satisfied as m_R^2, where m_R is the renormalized mass.

Thus, the interaction has shifted the value of the mass of the scalar field from m to m_R. In fact, only m_R will be physically observable because in the real world interactions cannot be switched off. It must be stressed that renormalization of physical parameters is a direct consequence of interactions and has *nothing to do, a priori*, with any divergence. For example, an electron in a crystal lattice will have an effective mass which is different from the mass of the free electron; both values will be finite. In field theory, however, we can *use* the process of renormalization to remove the divergences. For example, any infinite part in $i\Pi(p)$ can be absorbed into the bare mass m^2, in (A.9) and (A.10), thereby keeping renormalized value m_R finite.

As the next illustration let us consider the renormalization of the coupling constant λ. We anticipate the interactions to modify λ to some value λ_R. Also, it is likely that (the original) λ should have a divergent part to cancel out any divergences that arise due to interactions. With these considerations in mind, we shall first write our Lagrangian in (A.1) in the form

$$L = \tfrac{1}{2}(\varphi^i\varphi_i - m^2\varphi^2) - \frac{\lambda}{4!}\varphi^4 + \frac{A\lambda}{4!}\varphi^4. \tag{A.11}$$

Clearly, we have done nothing but redefine λ in (A.1) as $\lambda(1 - A)$. However, treating the last two terms as *separate* interactions we are led to one extra diagram of the following kind:

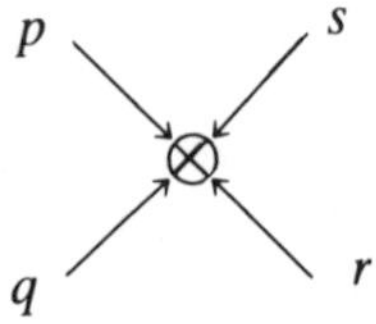

$$= iA\lambda. \tag{A.12}$$

(This should be obvious from a comparison of (A.1), (A.11), and Table A.1).

We shall now proceed to look at higher order corrections that modify the value of λ to a renormalized value λ_R. To the lowest order, of course, λ arises in the diagram

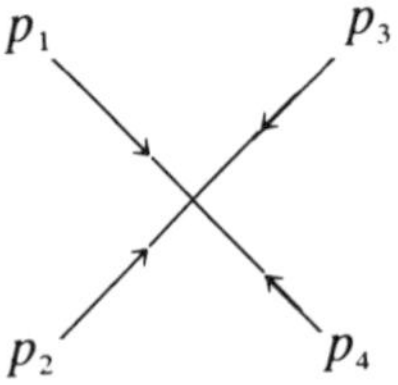

$$= -i\lambda. \tag{A.13}$$

In the next order (to obtain the renormalized value $(-i\lambda_R)$), we have to sum up the following diagrams:

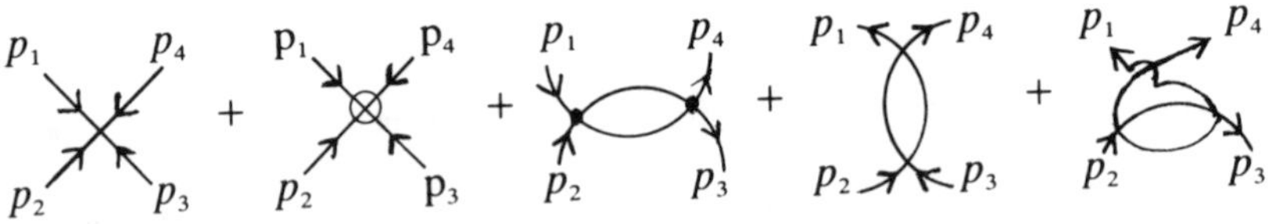

(A.14)

As a prototype consider the evaluation of the third diagram:

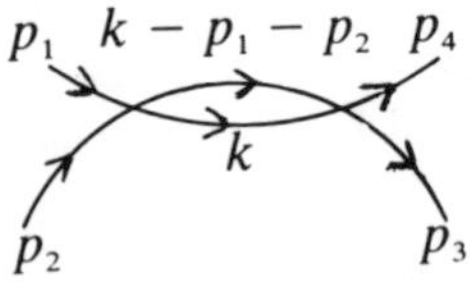

$$;\quad p_1 + p_2 = p_3 + p_4. \tag{A.15}$$

Using our 'dictionary' the loop part of this diagram can be translated into the integral

$$\frac{\lambda^2}{(2\pi)^4}\int d^4k \frac{1}{\{k^2 - m^2 + i\varepsilon\}\{(k - p_1 - p_2)^2 - m^2 + i\varepsilon\}} \equiv I. \qquad (A.16)$$

We now use the identity

$$\frac{1}{X_1X_2 \ldots X_n} = \Gamma(n)\int_0^1 \frac{d\alpha_1\, d\alpha_2 \ldots d\alpha_n}{[\alpha_1X_1 + \ldots \alpha_nX_n]^n}\,\delta\left(\sum_{i=1}^{n} \alpha_i - 1\right) \qquad (A.17)$$

and obtain, after some algebra,

$$I = \frac{\lambda^2}{(2\pi)^4}\int \frac{d^4k\, d\alpha}{[\{k + (1-\alpha)(p_1 + p_2)\}^2 - m^2 + (1-\alpha)s - (1-\alpha)^2 s]^2}, \qquad (A.18)$$

where $s = (p_1 + p_2)^2$. We shall now change variables from k to $\{k + (1 - \alpha)(p_1 + p_2)\}$, to get

$$I = \frac{\lambda^2}{(2\pi)^4}\int \frac{d^4q\, d\alpha}{[q^2 + a^2]^2}; \quad a^2 = -m^2 + \alpha(1-\alpha)s. \qquad (A.19)$$

Using the definite integral

$$\int \frac{d^4q}{(q^2 + a^2)^n} = i\pi^2 \frac{\Gamma(n-2)}{\Gamma(n)} \frac{1}{(a^2)^{n-2}}, \qquad (A.20)$$

we see that our integral in (A.19) is divergent.[2] The reader can convince himself that the last two diagrams in (A.14) will give contributions similar to (A.19) with s replaced by t $(p_1 - p_4)^2$ and u $(p_1 - p_3)^2$. Thus, the series in (A.14) has the (divergent) value

$$-i\lambda_R = -i\lambda + iA\lambda + \frac{\lambda^2}{(2\pi)^4}\int \frac{d^4q\, d\alpha}{(q^2 + a^2)^2} + \frac{\lambda^2}{(2\pi)^4}\int \frac{d^4q\, d\alpha}{(q^2 + b^2)^2} + $$

$$+ \frac{\lambda^2}{(2\pi)^4}\int \frac{d^4q\, d\alpha}{(q^2 + c^2)^2}, \qquad (A.21)$$

with

$$a^2 = -m^2 + \alpha(1-\alpha)s;\; b^2 = -m^2 + \alpha(1-\alpha)t;\; c^2 = -m^2 + \alpha(1-\alpha)u \qquad (A.22)$$

$$s = (p_1 + p_2)^2;\; t = (p_1 - p_4)^2;\; u = (p_1 - p_3)^2. \qquad (A.23)$$

[2] This fact, of course, makes the legitimacy of the change of variables from k to $k + (1 - \alpha)(p_1 + p_2)$ suspect; we shall not worry about this point. A more sophisticated method called 'dimensional regularization' is required to justify this procedure.

We shall now see how finite results are extracted out of these expressions by choosing A carefully to cancel the infinities.

Let us again look at the prototype integral in (A.19). We expand this integral about $s = 0$, obtaining

$$I(s) = \frac{\lambda^2}{(2\pi)^4} \int \frac{dq\, d\alpha}{[q^2 - m^2 + \alpha(1-\alpha)s]^2}$$

$$= \frac{\lambda^2}{(2\pi)^4} \int \frac{dq\, d\alpha}{(q^2 - m^2)^2} +$$

$$+ \left\{ \frac{-2\lambda^1}{(2\pi)^4} \int \frac{\alpha(1-\alpha)\, dq\, d\alpha}{[q^2 - m^2 + \alpha(1-\alpha)s]^3} \right\} s + \cdots \tag{A.24}$$

Clearly, *only* the first term in the expansion of $I(s)$ is infinite. The second and higher order terms lead to integrals of the type in (A.20) with $n \leq 3$, which are finite. Thus we have isolated the divergence in the first term

$$I(0) = \frac{\lambda^2}{(2\pi)^4} \int \frac{d^4q\, d\alpha}{(q^2 - m^2)^2} = \frac{\lambda^2}{(2\pi)^4} \int \frac{d^4q}{(q^2 - m^2)^2}\,, \tag{A.25}$$

which is independent of s. Since the last two terms in (A.21) are obtained by replacing s by t and u, the nature of divergences in these terms must be the same as that of $I(0)$.

We shall evaluate $I(0)$ by cutting off the integration at $q^2 = \Lambda^2$ (say). The angular integrations can be performed by using the formula

$$\int d^np\, f(p^2) = \frac{2\pi^{n/2}}{\Gamma(n/2)} \int_0^\infty dp\, p^{n-1} f(p^2), \tag{A.26}$$

leading to

$$I(0) = \frac{i\lambda^2}{(2\pi)^4} \frac{2\pi^2}{\Gamma(2)} \int_0^\Lambda \frac{q^3\, dq}{(q^2 - m^2)^2}$$

$$= \frac{i\lambda^2}{(2\pi)^4} \pi^2 \left\{ 1 + \frac{1}{(1 - (\Lambda^2/m^2))} + \ln\left(1 - \frac{\Lambda^2}{m^2}\right) \right\}. \tag{A.27}$$

The expression is seen to be logarithmically divergent as $\Lambda \to \infty$. We can eliminate this divergence by choosing the hitherto unspecified constant A to be

$$A = -\frac{\lambda\pi^2}{(2\pi)^4} \left\{ \frac{1}{(1 - (\Lambda^2/m^2))} + \ln\left(1 - \frac{\Lambda^2}{m^2}\right) \right\}. \tag{A.28}$$

(Actually, since three diagrams lead to the same divergent contribution, we should choose A to be three times the value given above.)

Notice that, as $\Lambda \to \infty$, $A \to \infty$. Thus the counter term we have added is actually infinite. However, A is supposed to be so adjusted that the infinities due to interactions cancel with the infinities in the counter term. Our sympathies are entirely with the reader who feels that the process is ugly, to say the least.

After 'cancelling' the infinities, we are left with the task of evaluating the finite part of $I(s)$ in (A.24). The simplest procedure is: (a) to evaluate $(\partial I/\partial s)$ and then (b) to integrate the result with respect to s. This process naturally eliminates the s-independent part of I. From $I(s)$ defined in (A.19), we get, (using (A.20))

$$\begin{aligned}\frac{\partial I}{\partial s} &= -\frac{2\pi^2}{(2\pi)^4}\int \frac{\alpha(1-\alpha)\,\mathrm{d}^4q\,\mathrm{d}\alpha}{[q^2 - m^2 + \alpha(1-\alpha)s]^3} \\ &= -\frac{2i\pi^2\lambda^2}{(2\pi)^4\Gamma(3)}\int_0^1 \frac{\alpha(1-\alpha)\,\mathrm{d}\alpha}{[\alpha(1-\alpha)s - m^2]}. \end{aligned} \tag{A.29}$$

Therefore,

$$\begin{aligned} I(s) &= \int \frac{\partial I}{\partial s}\,\mathrm{d}s = -\frac{2i\pi^2\lambda^2}{(2\pi)^4\Gamma(3)}\int_0^1 \mathrm{d}\alpha \int \mathrm{d}s\, \frac{\alpha(1-\alpha)}{[\alpha(1-\alpha)s - m^2]} \\ &= -\frac{i\pi^2\lambda^2}{(2\pi)^4}\int_0^1 \ln\left[\alpha(1-\alpha)s - m^2\right]\mathrm{d}\alpha + C. \end{aligned} \tag{A.30}$$

The interchanging of orders of integration between α and s is allowed because these expressions are finite. We notice that the indefinite integral over s has produced a constant of integration C, which is not yet determined. Performing the α integral we get finally

$$I(s) = -\frac{i\lambda^2\pi^2}{(2\pi)^4}\left[\left(1 - \frac{4m^2}{s}\right)^{1/2} \ln\left\{\frac{1 + (1 - (4m^2/s))^{1/2}}{1 - (1 - (4m^2/s))^{1/2}}\right\} + 2\right] + C. \tag{A.31}$$

What about the constant of integration C? The value of C can be specified once $I(s)$ is known for any given value of s. This arbitrariness is related to the concept of a 'running coupling constant' mentioned in Chapter 4. (The quantities s, t, u are related in a simple manner to the energy scales associated with p_1, p_2, p_3, p_4.) Given the value of I at any chosen energy scale, (A.31) will give a (finite) value to $I(s)$ for all s.

Putting all these together, we find that the diagrams in (A.15) give the following contribution to the renormalized coupling constant λ_R:

$$-i\lambda_R = -i\lambda + I(s) + I(t) + I(u).$$

Or, in other words, λ_R as a function of (s, t, u) is given by

$$\lambda_R(s, t, u) = \lambda + iI(s) + iI(t) + iI(u).$$

We conclude our discussion of renormalization at this point. More exhaustive discussion of this subject can be found in the texts cited in Chapter 4. Judged by mathematical standards, the renormalization programme is certainly illegitimate. Aesthetically it is ugly and *ad hoc*. It has, however, two points to commend it. Its rules, once specified, are unambiguous and can be applied at all orders of perturbation in a renormalizable field theory. Second, its predictions in terms of experimentally verifiable numbers have been vindicated in the only experimentally tested renormalizable theory, viz. the electro-weak gauge theory.

One is reminded of Ptolemy's model for the universe. A series of unambiguous rules (based on 'epicycles') were given, using which one could calculate planetary positions. The predictions agreed reasonably well with observation. Nevertheless, the basic premises on which the model was based turned out to be wrong.

Appendix B

Basic Group Theory

In this appendix we shall summarize some basic results in group theory which are of relevance to Chapter 5. Our discussion will be neither exhaustive nor rigorous.

B.1. Definition of a Group

A Group G is a set $g_1, g_2, \ldots, g_n \varepsilon G$ together with an operation (which will be indicated by a dot '$\cdot$') such that

(i) $g_i \varepsilon G, g_j \varepsilon G, \quad g_i \cdot g_j \varepsilon G$; (B.1)

(ii) $g_i \cdot (g_j \cdot g_k) = (g_i \cdot g_j) \cdot g_k$; (B.2)

(iii) there exists an element ('identity element') g_1 such that

$$g_1 \cdot g_i = g_i \cdot g_1 = g_i; \tag{B.3}$$

(iv) to every element g_i, there exists another element g_i^{-1} such that

$$g_i \cdot g_i^{-1} = g_i^{-1} \cdot g_i = g_1. \tag{B.4}$$

We note that the number of elements in a group may be finite or infinite.[1]

Example B.1. (a) A set of real $n \times n$ matrices which are nonsingular form a group under the operation of matrix multiplication. The properties (i) to (iv) are easily verified to be true; (i) and (ii) are trivially satisfied under matrix multiplication. The n-dimensional identity matrix can be taken to be g_1 and will satisfy (iii). As long as the matrices are nonsingular, the inverse matrix is well defined and will satisfy (iv).

This group is usually denoted as GL(n, r). It is not necessary to keep to real matrices. The complex extension of this group is denoted by GL(n, c).

(b) A subset of GL(n, r) consisting of matrices with determinant (+1) forms another group. Since

$$\det(AB) = (\det A)(\det B)$$

it is easy to verify that all the above conditions are once again satisfied. This group is denoted by SL(n, r) and its complex extension by the symbol SL(n, c).

(c) All the ($n \times n$) unitary matrices form a group under multiplication. The verification is once

[1] Note that we have not demanded, in general, $g_i \cdot g_k = g_k \cdot g_i$. If this additional requirement is satisfied for all i, k, then the group is called *abelian*.

again straightforward since the product of unitary matrices is unitary. This group is denoted by U(n). A subset which consists of unitary matrices of unit determinant forms the Special Unitary group denoted by SU(n). □

The matrix groups discussed in the above example belong to a class of groups called 'transformation groups', which we shall now discuss. Consider a set of N real numbers $(x^1, x^2, \ldots, x^N)$ which are transformed to another set $(y^1, \ldots, y^N)$ by the 'rule':

$$y^i = f^i(\alpha^1, \alpha^2, \ldots, \alpha^P; x^1, x^2, \ldots, x^N). \tag{B.5}$$

Here the functions f^i are taken to depend on a set of parameters α^A. We shall write (B.5) in a condensed notation as

$$y = f(\alpha; x). \tag{B.6}$$

Now consider two successive transformations:

$$x \to x' = f(\alpha; x) \to x'' = f(\beta; x') = f(\beta; f(\alpha, x)), \tag{B.7}$$

In order for these transformations to form a group, we must demand the existence of a function $\varphi = \varphi(\alpha, \beta)$ such that

$$x'' = f(\beta; f(\alpha, x)) = f(\varphi(\alpha, \beta); x). \tag{B.8}$$

The function $\varphi = \varphi(\alpha, \beta)$ associates one given element (φ) of the group with two elements (α, β) thereby defining the group operation. The conditions (B.2) to (B.4) now translate into the conditions

$$\varphi(\gamma; \varphi(\beta, \alpha)) = \varphi(\varphi(\gamma, \beta); \alpha), \tag{B.9}$$

$$\varphi(\varepsilon, \alpha) = \alpha = \varphi(\alpha, \varepsilon), \tag{B.10}$$

$$\varphi(\alpha, \alpha^{-1}) = \varepsilon = \varphi(\alpha^{-1}, \alpha), \tag{B.11}$$

Here ε^A denotes the parameters for the identity element of the group; $(\alpha^{-1})^A$ denotes the parameters for the group element that is the inverse of the one with parameters α^A. Conditions (B.9) to (B.11) assure associativity, the existence of identity, and the existence of the inverse. For our purpose, we shall class the group of transformations as a *Lie group* if the function φ is an analytic function of its arguments.

Example B.2. A simple two-parameter transformation group is given by the transformation rule

$$f(\alpha^1, \alpha^2; x) = \alpha^1 x + \alpha^2 \equiv x'; \qquad \alpha^1 \neq 0.$$

This produces a mapping from the set of all real numbers x to itself. The group operation rule can be easily found:

$$\begin{aligned} x'' &\equiv f(\beta^1, \beta^2; x') = \beta^1 x' + \beta^2 = \beta^1(\alpha^1 x + \alpha^2) + \beta^2 \\ &= (\beta^1\alpha^1)x + (\beta^1\alpha^2 + \beta^2) \\ &= \varphi^1 x + \varphi^2. \end{aligned}$$

Clearly $\varphi = \varphi(\alpha, \beta)$ is given by the relation,

$$\varphi^1 = \beta^1\alpha^1; \qquad \varphi^2 = \beta^1\alpha^2 + \beta^2.$$

It is straightforward to verify that (B.9) to (B.11) are satisfied. □

B.2. Generators

We shall now motivate the concept of 'generators' for the Lie group. Let us consider an N-dimensional space in which a coordinate system S is chosen. Any particular point P of the manifold will be described by the N-tuple $(x^1, \ldots, x^N)$. Let $F(P)$ be some function defined on the manifold. In the coordinate system S, F will have a particular structural form, say:

$$F = F[x^i] \equiv F(P). \tag{B.12}$$

Suppose we had chosen a different coordinate system S'. In order to have the same numerical value $F(P)$, the function must have a different form in terms of the new coordinates x'. Let

$$F' = F'[x'] \equiv F(P). \tag{B.13}$$

We shall assume that the two coordinate systems are related by a transformation which is an element of a Lie group of transformations:

$$x'^i = f^i[\alpha; x]. \tag{B.14}$$

How are F and F' related? It is easy to see that

$$F'(x') = F[f(\alpha^{-1}; x')]. \tag{B.15}$$

This equation can be cast into a much more useful form by concentrating on infinitesimal coordinate transformations. Let us parametrize the group so that $\alpha = 0$ is the identity element of the group. For an infinitesimal transformation which parameters $\delta\alpha^A$ $(A = 1, 2, \ldots)$ we can write

$$x^i = f^i\,[(\delta\alpha^A)^{-1}, x'\,] \cong x'^i - \delta\alpha^A \left.\frac{\partial f^i[\beta, x']}{\partial\beta^A}\right|_{\beta=0} \tag{B.16}$$

Note that the inverse of $\delta\alpha^A$ is $-\delta\alpha^A$. From (B.15) and (B.16) we get

$$F'(x') - F(x') = \delta\alpha^A T_A(x')F(x'), \tag{B.17}$$

where the operators

$$T_A(x') = -\left.\frac{\partial f^i(\beta, x')}{\partial\beta^A}\right|_{\beta=0}\left(\frac{\partial}{\partial x'^i}\right), \quad A = 1, 2, \ldots \tag{B.18}$$

are called the *generators* for the transformation.

It is clear from (B.6) and the definition of the group operation $\varphi = \varphi(\alpha, \beta)$

that a Lie group can be considered to be a transformation group acting on itself. Thus, given the functions

$$\varphi^A = \varphi^A(\alpha; \beta) \tag{B.19}$$

the generators can also be defined as

$$T_A = -\left.\frac{\partial \varphi^B(\beta, \chi)}{\partial \beta^A}\right|_{\beta=0} \left(\frac{\partial}{\partial \chi^B}\right) . \tag{B.20}$$

Example B.3. As an illustration, consider the group in Example B.2. Since we want the identity element to correspond to zero values of the parameter we shall first write

$$\alpha^1 = \exp(a^1); \quad \beta^1 = \exp(b^1); \quad \alpha^2 = a^2, \ \ \beta^2 = b^2.$$

Then the group operation rule corresponds to

$$\varphi^1 = a^1 + b^1; \quad \varphi^2 = e^{b^1} a^2 + b^2.$$

The generators are

$$T_1 = -\frac{\partial}{\partial a^1} - a^2 \frac{\partial}{\partial a^2}; \quad T_2 = -\frac{\partial}{\partial a^2} .$$ □

Consider two elements of the group α, β which are close to the identity. We represent them in terms of the generators, obtaining

$$\alpha: \ 1 + \delta\alpha^A T_A + \tfrac{1}{2}\delta\alpha^A\, \delta\beta^B T_A T_B,$$

$$\beta: \ 1 + \delta\beta^A T_A + \tfrac{1}{2}\delta\alpha^A\, \delta\beta^B T_A T_B. \tag{B.21}$$

Here $\delta\alpha^A$ and $\delta\beta^A$ are the parameters that specify the group elements α and β and the 1 in the right-hand side represents the identity element. Clearly the group operation $(\alpha\beta)\,(\beta\alpha)^{-1}$ will correspond to

$$(\alpha\beta)(\beta\alpha)^{-1} \rightarrow 1 + \delta\alpha^A\, \delta\beta^B\, [T_A, T_B]. \tag{B.22}$$

Since $[T_A, T_B]$ is another element of the group we can expand it in terms of the generators as

$$[T_A, T_B] = C^D{}_{AB}\, T_D. \tag{B.23}$$

where $C^D{}_{AB}$ are called the *structure constants* of the group.

Any general element of the group can be obtained by 'exponentiating' the infinitesimal generators. An arbitrary element U may be expressed as

$$U(\alpha) = \exp(\alpha^A T_A). \tag{B.24}$$

For physical applications it is often convenient to use (in place of T_A) the generators

$$L_A \equiv iT_A; \tag{B.25}$$

clearly,

$$[L_A, L_B] = -[T_A, T_B] = -C^D{}_{AB}T_D = iC^D{}_{AB}L_D. \tag{B.26}$$

In fact, for most physical purposes, we may consider a Lie group to be a continuous group generated by a set of generators L_A. An arbitrary member of the group now becomes

$$U = \exp(-i\alpha^A L_A). \tag{B.27}$$

The structure constants of the Lie group satisfy two important conditions. It is obvious from (B.23) that

$$C^A{}_{MN} = - C^A{}_{NM}. \tag{B.28}$$

Further, straightforward algebra using (B.20) and (B.9) will show that the generators satisfy the relation called the Jacobi identity:

$$[[L_A, L_B], L_C] + [[L_B, L_C], L_A] + [[L_C, L_A], L_B] = 0, \tag{B.29}$$

implying

$$C^A{}_{MN}\, C^D{}_{AP} + C^A{}_{NP}\, C^D{}_{AM} + C^A{}_{PM}\, C^D{}_{AN} = 0. \tag{B.30}$$

The classification of Lie groups essentially boils down to determining the class of all $C^A{}_{MN}$ that satisfy (B.28) and (B.30).

B.3. Representations

For practical calculations, we often work with what is called a 'matrix representation' of the generators of a group. If $\{L_A\}$ is a set of matrices with the same commutation law as the generators in (B.26), we say that the matrices provide a 'representation' for the generators. If $\{L_A\}$ satisfy (B.26) it is clear that $\{L_A^\dagger\}$ also satisfy (B.26). Thus, we can always choose the matrices of the representation to be Hermitian. The group elements expressed as in (B.27) are then represented by unitary matrices.

The representation of the lowest dimensionality is called the *fundamental representation*. There exists another representation (used very often) which is completely determined by the structure constants themselves. This is called the *adjoint representation*, in which the matrix corresponding to L_A will have elements

$$(L_A)_{BD} = -iC^D{}_{AB}. \tag{B.31}$$

In order to show that this forms a representation we calculate

$$\begin{aligned}[L_A, L_B]_{ED} &= (L_A)_{EM}(L_B)_{MD} - (L_B)_{EN}(L_A)_{ND} \\ &\quad C^M{}_{EA}C^D{}_{BM} + C^M{}_{BE}C^D{}_{AM}\end{aligned} \tag{B.32}$$

Using the Jacobi identity (B.30) we get

$$[L_A, L_B]_{ED} = -C^M{}_{AB}\, C^D{}_{ME} = C^M{}_{AB}\, i(L_M)_{ED}, \tag{B.33}$$

which proves that matrices in (B.31) do form a representation.

Suppose the group index A can be divided into two sets such that $C^D{}_{AB} = 0$ whenever A and B fall on different sets; then we say that the group has been factored into two. A *nonabelian* group that cannot be so factored is called 'simple'. For a simple Lie group we can define a 'metric' by

$$g_{AB} = \mathrm{Tr}(L_A L_B)_{\text{adjoint representation}} \tag{B.34}$$

$$= -C^M{}_{AN}\, C^N{}_{BM}. \tag{B.35}$$

As the L's are Hermitian, the norm of L_A is positive definite. Thus we can choose a basis in which g_{AB} is diagonalized to δ_{AB}.

The above analysis was done in the adjoint representation. In *any* representation we can take

$$\mathrm{Tr}(L_A\, L_B) = F\, \delta_{AB}, \tag{B.36}$$

where F depends only on the representation. This can be proved as follows:

In a suitable basis $\mathrm{Tr}(L_A\, L_B)$ can always be diagonalized to the form

$$\mathrm{Tr}(L_A\, L_B) = \begin{cases} 0 & \text{for } A \neq B, \\ F_A & \text{for } A = B. \end{cases} \tag{B.37}$$

We only have to show that F_A is independent of A. Consider the tensor

$$\begin{aligned} K_{ABD} &\equiv \mathrm{Tr}(L_A L_B L_D) - \mathrm{Tr}(L_B L_A L_D) \\ &= iC_{ABN}\, \mathrm{Tr}(L_N L_D) \\ &= iC_{ABD} F_D \quad \text{(no summing on } D). \end{aligned} \tag{B.38}$$

Here we have defined the completely antisymmetric tensor

$$C_{ABD} \equiv C^N{}_{AB} g_{ND}. \tag{B.39}$$

Interchanging B, D we have

$$K_{ADB} = iC_{ADB} F_B \quad \text{(no summing on } B). \tag{B.40}$$

Comparison of (38) and (40) shows that

$$F_D = F_B \tag{B.41}$$

as long as $[L_D, L_B] \neq 0$. Since the group is assumed to be simple it immediately follows that F_A is independent of A. This result was used in Chapter 5.

More detailed discussion of group theoretical methods can be found in the books referred to at the end of Chapter 5.

Appendix C

Differential Geometry

In this appendix we shall discuss the concepts of tensor analysis, covariant differentiation, curvature, etc., from the point of view of differential geometry. The aim is limited to the extent of introducing the reader to the basic concepts without being unduly rigorous or abstract.

C.1. Basic Concepts

In elementary calculus we deal with functions of, say, n-variables $(x^1, x^2, \ldots, x^n)$. These functions are mapping from R^n to R^1. Here, and in what follows, we denote by R^m the Euclidean space of m-dimensions. The idea of continuity, differentiability, etc., are well understood in this particular context.

The central object in differential geometry is a 'manifold.' Manifolds are essentially spaces which behave locally like a Euclidean space. Since the concepts of differential calculus are essentially local, we can use the known concepts of calculus on R^n to develop the calculus on an arbitrary manifold. To do this, let us first make precise the notion of a manifold.

Suppose R^n is the n-dimensional Euclidean space with the neighbourhoods (i.e. open and closed sets) defined in the usual way. An n-dimensional *manifold M* is a union of neighbourhoods u_α, where we have associated with each u_α a particular one–one map φ_α, which images every $p \varepsilon u_\alpha$ to a point in an open neighbourhood of R^n. We also impose the following differentiability condition on the set of mappings $\{\varphi_\alpha\}$. Consider two neighbourhoods u_α, v_α with *non*-vanishing intersection. Let φ_α and ψ_α be the corresponding maps from M to R^n. Now consider the map $\varphi_\alpha \circ \psi_\alpha^{-1}$ for the intersection region $(u_\alpha \cap v_\alpha)$. Clearly, this is a map from R^n to R^n. We demand that this map must be differentiable in the usual sense of the word.

The above discussion is illustrated schematically in Figure C.1. The existence of the maps φ_α allows us to introduce 'coordinates' on the manifold M. With every point $p \varepsilon M$ we have now associated a point $\{x^1, x^2, \ldots, x^n\} \equiv \{x^i\}$ of R^n. Notice that there is nothing unique about the set of mappings

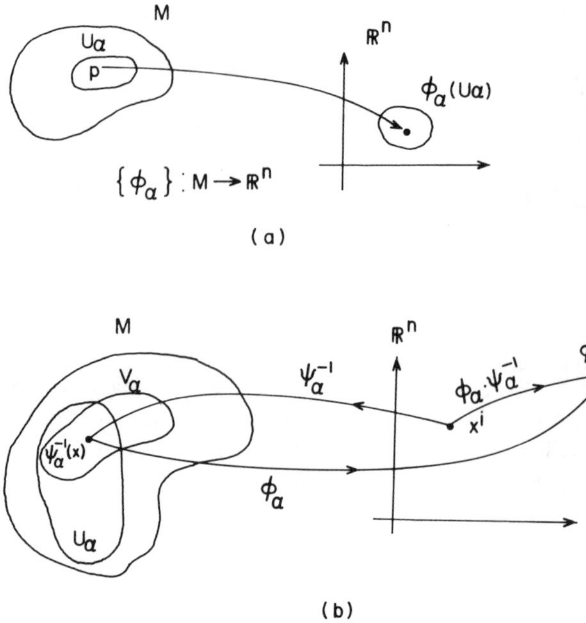

Fig. C.1. (a) Concept of a manifold. In the neighbourhood U_α around $p \; \varepsilon M$, there is a mapping φ_α which assigns to p an n-tuple $(x^1, \ldots, x^n$ of R^1. (b) The concept of differentiability of mappings. We demand that when two neighbourhoods have a nonvanishing overlap the mappings must be compatible in that region.

$\{\varphi_\alpha\}$. We could have used a different set $\{\varphi'_\alpha\}$ which would have associated with the *same* $p \; \varepsilon \; M$ a *different* $\{x'^i\}$ of R^n. In other words, the change from $\{\varphi_\alpha\}$ to $\{\varphi'_\alpha\}$ is what we have called a 'general coordinate transformation' in Chapter 6. The coordinate transformation merely relabels the points of the manifold.

The existence of the maps $\{\varphi_\alpha\}$ allows us to translate known concepts in R^n to an arbitrary manifold M. We have already introduced the concept of differentiability on a manifold. As illustrated in Figure C.1(b) we have used the composition $\varphi_\alpha \circ \psi_\alpha^{-1}$, which is a map from R^n to R^n, to achieve this goal. We shall repeatedly use this 'trick' in what follows.

A *function* on the manifold M is defined by a map f: $M \rightarrow R^1$. We shall assume that the composition $f \circ \varphi_\alpha^{-1}$ which maps R^n to R^1 is a smooth function.[1]. A *smooth curve* C on M is again defined to be a map C from some interval of R^1 to M such that $(\varphi_\alpha \circ C)$, which is a map from R^1 to R^n, is smooth. It is customary to choose the interval in R^1 to be (0, 1). For all t with $0 < t < 1$, $C(t)$ gives a point $p \; \varepsilon \; M$. Further, $\varphi_\alpha \cdot C(t) = \varphi_\alpha(p)$ gives an

[1] By 'smooth' we mean the existence of derivatives of sufficiently high order to warrant the ensuing manipulations.

n-tuple $\{x^1(t), \ldots, x^n(t)\}$. Note that the curve C defined on M is a coordinate independent entity, while a particular choice of φ_α allows a particular parametrization $\{x^i(t)\}$ via the map $(\varphi_\alpha \cdot C)$. Now we can easily define a function on the curve C by the composition $(f \circ C)$ (clearly, C: $R^1 \to M$, f: $M \to R^1$ so that $(f \circ C)$: $R^1 \to R^1$). It is usual to denote this function on C by $f(t) \equiv f[x^1(t), \ldots, x^n(t)]$.

C.2. Vectors and 1-Forms

A tangent vector to a curve C on a manifold M, at a point p, is defined to be an *operator* which maps each function f (defined at p) to the number

$$\left(\frac{\partial f}{\partial t}\right)_p \equiv \lim_{s \to 0} \frac{1}{s} \{(f \circ C)(t+s) - (f \circ C)(t)\} . \tag{C.1}$$

Note that $(f \circ C)$ is a map from $(0, 1)$ of R^1 to R^1. Thus, using a particular set of $\{\varphi_\alpha\}$, we can write $(f \circ C) = f(x^i(t))$, where $\{x^i(t)\} = \varphi_\alpha \cdot C(t)$. Then

$$\left(\frac{\partial f}{\partial t}\right)_p = \left.\frac{\mathrm{d}x^i}{\mathrm{d}t}\right|_{t=t_0} \left(\frac{\partial f}{\partial x^i}\right)_{C(t_0)} \equiv Q^i \frac{\partial}{\partial x^i} f. \tag{C.2}$$

It is thus clear that every tangent vector at p can be expressed as a linear combination of the operators $\{\partial/\partial x^i\}$; conversely, given an operator $\{V^i(\partial/\partial x^i)\}$ with a set of numbers $\{V^i\}$ there exists the curve, $x^i(t) = x_0^i + tV^i$ with $V^i(\partial/\partial x^i)$ as the tangent vector.

The set of all tangent vectors at a point p forms a linear vector space of n dimensions. It is usual to call this space the tangent space at p and denote it by T_p.

This definition of vectors as differential operators will be the first of a series of shocks an uninitiated student will suffer in this appendix. It should be borne in mind that 'vector' is a geometric entity and is independent of the coordinates assigned to M by the choice of $\{\varphi_\alpha\}$. In the usual definition (as in Chapter 6), a set of n numbers $\{V^i\}$ which transforms in a particular way under coordinate transformations is called a vector. Rigorously speaking this definition only specifies the *components* of the vector and not the vector itself!

The transformation property of the components of a vector under a coordinate transformation follows trivially from the above definition. Suppose $\{\varphi_\alpha\}$ and $\{\varphi'_\alpha\}$ are two sets of coordinate assignments on M. We assume that $\varphi_\alpha \circ \varphi'^{-1}_\alpha$ and $\varphi'_\alpha \circ \varphi_\alpha^{-1}$ (which are maps from R^n to R^n) are smooth. Then it is clear that, for function f defined on M,

$$\left.\frac{\partial f}{\partial x^i}\right|_p = \left.\frac{\partial x^{j'}}{\partial x^i} \frac{\partial f}{\partial x^{j'}}\right|_p , \tag{C.3}$$

where, of course, $\varphi_\alpha(p) = \{x^i\}$ and $\varphi'_\alpha(p) = \{x^{i'}\}$. Using our definition of vectors, it follows that

$$V^{i'} = \frac{\partial x^{i'}}{\partial x^j} V^j. \tag{C.4}$$

This is the usual transformation law for the components of a vector.

In the tangent space T_p we can choose n linearly independent basis vectors $\mathbf{e}_i$ which span T_p. Then any arbitrary vector $\mathbf{V} \, \varepsilon \, T_p$ can be expressed as a linear combination:

$$\mathbf{V} = V^i \mathbf{e}_i. \tag{C.5}$$

(Hereafter, we shall denote vectors by boldface: $\mathbf{V}$, $\mathbf{e}_i$, etc.) Thus, V^i is the ith component of vector $\mathbf{V}$ in a given basis $\mathbf{e}_i$. (The subscript i on $\mathbf{e}_i$ denotes 'which' basis vector and *not* a component.) One natural choice for $\mathbf{e}_i$ is (with $\varphi_\alpha(p) = \{x^i\}$)

$$\mathbf{e}_i \bigg|_p = \frac{\partial}{\partial x^i}\bigg|_p . \tag{C.6}$$

This is called the 'coordinate basis'. Of course, we are free to choose any other basis. In fact, it is often convenient to choose a noncoordinate basis.

Example C.1. Consider the manifold R^3 in which we have made the Cartesian coordinate assignments (x, y, z). The vectors

$$\mathbf{e}_1 = \frac{\partial}{\partial x}, \quad \mathbf{e}_2 = \frac{\partial}{\partial y}, \quad \mathbf{e}_3 = \frac{\partial}{\partial z}$$

form a coordinate basis. On the other hand, we can assign to points on R^3 the spherical coordinates (r, θ, φ). Then, the coordinate basis will be

$$\mathbf{e}_1 = \frac{\partial}{\partial r}, \quad \mathbf{e}_2 = \frac{\partial}{\partial \theta}, \quad \mathbf{e}_3 = \frac{\partial}{\partial \varphi}$$

However, it is usual to work with the noncoordinate basis

$$\mathbf{e}_1 = \frac{\partial}{\partial r}, \quad \mathbf{e}_2 = \frac{1}{r}\frac{\partial}{\partial \theta}, \quad \mathbf{e}_3 = \frac{1}{r \sin\theta}\frac{\partial}{\partial \varphi}.$$ □

In Chapter 6, we introduced two kinds of vectors: contravariant and covariant. So far, we have only identified the contravariant components of the vector. What about the covariant vectors? It turns out that these actually correspond to a set of creatures called '1-forms', which we shall now introduce.

A 1-form[2] $\boldsymbol{\omega}$ at p is real valued linear function from T_p to R^1. The number in

[2] We shall reserve *Greek* boldface characters for 1-forms to distinguish them from vectors.

R^1 to which a 1-form $\boldsymbol{\omega}$ maps a vector $\mathbf{V} \varepsilon T_p$ will be denoted by $\langle \boldsymbol{\omega}, V \rangle$. Linearity implies that

$$\langle \boldsymbol{\omega}, \alpha \mathbf{V} + \beta \mathbf{U} \rangle = \alpha \langle \boldsymbol{\omega}; \mathbf{V} \rangle + \beta \langle \boldsymbol{\omega}, \mathbf{U} \rangle; \quad \mathbf{U}, \mathbf{V} \ \varepsilon \ T_p. \tag{C.7}$$

We define the linear combinations of 1-forms by the rule:

$$\langle \alpha \boldsymbol{\omega} + \beta \boldsymbol{\theta}, \mathbf{V} \rangle = \alpha \langle \boldsymbol{\omega}, \mathbf{V} \rangle + \beta \langle \boldsymbol{\theta}, \mathbf{V} \rangle. \tag{C.8}$$

Now consider a particular basis $\mathbf{e}_i$ for T_p. We define a unique set of 1-forms $\boldsymbol{\omega}^j$, corresponding to a particular choice $\mathbf{e}_i$, by the following rule. For any $\mathbf{X} \ \varepsilon \ T_p$,

$$\langle \boldsymbol{\omega}^j, \mathbf{X} \rangle = X^j, \tag{C.9}$$

where $\mathbf{X} = X^i \mathbf{e}_i$. It immediately follows that

$$\langle \boldsymbol{\omega}^j, \mathbf{e}_i \rangle = \delta_i^j \, . \tag{C.10}$$

We can regard $\boldsymbol{\omega}^j$ as forming a basis for any 1-form at p, since we can write $\boldsymbol{\theta} \equiv \theta_i \boldsymbol{\omega}^i \equiv \langle \boldsymbol{\theta}, \mathbf{e}_i \rangle \boldsymbol{\omega}^i$ for any $\boldsymbol{\theta}$. The set of all 1-forms at p forms a linear vector space denoted by T_p^*. The basis $\boldsymbol{\omega}^i$ for this vector space is called the 'dual basis' to $\mathbf{e}_i$.

To every function f defined on M we can associate a 1-form (**df**) by the rule

$$\langle \mathbf{df}, \mathbf{V} \rangle \equiv \mathbf{V}(f); \tag{C.11}$$

in other words, **df** associates with every vector $\mathbf{V}$ the real number $\mathbf{V}(f)$. The notation **df** is motivated by the fact that, when $\mathbf{V} = \mathbf{e}_i$ in a coordinate basis,

$$\langle \mathbf{df}, \mathbf{e}_i \rangle = \mathbf{e}_i(f) = \frac{\partial f}{\partial x^i} \, . \tag{C.12}$$

Since $\{\varphi_\alpha\}$ themselves are functions on M mapping a point $p \ \varepsilon \ M$ to $\{x^i\}$ of R^n, we can also consider the coordinate 1-forms $\mathbf{dx}^i$; we have

$$\langle \mathbf{dx}^i, \mathbf{e}_j \rangle = \mathbf{e}_j(x^i) = \frac{\partial x^i}{\partial x^j} = \delta_j^i \, . \tag{C.13}$$

Thus $\mathbf{dx}^i$ forms the dual for the coordinate basis $\mathbf{e}_j = \partial / \partial x^j$, allowing us to write

$$\mathbf{df} = \frac{\partial f}{\partial x^i} \, \mathbf{dx}^i. \tag{C.14}$$

The reader will recall that the 'covariant vector' is defined by considering the gradient of a scalar function, like $(\partial f / \partial x^i)$. It can easily be seen that the components of a 1-form transform under coordinate transformations as

$$\boldsymbol{\theta}_i' = \frac{\partial x^j}{\partial x'^i} \, \theta_j. \tag{C.15}$$

Thus the 1-form is the proper mathematical characterization of the covariant vector defined in Chapter 6.

We therefore see that contravariant and covariant vectors introduced in Chapter 6 are actually elements of T_p (space of tangent vectors) and T_p^* (space of 1-forms). At this stage, they are separate entities. (For example, we cannot 'raise or lower indices' and obtain covariant vectors from contravariant vectors etc.) We shall later show how, by introducing a metric on M, we can set up a one–one correspondence between vectors and 1-forms.

Given the linear vector spaces T_p and T_p^*, we can construct the usual Cartesian product space

$$Q_r^s \equiv \underbrace{T^*{}_p \otimes T^*{}_p \ldots T^*{}_p}_{r \text{ factors}} \otimes \underbrace{T_p \otimes T_p \ldots T_p}_{s \text{ factors}}. \tag{C.16}$$

A tensor of type $\binom{S}{r}$ at p is a linear function from Q_r^s to R^1. It is trivial to construct the direct product basis for Q_r^s from the basis vectors $\mathbf{e}_i$ and basis forms $\boldsymbol{\omega}^i$ and show that these objects have components which transform as a mixed tensor with s contravariant and r covariant indices. If $\mathbf{T}$ is a tensor of type $\binom{S}{r}$ then the number in R^1 to which $\mathbf{T}$ maps the element $(\boldsymbol{\theta}^1, \boldsymbol{\theta}^2, \ldots, \boldsymbol{\theta}^r; \mathbf{V}_1, \mathbf{V}_2, \ldots, \mathbf{V}_s)$ of Q_r^s will be denoted by $\mathbf{T}(\boldsymbol{\theta}^1 \ldots \boldsymbol{\theta}^r; \mathbf{V}_1 \ldots \mathbf{V}_s)$. All *algebraic* operations with tensors like addition, contraction, etc., can now be performed locally at any point p.

C.3. Lie Derivative and Covariant Derivative

We have now endowed the manifold M with a tangent space T_p and its dual space T_p^* *at each point* p. The space Q_r^s constructed out of (T_p, T_p^*) is also strictly local and is assigned to each point p. By this construction we have vectors, 1-forms, and tensors residing at each point p.

From these rudiments, it is easy to construct a vector field on M. A *vector field* may be looked upon as an operator which assigns to each $p \; \varepsilon \; M$ a unique vector $\mathbf{V} \; \varepsilon \; T_p$. Similarly a *tensor field* of type $\binom{S}{r}$ is an assignment of an element of Q_r^s to each $p \; \varepsilon \; M$.

Once such a vector (or tensor) field is defined it is meaningful to think of comparing the vectors at different points. We ask how the vector field changes with $p \; \varepsilon \; M$ and are led to the concept of differentiation of a vector (or tensor) field.

It turns out that three different kinds of 'differential structures' are of importance in differential geometry. Two of these, called 'exterior differentiation' and 'Lie differentiation' can be defined without introducing any other external structure on M. The third, called 'covariant differentiation', requires an extra structure, called 'affine connection'. In what follows, we shall *not* discuss exterior differentiation and shall begin with the concpet of Lie differentiation.

Suppose **X**, **Y** are two vector *fields* on M. We define the *Lie bracket* of **X** and **Y** by its action on any function f:

$$[\mathbf{X}, \mathbf{Y}]f \equiv (\mathbf{XY} - \mathbf{YX})f \equiv \mathbf{X}(\mathbf{Y}f) - \mathbf{Y}(\mathbf{X}f). \tag{C.17}$$

It is easy to verify that

$$[\mathbf{X}, \mathbf{Y}]\,(\alpha f + \beta g) = \alpha[\mathbf{X}, \mathbf{Y}]f + \beta[\mathbf{X}, \mathbf{Y}]g, \tag{C.18}$$

$$[\mathbf{X}, \mathbf{Y}](fg) = f[\mathbf{X}, \mathbf{Y}]g + g[\mathbf{X}, \mathbf{Y}]f, \tag{C.19}$$

$$\Big[[\mathbf{X}, \mathbf{Y}], \mathbf{Z}\Big] + \Big[[\mathbf{Z}, \mathbf{X}], \mathbf{Y}\Big] + \Big[[\mathbf{Y}, \mathbf{Z}], \mathbf{X}\Big] = 0. \tag{C.20}$$

Here f, g are functions on M and α, β are constants. The property of differentiation exhibited in (C.19) motivates us to define the *Lie derivative* of **Y** in the direction of **X** as

$$L_x\mathbf{Y} \equiv [\mathbf{X}, \mathbf{Y}] = -[\mathbf{Y}, \mathbf{X}] = -L_y\mathbf{X}. \tag{C.21}$$

More formally, it can be noticed that, given any vector field **V** on M, there exists a *unique* set of integral curves to **X** which satisfy (in a chosen coordinate system) the equations

$$\frac{\mathrm{d}x^i}{\mathrm{d}t} = X^i(x^i). \tag{C.22}$$

Here the curve is parametrized by t: i.e. $x^i = x^i(t)$. These curves define a family of diffeomorphisms φ_t: $M \to M$ and thus assigns to every tensor **T** of type $\binom{S}{r}$ another tensor, usually denoted by $(\varphi_f * \mathbf{T})|_{\varphi_f(p)}$. The Lie derivative of an *arbitrary* tensor field may be defined as

$$L_\mathbf{x}\mathbf{T}|_p = \lim_{t\to 0} \frac{1}{t}\left\{\mathbf{T}|_p - \varphi_t * \mathbf{T}|_p\right\}. \tag{C.23}$$

Using either of these definitions, we can show that for any 1-form $\boldsymbol{\omega}$

$$(L_\mathbf{x}\boldsymbol{\omega})_i = \omega_{i,k}X^k + \omega_k X^k{}_{,i}, \tag{C.24}$$

where we have used (some) coordinate assignment on M. Similarly, definition (C.21) can be written in component notation as

$$(L_\mathbf{x}\mathbf{Y})^i = X^k Y^i{}_{,k} - Y^k X^i{}_{,k}. \tag{C.25}$$

We shall now proceed to the concept of *covariant differentiation*. We define a connection ∇ on M to be a rule that assigns to each vector field **X**, at a point p, a differential operator $\nabla_\mathbf{x}$ which maps an arbitrary vector field **Y** in M, to a vector field $\nabla_\mathbf{x}\mathbf{Y}$ such that,

$$\text{(i)}\quad \nabla_{f\mathbf{x} + g\mathbf{y}}\mathbf{Z} = f\nabla_\mathbf{x}\mathbf{Z} + g\nabla_\mathbf{y}\mathbf{Z}, \tag{C.26}$$

$$\text{(ii)}\quad \nabla_\mathbf{x}(\alpha\mathbf{Y} + \beta\mathbf{Z}) = \alpha\nabla_\mathbf{x}\mathbf{Y} + \beta\nabla_\mathbf{x}\mathbf{Z}, \tag{C.27}$$

(iii) $\nabla_{\mathbf{x}}(f\mathbf{Y}) = \mathbf{X}(f)\mathbf{Y} + f\nabla_{\mathbf{x}}\mathbf{Y}.$ (C.28)

Here t, g are functions on M and α, β are constants. More generally, we define $\nabla\mathbf{Y}$ to be a (1-1) tensor which, on contraction with $\mathbf{X}$ (at p), produces $\nabla_{\mathbf{x}}\mathbf{Y}$. In component notation, we write

$$\nabla\mathbf{Y} = Y^a{}_{;b}\,\boldsymbol{\omega}^b \otimes \mathbf{e}_a.$$

Because of linearity, the action of ∇ on any tensor $\mathbf{T}$ is known once the action on basis vectors and 1-forms are given. If $\boldsymbol{\omega}^a$ and $\mathbf{e}_b$ are the basis vectors chosen in T_p^* and T_p, we have the functions $\Gamma^a{}_{bc}$:

$$\Gamma^a{}_{bc} = \langle \boldsymbol{\omega}^a, \nabla_{\mathbf{e}_b}\mathbf{e}_c\rangle\ ; \tag{C.29}$$

which completely determine the connection ∇. It is easy to verify that

$$Y^a{}_b = Y^a{}_{,b} + \Gamma^a{}_{bc}Y^c, \tag{C.30}$$

which is the rule for covariant differentiation derived in Chapter 6.

We can generalize the action of ∇ on an arbitrary tensor field of type $\binom{r}{s}$ by the following additional rules:

(i) if $\mathbf{T}$ is of type $\binom{r}{s}$ then $\nabla\mathbf{T}$ is of type $\binom{r}{s+1}$.

(ii) ∇ operates linearly and commutes with contraction.

(iii) For arbitrary $\mathbf{S}$ and $\mathbf{T}$

$$\nabla(\mathbf{S}\otimes\mathbf{T}) = \nabla\mathbf{S}\otimes\mathbf{T} + \mathbf{S}\otimes\nabla\mathbf{T}. \tag{C.31}$$

(iv) For any function f, $\nabla f = \mathbf{df} = \dfrac{\partial f}{\partial x^a}\,\mathbf{dx}^a$.

Using these rules, we can calculate the action of ∇ on the basis 1-forms $\boldsymbol{\omega}^a$. We have

$$0 = \nabla_{\mathbf{e}_c}(\delta^a_b) = \nabla_{\mathbf{e}_c}\langle\boldsymbol{\omega}^a, \mathbf{e}_b\rangle = \langle\nabla_{\mathbf{e}_c}\boldsymbol{\omega}^a\ \mathbf{e}_b\rangle + \langle\boldsymbol{\omega}^a, \nabla_{\mathbf{e}_c}\mathbf{e}_b\rangle$$

$$= \langle\nabla_{\mathbf{e}_c}\boldsymbol{\omega}^a, \mathbf{e}_b\rangle + \Gamma^a{}_{cb}. \tag{C.32}$$

Therefore,

$$\nabla_{\mathbf{e}_c}\boldsymbol{\omega}^a = -\,\Gamma^a{}_{cb}\boldsymbol{\omega}^b. \tag{C.33}$$

Linearity of ∇ allows us to construct the component form of $\nabla\mathbf{T}$ for any tensor $\mathbf{T}$. The reader can easily convince himself that these rules are the same as those given in Chapter 6.

C.4. Curvature and Metric

We define a type $\binom{1}{3}$ tensor, called a *curvature tensor*, by the following mapping:

$$\mathbf{R}(\boldsymbol{\theta};\ \mathbf{X},\ \mathbf{Y},\ \mathbf{Z}) \equiv \langle \boldsymbol{\theta},\ \nabla_{\mathbf{x}}\nabla_{\mathbf{y}}\mathbf{Z} - \nabla_{\mathbf{y}}\nabla_{\mathbf{x}}\mathbf{Z} - \nabla_{[\mathbf{x},\mathbf{y}]}\mathbf{Z}\rangle \tag{C.34}$$

Here $\boldsymbol{\theta}$ is an arbitrary one form and **X**, **Y**, **Z** are vectors. It is somewhat more convenient to work with the vector field $\mathbf{R}(\ ,\ \mathbf{X},\ \mathbf{Y},\ \mathbf{Z}) \equiv R^a{}_{bcd}X^cY^dZ^b\mathbf{e}_a$. From the basic definition (C.34) one sees that

$$\begin{aligned} R^a{}_{bcd}X^cY^dZ^b &= (Z^a{}_{;d}Y^d)_{;c}X^c - (Z^a{}_{;d}X^d)_{;c}Y^c \\ &= (Z^a{}_{;dc} - Z^a{}_{;cd})X^cY^d. \end{aligned} \tag{C.35}$$

Thus,

$$R^a{}_{bcd}Z^b = Z^a{}_{;dc} - Z^a{}_{;cd}, \tag{C.36}$$

which is the definition for curvature tensor given in Chapter 6 (see Equation (6.62)).

The *metric tensor* is defined to be a type $\binom{0}{2}$ tensor on M, which is symmetric. We denote the components of the metric by g_{ab}; i.e.

$$g_{ab} \equiv \mathbf{g}(\mathbf{e}_a,\ \mathbf{e}_b); \tag{C.37}$$

so that

$$\mathbf{g} = g_{ab}\ \mathbf{dx}^a \otimes \mathbf{dx}^b. \tag{C.38}$$

The introduction of the metric allows a natural identification between vectors and 1-forms. Suppose **V** is a vector in T_p. We assign to it a 1-form $\tilde{\mathbf{V}}$ by the following choice:

$$\langle \tilde{\mathbf{V}},\ \mathbf{U}\rangle = \mathbf{g}(\mathbf{V},\ \mathbf{U}). \tag{C.39}$$

In other words, $\tilde{\mathbf{V}}$ maps an arbitrary vector **U** to a real number in R, which is numerically the same as the one given by $g(\mathbf{V},\ \mathbf{U})$. Thus, the metric gives a linear map between T_p and T^*_p. Let us work out the components of $\tilde{\mathbf{V}}$. We have

$$\langle \tilde{\mathbf{V}},\ \mathbf{U}\rangle = \tilde{V}_aU^a = g_{mn}V^mU^n. \tag{C.40}$$

Therefore,

$$\tilde{\mathbf{V}}_a = g_{an}V^n. \tag{C.41}$$

This is precisely the concept, introduced in Chapter 6 of 'raising and lowering' an index. The existence of a metric on M allows us to talk interchangeably between elements of T_p and T^*_p. This is the reason why we do not usually make a physical distinction between contravariant and covariant indices in dealing with tensors, in a spacetime manifold endowed with a metric.

So far we have introduced the metric tensor and the connection as separate structures on M. However, given a metric **g** on M, there exists a unique connection on M defined by the condition

$$\nabla \mathbf{g} = 0. \tag{C.42}$$

It is trivial to see that, in component notation, this translates to

$$g_{ab;c} = 0.$$

As shown in Chapter 6, this relation can be used to determine the connection (uniquely) in terms of the metric tensor.

In conclusion, we wish to point out the hierarchy of structures that we have developed. In any differential manifold structures like curves, tangent vectors, 1-forms, and tensors exist. This allows us to construct vector fields and Lie derivatives in a straightforward manner. (The existing structure also allows us to develop exterior differentiation – a topic which we have not discussed.) In the next level of the hierarchy, we introduce a *connection* on M making it (what is called) an *affine manifold*. This allows us to differentiate covariantly, vector and tensor fields. Further, the noncommutative nature of the covariant differentiation allows us to define the *curvature tensor* in a natural fashion. The third level of hierarchy is reached when a metric (which is compatible with the connection) is introduced on M, making it a *Riemannian manifold*. The metric allows us to define the concepts of length of vector, etc., and also gives a mapping between vectors and 1-forms.

Appendix D

Spacetime Symmetries

D.1. Displacement of Spacetime

The notions of homogeneity and isotropy of spacetime which were outlined in an intuitive manner in Chapter 8 can be given formal definitions. To understand these definitions we first introduce the concepts of Killing vectors and spacetime displacement. Again, to this end it helps to visualize concrete examples. Take for instance the circle of radius unity which is a compact one-dimensional space. Suppose we cut out a small arc of this circle and slide along it. At each stage the arc will lie plush on the curve underneath it. It is this example that generalizes to arbitrary displacements of spacetime.

Suppose x^i_P are the coordinates of a point P in a spacetime manifold $\mathcal{M}$. Suppose we 'cut out' a small neighbourhood $N(P)$ surrounding P and 'slide' it along by a small amount. In this process we imagine every point of $N(P)$ to retain its old coordinates. Thus, in this process of displacement P falls on a point P' of $\mathcal{M}$ with coordinates $x^i_p + \xi^i_p$, say. Likewise we may associate each point X with coordinates x^i in $N(P)$ to a point X' of $\mathcal{M}$ with coordinates $x^i + \xi^i(x^k)$. In this way we have a vector field defined on $N(P)$ (see Figure D.1).

The question is: "Is it possible for $N(P)$ to cover a corresponding portion of $\mathcal{M}$ in a congruent manner?" If it is, then we say that the spacetime is invariant under the displacement ξ^i in the neighbourhood of P.

To settle this issue we first introduce the coordinate transformation at and near P' given by

$$x'^i = x^i - \xi^i(x^k). \tag{D.1}$$

This achieves equality of coordinates at the corresponding points of $N(P)$ and $\mathcal{M}$. But coordinates alone do not ensure that the neighbourhoods of P and P' will exactly match each other. To achieve complete matching the spacetime metrics at the corresponding points must be the same.

Let the metric tensor at X of $N(P)$ be $g_{ik}(x^k)$. What is the metric tensor at X'? We know that it was

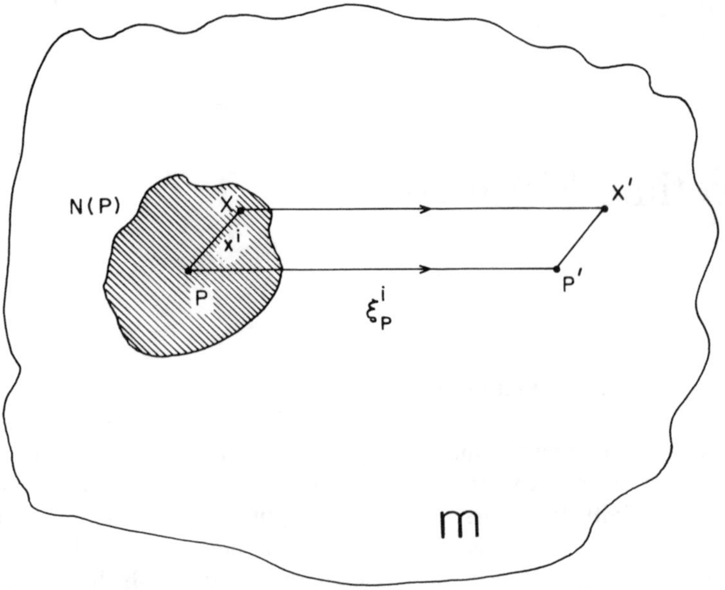

Fig. D.1. The hatched region denotes a neighbourhood $N(P)$ of point P. When $N(P)$ is displaced, the point P falls on P' and X on X'.

$$g_{ik}(x^l + \xi^l)$$

in the *original* coordinate system. We also know that to ensure congruence it must be $g_{ik}(x^k)$ in the *new* coordinate system given by (1). Using the tensor transformation law (6.19) we get

$$g_{ik}(x^k) = \frac{\partial x^l}{\partial x'^i} \frac{\partial x^m}{\partial x'^k} g_{lm}(x^k + \xi^k).$$

So far we have not made any approximations. We now assume that the displacement is infinitesimal and so neglect terms of second and higher orders in ξ^i, $\xi^i{}_{,k}$, etc. A simple computation gives the relation

$$g_{il}\xi^l{}_{,k} + g_{kl}\xi^l{}_{,i} + g_{ik,l}\xi^l = 0. \tag{D.2}$$

These equations, known as the *Killing equations* are the necessary and sufficient conditions for the spacetime to be invariant under the displacement ξ^k. The vector ξ^k itself is called the *Killing* vector (for more details see Chapter 6, Note 1 (Eisenhart) and Note 3 (Weinberg)).

A more compact form of (D.2) is

$$\xi_{i;k} + \xi_{k;i} = 0. \tag{D.3}$$

It is, of course, not always possible for these equations to have a solution. Spacetime is said to have a *symmetry of motion* or an *isometry* when a solution does exist.

D.2. Killing Vectors

Let us briefly study some properties of the Killing vectors.

D.2.1. *Maximum Number of Independent Killing Vectors*

Consider the consequences of Equation (D.3), using Equation (6.62). We have

$$\begin{aligned} 2\xi_{m;np} &= (\xi_{m;np} - \xi_{m;pn}) + (\xi_{p;nm} - \xi_{p;mn}) + (\xi_{n;pm} - \xi_{n;mp}) \\ &= -R^{l}{}_{pmn}\xi_l. \end{aligned} \tag{D.4}$$

Thus, if ξ_l and $\xi_{l;m}$ are known at any given point, we can evaluate all higher derivatives of ξ_l at that point from (D.4) and from differentiating (D.4) further. In other words, ξ_l as a vector field is completely determined from the four quantities ξ_l and $4\times3/2 = 6$ quantities $\xi_{l;m}$. (From (D.3), $\xi_{l;m}$ is an antisymmetric tensor.) Thus there can be at most 10 linearly independent Killing vector fields on $\mathcal{M}$.

D.2.2. *Integrability*

Although their maximum number is 10, there need not be *any* vector fields satisfying (D.3). The condition for (D.3) to have a solution can be found with the help of (D.3) and (D.4). Differentiate (D.3) and use Equation (6.62) to get

$$\xi_{m;npq} = -R^{l}{}_{pmn;q}\xi_l - R^{l}{}_{pmn}\xi_{l;q}.$$

Using this result and some simple manipulations we get

$$\begin{aligned} \xi_l(R^{l}{}_{qmn;p} - R^{l}{}_{pmn;q}) + R^{l}{}_{qmn}\xi_{l;p} - R^{l}{}_{pmn}\xi_{l;q} \\ - R^{l}{}_{mpq}\xi_{l;n} - R^{l}{}_{npq}\xi_{m;l} = 0. \end{aligned} \tag{D.5}$$

These are the integrability conditions which relate ξ_l and $\xi_{l;m}$ at any given point. If no ξ_l and $\xi_{l;m}$ at a point can be found satisfying (D.5), no Killing vector exists. The proof that these conditions are sufficient for the existence of a Killing vector is more involved and will not be given here (see Weinberg (1972) in Note 3, Chapter 6).

D.2.3. *Finite Displacement*

In the special case where the g_{ik} are independent of a particular coordinate it is possible to write down a Killing vector: ξ^i has only one component in the direction of that coordinate and that component is constant. By successive application of ξ^i we can readily generate a finite displacement which leaves the spacetime unchanged.

For example, for the line element of the Schwarzschild spacetime, viz. (7.19) the timelike Killing vector $(1, 0, 0, 0)$ generates a finite time-translation that leaves the spacetime unchanged. Likewise the line element (7.87) for the Kerr spacetime is unchanged by a time translation and a finite φ-rotation.

D.2.4. *Integrals of Geodesics*

Since ξ^i represents a spacetime symmetry it leads to a conservation law. This law is the first integral of the geodesic equation

$$\frac{du^i}{ds} + \Gamma^i_{kl} u^k u^l = 0. \tag{D.6}$$

It is easy to verify that

$$u^i \xi_i = \text{const} \tag{D.7}$$

is a first integral of (D.6). Spacetime symmetries thus give solutions of geodesic equations in a simple way. It can be easily verified, for example, that the first integrals (7.35) follow immediately from the relation (D.7) above.

D.3. Homogeneity

We are now able to define the concept of a homogeneous spacetime $\mathscr{M}$ in a rigorous way. $\mathscr{M}$ is said to be homogeneous if there exist isometries which carry any point P in it to any point P' in its neighbourhood. To this end it is sufficient to have, at P, *four* linearly independent Killing vectors.

Example D.1. The Gödel spacetime given by the line element

$$ds^2 = (dx^0)^2 + 2e^{x^1} dx^0 dx^2 - (dx^1)^2 + \tfrac{1}{2} e^{2x^1} (dx^2)^2 - (dx^3)^2$$

is easily seen to be homogeneous. First, because the metric tensor is independent of x^0, x^2, and x^3, we get *three* linearly independent Killing vectors:

$$(1, 0, 0, 0), (0, 0, 1, 0), (0, 0, 0, 1).$$

By setting up the Killing equations we can verify that

$$(0, 1, -x^2, 0), (-4e^{-x^1}, 2x^2, -(x^2)^2 + 2e^{-2x^1}, 0)$$

are also Killing vectors of which we can choose the first one to be the fourth member of the quartet of linearly independent vectors. □

Note that like Gödel's universe there are other spacetime manifolds which are homogeneous with more than four Killing vector fields.

D.4. Isotropy

We approach the notion of isotropy with the example of the hypersphere in four dimensions. Suppose we take at a typical point P a locally inertial coordinate system $\{y^i\}$ with origin at P. Let P' be a neighbouring point of P with coordinates y^i. Suppose we 'rotate' the neighbourhood $N(P)$ of P around P so that P is kept fixed while P' moves to a point P'' which originally had the coordinates $\xi^i + y^i$ (see Figure D.2).

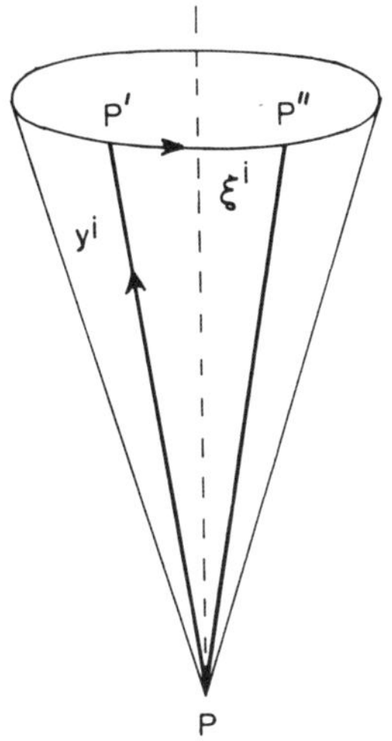

Fig. D.2. Infinitesimal rotation about an axis through P (shown by the dashed line). A typical point P' moves to P''. The displacement is shown on an exaggerated scale.

In rotating $N(P)$ we of course keep the 'distance' PP' fixed, i.e. $PP' = PP''$. To first order in ξ^i this implies

$$y^i \xi_i = 0. \tag{D.8}$$

(Note that since $N(P)$ itself is small, we tacitly assume that $|\xi^i| \ll |y^i|$.)

The most general solution of Equation (D.8) is

$$\xi_i = \varepsilon_{ijkl} y^j A^{kl}, \tag{D.9}$$

where A^{kl} is an arbitrary antisymmetric tensor. We may write A^{kl} in terms of another arbitrary antisymmetric tensor B_{mn}:

$$A^{kl} = \tfrac{1}{2} \varepsilon^{klmn} B_{mn}. \tag{D.10}$$

From (D.9) and (D.10), therefore, we get

$$\xi_{i,j} \equiv \frac{\partial \xi_i}{\partial y_j} = (\delta_i^m \delta_j^n - \delta_i^n \delta_j^m) B_{mn}. \tag{D.11}$$

Expressed thus, we see that for an arbitrary antisymmetric tensor B_{mn} there are basically *six* independent values of $\xi_{i,j}$ that specify arbitrary rotation about P.

These ideas can be incorporated in the following definition of isotropy: the spacetime $\mathscr{M}$ is isotropic at a given point P if there are Killing vector fields ξ^i in the neighbourhood $N(P)$ of P such that $\xi^i = 0$ at P and $\xi_{i;k}$ span the space of second-rank antisymmetric tensors at P. In such a case we can choose vector fields $\xi^i_{[mn]}(P', P)$ in $N(P)$ such that

$$\left.\begin{aligned} \xi^i_{[mn]}(P', P) &= \xi^i_{[nm]}(P, P'), \\ \xi^i[P, P] &= 0, \\ \lim_{P' \to P} \xi_{i;k[mn]}(P) &= \delta^m_i \delta^n_k - \delta^n_i \delta^m_k . \end{aligned}\right\} \tag{D.12}$$

It can be shown that a spacetime $\mathscr{M}$ which is isotropic about every point in it is homogeneous. Such a spacetime is called *maximally symmetric* since it has the maximum number (D.10) of independent Killing vector fields that is possible. It can be shown that a maximally symmetric *three-dimensional space* (which has six independent Killing vector fields) has the line element

$$\mathrm{d}l^2 = Q^2 \left[\frac{\mathrm{d}r^2}{1 - kr^2} + r^2(\mathrm{d}\theta^2 + \sin^2\theta\, \mathrm{d}\varphi^2) \right], \tag{D.13}$$

where Q is a constant and k can take values 0, 1, or -1. We refer the reader to references cited previously for proofs of these results.

Index